# COGENERATION

**CHARLES H. BUTLER**

# COGENERATION: Engineering, Design, Financing, and Regulatory Compliance

**MCGRAW-HILL BOOK COMPANY**

*New York St. Louis San Francisco Auckland Bogotá Hamburg Johannesburg London Madrid Mexico Montreal New Delhi Panama Paris São Paulo Singapore Sydney Tokyo Toronto*

**Library of Congress Cataloging in Publication Data**

Butler, Charles H.

Cogeneration: engineering, design, financing, and regulatory compliance.

Includes index.

1. Cogeneration of electric power and heat.

I. Title.

TK1041.B88 1984 621.1′9 84-4437

ISBN 0-07-009364-4



1 2 3 4 5 6 7 8 9 0 DOC/DOC 8 9 8 7 6 5 4

ISBN 0-07-009364-4

The editors for this book were Harold Crawford and Georgia Kornbluth, the designer was Elliot Epstein, and the production supervisor was Thomas G. Kowalczyk. It was set in Caledonia by Achorn Graphics.

Printed and bound by R. R. Donnelley & Sons Company.

# CONTENTS

# PREFACE

This book provides a comprehensive treatment of all areas critical to obtaining permits, meeting regulatory requirements, planning, financing, engineering, and allocating risks in the development of a qualified cogeneration facility. Basic concepts of cogeneration, a site design data checklist, and applications complete with drawings, diagrams, figures, and tables show how to size, select, and implement cogeneration projects. Fundamental definitions and engineering design criteria are included. Electric utility and industrial concerns, common goals, and interactions are presented. The integration of financial, regulatory, and environmental considerations is illustrated by examples. The technical feasibility of power-cycle alternative for qualified topping- and bottoming-cycle cogeneration facilities is illustrated. Descriptions of equipment and components necessary to satisfy best available control technology are presented. An overview and update of the impact the controversial PURPA and FERC government policies, permits, and regulatory status have on the development of cogeneration viability are included. The economic benefits of cogeneration associated with financing and tax incentives are described in relation to the development of a successful cogeneration project. An example shows how to calculate and perform a cash-flow analysis of capital and operating costs and revenues.

# COGENERATION

# 1

# INTRODUCTION TO COGENERATION

Corporate financial survival during periods of high fuel and operating costs requires the control of energy costs to produce competitive products and services. Cogeneration is a direct approach to conserving fuel through improved energy utilization and higher overall system efficiency and is indeed a proven technology offering energy savings with attractive cost payback periods.

Cogeneration facilities have been installed and are operating at chemical, petrochemical, refinery, mining and metals, paper and pulp, food-processing, and utility complexes. The three most significant factors affecting a cogeneration project are availability of financing, acceptable economic payback period, and ability to satisfy regulatory requirements. Most cogeneration projects providing their base process steam and/or heating and drying thermal loads, in addition to satisfying internal electric energy requirements except during peak demand periods, are easily proved to have an economically attractive payback period. Each of these items is considered in the following discussion.

## HISTORY

Cogeneration is a new name for an old and proven method of producing power. During the late 1800s and early 1900s, industry generated close to half the energy it used. As late as 1950, cogeneration supplied approximately 15 percent of U.S. energy needs. The use of cogeneration declined as fuel became less expensive during the 1950s and 1960s and electric power generated by utilities became more reliable and less expensive. By 1977, only 4 percent of the nation's energy came from cogeneration, although in Europe higher fuel costs spurred industries to continue and to expand the use of this technology.

As an energy conservation technology, cogeneration uses oil, gas, and other fuel resources more efficiently than either typical industrial processes or traditional power plants. Increased efficiency reduces pollution, fuel consumption, and reliance on imported energy sources.

In 1978 the stage was set for private investment in cogeneration by enactment of the federal Public Utility Regulatory Policy Act (PURPA); and as a result cogeneration became much more profitable for industry. PURPA exempted most cogenerators from regulation as utilities and directed utilities to set reasonable rates for buying power from and selling power to cogenerators. States were required to set rates that utilities must pay cogenerators for electricity by March 1981.

## DEFINITIONS

A wide range of definitions of cogeneration exist. Several of these and their source are given below.

> The sequential production of electricity and heat, steam, or useful work from the same fuel source. *(California Public Utilities Commission OIR 2)*
>
> The sequential production of useful shaft power and thermal energy from a single primary energy source. *(Southern California Gas Company)*
>
> Equipment used to produce electric energy and forms of useful thermal energy (such as heat or steam), used for industrial, commercial, heating and cooling purposes, through the sequential use of energy. *(Natural Gas Policy Act)*
>
> "Cogeneration Facility" means an electric power plant or a major fuel burning installation that produces: (1) Electric power; and (2) Any other form of useful energy (such as steam, gas, or heat) that is, or will be, used for industrial, commercial, or space heating purposes. In addition, for purposes of this definition, electricity generated by the cogeneration facility must constitute more than ten (10) percent and less than ninety (90) percent of the useful energy output of the facility. *(Economic Regulatory Administration, Department of Energy)*
>
> "Cogeneration Facility"—Equipment used to produce electric energy and forms of useful thermal energy (such as heat or steam), used for industrial, commercial heating, or cooling purposes, through the sequential use of energy. *(California Public Utilities Commission OIR 2)*

## INDUSTRIAL CONCERNS

The items listed below should be of concern to an industrial facility considering the development of a cogeneration project.

- Benefits of quick return on cogeneration capital investment compared to the following disadvantages:

    An operation removed from the main product line of the business

    The availability of existing staff to operate and maintain the cogeneration facility

    Added environmental considerations

- States in which oil or gas is burned as the major fuel generally offer higher electric rates to cogenerators.
- The electric utility is a guaranteed purchaser of electricity output.
- The cogenerator can shop around for the right customer, "wheeling" its power through a local utility to a higher-paying utility, provided the transfer costs are not too high and the utility agrees to wheel the power from its transmission systems.
- Profit and energy self-sufficiency.
- Investment decisions by management in the face of regulatory and tax incentive uncertainty.
- A noncogenerator must compete against cogeneration firms with lower energy costs.
- Backup power needs:

  Purchase of redundant generating capacity (an expensive option)

  Purchase of backup power from a utility
- Federal, state, and local air pollution control district ambient air quality standards and permit requirements.
- Air and noise emission controls required by regulatory agencies:

  Size-dependent (larger facilities emit greater emissions)

  Determined on a case-by-case basis

  Exemption from emission standards for nitrogen oxides, sulfur oxides, and particulates

  For existing facilities, reduced emissions from nonoperating boiler(s) should be taken into consideration in estimating the emissions from the cogeneration system.

**UTILITY CONCERNS**

An electric utility should consider the following factors when investigating cogeneration.

- Need for additional capacity:

  Cogeneration can help a utility defer or cancel construction of a new power station, or reduce the need to buy power from another utility.

  Cogeneration helps the utility avoid the burning of expensive gas or oil to produce power (or reduces the use of fossil fuels).

- Availability of power during peak load periods.
- Reduction in line losses.
- States can provide rates offering additional incentives for cogeneration, even to the point of subsidy by the electric utility.
- Dealing with potential capacity demands by cogenerators.
- Impact of a large group of small cogenerators on the reliability of power to their grid.
- Whether or not utilities can rely on cogeneration for

  Supplying additional power blocks of electricity with a shorter lead time than conventional power plants

  Easier siting

  Advantageous fuel and air pollution control regulations

## ENGINEERING AND DESIGN OF FACILITIES

Full-scope services are normally required to complete cogeneration projects. These capabilities include performing technical cost-benefit studies, conceptual engineering, preliminary design, detailed engineering and design, construction, and start-up services. Cogeneration projects are currently in use in proven operating applications. These projects use alternative fuels including gas, oil, coal, lignite, petroleum pitch, wood waste, biomass, and refuse.

# 2

# PRELIMINARY STUDIES, ENGINEERING AND CONSTRUCTION PLANNING, AND DESIGN DOCUMENTATION

This chapter outlines the tasks required to successfully perform studies, engineering, design, and construction of cogeneration projects.

## PRELIMINARY STUDIES AND CONCEPTUAL ENGINEERING

Cogeneration studies are initiated to evaluate cost-effective alternatives and to select the most appropriate options. This is achieved by performing a technical feasibility and economic cost-benefit study to rank and recommend the alternatives. Determination of technical feasibility includes a realistic assessment of each application with respect to environmental impact, regulatory compliance, and interfaces with a utility. An economic payback analysis determines the payback period which is then compared to client financial goals. The selected design can be further developed during a subsequent detailed feasibility study to obtain a more comprehensive technical definition of the project and improved accuracy of cost estimates. Many clients elect to proceed directly to the engineering and design phase upon comparison of the initial technical and economic evaluations. After obtaining client authorization to proceed with a project, detailed engineering, design, procurement, construction, and start-up activities can be initiated.

The various types of analyses and their objectives are listed below.

- TECHNICAL FEASIBILITY STUDY

  Provide a realistic assessment of cogeneration potential.

  Identify available cogeneration alternatives.

  Establish fuel availability and priority classifications.

  Prepare concept design criteria.

  Determine modes of operation.

Optimize the power cycle.

Utilize existing proven cogeneration equipment and extensive system data base experience.

Match systems and equipment to the generation of electricity and process steam or exhaust heat demands.

Plan for the efficient utilization of fuel.

- ECONOMIC PAYBACK ANALYSIS

  Compare economically competitive cogeneration options with noncogeneration alternatives.

- ENVIRONMENTAL ASSESSMENT

  Select the site.

  Assist in the preparation of permit applications.

  Examine air quality problems and compliance with regulations.

  Select appropriate emission control equipment.

- IDENTIFICATION OF REGULATORY REQUIREMENTS

  Clarify applicable regulatory requirements.

  Determine the exemption status.

  Advise the client regarding regulatory applications.

  Assist the client in dealing with regulatory agencies.

  Identify permit requirements.

- ESTABLISHMENT OF ELECTRIC UTILITY INTERFACE

  Clarify the avoided cost rate.

  Define the buy-and-sell contract and ownership options.

  Determine the technical electrical interconnection interface.

- IDENTIFICATION OF FUNDING SOURCES

## ENGINEERING AND CONSTRUCTION PLANNING

Preliminary or detailed engineering is initiated by establishing the scope of the project. Meetings are held with the client to identify the design and operating requirements for a new or an existing facility. In developing the design criteria the following should be considered.

- Specific information regarding the site and conditions must be obtained.
- The system must be clearly defined.
- Individual components or combinations of components must be identified.
- The specific mode or modes of operation must be determined.
- The operating modes must be satisfied by selection of the proper cogeneration equipment, systems components, and controls.
- A thorough review of the existing facility operating modes and the existing systems interconnection interface points with the cogeneration facility must be conducted.
- The assumptions and design criteria to be used in the analysis or study must be defined.
- Detailed description of the system design, installation, operations, and environment must be obtained.
- Complete system and equipment data must be identified.
- All the alternatives or combinations of cycles should be examined.
- The systems and equipment must be sized.

Steps listed below should be followed in choosing engineering systems, selecting equipment, and utilizing the latest cogeneration technology to match electrical and process steam or heating applications to provide the highest possible efficiency and lowest total installed cost.

Technical evaluation of a project should

- Establish design criteria.
- Identify fuel considerations.
- Verify the fuel supply.
- Match varying client requirements to the cogeneration application: process steam, exhaust heat utilization or drying, hot water, generation of electricity.
- Select the most suitable plant, systems, equipment, and control systems for the facility.
- Optimize the power cycle.
- Size the plant equipment and systems to match the application.

- Apply experience with the development of compact and dependable heat recovery equipment.

Major equipment and component applications should be reviewed to optimize overall system efficiency and operating characteristics in a comparison of technical performance with cost-effectiveness. Feedwater quality has a direct affect on plant design and cost. The makeup and feedwater for existing facilities should be reviewed relative to equipment water quality requirements.

Efficient cogeneration energy systems design, equipment selection, and construction have become increasingly important because of the rising price and scarcity of fuel. More efficient cogeneration systems conserve energy. The following effective energy conservation and heat recovery practices should be integrated into the power cycle and system design.

- PROJECT MANAGEMENT

  Planning.

  Scheduling.

  Cost control.

- ENERGY CONSERVATION

  Improve energy efficiency by implementing an energy conservation plan.

  Obtain more efficient utilization of fuel by

    Increasing the overall efficiency of electrical generation.

    Providing energy at the lowest possible cost.

  Economically recover heat rejected to exhaust and/or water systems.

    Utilize lowest technically practical temperature relative to ambient conditions.

  Obtain maximum economical heat recovery while ensuring reliability and operability of the equipment.

  Reduce exhaust losses.

  Reduce cooling system losses.

  Minimize the energy demand.

  Utilize available energy for internal process applications and displace less efficient sources.

  Convert existing standby power generator sets to continuous profit producers.

• ENGINEERING ASSURANCE

Assure client of technical quality and economical project.

Preclude cost of nonconformance and schedule delays.

Review project for compliance with specifications and industry standards.

• FINANCIAL CONSIDERATIONS

Performing the activities listed below will help to identify the economic impact of the project.

- Establish economic evaluation criteria.
- Prepare and evaluate economic factors:

  Determine fuel costs and consumption.

  List buy or sell prices for electricity.

  Estimate electric energy and capacity purchase prices.

  Assist client in preparing the power agreement.

  Prepare an equipment cost estimate.

  Determine operating and maintenance expenditures.
- Perform a cost-benefit and cost-effectiveness analysis.
- Estimate the total installed cost of the project.
- Prepare estimated cash flow requirements.
- Provide liaison services with local economic development officials.
- Advise client on financial assistance programs.

Cogeneration projects should be designed to satisfy applicable environmental requirements to ensure that the facility will be permitted by regulatory agencies. The following factors should be taken into account.

• ENVIRONMENTAL CONSIDERATIONS

Indentify applicable air quality regulations.

Coordinate the activities of local, state, and federal regulatory agencies.

Evaluate tradeoffs for air quality.

Apply best available control technology (BACT).

Determine the requirements of the Environmental Protection Agency and the Air Resource Board concerning emission offsets.

- REGULATORY REQUIREMENTS

  Identify applicable governmental regulatory policies.

  Determine the Fuel Use Act exemption status of project.

- PROJECT PROCUREMENT

- CONSTRUCTION SERVICES

  Preconstruction planning.

  Construction management.

  Construction.

  Start-up.

- SUPPORT SERVICES

  Quality assurance.

  Maintenance.

  Training.

## DESIGN DOCUMENTATION

Design documentation includes the preparation of project flow diagrams, piping and instrument diagrams, general arrangement drawings, equipment layouts, building and structural drawings, foundation drawings, and electrical one-line diagrams.

Integrating cogeneration projects into existing facilities requires a review of the site areas available and considerations of planned expansion. This area is sensitive to the type of fuel selected. Solid fuels require more space for delivery, storage, handling, and ash removal facilities. The following objectives should be considered in designing a project.

- Design for a specific base fuel.
- Design the facility to integrate with existing client plant operating characteristics:

  Define system and control philosophy.

  Satisfy the electric utility interconnection requirement.

- Design for existing space allocations or limitations.
- Select equipment suitable for the application.
- Arrange equipment for easy access and maintenance.
- Assist in finding a site for the facility.

Fuel availability and alternative fuel possibilities should be analyzed. Potential alternate fuels to be considered for cogeneration applications include natural gas, fuel oil, petroleum-based fuels, wood waste, biomass, coal, lignite, and refuse-derived fuel. Fuel selection should be based on the following considerations.

- AVAILABILITY

  Reliability of equipment and system

  Delivery time

- SITE LOCATION

  Fuel delivery alternatives available

  On-site versus off-site storage

  Delivery costs

- SITE AREA REQUIREMENTS
- AIR EMISSIONS
- EQUIPMENT DESIGN, SIZE, AND COST
- OPERATIONS

  Maintenance

  Training of employees

  Personnel staffing

  Control response time

- GOVERNMENT REGULATIONS

  Exemptions

  Air quality standards

  Tax credits

The equipment and systems listed below are generally considered major factors in the design of a cogeneration facility.

- Combustion turbine generators
- Waste heat recovery steam generators
- Steam turbine generators
- Condensers
- Cooling towers
- Chiller-heaters
- Fuel systems
- Instrumentation and control systems
- Electric switchgear and motor control centers
- Condensate and feedwater systems
- Water treatment systems
- Heat recovery components

# POWER CYCLES AND APPLICATIONS

## POWER CYCLES

Cogeneration projects generally employ topping or bottoming power cycles. The topping cycle has the widest industrial application. The bottoming cycle is of limited application because of its high cost and the probability that more efficient use of the waste energy stream can be made.

TOPPING CYCLE The primary energy source is used to produce useful electric or mechanical power output. Reject heat from power production is then used to provide useful thermal energy.

BOTTOMING CYCLE The primary energy source is applied to a useful heating process, and the reject heat emerging from the process is then used for power production.

These power cycles are shown in Figures 3-1 to 3-4. The efficiency and heat utilization standards established by the Public Utility Regulatory Policy Act and the Federal Energy Regulatory Commission are indicated. A topping cycle for a simplified combined cycle cogeneration facility is shown in Figure 3-1. A topping cycle for a simplified combustion turbine exhaust heat cogeneration facility is shown in Figure 3-2. A topping cycle for a simplified internal-combustion diesel or gas exhaust heat cogeneration facility is shown in Figure 3-3. A bottoming cycle for a cogeneration facility is shown in Figure 3-4.

Total overall thermal efficiencies of up to 80 percent are possible for energy cogeneration systems. This efficiency is approximately twice that of a conventional power plant application. Conventional power plants are approximately 30 to 40 percent efficient. Boilers for conventional plants are approximately 75 to 85 percent efficient, as compared to unfired waste heat recovery boilers for cogeneration applications which can be 98+ percent efficient. Because cogeneration cycles are more efficient they use 30 to 40 percent less total energy. The gross heat rate (Btu

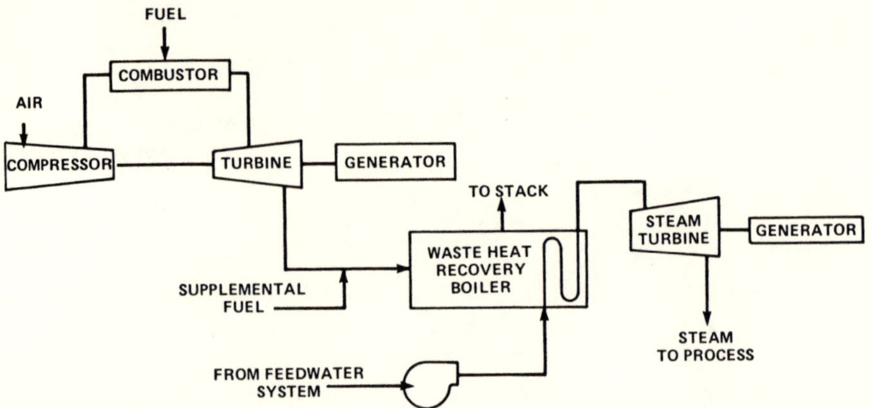


**Figure 3-1 Simplified combined cycle cogeneration. Topping cycle: fuel → electricity → process.**

*PURPA and FERC Efficiency and Heat Utilization Standards*

1. Useful thermal energy is the steam to process which must be greater than 5 percent of the total energy.
2. Electricity generated plus one-half the useful thermal energy must be equal to or greater than 42.5 percent of the fuel energy input.

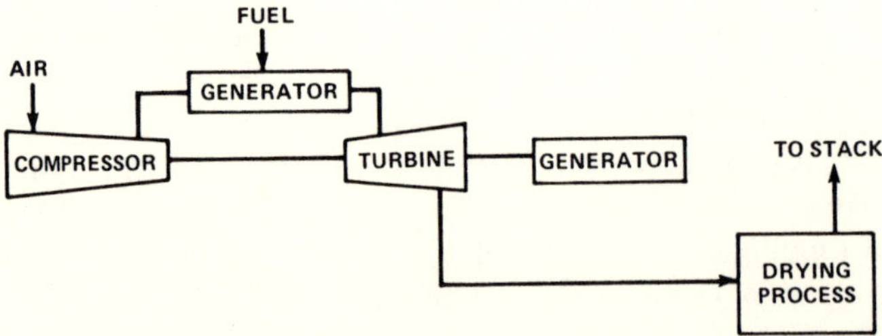


**Figure 3-2 Simplified combustion turbine exhaust heat cogeneration facility. Topping cycle: fuel → electricity → process.**

*PURPA and FERC Efficiency and Heat Utilization Standards*

1. Useful thermal energy is the exhaust heat to drying process which must be greater than 5 percent of the total energy.
2. Electricity generated plus one-half the useful thermal energy must be equal to or greater than 42.5 percent of the fuel energy input.

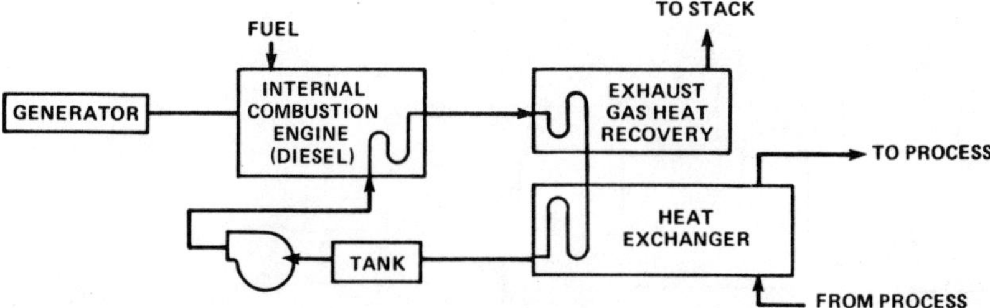


**Figure 3-3 Simplified internal-combustion diesel exhaust heat cogeneration facility cycle. Topping cycle: fuel → electricity → process.**

*PURPA and FERC Efficiency and Heat Utilization Standards*

1. Useful thermal energy is the exhaust heat recovered which must be greater than 5 percent of total energy.
2. Electricity generated plus one-half the useful thermal energy must be equal to or greater than 42.5 percent of the fuel energy input.

| Process | | Engine | Exhaust to stack | Engine cooling system | Radiation to ambient air |
|---|---|---|---|---|---|
| Overall engine thermal efficiency without | | | | | |
| Heat recovery | 33% | 33 | | | |
| Losses | 67% | | −30 | −30 | −7 |
| Total fuel input | 100% | | | | |

| Process | | Radiation to ambient air | Engine cooling system | Exhaust to stack | Engine |
|---|---|---|---|---|---|
| Overall engine thermal efficiency with | | | | | |
| Heat recovery | 75% | | 30 | +12 | +33 |
| Losses | 25% | +7 | | +18 | + 0 |
| Total fuel input | 100% | | | | |

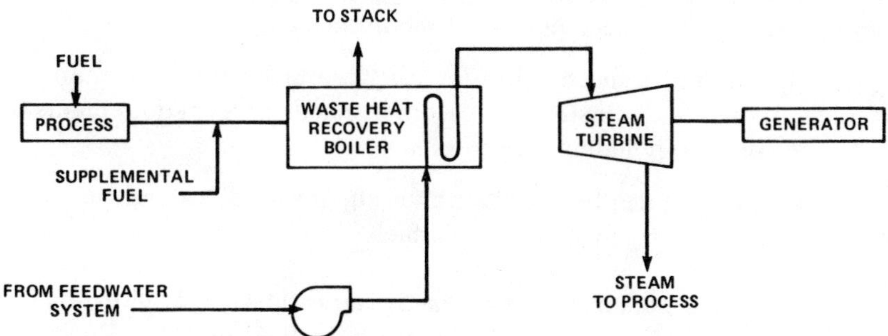


**Figure 3-4 Cogeneration bottoming cycle: fuel → process → electricity.**

*PURPA and FERC Efficiency Standards*

Electricity generated must be equal to or greater than 45 percent of the fuel energy input.

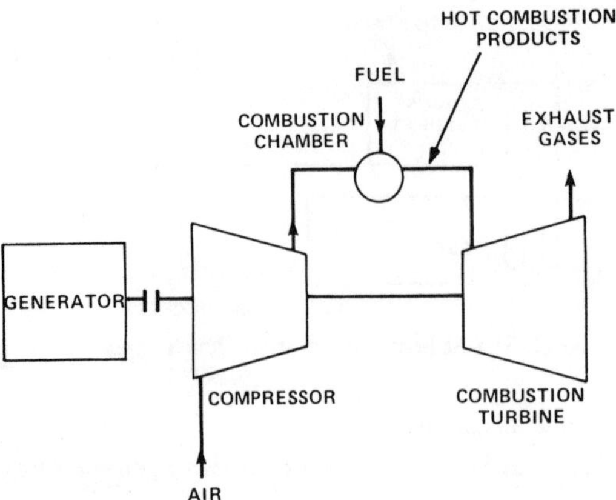


**Figure 3-5 Simple cycle gas turbine generator for power generation.**

per kilowatthour) is the energy required by the prime mover without considering the use of its exhaust heat. The net heat rate is the gross heat rate less the heat used in the process. The lower the heat rate, the more power produced for a given amount of heat and the more efficient the facility.

## A BASIC COMBUSTION TURBINE GENERATOR FOR POWER GENERATION

The basic combustion turbine generator configuration shown in Figure 3-5 is simple and compact. It has the following advantages.

- A minimum cost investment in a combustion turbine generator plant can be made when emergency or peak power demand periods exist.
- The combustion turbine generator plant can be installed at any site and remote-controlled to reduce personnel requirements.
- The installation time is shorter than for other thermal plants. It is therefore possible to sequence the installation cost to planned energy requirements by purchasing completely independent units.
- Combustion turbines can be fired by various fuels such as natural gas, no. 2 fuel oil, kerosene, and low-cost residual fuels.

A combustion turbine generator power package plant is shown in Figure 3-6. The general arrangement of this plant is shown in Appendix B, Figure B-1. Engineering data and standard performance are given in Table 3-1. Figure 3-7 illustrates a typical combustion turbine cross-sectional view.

Figure 3-6 W191 Gas turbine power package plant. (*Westinghouse Standard Proposal for the W191 Econopac Gas Turbine Power Plant*)

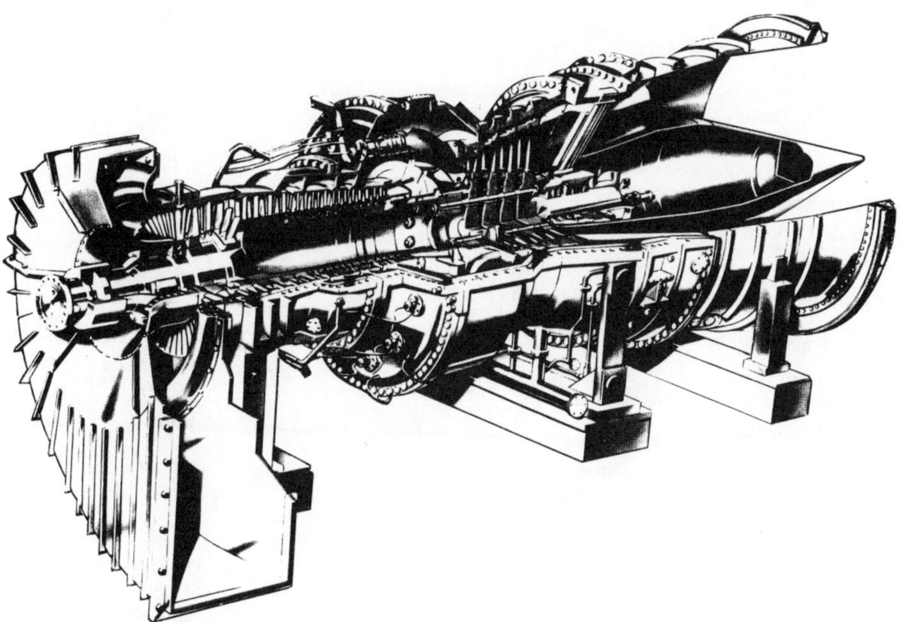

Figure 3-7 Combustion turbine cross section. (*Westinghouse Combustion Turbines, "The Ready Source of Power Worldwide," Bulletin SA 11152*)

**Table 3-1 W191G Gas Turbine Generator Econo-Pac Power Plant: Engineering Data***

| Type of operation | Plant net rating, kW | Minimum exhaust flow, lb/hr | Minimum exhaust temperature, °F | Plant performance: Heat rate, Btu/kWh | Plant performance: Efficiency, % |
|---|---|---|---|---|---|
| **Fuel—Standard natural gas: Heating value 1,000 (HHV) and 900 (LHV) Btu/scf** | | | | | |
| Base | 17,700 | 985,000 | 780† | 13,390 | 25.49 |
| Peak | 19,000 | 985,000 | 813† | 13,210 | 25.84 |
| **Fuel—Standard no. 2 distillate oil: Heating value 19,300 (HHV) and 18,300 (LHV) Btu/lb** | | | | | |
| Base | 17,300 | 985,000 | 781† | 13,680 | 24.95 |
| Peak | 18,600 | 985,000 | 815† | 13,510 | 25.26 |

| | 60 Hz | 50 Hz |
|---|---|---|
| **Gas turbine:** | | |
| Axial compressor stages | 15 | 15 |
| Turbine stages | 5 | 5 |
| Unit pressure ratio | 7.47–1 | 7.37–1 |
| Combustor baskets | 6 | 6 |
| Rated speed, rpm | 4912 | 4830 |
| **Equipment weight (approximate), lb:** | | |
| Turbine (erected including piping and lube oil) | 169,500 | 169,500 |
| Gear | 43,500 | 46,770 |
| Generator and exciter | 89,300 | 101,000 |
| Auxiliary assembly | 20,100 | 20,100 |
| Complete turbine generator | 322,400 | 337,370 |
| Heaviest piece to be handled | | |
| During installation (turbine) | 145,000 | 145,000 |
| After installation (generator rotor) | 40,800 | 46,000 |
| Turbine rotor | 26,000 | 26,000 |
| Turbine cylinder cover | 29,000 | 29,000 |
| Starting diesel | 12,000 | 12,000 |
| $WR^2$ (turbine generator unit referred to turbine speed), lb ft$^2$ | 26,600 | 27,500 |
| **Lubrication system:** | | |
| Oil reservoir capacity, gal (U.S.) | 1,750 | 1,750 |
| Lube system heat load, Btu/min | 30,000 | 30,000 |

**Table 3-1 (Cont.)**

| | 60 Hz | 50 Hz |
|---|---|---|
| **Lubrication system (cont.):** | | |
| Oil pump, capacities, gal/min (U.S.) | | |
| Shaft-driven, main | 370 | 364 |
| Motor (ac)-driven, primary auxiliary | 300 | 300 |
| Motor (dc)-driven, secondary auxiliary | 150 | 150 |
| **Starting device:** | | |
| Diesel engine (rating at 100% speed), hp | 430 | 430 |
| Fuel oil tank capacity (U.S.) | 50 | 50 |

| | 60 Hz | | 50 Hz | |
|---|---|---|---|---|
| | Generator | Exciter | Generator | Exciter |
| **Type of unit: salient pole** | | | | |
| Rated | | | | |
| Kilovoltamperes | 20,820 | 111 | 20,850 | 111 |
| Power factor | 0.85 | 0.90 | 0.85 | 0.90 |
| Kilowatts | 17,700 | 100 | 17,720 | 100 |
| Air flow, cfm | 40,000 | 1500 | 37,500 | 1500 |
| Ambient air, °C | 15 | 15 | 15 | 15 |
| Rpm | 900 | 900 | 750 | 750 |
| Phase | 3 | 3 | 3 | 3 |
| Cycle | 60 | 75 | 50 | $62\frac{1}{2}$ |
| Voltage | 13,800 | 125 | 11,500 | 125 |
| Short circuit ratio | 0.645 | | 0.60 | |
| F.L. amps | 870 | | 1050 | |
| Low ambient maximum capability, KVA | 22,940 | | 22,940 | |
| Class insulation | B | B | F | B |
| Speed of response | | 0.5 | | 0.5 |
| Reactances | | | | |
| Transient (sat., $Xd'$) | 0.363 | | 0.40 | |
| Subtransient (sat. $Xd''$) | 0.259 | | 0.27 | |
| Synchronous ($Xd$) | 2.03 | | 2.06 | |
| Zero sequence ($X_0$) | 0.159 | | 0.13 | |
| Negative sequence ($X_2$) | 0.248 | | 0.26 | |
| Time constant, sec | 6.8 | | 8.0 | |
| Generator and exciter $WR^2$ at generator speed, lb/ft² | | 80,000 | | 94,100 |
| Cooling water required, gal/min at °C | None | | None | |

Table 3-1 (Cont.)

| | Horsepower at 50 Hz | Horsepower at 60 Hz | Operation Category§ |
|---|---|---|---|
| **Plant auxiliaries** | | | |
| Ac drives—440-V, three-phase, 60-Hz or 380-V, three-phase, 50-Hz | | | |
| Primary auxiliary lube oil pump | 30 | 25 | BA |
| Control and atomizing compressor | 5 | 5 | AR |
| Lube oil cooler fan | 2–20 | 2–20 | C |
| Enclosure exhaust fans (total) | 20 | 20 | AR |
| Optional arrangements, ac motors | | | |
| Inlet air evaporative cooler pump | 20 | 20 | C |
| Inlet air filter screen (approximate total) | ½ | ½ | C |
| Fuel oil (main) pump | 50 | 40 | BC |
| Fuel oil (tank-to-unit) transfer pump | 3 | 3 | BC |
| Air conditioner | 16 | 8 | AR |
| Fuel oil filter | ⅛ | ⅛ | BC |
| Glycol pump (option) | 20 | 20 | AR |
| Dc drives—125-V | | | |
| Secondary auxiliary lube oil pump | 3 | 3 | D |
| Turning gear | 5 | 5 | BD |
| Clutch air compressor | 1–5 | 1–5 | B |
| Controls, kW | 1 | 1 | ABCD |
| Heaters LV, three-phase | | | |
| Generator space heaters, kW | 5.6 | 5.6 | AR |
| Building unit heaters, kW | 60 | 60 | AR |
| Lube oil heater, kW | 18 | 15 | AR |
| Exciter space heater, kW | 0.6 | 0.6 | AR |
| Controls and lighting, kV | 32 | 32 | AR |

*ISO ambient conditions: Temperature 59°F (15°C); Barometric pressure 14.7 psia (sea level)

†Exhaust temperatures shown are for a 50-Hz plant or 776 base and 809 peak at 60 Hz.

‡Exhaust temperatures shown are for a 50-Hz plant or 777 base and 810 peak at 60 Hz.

Performance based on lower heating value of the fuel. Performance based on unfiltered inlet and exhaust systems with level A silencers (4-in. $H_2O$ pressure loss).

§A, Standby; B, start-up; C, unit operating normally; D, emergency unit shutdown; AR, as required.

SOURCE: Westinghouse Standard Proposal for the W191 Econopac Gas Turbine Power Plant.

Fig. 3-8 shows a cutaway section of a typical gas turbine generator power plant. A compact arrangement requiring minimum platform space for offshore combustion turbine generator applications is shown in Figure 3-10. A cutaway view of a boiler and a condensing steam turbine generator marine propulsion power system is given in Figure 3-9. A modular combustion turbine designed for high availability, good serviceability, and easy maintenance and overhaul at the site is shown in Figure 3-11. Materials and temperatures for industrial applications are indicated on Figure 3-12. A cutaway drawing of a combustion turbine generator power

Figure 3-8 Combustion turbine cross section. (*Stal-Laval, Inc., "Turbine Power," Bulletin 572E9.79.4.000*)

Figure 3-9 Marine propulsion power system. (*Stal-Laval, Inc., "Turbine Power," Bulletin 572E.9.79.4.000*)

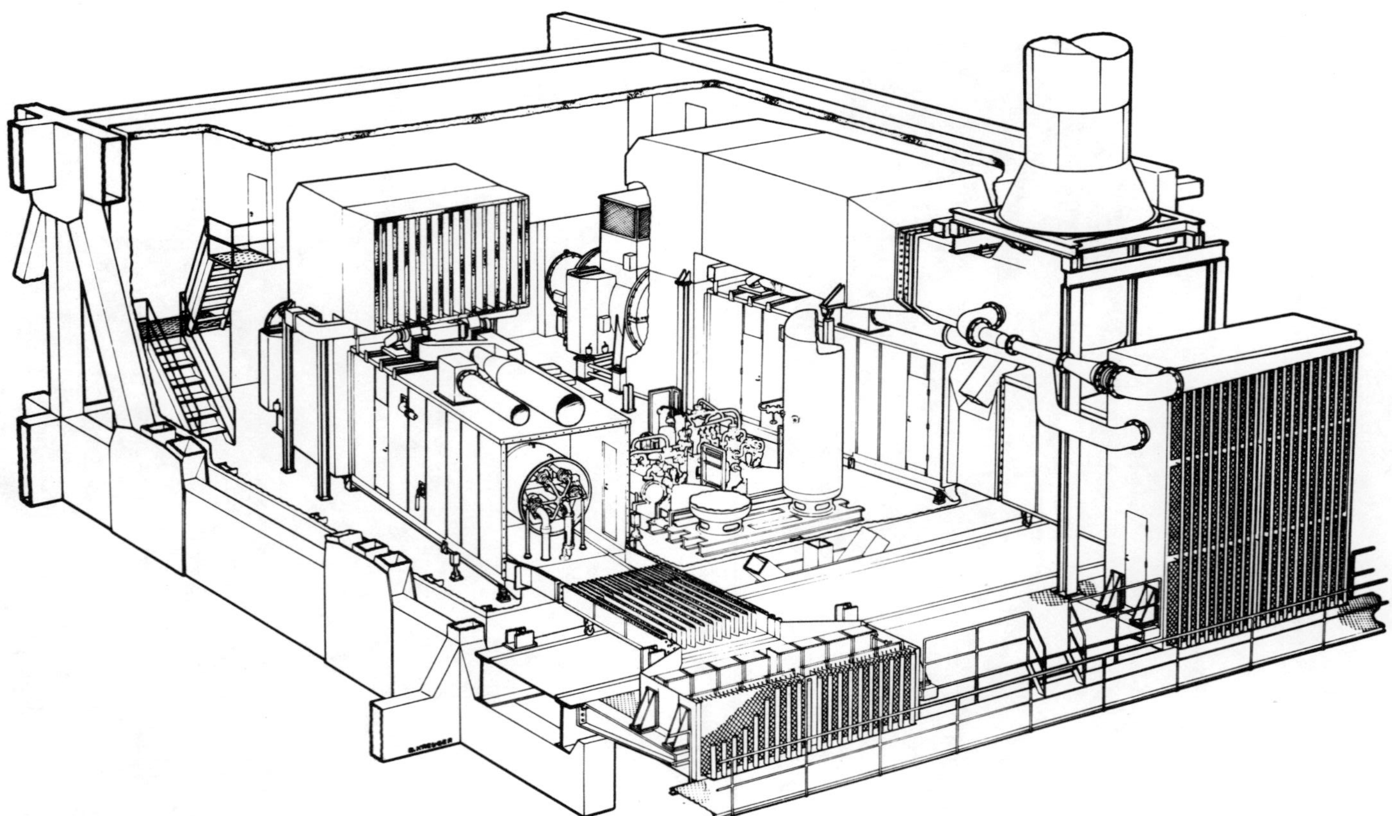

**Figure 3-10** Offshore platform combustion turbine generator application. (*Stal-Laval, Inc., "The GT 35 Gas Turbine 10-15MW," Bulletin 588E02.801.000*)

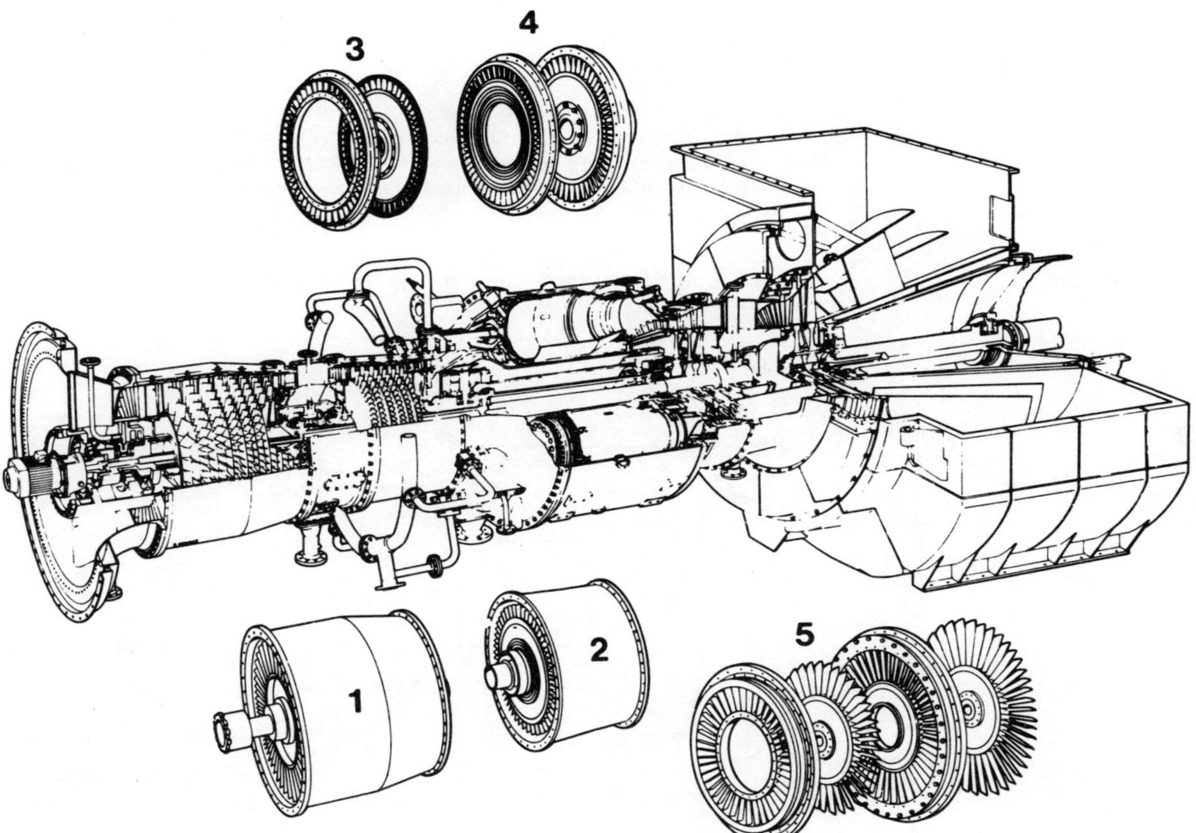


**Figure 3-11** Modular combustion turbine. GT 35 and modules: (1) LP compressor, (2) HP compressor, (3) HP turbine, (4) LP turbine, (5) power turbine. (*Stal-Laval, Inc.*, *"The GT 35 Gas Turbine 10-15MW,"* *Bulletin 588E02.80.1.000*)

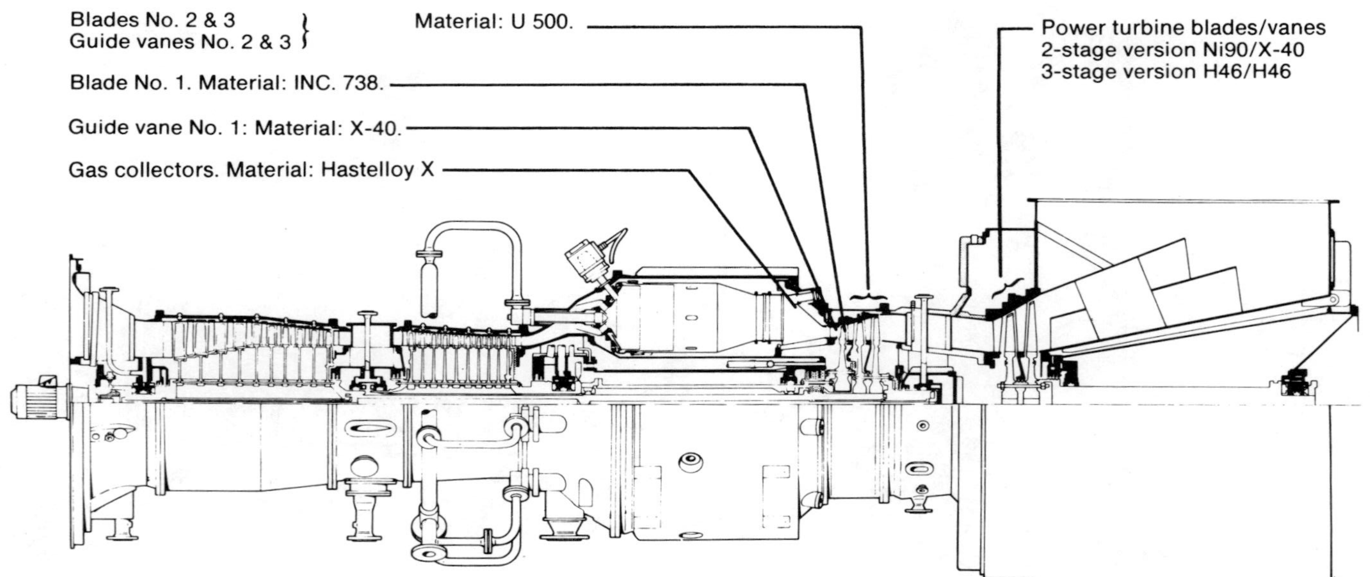


**Figure 3-12** **Industrial combustion turbine material and temperature. Temperatures, materials, and coatings: At base load, ISO conditions, turbine inlet temperature is 810°C (1490°F). At maximum continuous load, ISO conditions, turbine inlet temperature is 850°C (1562°F). Coatings are available for hostile environments. (*Stal-Laval, Inc., "The GT 35 Gas Turbine 10-15MW," Bulletin 588E02.801.000*)**

Figure 3-13 Combustion turbine generator power station. (*Stal-Laval, Inc., "The GT 35 Gas Turbine 10-15MW," Bulletin 588E02.801.000*)

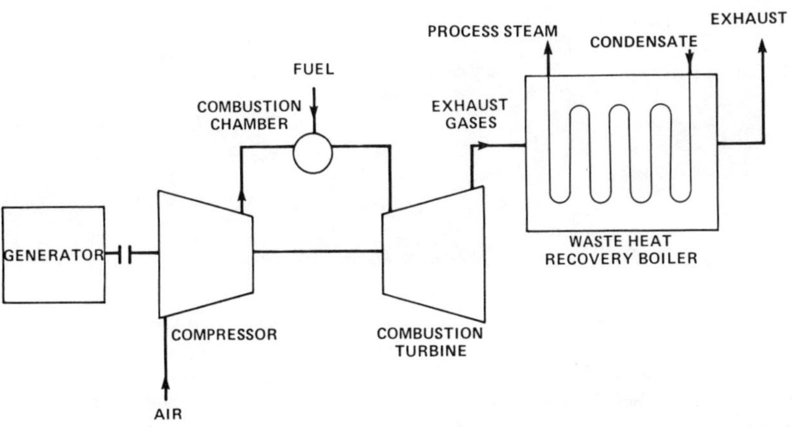


Figure 3-14 Simple cycle gas turbine generator for power and process heat generation.

station is shown in Figure 3-13. The basic arrangement layout for this power station is shown in Appendix B, Figure B-3.

## A HEAT RECOVERY COMBUSTION TURBINE GENERATOR CYCLE FOR GENERATION OF BOTH POWER AND PROCESS HEAT

The heat content of the turbine exhaust gas is recovered through a waste heat recovery steam generator as shown in Figure 3-14. Typical combustion turbine generator and waste heat recovery steam applications are shown in Figure 3-15. The high temperature of the turbine exhaust gas and the heat transferred to the environment make the combustion turbine more suitable than other thermal engines because of the ratio between the useful power, both as electricity and hot

(a)

(b)

(c)

**Figure 3-15 Typical cogeneration installations. (*a*) Repowering application in Canada. Two W251 (35-megawatt) combustion turbines with waste heat making steam for existing steam plant. (*b*) Cogeneration in the United States. W501D combustion turbine generator and waste heat boiler making steam for process. (*c*) Combined cycle in Mexico. A Westinghouse preengineered, packaged 520-megawatt PACE plant. (*Westinghouse Combustion Turbines, "The Ready Source of Power Worldwide," Bulletin SA 11152*)**

fluid, and the power supplied to the plant by the fuel. This heat utilization ratio can be as high as 80 percent, a value that can be achieved with installations of no particular complexity. Heat utilization varies from process steam for industry requirements to hot water or steam for district heating. Installation of a combustion turbine with a waste heat recovery boiler is particularly simple. Installation time and the number of operating personnel are considerably reduced.

The general arrangement of a waste heat recovery steam generator and a combustion turbine generator is illustrated in Appendix B, Figure B-4. Figure B-5 shows the layout of a waste heat recovery steam generator, indicating the dimensional relationships of the economizer, boiler drum, superheater, combustion turbine generator–exhaust duct interface, bypass, and isolation damper locations.

## A COMBINED CYCLE FOR POWER GENERATION

The combined cycle shown in Figure 3-16 utilizes the steam produced from the exhaust gas heat recovery steam generator to drive a steam turbine generator set. The steam turbine is the condensing type. It can either be manufactured for the application or be an existing unit repowered by replacing an existing boiler with a waste heat recovery boiler. In plants of this kind the output of the steam turbine may be approximately half that of the gas turbine, and efficiency can reach 45 percent, exceeding that of all other thermal plants.

Combining a gas turbine generator and a waste heat recovery boiler with an existing steam turbine-generator gives new life to and increases the output of

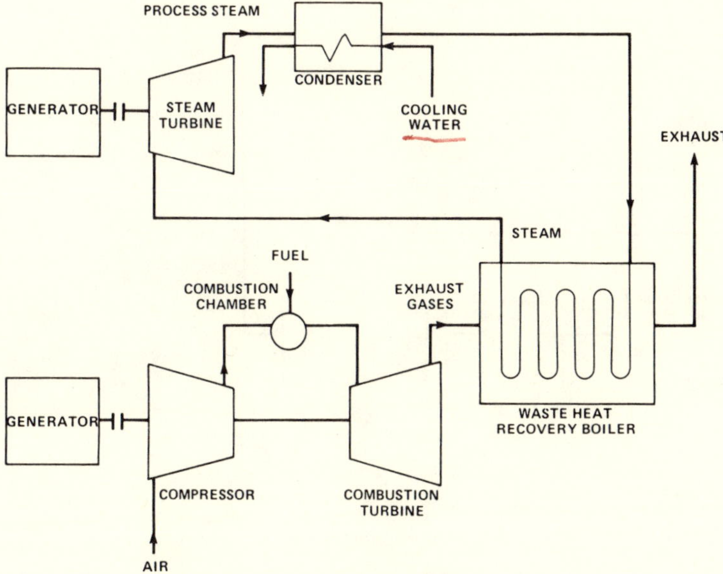


**Figure 3-16 Combined cycle for power generation.**

plants that, because of obsolescence of the conventional boiler, would have been dismantled.

A cross-sectional illustration of a gas turbine generator, steam turbine genera-

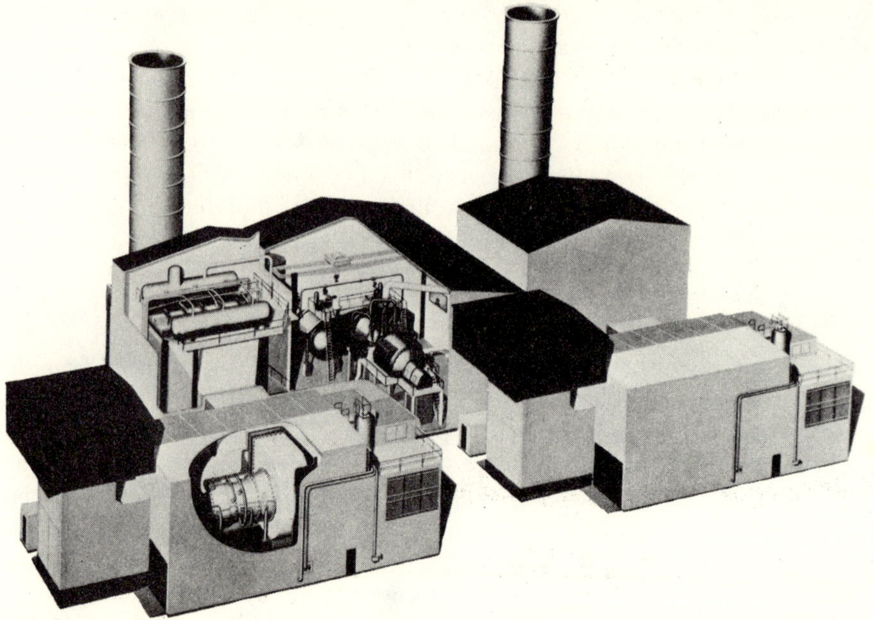

**Figure 3-17 Combined cycle power plant.** (***Stal-Laval, Inc., "Turbine Power," Bulletin 572 E.9.79.4.000***)

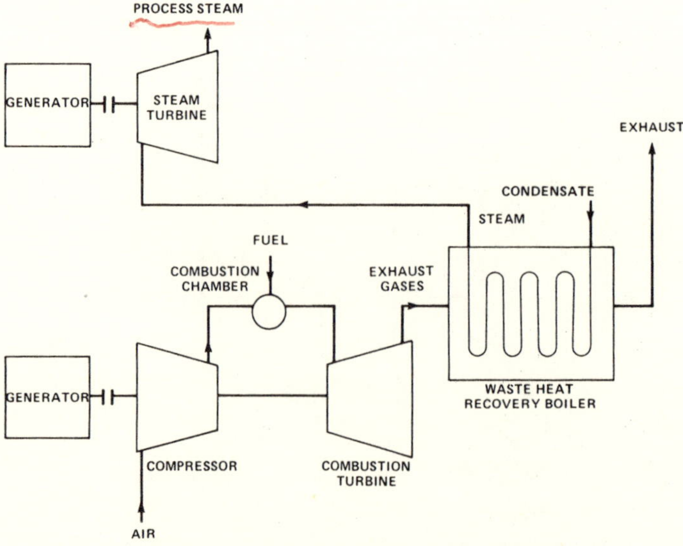


**Figure 3-18 Combined cycle for power and process heat generation.**

tor, and waste heat recovery steam generator combined cycle facility is shown in Figure 3-17.

## A COMBINED CYCLE FOR GENERATION OF BOTH POWER AND PROCESS HEAT

A gas turbine generator and waste heat recovery boiler can be combined with a condensing or extracting or a back-pressure steam turbine generator, as shown in Figure 3-18, to provide electric power and process steam. Multidisk radial steam turbine cross sections and steam flow paths are shown in Figures 3-19 and 3-20.

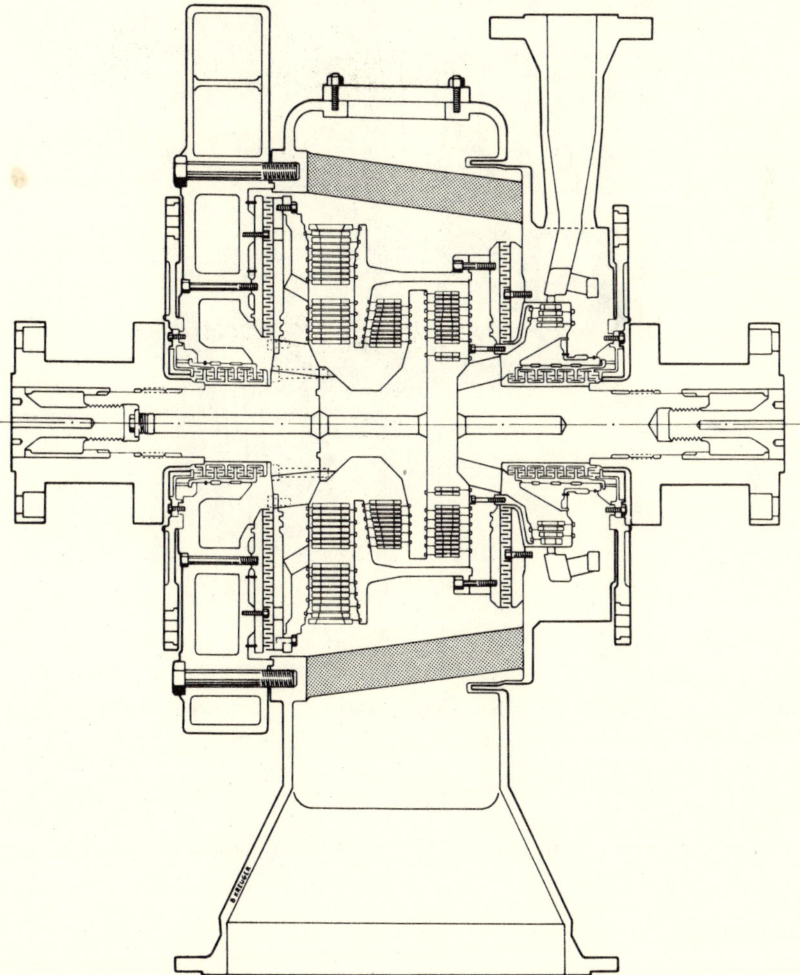

**Figure 3-19** **Steam turbine cross-sectional view.** (***Stal-Laval, Inc., "Turbine Power," Bulletin 572E.9.79.4.000***)

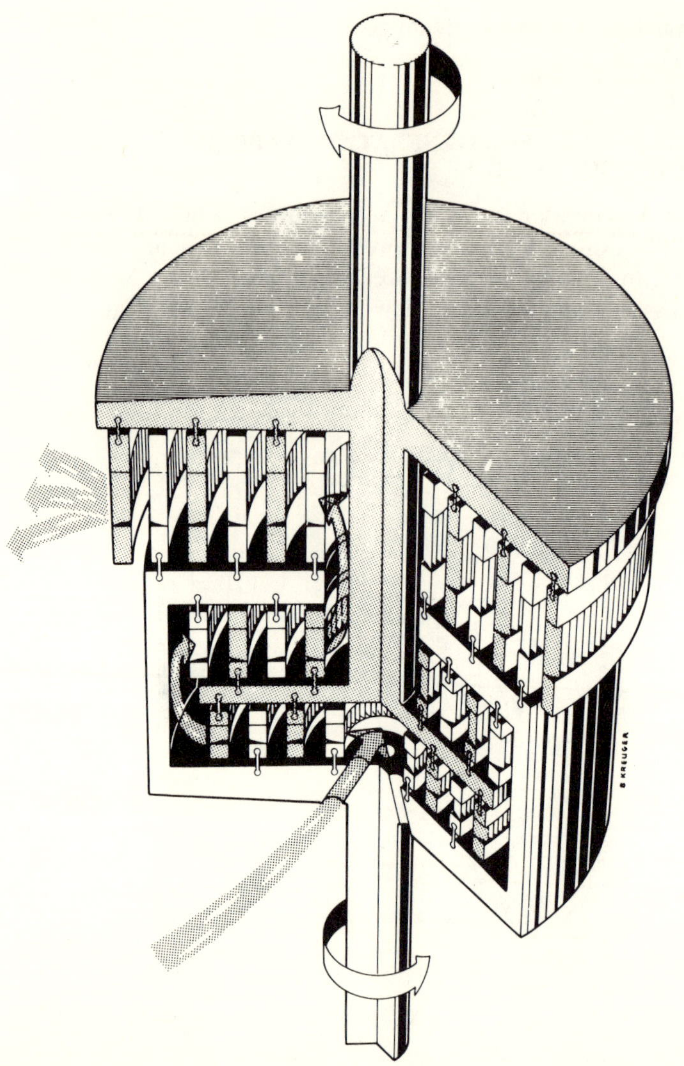

**Figure 3-20 Radial steam turbine steam flow path.** (*Stal-Laval, Inc., "Turbine Power," Bulletin 572E9.79.4.000*)

## APPLICATIONS

Cogeneration is most feasible and financially rewarding for the following types of applications requiring a large consumption of natural gas and energy.

- INDUSTRIAL FACILITIES

    Paper and pulp.

    Food processing.

Chemical manufacturing.

Petroleum refining.

Oil processing.

Secondary recovery in low-gravity oil fields. Extracting heavy crude oil from the ground is difficult and expensive. For every 4 barrels of oil produced, more than 1 barrel is burned to generate steam. The steam is injected into wells to heat the oil in the ground so it can be pumped.

Lumber.

Canning.

- HOSPITALS
- HOTELS
- INSTITUTIONS

  Correctional facilities

  Educational facilities and schools
- OFFICE BUILDINGS
- APARTMENT COMPLEXES
- SHOPPING CENTERS
- UNIVERSITIES AND COLLEGES

Some simple cycle diagrams are included here to help define cogeneration.

A process steam plant consisting simply of a boiler providing steam to a process is shown in Figure 3-21. An ideal temperature-entropy diagram is shown for this cycle. 1′ is the boiler inlet, and 2′ is the superheater outlet. Both the work out (2′ to 3′) and the heat of condensation (3′ to 4′) are received by the process. Ideally this is an efficient cycle with an actual plant efficiency in the low 70 percent range.

The conventional power plant cycle, shown in Figure 3-22, has additive losses in the mechanical equipment plus heat loss to the condenser. The station heat rate is approximately 10,000 Btu per kilowatthour. The ideal temperature-entropy diagram for this cycle shows boiler work in from 1 to 2, assuming no loss in the main steam piping, and turbine work out from 2 to 3. The heat removed to condense the turbine exhaust steam (3 to 4) is wasted in this case. This heat is proportional to the area under line 3-4. Typical enthalpy values are shown, resulting in a calculated theoretical efficiency of 43.5 percent which converts to an actual efficiency of 34.8 percent.

The cogeneration cycle depicted in Figure 3-23 is very much like the conventional power cycle. It can vary in numerous ways, but for this example a single process steam requirement of 60 pounds per square inch gauge was chosen. A noncondensing turbine is used exhausting at 60 pounds per square inch gauge. The temperature-entropy diagram is much like the one for power, except that the bottom line is higher because less energy was removed from the steam in the turbine. The energy passed on to the process via the exhaust steam is 1112-262 Btu per pound. There is less turbine workout and the work-in by the boiler is slightly reduced, but the large loss in the condenser is eliminated, giving an actual plant efficiency in the 72 percent range. Normal heat rates of 4500 to 5000 Btu per kilowatthour are possible.

Various options must be studied when optimizing cogeneration cycles. The cycle must be designed to fit the requirements of the process. Examples of various methods of providing steam for process requirements are given in Figures 3-24 to 3-26, and the power cycle optimization and evaluation are shown in Figure 3-27. The following factors must be considered.

- Are the steam or power supplies to be continuous? What is their ratio?
- Additional types of turbines (i.e., back pressure, condensing).
- Process steam conditions.
- Initial steam conditions.
- Number of feedwater heaters (if any).
- Quantity and quality of condensate.

The evaluation should

- List all the operating condition options.
- Select the system best suited for each operating condition.
- Apply the systems chosen to each operating condition for evaluation of the best overall system.

Higher temperature and pressure steam can be produced at the turbine throttle. This additional temperature and pressure produce an amount of electricity many times greater in value than the additional fuel cost. The cycles of some typical cogeneration projects are depicted in Figures 3-28 to 3-31.

## SUPPLEMENTAL FIRING

A high oxygen content in the exhaust gases permits supplemental firing in the steam generator furnace or the ducting between the combustion turbine exhaust and the waste heat recovery boiler, as shown in Figure 3-32. Supplemental firing can be used if the exhaust gas heat from the combustion turbine or cycle is insufficient for the desired operating mode. Supplemental firing is one approach to meeting peak steam or hot water demands while allowing wide flexibility of application. Temperature limitations must not be exceeded during supplemental firing.

Supplemental firing increases the steam generation capacity through higher gas turbine exhaust temperatures. The steam generation capacity can also be maintained through supplemental firing when combustion turbine generator loads are reduced. Where standby boilers do not exist, a 100 percent back-up burner system should be considered for the waste heat recovery boiler. Environmental considerations limit a 100 percent back-up burner system to 35 × $10^6$ Btu per hour. Selective catalytic reduction and low nitrogen oxide burners are required for waste heat recovery boiler auxiliary firing systems larger than 35 × $10^6$ Btu per hour.

Supplemental fuel-fired waste heat recovery boilers require less fuel consumption than conventional boilers providing the same steam generation.

Supplemental firing is usually done in the inlet ducting, as shown in Figure 3-33. A low–nitrogen oxide duct burner system normally fires either natural gas or no. 2 fuel oil. Ducting from the combustion turbine to the duct burner may be fabricated of alloy steel or carbon steel plate internally lined with a ceramic fiber blanket and covered with a stainless steel sheet to prevent erosion of the insulation. The ducting from supplemental firing burners and the transition duct from the superheater to the boiler may be internally lined with ceramic fiber blanket thermal insulation covered with a rigidized wet felt ceramic liner to withstand temperatures up to 2000°F resulting from supplemental firing. This duct is sized for uniform temperature distribution into the waste heat recovery boiler.

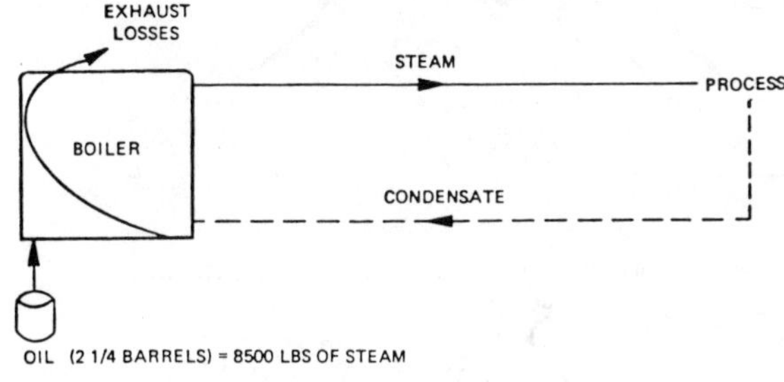


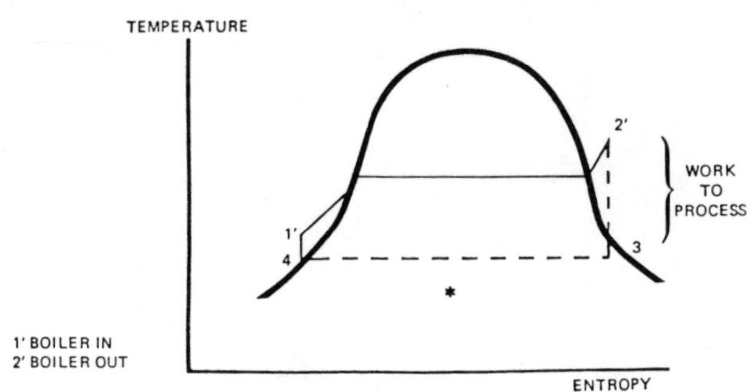


**Figure 3-21 Process steam plant.**

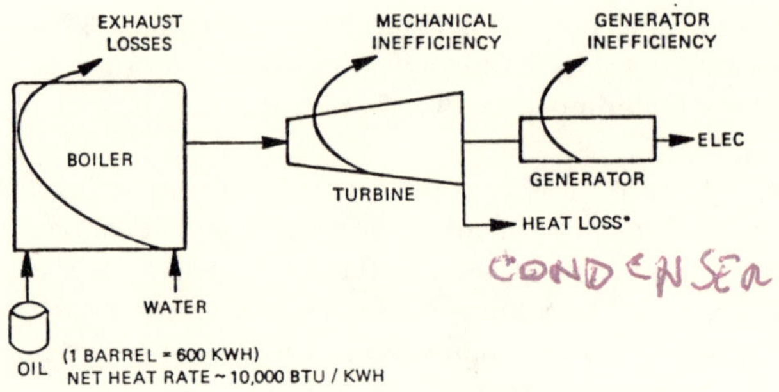


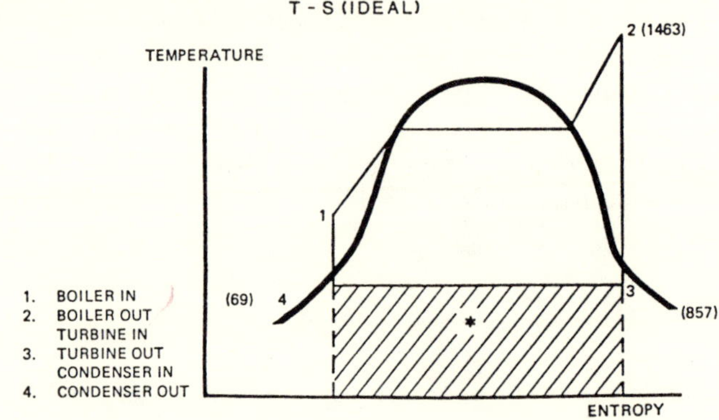


*REJECTED HEAT IS PROPORTIONAL TO THIS AREA

( ) TYPICAL ENTHALPYS

$$EFF_{TH} = \frac{\text{WORK OUT}}{\text{WORK IN}} = \frac{1463 - 857}{1463 - 69} \times 100 = 43.5\%$$

$$EFF_{(ACT)} = 0.80 \times 43.5 = 34.8\%$$

**Figure 3-22 Conventional plant.**

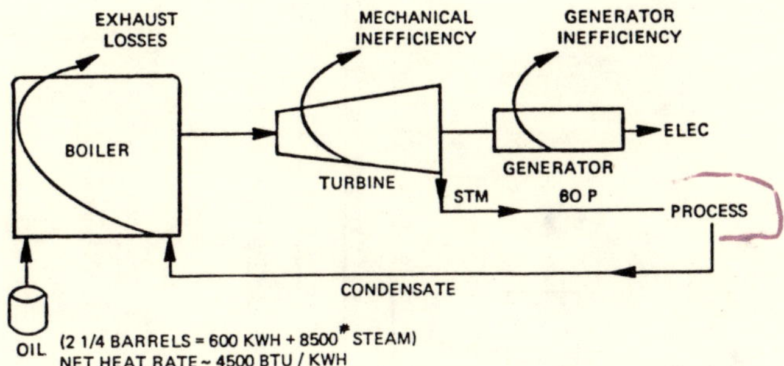


T - S (IDEAL)

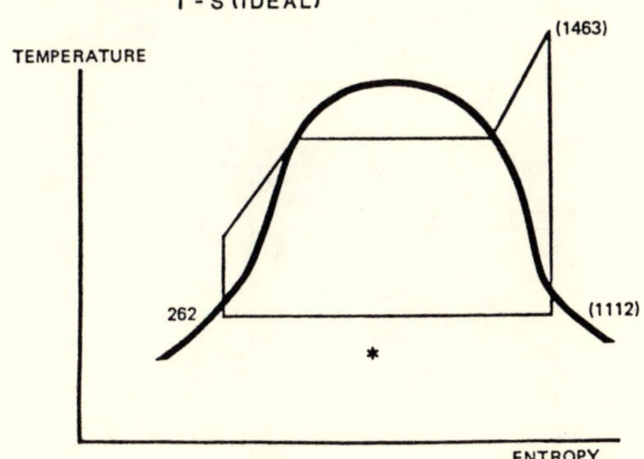


*HEAT LOSS CHARGED TO PROCESS
( ) TYPICAL ENTHALPYS
TURBINE WORKOUT = 1463 - 1112 = 351 BTU/lb
STEAM WORKOUT = 1112 - 262 = 850 BTU/lb
PLANT $EFF_{(ACT)}$ ~ 72%

**Figure 3-23 Cogeneration plant.**

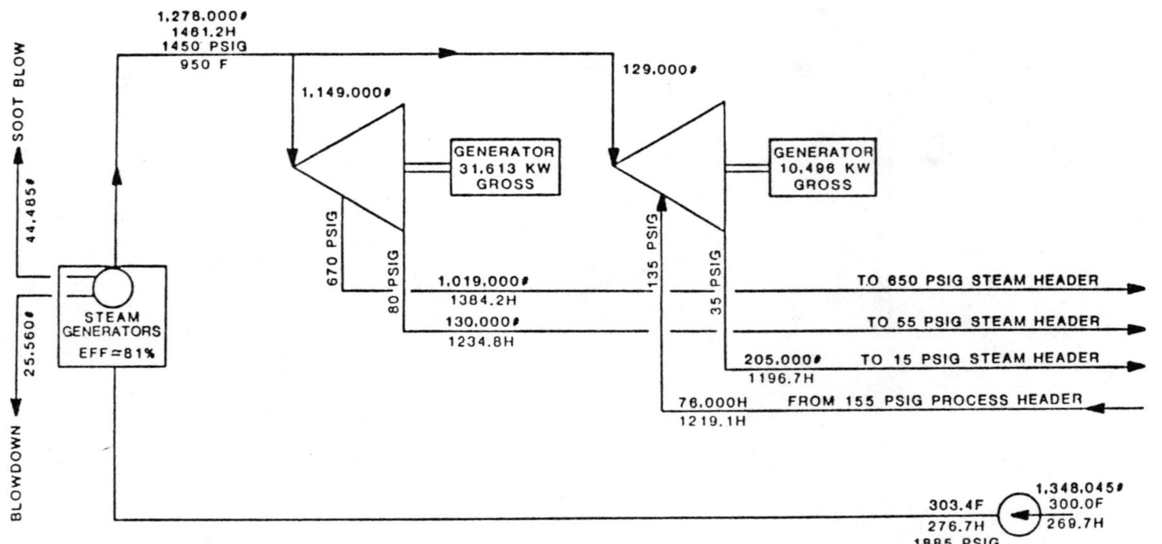


Figure 3-24 Steam generator and extraction steam turbine generator heat balance.

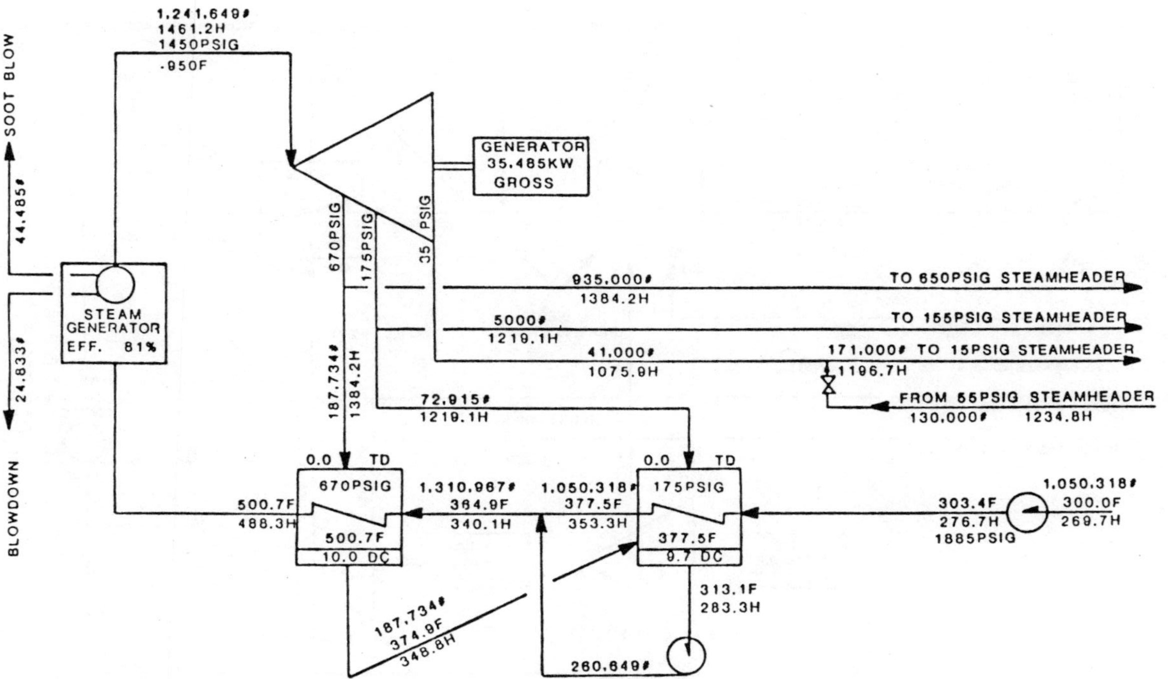


Figure 3-25 Regeneration cycle heat balance.

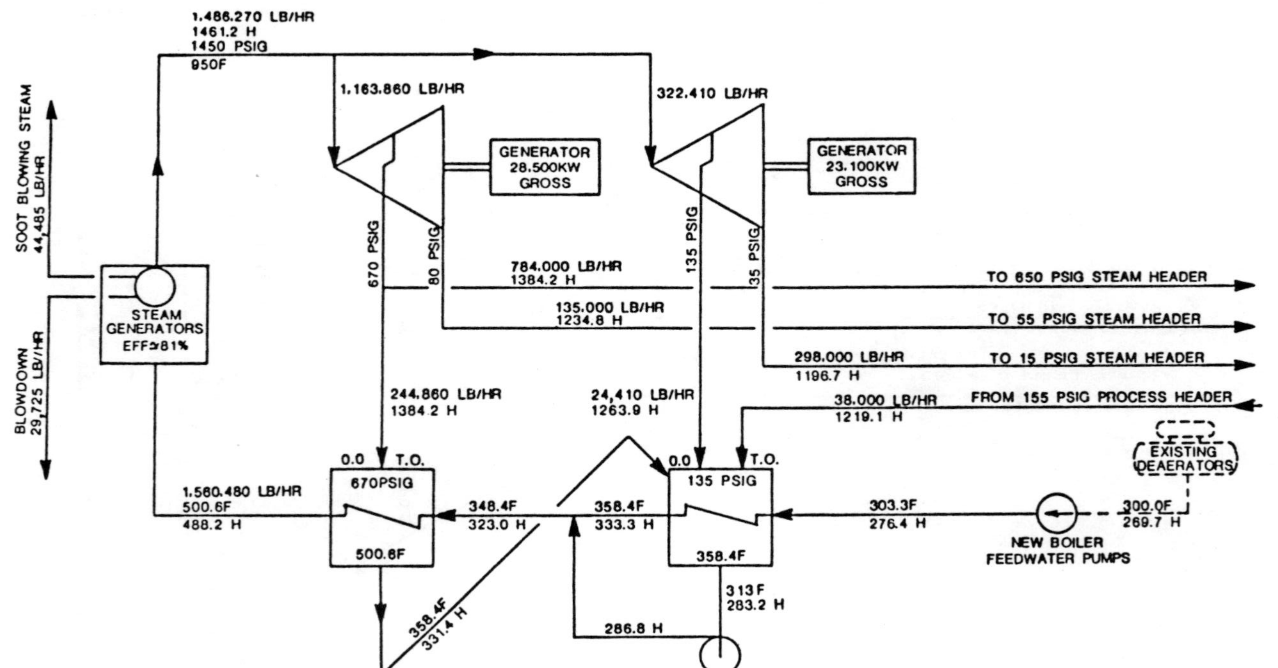


Figure 3-26 Dual steam turbine generator regenerative cycle heat balance.

**OPTIMIZATION CONSIDERATIONS**

- Ratio of power to process steam
- Type of turbine
- Process steam temperature and pressure
- Initial steam temperature and pressure
- Number of heaters
- Quantity and quality of condensate from process

**Figure 3-27** Turbine thermal cycle.

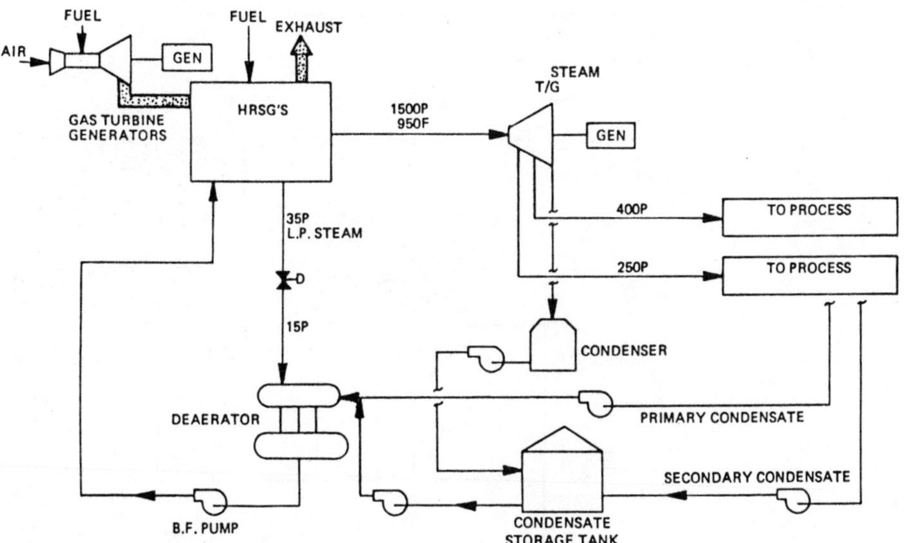


**Figure 3-28** Extraction condensing steam turbine cycle.

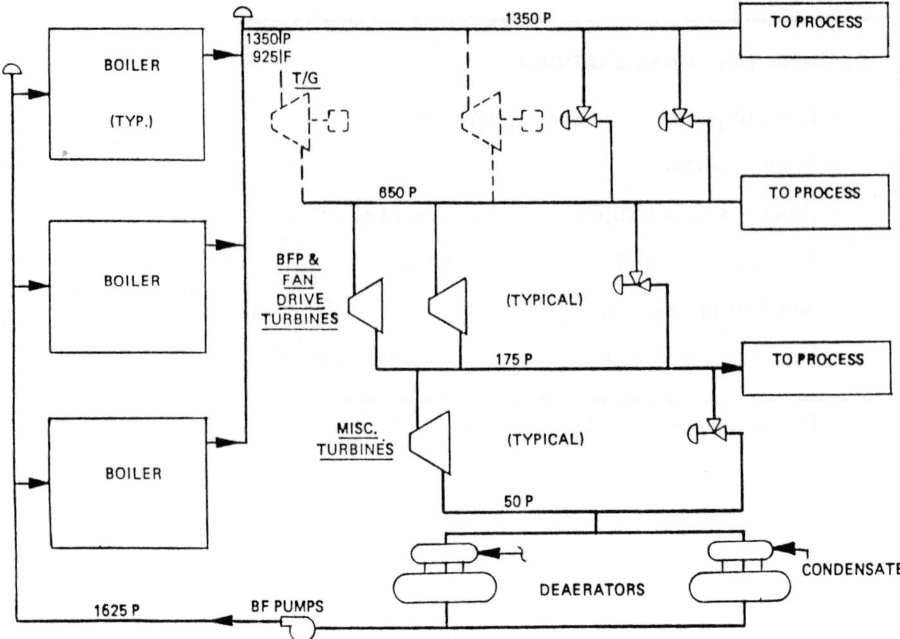


Figure 3-29 Topping turbine power cycle.

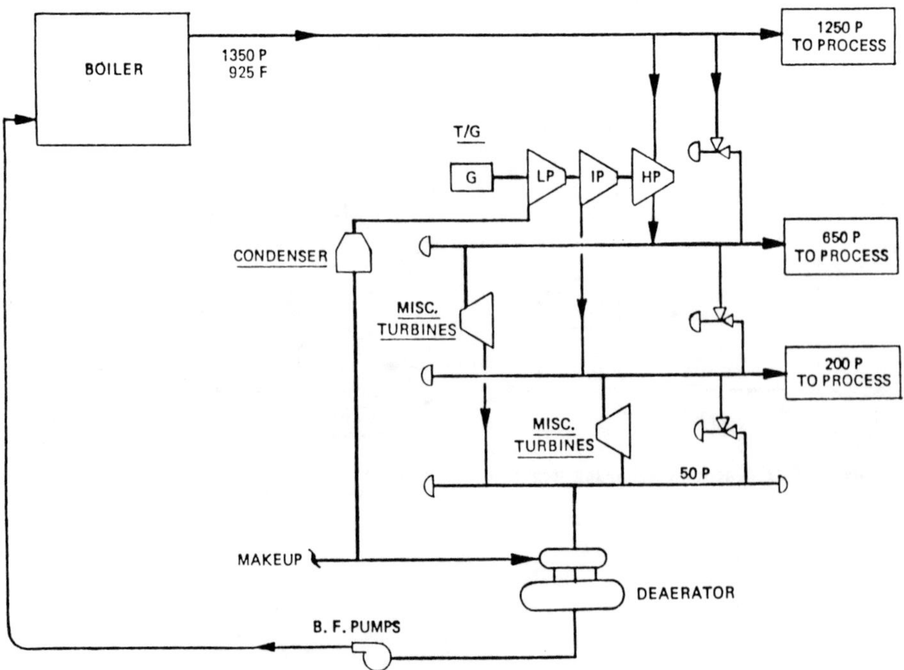


Figure 3-30 Tandem steam turbine generator cycle.

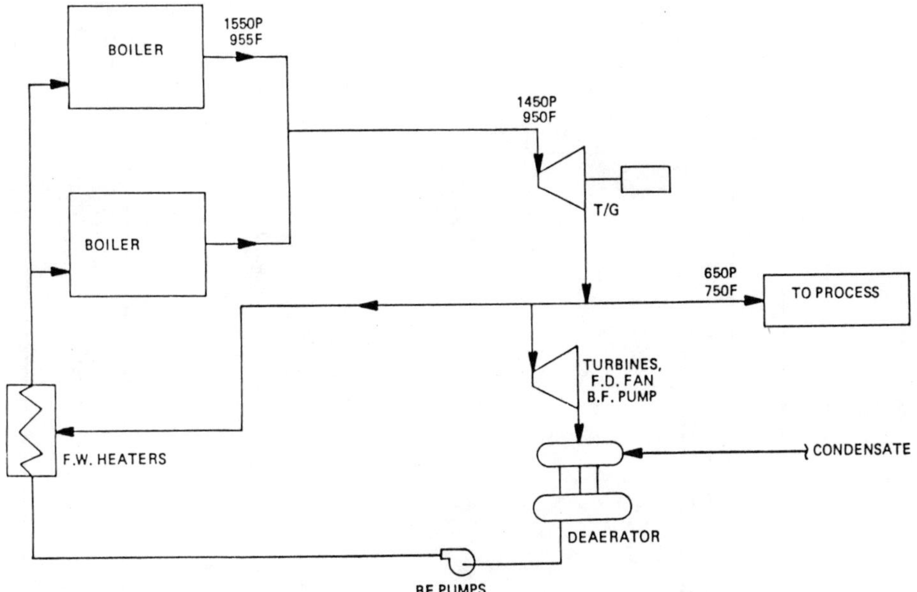


**Figure 3-31 Power and process steam regenerative power cycle.**

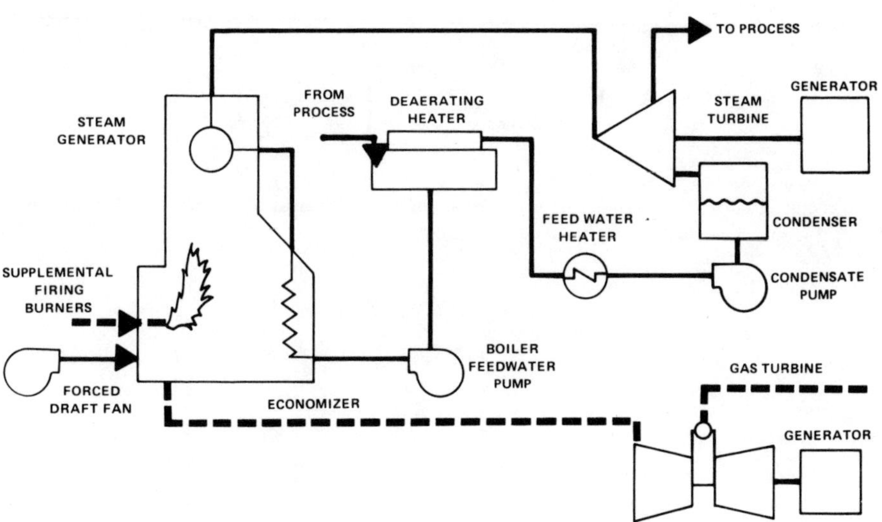


**Figure 3-32 Supplemental fired heat recovery power cycle.**

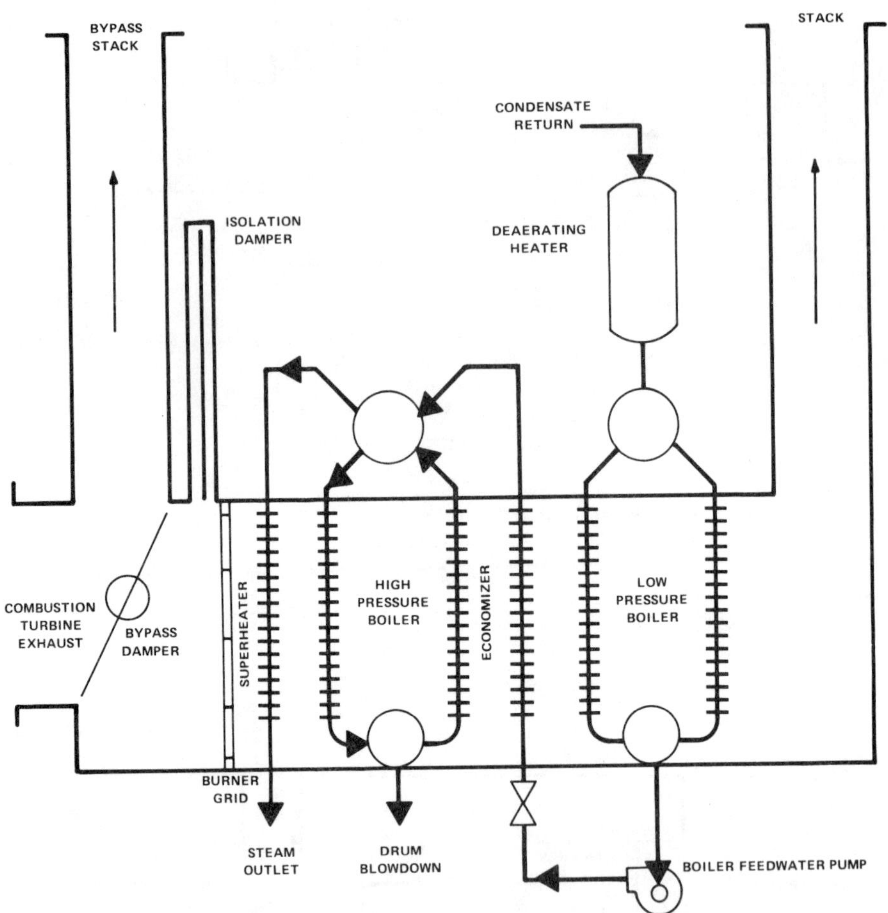


Figure 3-33 Combustion turbine exhaust heat recovery multipressure steam generator system.

# 4

## COGENERATION IN HEATING AND COOLING SYSTEMS FOR BUILDINGS AND DISTRICTS

Cogeneration for building and district space heating and cooling purposes consists of producing electricity and sequentially utilizing useful energy. Useful energy in the form of steam, hot water, or direct exhaust gases is provided for building or district heating and air-conditioning loads as shown in Figure 4-1. The benefits from combined district heating and production of electric power are illustrated in Figure 4-2. A district heating power station and control room are shown in Figures 4-3 and 4-4. An integrated district heating and power-producing complex flowchart is given in Figure 4-5. A typical district heating power station plant layout is shown in Appendix B, Figure B-2.

The two most common heating, ventilation, and air conditioning (HVAC) cycles are the vapor compression cycle and the absorption cycle.

### VAPOR COMPRESSION CYCLE

The vapor compression cycle shown in Figure 4-6 consists of a compressor, a condenser, an expansion valve, and an evaporator. The compressor driver can be an electric motor, steam turbine, or combustion turbine. A single fluid used as the refrigerant vaporizes in the evaporator as a result of heat transfer from the refrigerated space, is compressed by applying work to the compressor, and condenses in the condenser as a result of heat transfer to cooling water or to the surroundings. The refrigerant leaves the condenser as a high-pressure liquid. The pressure of the liquid is decreased as it flows through the expansion valve, and some of the liquid flashes into vapor.

Centrifugal chillers are the most commonly used chillers in the United States, especially for office building air-conditioning systems. They are simple and clean. The chiller compressor drive is usually an electric motor, although either a gas turbine or a steam turbine drive can be used. The centrifugal chiller HVAC application is shown in Figure 4-7. This arrangement is essentially the same as the cogeneration arrangement shown in Figure 4-1 for the absorption chiller cycle, except that the chiller-equipment-related pumps can be driven by electric motors, combustion turbines, diesel or gas engines, or steam turbines. The cost of power or energy consumption for an electric-motor-driven centrifugal chiller is approximately 30 to 50 percent more than that for an absorption chiller system.

## ABSORPTION REFRIGERATION CYCLE

The absorption refrigeration cycles for cogeneration air-conditioning applications primarily utilize water as the refrigerant and lithium bromide solution as the absorbent. Lithium bromide solution has a strong affinity for water vapor, thus giving the absorbent the ability to compress the refrigerant vapor.

The absorption cycle shown in Figure 4-1 consists of an absorber generator (concentrator), condenser, expansion valve (pressure-reducing device), evap-

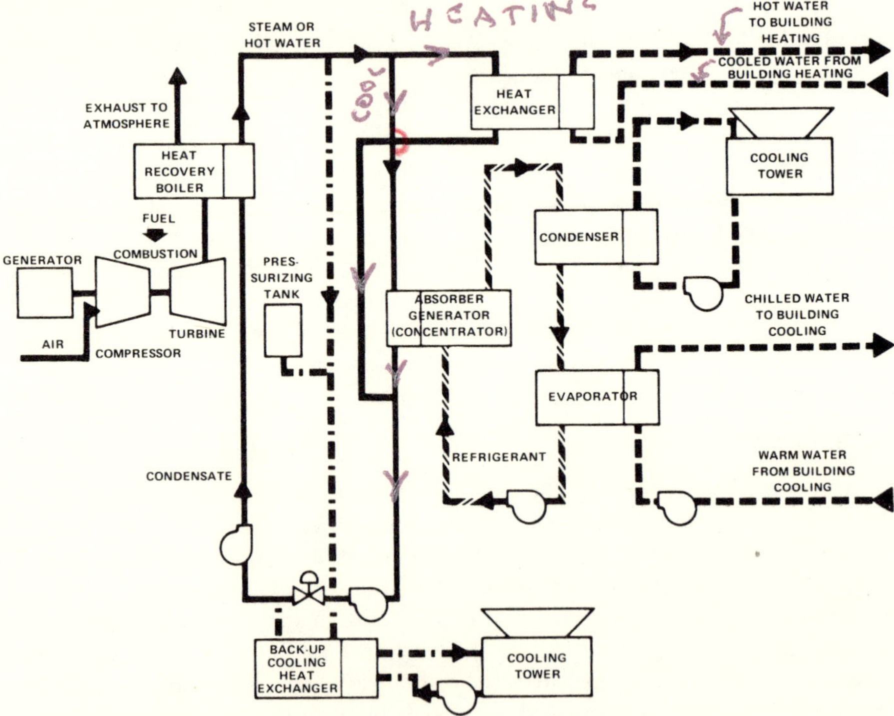


**Figure 4-1 Cogeneration absorption district heating and cooling cycle: cogeneration → electricity → process topping.**

*PURPA and FERC Efficiency and Heat Utilization Standards*

1. Useful thermal energy is the steam to process which must be greater than 5 percent of the total energy.

2. Electricity generated plus one-half the useful thermal energy must be equal to or greater than 42.5 percent of the fuel energy input.

orator, and pump. The absorber functions as a compressor. The pump circulating the absorbent requires less energy than that to drive the compressor in the vapor compression cycle.

This cycle, combined with one or more combustion turbine generators or internal-combustion engines exhausting to one or more waste heat recovery steam or hot water generators, provides a common cogeneration system.

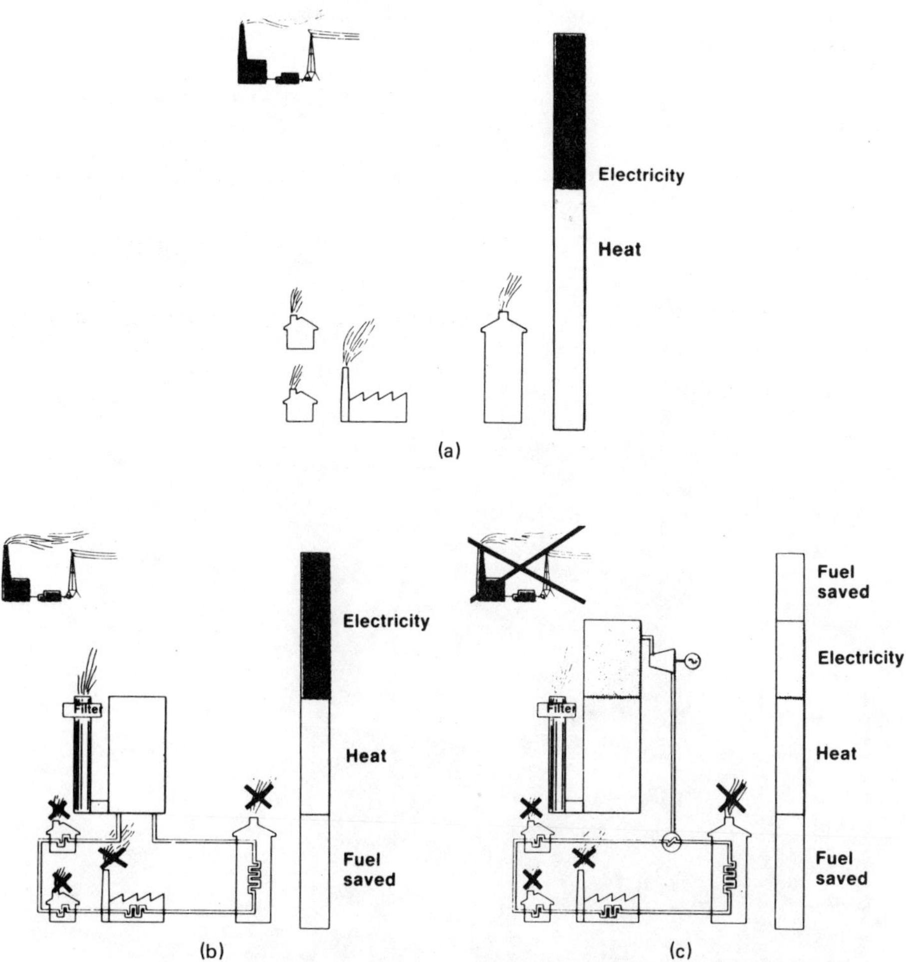


**Figure 4-2 District heating and electric power production. (*a*) When the owners of houses, flats, and industrial premises make their own arrangements for heating and buy their electric power, there is a high rate of fuel consumption. (*b*) If there is a district heating boiler plant which can supply hot water for heating, a great deal of fuel can be saved and, in addition, there is less atmospheric pollution. (*c*) If a district heating power station is built, fuel is used more effectively and large amounts of electric power are obtained at the same time. (*Stal-Laval, Inc.*, "*Turbine Power*," *Bulletin 572E9.79.4.000*)**

## BUILDING HEATING AND COOLING SYSTEMS

Typical cogeneration systems supply power and steam or hot water for heating and air-conditioning loads. The following systems are used.

- Combustion turbine generator exhausting into a heat recovery steam generator (HRSG) supplying steam to a back-pressure turbine generator exhausting steam to an air-conditioning absorption chiller and a steam heating system
- Combustion turbine generator exhausting directly into a HVAC absorption chiller-heater
- Combustion turbine generator exhausting into a HRSG supplying hot water for HVAC applications
- Combustion turbine generator exhausting directly into a HRSG supplying steam for HVAC applications

**Figure 4-3 District heating and power station.** (*Stal-Laval, Inc., "Turbine Power," Bulletin 572E9.79.4.000*)

Figure 4-4 District heating and power station control room. (*Stal-Laval, Inc., "Turbine Power," Bulletin 572E.9.79.4.000*)

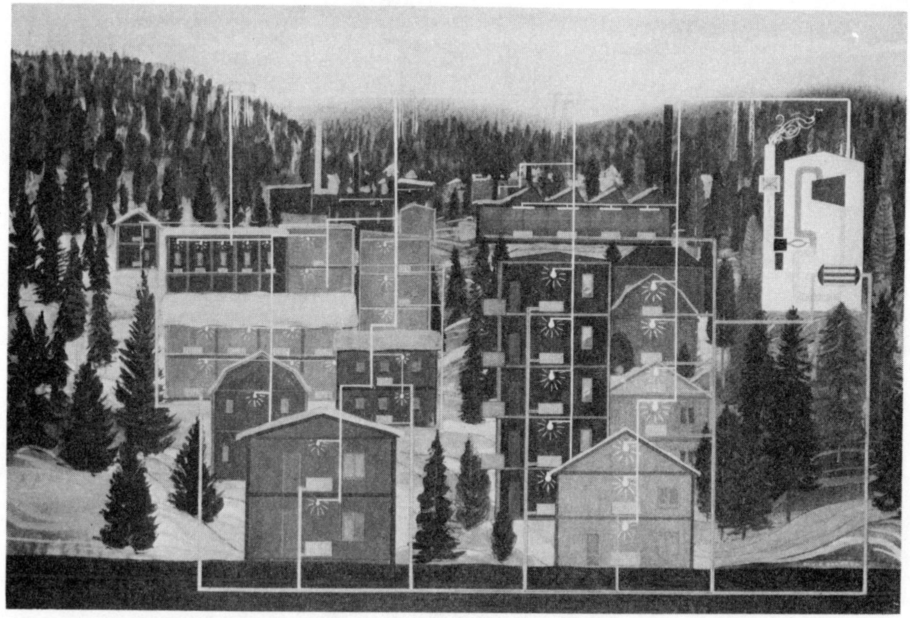

Figure 4-5 District heating and power simplex flowchart. (*Stal-Laval, Inc., "Turbine Power," Bulletin 572E9.79.4.000*)

- Internal-combustion engine generator set exhausting into HVAC absorption chiller-heaters
- Internal-combustion engine generator set exhausting into a HRSG supplying steam or hot water for HVAC applications

The above-mentioned system, consisting of a combustion turbine generator exhausting directly into an absorption chiller-heater, offers benefits such as a minimum number of components, lower associated capital costs, and higher efficiency associated with direct energy conversion of exhaust heat, lower fuel consumption per HVAC output, lower maintenance and spare parts costs, a requirement for less space, flexibility to be used for heating, and simplicity of operation. Absorption chiller-heaters can be supplied with hot water or steam from a HRSG or provided with optional direct gas-fired burners to supplement or use when the combustion turbine is not in service.

## AIR-CONDITIONING SYSTEMS

Typical air-conditioning system alternatives are steam absorption chiller-heaters, direct-fired absorption chiller-heaters, and centrifugal chillers. (Drivers that may be used include combustion turbines, steam turbines, electric motors, and diesel or gas engines.)

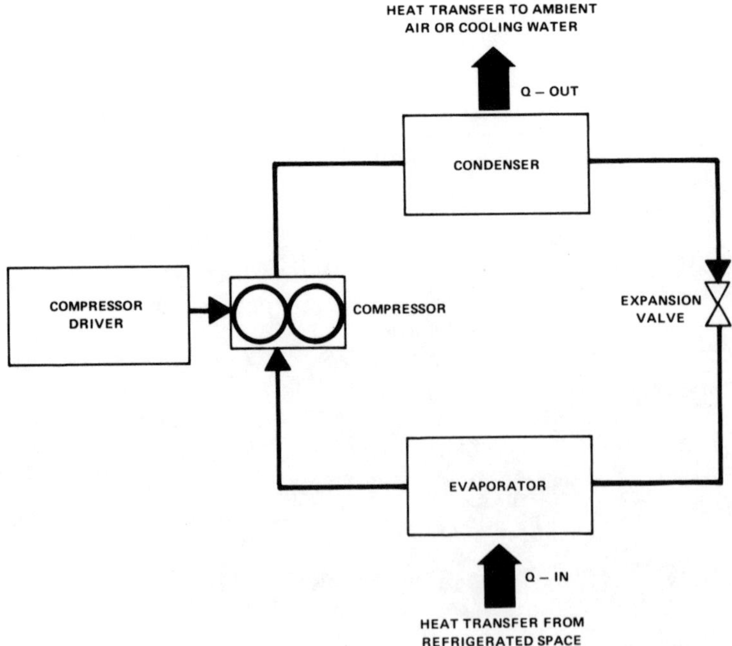


**Figure 4-6 Basic vapor compressor cycle.**

## HEATING SYSTEMS

The two heating systems commonly used are steam heating systems and hot-water heating systems. Hot-water heating systems are widely used in medium to large centralized heating systems because of their higher reliability, lower maintenance requirements, lower operating costs, and longer component life, as compared with steam heating systems. The initial cost of a hot-water heating system, however, is slightly higher than that of a steam heating system.

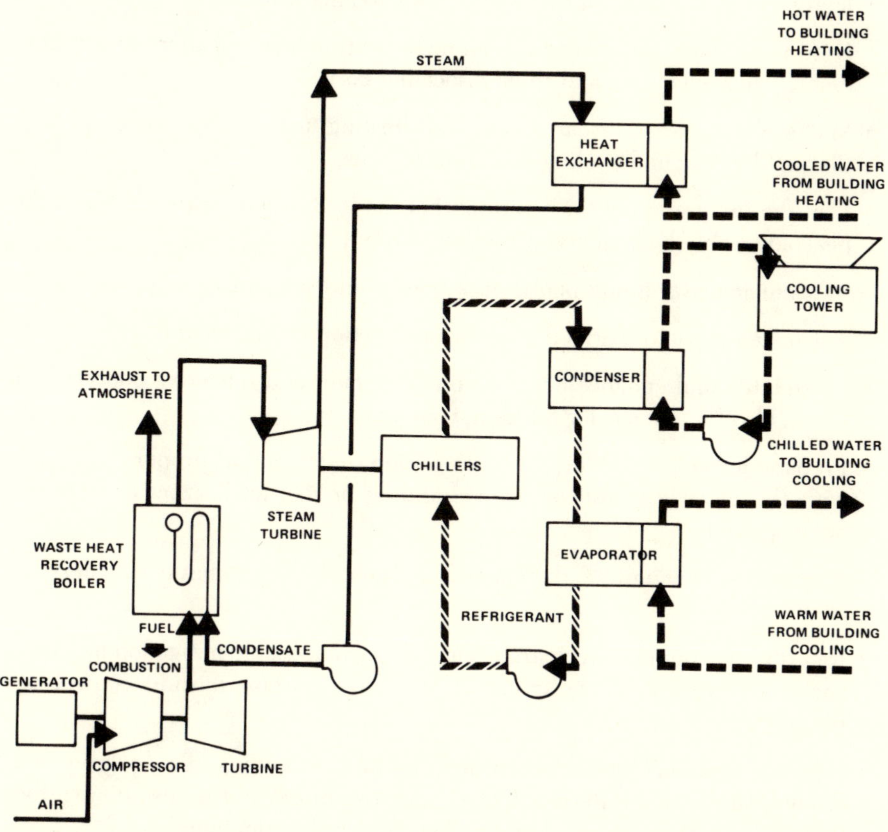


**Figure 4-7 Cogeneration centrifugal chiller district heating and cooling cycle: Cogeneration fuel → electricity → process topping.**

*PURPA and FERC Efficiency and Heat Utilization Standards*

1. Useful thermal energy is the steam to process which must be greater than 5 percent of the total energy.

2. Electricity generated plus one-half the useful thermal energy must be equal to or greater than 42.5 percent of the fuel energy input.

## PLANT DESIGN

The cogeneration facility should be thoroughly examined to confirm the design basis for the plant. The list below includes guidelines for the major steps that must be covered in the preliminary study.

- Determine the suitability of the identified site for the cogeneration facility. Factors to consider include the physical site available, legal matters, and other pertinent features.
- Determine the energy utilization factors.
- Determine the nature of the heating and cooling loads.
- Determine the technical and economic benefits of practical alternative cogeneration thermal cycles and their associated costs.
- Match alternative thermal cycles with heating and cooling loads and with thermal heating and cooling heat-sink options.
- Plan the integration of the building thermal load cycling with additional off-peak accessible heat-sink and cooling sources, such as

  Municipal swimming pools, park ponds, and so forth

  Cooling capacity derived from fire protection storage water

  Architectural enhancements such as fountains and reflecting pools (which can also be utilized for building cooling)

- Prepare design criteria for the cogeneration facility and its proper integration with the electric system and the building or district heating and cooling development.
- Prepare an economic and financial analysis of the proposed cogeneration facility.
- Identify environmental constraints and determine viable mitigation measures such as nitrogen oxide, sulfur dioxide, and carbon dioxide control applications.
- Prepare a project implementation schedule containing the analyses conducted, the design criteria selected, and a summary of the results, together with conclusions and recommendations for implementation.

## INTEGRATION OF COGENERATION WITH BUILDING HEATING AND COOLING SYSTEMS

Efficiently designed cogeneration projects provide technical and economic features which can be optimized to meet both electrical and thermal heating and cooling requirements.

It is important to assess load requirements on a seasonal basis so that heat-utilization factors for the cogeneration plant can be accurately determined. A comprehensive knowledge of the nature of the plant heating and cooling requirements, as well as the cogeneration cycle characteristics, is essential to creation of an optimal overall design for both the building and the cogeneration facility.

The engineer or project manager should work closely with the client to assess present and future electrical and thermal load requirements. The steps involved in performing this assessment include

- Identify and quantify the overall energy requirements.
- Identify the peak heating and cooling requirements.
- Establish daily and seasonal variations.
- Develop daily load duration curves to characterize heating and cooling requirements.

The most cost-effective integrated system possible for the particular facility may be designed by

- Integrating cogeneration energy systems into the entire pattern of energy consumption associated with the building heating and cooling systems
- Designing the system to compensate for either heat loss or heat gain in a building, and taking into consideration factors such as infiltration of outside air, thermal transmission, ventilation, lighting, solar heat gain, equipment for heat dissipation, and number of occupants and type of activity.

The cogeneration concept—combining heating and cooling in one energy system—offers tenants

- Cost-saving incentives relative to energy consumption
- Satisfactory lighting levels
- Satisfactory heating and cooling

The heating-and-cooling system must be properly balanced to achieve the airflow rates necessary for the safety and comfort of building occupants. Listed below are some factors which must be considered in the design phase.

- Systems must be adjusted and balanced to minimize improper control causing

    Overcooling

    Overheating

Poor zoning

Poor distribution

- For proper air balance and to retard infiltration-caused heat losses and heat gains, a slight positive pressure should be maintained.
- Both positioning of dampers and selection and maintenance of filters are important in creation of a properly balanced system.
- Damper sequencing of outdoor air and fan control adjustments should be programmed to take into account hours of occupancy and equipment utilization.
- Total energy management can substantially reduce energy consumption and provide dramatic savings. These ends can be achieved by combining programmed start-stop and night setback adjustments. Demand forecasting and load shedding can be combined with more sophisticated computerized control techniques, such as

  Enthalpy switchover and supply air reset control

  Optional morning start-up

  Chiller plant control
- Centralized control also means that fewer personnel are needed, and yet constant surveillance over all building systems can be maintained.

Cogeneration savings go hand in hand with building temperature control and energy management systems. All the factors listed below should be evaluated in relation to possible energy-saving improvements which can lead to improved cost-effectiveness and increased return on investment.

- The climatic conditions of the building
- The working environment
- Business or other equipment required by tenants in addition to basic heating, cooling, lighting, and ventilation
- Building complex envelope or boundaries consisting of light, insulation, glazing, alternate fuels, thermal storage, spot cooling or heating, and the power factor correction

# 5

## RECIPROCATING INTERNAL-COMBUSTION ENGINES

A reciprocating internal-combustion diesel engine and a gas Otto cycle engine can be used as the prime mover to drive an electric generator, as shown in Figure 5-1. Waste heat can be recovered from the engine exhaust gases, the engine cooling water, and the engine lubricating oil to produce steam, hot water, or both, to be used for building heating and cooling applications, as shown in Figure 5-2.

At reduced loads, the heat rates of gas engines and combustion turbine engines increase more significantly than do the heat rates of diesel engines.

Cogeneration facilities which use a reciprocating engine operate at lower temperatures than alternative cogeneration systems because of engine temperature limits. Overall engine thermal efficiency can be increased up to approximately 75 percent by waste-heat recovery.

### OTTO CYCLE ENGINES

With minor modifications, gas engines can burn fuels such as gasoline, propane, butane, natural gas, sewage gas, alcohol, and gasoline and alcohol mixtures. Gas engine fuel-to-air ignition is achieved by a spark plug or magneto-type ignition system.

The Otto cycle gas engine can be used to drive an electric generator. Engine heat rejected to the cooling water, lubricating oil, and engine exhaust can be recovered for project heating. Cooling water can be used to generate low-pressure steam or hot water; however, lubricating oil heat recovery is sufficient only to produce hot water. The engine exhaust represents about 30 percent of the recoverable heat at temperatures sufficient to produce steam. The conversion efficiency for fuel to electricity is approximately 30 to 33 percent. Total system efficiencies of 75 to 85 percent can be achieved by rejected heat recovery in the conversion of fuel energy to usable energy.

The recovered heat temperature is limited by the permissible temperature of the engine and is in the range of approximately 180 to 250°F. Low-temperature

Figure 5-1 Engine generator sets. (*Colt Industries, Fairbanks Morse Engine Division*)

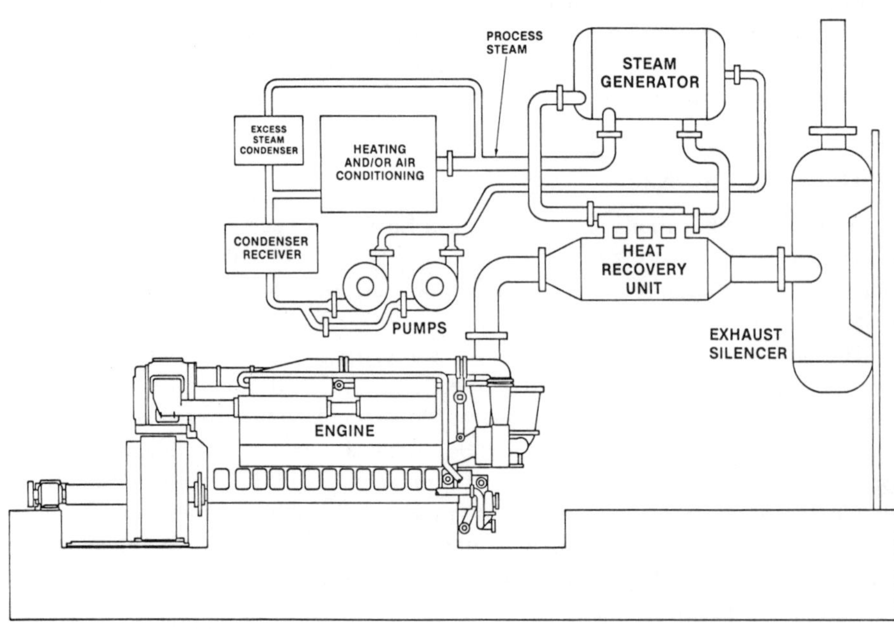


Figure 5-2 Internal combustion engine heat recovery system. (*Colt Industries, Fairbanks Morse Engine Division*)

single-stage absorption chiller machines should be used because of the low temperature of the recovered heat. These chillers are approximately 65 percent as efficient as steam turbine centrifugal chillers and high-temperature two-stage absorption machines.

The maximum hot-water temperature from the heat recovery system is limited to approximately 250°F. Each engine can have its own generator and heat recovery silencer, and heat can be recovered as hot water or as low-pressure steam.

Cooling is produced by means of low-temperature (250°F) absorption chillers with cooling towers. These towers can also be used to cool the engines when the load requires less cooling or less heat than is being produced and operation of the system for electricity generation alone is desired.

The Otto cogeneration cycle is essentially identical to the diesel cycle shown in Figure 3-3.

## DIESEL GENERATOR SETS

Diesel generator sets consist of small, high-speed and large, slow-speed two-stroke engines. Diesel engine fuel-to-air spontaneous ignition is achieved from the heat of compression.

Large, slow-speed units are more efficient and have a longer operating period between maintenance and overhauls than small, high-speed engines. Large engines generally cost more and require more space per unit electrical output.

Waste heat is recovered from diesel exhaust and cooling system. The thermal energy recovered from exhaust gases ranging from 1000 to 1400°F can be used to generate hot water or steam through a HRSG. Hot water can be recovered from the diesel cooling water and lubricating oil systems with chiller-heater equipment.

Diesel engine generator sets have a thermal efficiency of about 33 percent, as indicated in Figure 3-2. This means that about one-third of the energy content of the fuel is converted to work. Recovering the exhaust gas heat can increase the efficiency to about 45 percent. Additional capital equipment investment to recover the cooling water and lubricating oil heat content can increase the efficiency to about 75 percent.

Diesel engines have a higher thermal efficiency than gas turbines and steam turbines and thus greater electrical conversion efficiency. The diesel engine also has a higher fuel-to-electricity conversion ratio than the gas engine because of the diesel's higher compression ratio.

Engine performance remains essentially constant down to a 50 percent load. Engine reject heat can be recovered effectively from the exhaust gas, the jacket, and the turbocharger air coolers.

The recommended downtime for engine inspection and maintenance is 300 to 450 hours/year, which is equivalent to a plant availability of about 95 percent.

A diverting valve in the exhaust system passes the exhaust to the heat recovery boiler, or, if required, directly to the adjoining stack.

Diesels emit less sulfur dioxide but more oxides of nitrogen per kilowatthour of electricity than conventional oil-fired electric utility power plants. Nitrogen oxide emissions are a concern that may affect compliance with environmental regulations.

## HEAT RECOVERY

The heat requirements of a facility are usually the determining factor favoring one system of heat recovery over another. Heat recovery from engines may utilize the heat transferred from an engine radiator to a flow of air at temperatures ranging from 100 to 150°F. Air can be delivered essentially free of contaminants. The effect is a relatively efficient system converting approximately 33 percent of input fuel energy to work or power and 30 percent to recovered heat energy, resulting in an overall efficiency of approximately 63 percent. This percentage can be increased by approximately 12 percent by recovering a portion of the exhaust heat. Exhaust heat recovery equipment adds to the cost, except in processes where the exhaust can be used directly.

The primary responsibility of heat recovery equipment is to cool the engine. Secondary functions are to recover heat and to silence engine exhaust. These units include a heat recovery muffler, steam separator, and safety devices to protect the entire cooling system. Heat recovery units are constructed either as water tube (water inside a tube) or fire tube units. Pressure drops on both the water and gas sides must be considered when sizing the heat recovery unit to the engine.

Typical heat recovery systems include hot-water systems, steam systems, and ebullient systems, as shown in Figure 5.3

### Hot-Water System

A hot-water system utilizes a jacket water temperature of approximately 190 to 250°F recorded at the engine outlet. A shell-and-tube heat exchanger transfers rejected engine heat to a secondary loop. An exhaust heat recovery boiler or heater may also be included in the system. The primary coolant loop serving the engine jacket must be a closed system.

Pressure control provided in the engine coolant loop ensures sufficient pressure to preclude steaming or steam formation in the engine coolant loop. This pressure source may be designed as a static head from an elevated expansion tank, or controlled gas or air pressure in the expansion tank. Water-circulating pumps must be designed for the elevated temperatures and pressures.

### Steam System

A steam system incorporates features of a hot-water system plus a boiler for generating low-pressure steam. Steam is generated in the boiler and in the piping to the boiler as a result of the pressure differential existing by design between the engine outlet and the boiler. A lower pressure prevails in the boiler than at the

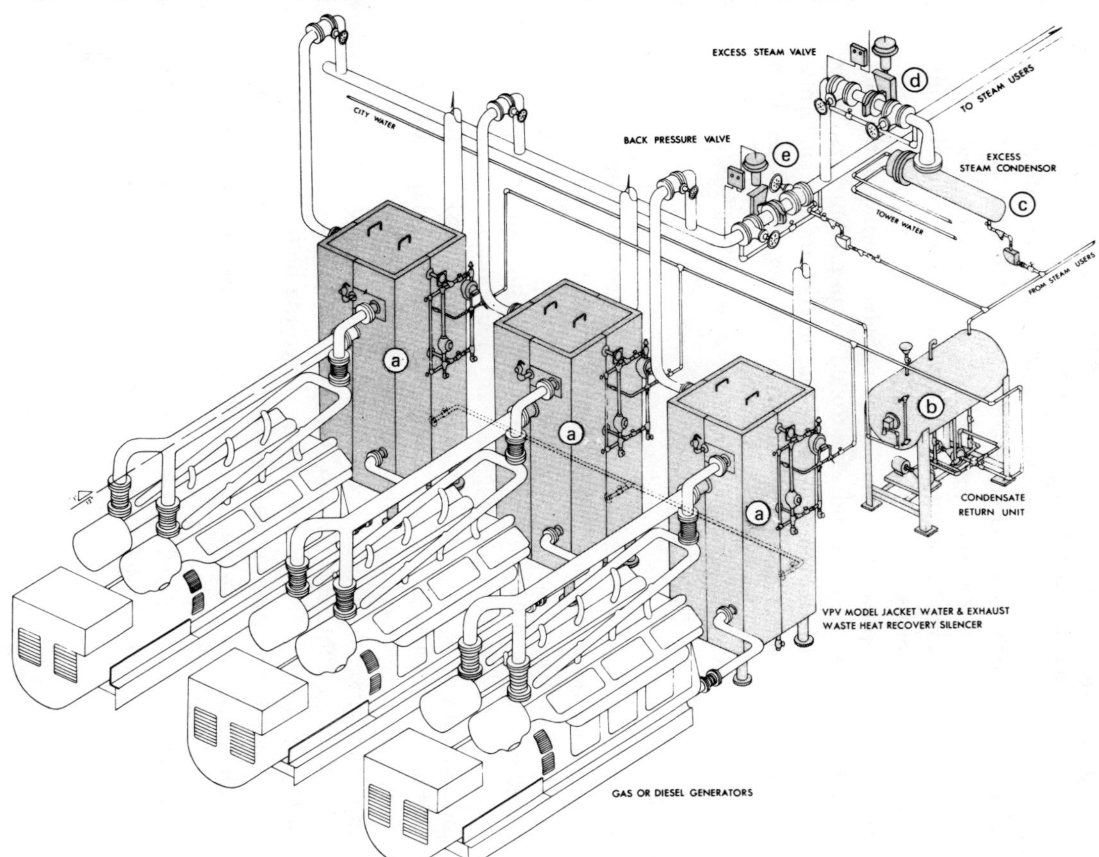


**Figure 5-3 Internal-combustion engine.** (*a*) **Vaporphase Model VPV packaged jacket water and exhaust heat recovery silencers.** (*b*) **Vaporphase Model ECCR condensate return unit.** (*c*) **Vaporphase Model ST excess steam condensor.** (*d*) **Vaporphase Model ESB pneumatic excess steam valve and pilot.** (*e*) **Vaporphase Model BPB pneumatic back-pressure valve and pilot.** (***Vaporphase packaged heat recovery units***)

engine outlet. As the high-temperature water from the engine approaches the boiler, the static head is reduced, as is the heat content of the liquid corresponding to the lower pressure. Heat of vaporization converts part of the water to steam. The temperature of both the steam and the remaining water adjust to the temperature corresponding to the controlled pressure in the boiler. The steam is supplied to the system and recycles through the process.

The total pressure in the engine cooling loop, consisting of the combined steam pressure and the static head, must be adequate to prevent boiling or flashing within the engine. For controlled boiler pressures, the static head required may be reduced but should always be adequate to prevent flashing in the engine cooling loop under normal pressure fluctuations.

A properly installed and operated pressure-reducing valve or orifice at the inlet to the boiler, instead of a static head and a controlled steam pressure system, is more susceptible to a pressure imbalance in the system and potential steam formation in the engine jacket water pump and in the engine, which can result in equipment damage.

### Ebullient System

An ebullient system utilizes the heat of vaporization to cool the engine. The engine must be protected at all times by maintaining adequate circulation of the coolant to absorb rejected heat. Steam is moved through the water passages, along with the high-temperature water, by natural circulation to a steam separator located above the engine, as shown in Figure 5-4. Circulation of the coolant is thermally induced and depends on the static head or finite density difference between the coolant water and the steam-water mixture. This is the simplest and least costly form of waste heat recovery, and a jacket water-circulating pump is not required. The temperature differential between the water inlet and outlet of the engine is normally 2 to 3°F. Flow through the engine is due to the change in coolant density as it gains heat from the engine. Since the higher-temperature coolant is lighter, a pressure differential is created between the water inlet and water outlet connections to the engine. Almost all the heat gain in the coolant is added in the form of heat of vaporization.

The exhaust gas boiler and the steam separator can be combined into a single unit, and a direct-fired section added to the exhaust boiler, eliminating the need for an auxiliary boiler.

## CONTROLS

The switchgear and controls used with gas and diesel engine generator sets vary in sophistication and cost. The type of facility and load served by the plant dictate the basic engineering sizing requirements for the switchgear.

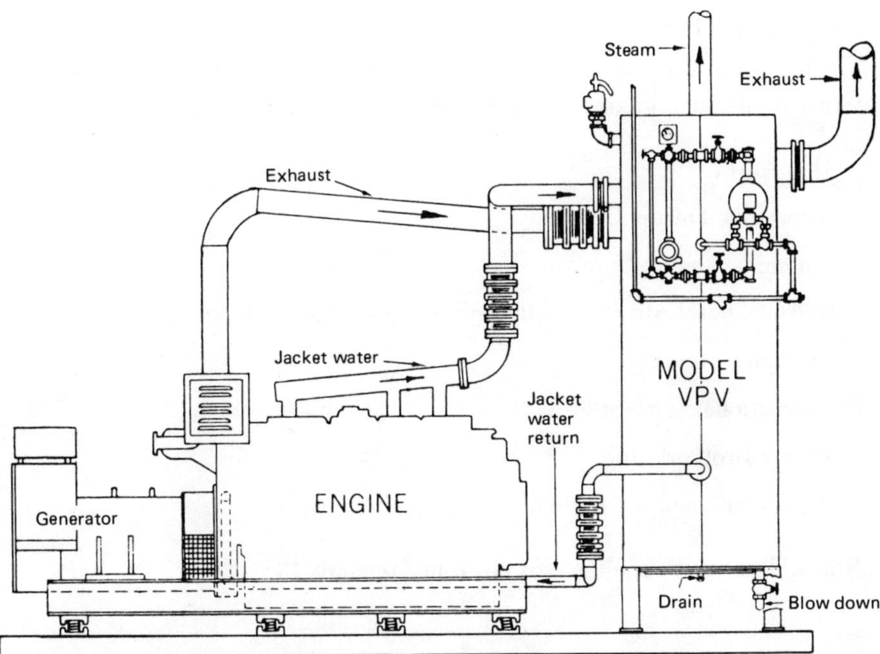


**Figure 5-4 Ebullient reciprocating internal-combustion engine and waste heat recovery system. The Vaporphase Model VP packaged heat recovery silencer is specifically designed to provide maximum economical recovery of waste heat from engine jacket water and exhaust while satisfying the many problems encountered in existing total energy and other waste heat recovery installations. Sizes available are standard 100 through 1500 horsepower. For larger sizes consult factory. (*Vaporphase packaged heat recovery units*)**

Cogeneration plants can be grouped into two basic categories:

1. MANUALLY CONTROLLED POWER PLANTS In manually controlled power plants, the operator sequences units in or out of service as the load profile demands. The units are generally paralleled and the load divided manually.

2. AUTOMATED POWER PLANTS An unattended or totally automated system utilizes an electronic governor to maintain control. Governors enable the units to be automatically paralleled and then divide the load proportionally. A signal-initiated device places units in operation or out of operation as the load profile demands. The electronic governor allows a group of engine-driven generator sets to be paralleled with reliability and ease. These units can be operated unattended by designing adequate monitoring and indicating devices to notify a distant or roving operator when abnormal operating conditions exist. Routine servicing and major maintenance is required.

## TYPICAL SUPPLIERS

The following is a listing of some representatives and suppliers of reciprocating internal-combustion engine generator set equipment.

Alco Power, Inc.

Caterpillar Tractor Company

Cummins Engine Company Inc.

Detroit Diesel Allison, A Division of General Motors Co.

Fairbanks Morse

International Harvester

Sultzer Brothers, Inc.

Waukesha Engine Division, Dresser Industries, Inc.

The addresses of suppliers are given in Appendix I.

# PERFORMANCE, TECHNOLOGY, AND COSTS OF COGENERATION EQUIPMENT

## PERFORMANCE COMPARISONS: ALTERNATIVE COMPONENTS AND CAPACITIES

The machine sizes and conservative technical performance data summarized in Table 6-1 are representative of the various major cycle components used in cogeneration projects. Applications cover simple cycle, combined cycle, and district heating and cooling alternatives. Additional application and detailed data should be obtained by consulting with the manufacturer to verify the latest technology and applicable thermal performance guarantees which vary widely, as shown in Table 6-1. By properly engineering the power cycle and optimizing the major equipment performance to the thermal and internal electrical demands of the cogeneration facility, these heat rates can be reduced and overall cycle efficiencies increased.

## PERFORMANCE SPECIFICATIONS AND TECHNICAL COMPARISONS: MAJOR EQUIPMENT

Selections of applications to match thermal loads to electric power generation requirements and comparisons of equipment performance characteristics and space requirements are provided by many manufacturers of gas and steam turbine generators.

A comparison of alternatives indicates which manufacturers have major equipment available in the size ranges required. Performance differences and technical evaluations of competing equipment can be made simply by comparing alternatives. The final equipment selection should be based on an engineered power cycle which matches the thermal and electric power output to the cogenerator's facilities to obtain the best overall efficiency, lowest capital installed cost, lowest operating cost, and lowest maintenance cost.

Specifications are prepared, and equipment manufacturers are asked to submit technical and commercial bids or proposals for each equipment item. The propos-

**Table 6-1 Alternate-Cycle Peformance Comparisons**

| Major cycle component | Unit size, kW | Range of heat rate, Btu per kWhr | Range of exhaust temperature, °F |
|---|---|---|---|
| Conventional combined cycle | 7,000–24,000 | 9,850–9,450 | — |
| | 38,000–75,000 | 7,650–7.650 | — |
| | 100,000–300,000 | 8,130–7,800 | — |
| | 400,000–750,000 | 8,080–7,210 | — |
| Combustion gas turbine generator | 800 | 16,250 | 860 |
| | 3,000–5,000 | 13,968–13,650 | 807–981 |
| | 7,000–10,000 | 15,140–10,520 | 860–668 |
| | 12,000–15,000 | 12,500–11,730 | 700–842 |
| | 15,000–20,000 | 12,140–9,950 | 871–855 |
| | 25,000–50,000 | 12,310–11,040 | 905–986 |
| | 65,000–105,000 | 11,790–10,620 | 985–953 |
| Diesel engine | 60–975 | 11,100–10,900 | 800–1,000 |
| Cogeneration | 7,000–25,000 | 4,700–5,500 | — |
| | 35,000–75,000 | 4,400–5,700 | — |
| | 100,000–300,000 | 4,700–4,000 | — |

als received are then evaluated for conformance to specifications, performance guarantees, warranties, terms of payment, prices, and delivery times. Finally, for each major equipment item, the supplier that offers the lowest price for equipment that satisfies the specifications is then selected. This approach will verify detailed manufacture design, performance data, latest technology developments, and successful or unsuccessful operating applications experience.

Manufacturers' specifications for and descriptions of gas and steam turbine generators are given in Appendix L. Performance specifications for simple cycle gas turbine generators and combined cycle plants are listed in Appendix M where manufacturers are listed alphabetically. The data given for gas turbines include the model, use or application, normal and maximum power rating and heat rate, compressor and power turbine stages, rpm, pressure ratio, number of combustors, exhaust mass flow rate, exhaust temperature, length, width, and height.

The data given for steam turbine generator sets include power rating ranges for electric power, industrial, process, marine, and other mechanical-drive applications.

Internal-combustion engine generator ratings appear in Table 6-2, and the corresponding drawings in Appendix B, Figure B-6.

**Table 6-2 Internal-Combustion Engine Ratings**

| No. of cycles | Kilowatts | Revolutions per minute |
|---|---|---|
| **Fairbanks Morse 38DD8-⅛ blower scavenged dual fuel** | | |
| 4 | 450–622 | 720–900 |
| 5 | 565–779 | 720–900 |
| 6 | 680–937 | 720–900 |
| 8 | 910–1250 | 720–900 |
| 9 | 1025–1410 | 720–900 |
| 10 | 1140–1566 | 720–900 |
| 12 | 1365–1882 | 720–900 |
| **Fairbanks Morse 38TDD8-⅛ turbocharged dual fuel** | | |
| 6 | 1203–1657 | 720–900 |
| 9 | 1807–2486 | 720–900 |
| 12 | 2412–3314 | 720–900 |
| **Fairbanks Morse 38DS8-⅛ blower scavenged spark ignition** | | |
| 6 | 770–1058 | 720–900 |
| 8 | 1025–1410 | 720–900 |
| 10 | 1282–1762 | 720–900 |
| 12 | 1480–2035 | 720–900 |
| **Fairbanks Morse 38TDS8-⅛ turbocharged spark ignition** | | |
| 6 | 940–1295 | 720–900 |
| 8 | 1255–1725 | 720–900 |
| 10 | 1570–2160 | 720–900 |
| 12 | 1885–2595 | 720–900 |
| **Colt-Pielstick PC-2 dual fuel** | | |
| 12 | 4300 | 514 |
| 14 | 5010 | 514 |
| 16 | 5730 | 514 |
| 18 | 6445 | 514 |

SOURCE: Colt Industries, Fairbanks Morse Engine Division.

## CAPITAL COSTS AND SIZING RELATIONSHIPS: MAJOR EQUIPMENT

Order-of-magnitude capital costs derived from a comprehensive data base in terms of 1982 dollars are given in Figures 6-1 to 6-3. Figure 6-1 indicates combustion turbine generator set equipment capital costs for electric power outputs ranging from 800 kilowatts to 75 megawatts, Figure 6-2 gives steam turbine generator set capital costs for electric power outputs ranging from 300 kilowatts to 10 megawatts, and Figure 6-3 compares the capital costs for reciprocating diesel and gas engine generator sets varying from 25 kilowatts to 2400 kilowatts. These costs are time-dependent, and therefore future costs may be forecast by extrapolation using known factors related to current changes in the cost of materials and labor.

The significance of these comparisons is seen in the curves showing the relationships between cost and unit electric power output. These curves represent the size-to-cost ratio and can be shifted as a function of time to compensate for changes in material and labor indexes.

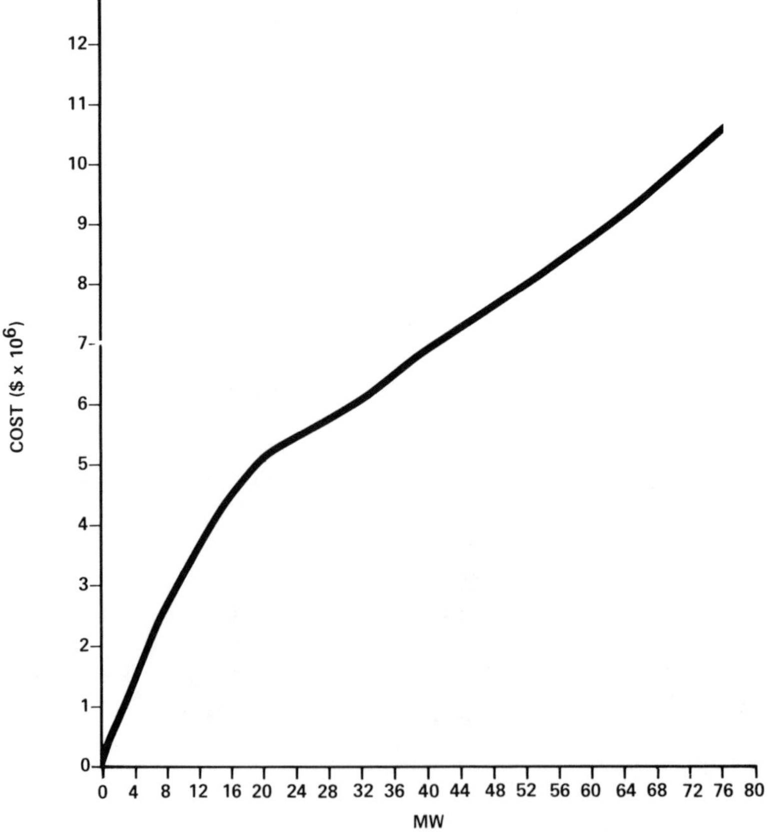


**Figure 6-1 Combustion turbine generator sets: capital cost comparisons.**

Note that reciprocating diesel and gas engine generator sets are the only major equipment alternatives which display a linear relationship with respect to capital equipment cost and electric power output rating.

The relative installed costs of a cogeneration project consisting of a combustion turbine-generator and a combined cycle facility are compared to those for a conventional electric utility fossil-fired steam electric plant in Figure 6-4.

Total installed costs of cogeneration projects can be plotted on similar curves. However, this information could be improperly interpreted because of the many combinations of major equipment components comprising each site-specific cogeneration project. Unique client requirements and preferences contribute to these nonstandard cogeneration project designs. Projects can range from large industrial cogeneration applications for process steam thermal loads utilizing a coal-fired steam generator, an extraction condensing steam turbine-generator, cooling towers, circulating water systems, and boiler emission control equipment

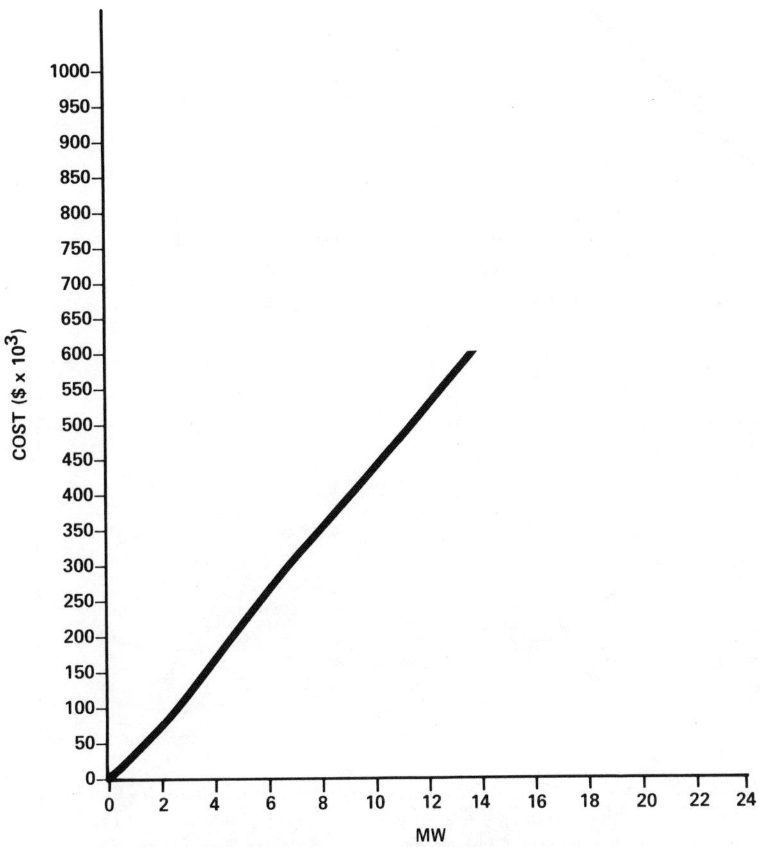


**Figure 6-2 Steam turbine generator sets: capital cost comparisons.**

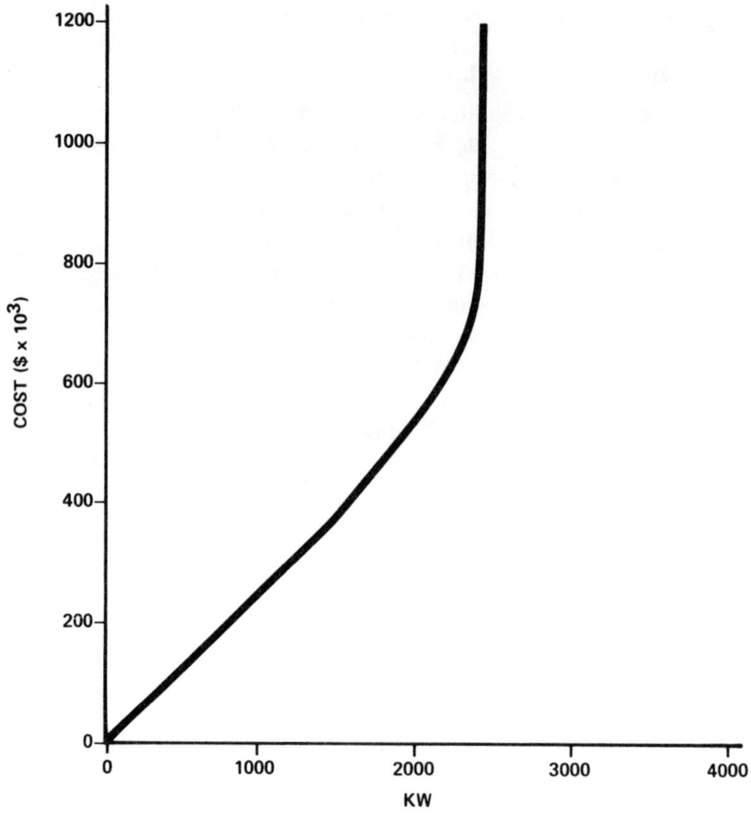


**Figure 6-3** Reciprocating diesel and gas engine generator sets: capital cost comparisons.

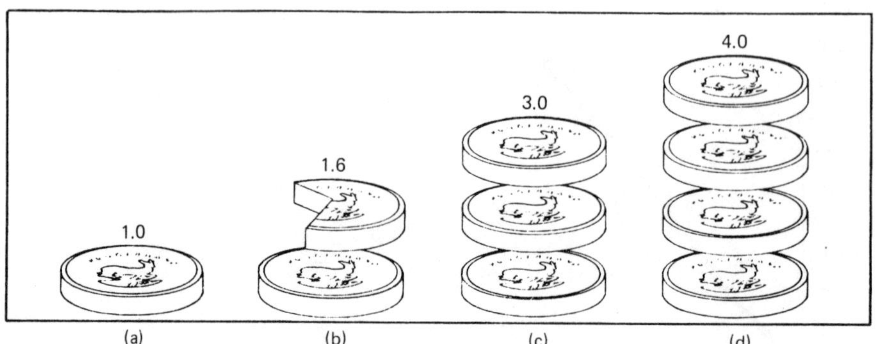


**Figure 6-4** Relative installed cost per kilowatt for various power cycle installations. (*a*) Combustion turbine (simple cycle), (*b*) combined cycle, (*c*) oil- or gas-fired steam plant, (*d*) coal-fired steam plant. (*Westinghouse Combustion Turbines, "The Ready Source of Power Worldwide," Bulletin SA 11152*)

to applications requiring a relatively small reciprocating engine generator set for heating and cooling a building. In addition to the multitude of technical alternatives other considerations that complicate comparisons of the total project cost are site ownership, real estate values, site development, and the impact of local building code and environmental restraints on overall project costs. Therefore, the estimated capital cost curves for cogeneration projects based on electric power output and thermal load are not given here but instead are left for the reader to estimate on a per-case basis.

# REGULATORY REQUIREMENTS

## UTILITIES INTERFACE

Recent state and federal legislation requires investor-owned utilities to

- Purchase electricity from cogeneration facilities at what it costs the utility to generate its most expensive electricity.
- Adopt preferential natural gas rates and high priorities for natural gas allocation to cogeneration customers.
- Assist potential users of cogeneration by preparing the financial, feasibility, and environmental analyses of potential projects.

All other utilities (primarily municipal utilities) must at least

- Purchase electricity from qualifying cogeneration facilities at "just and reasonable" prices.
- Supply backup electricity to qualifying cogeneration facilities at just and reasonable rates.
- Allow their transmission lines to be used to send electricity from a cogeneration facility to a different utility or to another facility in another part of the state.
- Make the connections necessary for cogeneration facilities to buy and sell electricity.

If a company would benefit by selling electricity to a distant utility, the utilities situated between the company and the buyer must agree to send electric power through their transmission lines for a fee. This is called wheeling electricity, and it offers an aggressive and innovative company an opportunity to obtain the best price for electricity.

A successful cogeneration project depends on careful negotiations between the company and the utility. The following crucial items should be negotiated as early as possible in the planning of a project.

- The price to be paid for the electricity
- The duration of the contract
- How the cogeneration equipment will be connected to the utility's grid and who will pay for the connection
- Who will operate and maintain the equipment

Alternative electric utility interconnection systems are listed below.

- A parallel operation in which the generating equipment is connected to a bus common to the utility's system. A technical interface between the cogenerator and the electric utility is required to ensure that design requirements for protective devices are met and are acceptable to the electric utility.
- A separate system in which the generating equipment cannot be connected in parallel with the utility's system. The cogenerator must be capable of transferring load between the two systems in an open transition in a nonparallel mode. The cogenerator provides protection to ensure adequate clearing of faults on his own system, and approval by the utility is required.
- An arrangement in which the cogenerator is liable for cogenerator-owned devices.
- An arrangement in which the utility is liable for utility-owned interconnection protection devices.

To make it easier for a company to negotiate a contract, each of the major utilities outlines contract terms that can serve as a starting point in negotiating the contract. The contract terms and additional information are available from the utility's cogeneration program manager.

## A Lower Price and a Guaranteed Supply of Natural Gas

The California Public Utilities Commission (PUC) has ordered investor-owned utilities to sell natural gas for use in cogeneration facilities at special rates. The rates may be no higher than those the utilities themselves pay when they purchase natural gas to generate electricity. Once these new rates are in effect, a company using cogeneration equipment will save approximately 2 to 5 cents a therm for natural gas.

In addition to the preferred rates, the PUC will assign cogeneration facilities a "3-A" priority for purchasing natural gas, the highest industrial category. This status ensures that cogeneration facilities will be able to purchase natural gas even if other industrial customers have their service curtailed in the event of a shortage. Only residential customers and commercial establishments have a higher priority.

## PERMIT REQUIREMENTS

Maintaining contact with local, state, and federal regulatory agencies will ensure that cogeneration studies, equipment selection, and systems meet efficiency, performance, and other standards. Applications should be optimized to satisfy process and electricity demands while meeting existing and pending regulatory requirements. A summary of the most significant regulatory policy acts and agencies, along with their specific impact on the entire future of cogeneration in fulfilling our national energy goals, is given in Table 7-1. PURPA and the Federal Power Plant and Industrial Fuel Use Act (FUA) are the two most significant regulatory acts affecting cogeneration applications and are therefore discussed in detail later in the chapter.

To help the cogenerator meet regulatory requirements, consultants offer the following services.

- Assist in securing necessary state and local permits.
- Determine when a permit is needed.
- Provide information on how to apply for a permit.
- Establish what criteria are used to evaluate the cogeneration project applications for permit approval.
- Identify the appropriate agency contact.
- Indicate the permit fee (if any).
- Assist in obtaining government agency approval of the project.
- Identify the permits required.
- Arrange meetings with the regulatory agencies.
- Prepare a schedule for filing applications and receiving decisions.
- Work with the cogenerator and the permit-granting agencies to ensure that the process goes smoothly and permits are issued.
- Work with the staff of the facility to resolve the problems and arrive at the most efficient design possible.
- Work closely with both the local government and the air pollution control district to ensure that the project meets their standards.
- Meet the staff of the local air pollution control district to ensure that the process of moving toward permit issuance goes smoothly.

**Table 7-1 Regulatory Policy Acts and Agencies**

**Public Utility Regulatory Policy Act of 1978 (PURPA)**

- Requires all utilities to buy whatever power a *new* cogenerator or independent small power producer offers and to pay either the avoided cost rate (the price the utility would have to pay to otherwise acquire or to produce the additional power) or a rate negotiated with the cogenerator.
- Exempts cogenerators from cumbersome state utility commission regulation and blocks utilities from charging *excessive* rates for power to back up the cogeneration systems.
- Allows each state to decide whether to apply this act to *existing* cogenerators also.
- Exempts cogenerators from the Public Utility Holding Company Act and from controls on rates and financial structure found in the Federal Power Act and in many state statutes.
- Provides exemption to cogenerator from incremental pricing of natural gas.

**Federal Energy Regulatory Commission (FERC)**

- An agency of the Department of Energy.
- Authorized by the Federal Power Act to regulate energy projects.
- Issues certifying status for qualifying cogeneration facilities.

**Natural Gas Policy Act (NGPA)**

- Exempts cogenerators from the incremental pricing program under which industrial boiler fuel users would absorb most increases in natural gas prices through 1985.

**Federal Power Plant and Industrial Fuel Use Act of 1978 (FUA)**

- Unless an exemption is granted by the U.S. Department of Energy, new power plants cannot use natural gas (or oil) and existing power plants cannot increase their natural gas use.
- A facility is exempt from the FUA if

  It uses an alternate fuel (wood waste, agricultural waste, refuse, coal, petroleum coke, or shale oil).

  It uses less than 100,000,000 Btu per hour (or generates approximately 8 to 10 megawatts) at a single facility.

  It uses less than 250,000,000 Btu per hour (or generates approximately 25 to 30 megawatts) at a combined facility.

  It uses equipment other than a boiler, such as an internal-combustion engine, a diesel engine, or a combustion turbine.

**Energy Tax Act (ETA)**

- Provides for a federal energy tax credit of 10 percent of the cost for qualified cogeneration equipment investments made by December 31, 1982. (The provisions of this act are described in more detail in Chapter 8.)

**Table 7-1 (Cont.)**

**Crude Oil Windfalls Profits Tax Act (COWPTA)**

- Provides a 10 percent tax credit for investment in certain cogeneration equipment.

  Waste heat boilers installed before December 31, 1982, and other equipment which increases the capacity of a system to produce its secondary energy product qualify for the credit, but only if oil and gas represent less than 20 percent of the fuel used in the system.

  The Secretary of the Treasury has the discretion to extend the tax credit to other cogeneration equipment.

  Credit is not available if electric utility ownership exceeds 50 percent. The utility may operate the facility.

**Public Utilities Commission (PUC)**

- State agency responsible for regulation of electric and gas utilities.
- Sets preferential natural gas rates and high priorities for natural gas allocation to cogenerators.
- Implements PURPA by requiring electric utilities to interconnect with cogenerators and purchase power from cogeneration facilities at the avoided cost.

**Clean Air Act of 1970 (CAA)**

- Requires states to develop a state implementation plan (SIP) describing an air pollution control plan they will implement to meet national ambient air quality standards (NAAQS).
- Each state is required to determine areas which satisfy the NAAQS for criteria pollutants subject to EPA approval. An area is then designated either attainment or nonattainment for that pollutant. SIP provisions are subject to EPA approval.
- An attainment area meets NAAQS for a pollutant. These areas are also referred to as prevention of significant deterioration (PSD) Areas. PSD is intended to ensure that air quality that is already better than that required by NAAQS does not deteriorate below that acceptable level. These areas require use of best available control technology (BACT).
- A nonattainment area is one in which NAAQS for one or more pollutants have not been met.

**United States Department of Energy (DOE)**

- Responsible for centralizing federal energy policies and operations.

**Economic Regulatory Administration (ERA)**

- An agency of the Department of Energy.
- Authorized to regulate the economic criteria for energy projects.

A permit from the city or county may be required for the installation of a cogeneration facility. If cogeneration equipment is being added to an existing plant, only a building permit may be needed, which local governments will routinely issue if the firm meets the building codes. If an entirely new facility that includes cogeneration equipment is being proposed, additional permits may be required.

Decisions on all local permit applications must be handed down within 1 year, otherwise the permit is automatically approved. Local governments may prepare either an environmental impact report (EIR) or a negative declaration. While an EIR is a detailed study identifying the likely environmental effects of a proposed project and ways to minimize them, a negative declaration consists of a short written statement describing why a proposed project will not have any significant environmental effect and therefore does not require further study.

An air quality permit is required for a cogeneration project. Early and frequent consultation with the local air pollution control district and the State Air Resources Board is important.

### Authority-to-Construct Permit

An authority-to-construct permit must be issued by the local air quality management district if the cogeneration facility

- Generates less than 50 megawatts of electricity.
- Uses state-approved pollution control devices.
- Offsets pollution from any other facilities in the same air basin. (If the firm does not own or operate any other facility in the basin this does not apply.)

In most instances, this permit can be obtained in as little as 6 months. But when the air pollution control district must prepare an EIR, it can take as long as a year. During this time, the district will study the cogenerator's proposal to see whether it will meet federal and state air pollution standards and whether the emissions control devices selected will be effective. The review period may be shortened by working with the staff of the local air pollution district in the early planning stages. Each district can help determine what equipment will best control emissions, given the cogeneration system to be installed.

### Permit to Operate

Once an authority-to-construct permit has been secured and building of the facility completed, a permit to operate must be obtained from the local air quality management district. A permit to operate certifies that the cogeneration facility has been built according to the conditions set forth in the authority-to-construct permit. The district also decides whether the cogenerator will be able to comply with the district's rules and regulations when operating the facility.

In addition to a permit from either the local air pollution control district or the state energy commission, one may also need a permit from the U.S. Environmental Protection Agency for Prevention of Significant Deterioration (PSD) which seeks to prevent clean air areas from deteriorating into dirty air areas.

A listing of typical permits, approving governmental agencies, estimated lead times, and application requirements is given in Table 7-2.

## PUBLIC UTILITY REGULATORY POLICIES ACT

Section 210 of PURPA, promulgated by Congress in March 1980 and administered by the Federal Energy Regulatory Commission (FERC), eliminates many regulatory and economic obstructions to the development of industrial cogeneration facilities by requiring that utilities purchase cogenerators' electricity at the utilities' avoided cost. In other words, a utility must pay the cogenerator what it saves by not generating the energy itself at its most expensive facility, either by increasing its power output or adding to its capacity. Another provision limits utility ownership of cogeneration facilities to 50 percent—anything above that eliminates eligibility for PURPA incentives.

Industrial cogenerators are also eligible for a 10 percent tax credit for all energy-saving equipment. Furthermore, under PURPA cogenerators are guaranteed fair rates for the electricity they sell, the backup power they buy, and interconnection charges.

### Rates

In cases where the cogeneration electrical facilities are industry-owned and the electrical output is applied to the industry's internal connected load, the utility's rates for service affect the economics of the cogeneration facility. If and to the extent that there is a need for capacity and energy during the time the industry's electrical generating facilities are out of service, the industry has the option of buying such capacity and energy from the utility under a contract for standby service. An alternative to consider in analysis of the economics of cogeneration and as a source of supplemental power in an industry-owned cogeneration project is the utility's rates for regular and supplemental service and industrial, regular, and time-of-use rates.

Time-of-use rates for cogeneration provide

- High rates for service during peak periods
- An improvement in the overall economics of cogeneration for an industrial customer
- An incentive for maximum reliability during periods when the capacity is greatest

Table 7-2 Regulatory Permits and Approvals

| Permit, license, or certification | Contact agency | Satisfying requirements | Lead time to obtain | Prerequisites | Self-certification or application required | Forms required |
|---|---|---|---|---|---|---|
| **Air emissions** | | | | | | |
| Permit to construct* | Local AQMD | Complete application: plans, specs, emissions data<br>BACT applied<br>NSPS achieved<br>Internal offsets obtained | 6–18 months before construction | Major equipment selections<br>Design plans and specs<br>Emissions data and/or calculations | Application | Application for permit to construct or operate |
| Permit to operate | Local AQMD | Notification of completion of construction<br>Local AQMD inspection | 3–6 months (construction permit used in interim) | Completion of construction<br>Emission testing | Certification | Same as above |
| Memorandum agreement* | EPA, regional office | AQMD complete application<br>Correspondence<br>Limitation on use of backup fuel oil | 6–18 months before construction | Completion of local AQMD application | Certifying correspondence | Same as above |
| **Wastewater** | | | | | | |
| Approval to increase wastewater flow | County, sanitation district | Request by correspondence | 1–3 months before construction | Water balance and schematic<br>Estimate of increase and characteristics | Possible application | Possible application |

**Table 7-2 (Cont.)**

| Permit, license, or certification | Contact agency | Satisfying requirements | Lead time to obtain | Prerequisites | Self-certification or application required | Forms required |
|---|---|---|---|---|---|---|
| **Water intake** | | | | | | |
| Approval to increase water intake | City water department | Request by correspondence | 1–3 months before construction | Water balance and schematic<br>Estimate of consumption increase | None | None |
| **Solid waste** | | | | | | |
| Notification of activity | Solid Waste Management Board | Correspondence verifying no increase in solid waste generation | Not applicable | None | None | None |
| Small power production qualification<br>*or* | FERC | Size, use, ownership | None | | Self-certification | None |
| Cogeneration facility qualification<br>*and/or* | FERC | Thermal efficiency 42.5% | Basis for criteria—compliance with FERC Order No. 70, 18 CFR Part 292.205(a) | | Self-certification unless required by gas or electric utility or by lending institution | None |
| Cogeneration facility qualification | FERC | Same as above | 90 days | | Application | None |
| Fuel Use Act exemption | ERA | Sell less than 50% of generation<br>Less than 100 MMBtu/hr single unit or 250 MMBTU/hr aggregate | Basis for criteria—Power Plant and Industrial Fuel Use Act of 1978, 10 CFR Part 500.2, 500.4, 500.5 | | Application | Yes |

Table 7-2 (Cont.)

| Permit, license, or certification | Contact agency | Satisfying requirements | Lead time to obtain | Prerequisites | Self-certification or application required | Forms required |
|---|---|---|---|---|---|---|
| **Solid waste (cont.)** | | | | | | |
| None | Public Service Commission | Review power sell contract | None | | Application | No |
| Building permit | City building or planning department | City codes | 30 days | Construction drawings | Application | Yes |
| High pressure boilder operation license | State license, bureau and/or insurance carrier | Pass test | 90 days | Training | Application | Yes |
| Thermal efficiency test | Gas company | Verify thermal efficiency to meet FERC Order No. 70 | | To qualify for cogeneration rate | Utility test | |
| Certification | State boiler inspector and/or insurance carrier | Certify welding of pipe and boiler | — | — | Inspection | Yes |
| Certification | Insurance, FM-FIA | Safety controls | | | Inspection and drawings | |

Before the enactment of PURPA in 1978, a cogenerator of electricity which sought to buy from or sell to an electric utility faced at least three major obstacles:

- A utility was not required to purchase a cogenerator's electricity output.
- Many utilities charged cogenerators discriminatory high rates for backup services.
- A cogenerator, in providing electricity to a utility grid, ran the risk of being considered an electric utility and thus being subjected to the complex state and federal regulations which apply to such firms.

Sections 201 and 210 of PURPA are designed to remove these three obstacles. An electric utility is required to offer to purchase electric energy from cogeneration facilities which obtain qualifying status under the legislation. For such purchases, utilities must pay rates which are considered just and reasonable by the rate payers of the utility, are in the public interest, and do not discriminate against cogenerators. Electric utilities must also provide backup power services at reasonable and nondiscriminatory rates. Finally, the commission may exempt qualifying facilities from rate regulations established under the Federal Power Act (which normally applies to utilities), from the requirements of the Public Utilities Holding Company Act, and from regulations based on similar state utility legislation. It is appropriate to consider whether and to what extent these exemptions might affect a proposed cogeneration project.

The Mississippi State Court found Titles I and II and Section 210 of PURPA unconstitutional in *Mississippi v. Federal Energy Regulatory Commission* on February 20, 1981. Title I of PURPA requires each state to assess the advisability of implementing 12 federal rate design and regulatory standards. Title II of Section 210 requires utilities to buy power from cogenerators and other alternate-energy producers, while Title III provides standards for gas utilities. *Mississippi v. Federal Energy Regulatory Commission* was appealed to the U.S. Supreme Court which upheld the constitutionality of all three challenged provisions of PURPA on June 1, 1982.

The case of *American Electric Power Service Corporation, Consolidated Edison, and the Colorado-Ute Electric Association v. Federal Energy Regulatory Commission,* filed in the U.S. Circuit Court of Appeals of the District of Columbia, challenged the following four aspects of FERC's regulations.

1. That power produced by qualifying facilities must be purchased by a utility at its avoided cost
2. That a qualifying facility may simultaneously sell all its power to a utility at the avoided cost rate and purchase its requirements back at standard tariff rates

3. That qualifying cogeneration facilities must be allowed to interconnect with utilities without meeting the requirements of Sections 210 and 212 of the Federal Power Act
4. That qualifying facilities need not comply with any fuel use criteria established under PURPA

On January 22, 1982, the U.S. Circuit Court of Appeals for the District of Columbia upheld FERC's regulations concerning the second and fourth issues but vacated the regulations involving the first and third issues.

Relative to the first issue, the court ruled that FERC must justify and explain its decision to prohibit across-the-board rates below the full avoided cost. Since it has failed to do so, this ruling has been vacated. The court concluded that the commission must justify and explain its decision on balancing the interests of cogenerators, the public, and customers of electric utilities. The court, on a remand, expects the commission to take a harder look at the percentage-of-avoided-cost approach in particular. The court suggested that FERC set a specific percentage or permit state commissions to set rates within a range of 80 to 100 percent of the full avoided cost authorized by the commission.

The third issue, involving the court's decision to vacate the rule regarding FERC's granting of blanket authority to cogenerators to interconnect with electric utilities without meeting the requirements of Sections 210 and 212 of the Federal Power Act is of more serious concern to cogenerators. This ruling could complicate scheduling uncertainties. FERC options are to establish a rule setting generic standards for interconnections or have Congress amend the Federal Power Act to allow easier interconnections. The court granted FERC's motion that the court stay its decision. Thus, FERC regulations will remain in effect until all appeals have been filed. The U.S. Supreme Court agreed on October 12, 1982, to review the Court of Appeals decision.

Despite the uncertainty of these two FERC regulations, some state PUCs are proceeding under their own authority to implement the FERC rules and structure their own regulatory program under state laws.

The U.S. Supreme Court decision on May 16, 1983, upheld the above two FERC issues. One rule requires electric utilities to purchase surplus electric energy from qualifying cogeneration and small power producer facilities at a rate equal to the utility's full avoided cost. The second rule requires utilities to make transmission system electric interconnections with cogenerators and small power producers as necessary for purchases or sales transactions of electricity authorized by PURPA without lengthy hearings.

Bill S. 1885, introduced in Congress by Senator Gordon J. Humphrey, is pending before the Senate Committee on Energy and Natural Resources. This bill would remove the regulation prohibiting utilities from owning more that 50 percent of a qualifying facility.

***Sales by Cogenerators to Electric Utilities*** On February 19, 1980, FERC issued Order No. 69, a final rule governing the sale of electricity by cogenerators to utilities. By Order No. 69, an electric utility is required to purchase at a fair rate all electric energy made available to it, either directly or indirectly, by a qualifying cogeneration facility. The rate for such purchases should generally reflect the utility's avoided cost. This is the cost the utility would have incurred, but for the purchase from the cogenerator, in acquiring a like quantity of energy or capacity, including the energy costs associated with the production of replacement energy and the capital costs of facilities to deliver the energy. Note that the emphasis is on the utility's and not the cogenerator's costs. The logic here is that rates based on utility costs, whether the electricity is supplied by the utility or by a cogenerator, do not work to the disadvantage of a utility's noncogeneration customers, who are arguably indifferent as to where their electricity comes from as long as it is at an equal cost.

It should be noted that the avoided cost mechanism does not exclude the possibility of an agreement negotiated between the utility and the cogenerator resulting in a rate for the sales and purchase of electricity which may differ from the avoided cost. The avoided cost mechanism may, however, provide the framework for such negotiations and will be applied by the state regulatory agencies and/or FERC on petition if the negotiations result in no agreement.

In arriving at the rate which reflects the utility's avoided cost, the agency will consider capital or capacity cost avoided and the operating or energy costs avoided by the utility through its purchase from the cogenerator.

Capital costs avoided represents an estimate of the extent to which the utility avoids the need for new capital expenditures by virtue of its purchases of firm or interruptible energy power from a qualifying cogenerator.

Operating costs avoided are calculated by comparing rates for qualifying facilities with rates for self-generation by the utility or rates for the sale of "wholesale power" between electric utilities. Where self-generation is the utility's most likely alternate to the purchase of a cogenerator's power, the operating costs of the highest-cost unit likely to be utilized determine applicable rates. Wholesale rates are generally less than rates for firm power sales, in which the seller is compensated for both energy and capacity costs, and are more than rates for the sale of interruptible power, where the buyer is charged only for the operating costs of the seller.

Consideration is also given to factors which bear on the availability of the cogenerator's energy and capacity during the electric utility's daily or seasonable peak periods, including

- The ability of the electric utility to dispatch or control the qualifying cogeneration facility.
- The expected or demonstrated reliability of the qualifying cogeneration facility.

- The terms of any contractual arrangement between the qualifying cogenerator and the utility. (Long-term contracts requiring notice before termination and providing remedies for failure to honor contractual obligations are considered more desirable by utilities and therefore worthy of higher purchase rates.)
- The extent to which the cogenerator's scheduled generator outages can be usefully coordinated with the scheduled outages of the utility's facilities.
- The usefulness of the cogenerated energy supplied during system emergencies, including the ability of the cogenerator to separate its load from the utility's overall generating load during such periods.
- The individual and aggregate value to the electric utility of energy and capacity supplied by qualifying facilities (for example, the overall value to the utility of a large number of small industrial cogenerators).
- The shorter lead times available to the utility as a result of using cogenerators for system expansion.

With these factors in mind, the appropriate rate is determined. In the case of cogeneration facilities with a generating capacity in excess of 100 kilowatts, the rate may be either a published rate based on cents per kilowatthour, or an individually negotiated contractual rate.

***Purchases of Electricity from Electric Utilities*** Demands for electricity from electric utilities can be significantly reduced by the conversion of selected plants to cogeneration. FERC has provided regulations to ensure that backup power is available from electric utilities at nondiscriminatory rates which are just and reasonable during cogenerator internal outages, maintenance, and overhaul periods. Four circumstances are identified in which such backup services may be needed:

- Backup power—to replace energy normally generated by a qualifying cogenerator for its own use
- Supplementary power—to supply power normally purchased from a utility in addition to a cogenerator's own self-generated power
- Maintenance power—to supply power during scheduled outages of the qualifying cogenerator
- Interruptible power—to provide power with specified limitations on use depending on the utility's peak load or other system requirements

The rates for these sales may not, in the absence of factual data to the contrary, be based on the assumption that forced outages or other reductions in output at a qualifying cogeneration facility will necessarily occur at times of utility system peak or that demands from a given class of qualifying cogenerators will occur simultaneously.

In times of utility system emergencies, the qualifying cogenerator is obliged to supply only as much power to the utility as fulfills its contractual obligations. Therefore, electricity generated in excess of its contractual commitments need not be diverted to the utility unless the President declares a national energy emergency under Section 202(c) of the Federal Power Act.

***Exemption of Cogenerators from Public Utility Regulations*** Under FERC and implementing state rules, qualifying cogeneration facilities are exempt from much public utility regulation, notwithstanding that electricity sale and purchase transactions might otherwise render them liable to regulation as utilities. A qualifying cogenerator is generally exempt from rate regulation under the Federal Power Act, from the Public Utility Holding Company Act, and from similar state laws and regulations regarding electric utility rates.

Under the provisions of PURPA, a qualifying cogenerator is exempt from regulation as a utility and is able to buy, sell, or exchange electricity with a utility either at independently negotiated rates or on the basis of the avoided cost mechanism.

## FEDERAL POWER PLANT AND INDUSTRIAL FUEL USE ACT

Congress enacted the Federal Power Plant and Industrial Fuel Use Act (FUA) November 9, 1978, to reduce the nation's dependence on foreign oil and petroleum-based gases by restricting industrial use of natural gas and oil in power plants and other major fuel-burning installations (MFBIs). An electric power plant or other MFBI is defined as one having

- A design capability of consuming any fuel at a fuel heat input rate of 100,000,000 Btu per hour or greater
- A combination of two or more units at the same site which in the aggregate have a design capability of consuming any fuel at a fuel heat input rate of 250,000,000 Btu per hour or greater

The act also generally requires conversion to coal use for existing electric power plants and MFBIs.

### Exemption from FUA

The act does not affect small cogeneration facilities having a fuel heat input rate of less than 100,000,000 Btu per hour (equivalent to approximately 8 to 10 megawatts), regardless of the fuel used. An exemption also exists for cogeneration facilities having a fuel heat input rate of less than 250,000,000 Btu per hour (equivalent to approximately 25 to 30 megawatts) at a combined facility. The act also exempts a cogeneration facility of any size that uses an alternative fuel, such as municipal, agricultural, or wood waste, industrial by-products, coal, petroleum coke, or shale oil.

A cogeneration facility larger than 8 to 10 megawatts using oil or natural gas as a fuel requires a permit from the Economic Regulatory Administration (ERA). In granting the permit, the administration uses two tests: whether the project will save fuel and whether it serves the public interest. The test of public interest offers the innovative user of cogeneration a wide range of possible locations and designs. For instance, a plant located in an urban area with high unemployment would qualify for a permit.

The purposes of the Fuel Use Act are to

- Reduce the importation of petroleum and increase the nation's capability to use indigenous energy resources
- Conserve natural gas and petroleum for uses other than electric utility or other industrial or commercial generation of steam or electricity
- Encourage and foster greater use of coal and other alternate fuels
- Encourage the use of synthetic gas derived from coal and other alternate fuels
- Encourage the rehabilitation and upgrading of railroads and the equipment necessary to transport coal
- Conserve gas and petroleum for the benefit of present and future generations
- Encourage the modernization or replacement of existing and new power plants to conserve gas and fuel oil where coal or other fuels cannot be utilized
- Ensure that all MFBIs which utilize natural gas, oil, coal, or other alternate fuels pursuant to the act comply with applicable environmental requirements
- Ensure that all federal agencies utilize their authority fully in furtherance of the objectives of the act
- Ensure that adequate supplies of natural gas are available for essential agricultural needs
- Reduce the vulnerability of the United States to energy supply interruptions
- Regulate interstate commerce

The act seeks to improve the U.S. balance of payments deficit while recognizing the realities of the U.S. natural resource situation.

A facility is exempt from the Fuel Use Act if it

- Uses an alternative fuel, such as

  Municipal wastes

  Agricultural wastes

  Wood wastes

  By-products from refineries or other industrial operations that are commercially unmarketable

  Coal

  Petroleum coke

  Shale oil

- Uses less than 100,000,000 Btu per hour (equivalent to approximately 8 to 10 megawatts) at a single facility
- Uses less than 250,000,000 Btu per hour (equivalent to approximately 25 to 30 megawatts) at a combined facility
- Uses equipment other than a boiler, such as

  Internal-combustion engine

  Diesel engine

  Gas turbine

## QUALIFYING AS A COGENERATION FACILITY

Qualifying facilities are exempt from state and federal regulation as utilities and from filing the kinds of reports with the Federal Securities Exchange Commission that utilities must. The sale of steam, heat, or electricity by a qualifying cogeneration facility is also exempt from state sales taxes.

Federal certification of a facility's status can be obtained early in the planning of a project by sending FERC a description of the facility. The commission assesses the proposed project against three criteria involving ownership and operating and efficiency standards. To pass the test of ownership, the facility must not be owned (defined as more than 50 percent of the equity) by a company primarily engaged in generating or selling power. In addition, the equipment must meet PURPA and FERC operating and efficiency standards.

If the project meets the three criteria, FERC will send a letter certifying that the facility is a qualifying cogeneration facility. Assistance in qualifying as a cogeneration facility can be obtained by filing for certification with the Federal Energy Regulatory Commission, 825 North Capitol Street, N.E., Washington, DC 20426.

The application for qualification requires the following information.

- The name and address of the facility
- A description of the facility in sufficient detail to indicate that it will meet the operating and efficiency standards
- Identification of the primary type of fuel
- A statement of the facility's capacity to generate electricity
- A statement of the percentage of equity in the facility held by others (if applicable)

## ENVIRONMENTAL AND AIR QUALITY CONSIDERATIONS

Air quality restrictions pose the single greatest technical problem for cogeneration facilities. Experienced personnel are required to satisfy various local, state, and federal air pollution requirements resulting from the Clean Air Act (CAA) of 1970 and the more stringent amendments adopted in 1977. Table 7-3 lists representative exhaust emissions obtainable with and without water injection for nitrogen oxide control for several combustion turbines of different sizes fueled by natural gas or No. 2 fuel oil. Table 7-4 summarizes air quality standards for one of the

**Table 7-3 Typical Exhaust Emissions**

| | Natural gas | No. 2 Fuel Oil |
|---|---|---|
| **With water injection for $NO_x$ control** | | |
| $NO_x$ (lb/hr) | 16 | 7.33 |
| CO (lb/hr) | 4.42 | 9.2 |
| Particulates | Negligible | Submicrometer size range |
| $SO_2$ (lb/hr) | Negligible | 43.2 |
| UHC (ppm volume) | 50 | 6 |
| Water/fuel ratio | 0.82 | 0.80 |
| Fuel rate (based on LHV) | 1621 cfm | 4209 lb/hr |
| **Without water injection for $NO_x$ control** | | |
| $NO_x$ (lb/hr) | 5.2 | 6.0 |
| CO (lb/hr) | 0.3 | 0.8 |
| Particulates | 0.1 | 0.5 |
| $SO_2$ (lb/hr) | Negligible | 4.25 |
| UHC (ppm volume) | 9 | 6 |
| Water/fuel ratio | No water injection | No water injection |
| Fuel rate (based on LHV) | 143 cfm | 439 lb/hr |

**Table 7-4 Regulatory Air Quality Standards**

| Item | Description | |
|---|---|---|
| **New source performance standards (NSPS) required by the Environmental Protection Agency (EPA) for each combustion turbine** | | |
| | **$NO_x$*** | **$SO_2$** |
| <10 MMBtu/hr | No standard | No standard |
| >10 MMBtu/hr <100 MMBtu/hr | 150 ppm† | 150 ppm *or* <0.8% sulfur fuel |
| >100 MMBtu/hr | 75 ppm | 150 ppm *or* <0.8% sulfur fuel |
| **Control equipment (AQMD/EPA)** | | |
| | **$NO_x$** | **$SO_2$** |
| Combustion turbine <1200 hp (gas fuel) | Air-to-fuel, ratio adjustment | Not applicable |
| >1200 hp (gas fuel) | Water or steam injection (70% $NO_x$ removal) *or* derating power output | Not applicable |
| >2000 hp (other fuel) | Same as above | <0.25% sulfur |
| <2000 hp (other fuel) | Air-to-fuel ratio adjustment | Not applicable |
| Boiler auxiliary firing system | | |
| <35 MMBtu/hr | Low-$NO_x$ burners, (30% $NO_x$ removal) | Natural gas, LPG, or <0.25% sulfur oil |
| >35 MMBtu/hr | Low-$NO_x$ burners (80% $NO_x$ removal) and selective catalytic reduction | Same as above |
| **Monitoring equipment for combustion turbine** | | |
| <10 MMBtu/hr | None required | |
| >10MMBtu/hr, with water injection for $NO_x$ reduction | Continuous monitoring and recording system of fuel consumption and ratio of water to fuel being fired in the turbine (accurate within ±5%) | |
| | Mechanisms for monitoring sulfur content and nitrogen content of fuel, including backup fuel oil—supplier analysis acceptable | |

*Nitrogen oxide adjusted upward for combustion turbines with thermal efficiencies >25% or firing fuels with a nitrogen content <0.015% by weight. Measured nitrogen oxide must be adjusted to International Standards Organization standard day conditions of 288 K, 60% relative humidity, and atmospheric pressure to determine compliance.

†No nitrogen oxide standard if construction commenced before October 3, 1982.

‡Continuous stack monitoring required for any piece of equipment which exceeds 99 tons per year of any pollutant.

more restrictive air quality management districts in the country. Specialized project services are available from experienced personnel to assist potential cogenerators in the following environmental areas.

- Assist clients in obtaining construction permits from regulatory agencies:

  Environmental Protection Agency (EPA)

  State Implementation Plan (SIP)

  Local Air Quality Management District (AQMD)

- Compare the energy, economic, and environmental impacts on alternative levels of emission control.
- Identify potential emission reduction offsets.
- Determine the most effective methods of air pollution control.
- Identify and evaluate alternate emission control systems.
- Develop a cost-effective emission control system satisfying permit requirements.
- Select equipment and an emission control system.
- Design systems for required emission control for major new facilities or modifications of existing plants:

  Attainment or prevention of significant deterioration (PSD) areas require use of best available control technology (BACT).

  Nonattainment areas must install technology yielding the lowest achievable emission rate (LAER).

  Application of new source performance standards (NSPS) review may be required for a cogeneration air-emission source.

- Assist in air quality monitoring and analysis.

# FINANCIAL CONSIDERATIONS

This chapter summarizes typical financial considerations associated with cogeneration installations. The following topics are discussed: tax credit and depreciation benefits, typical operating and maintenance costs, a method for determining the economic feasibility of a project by calculating the payback period necessary to obtain a return on the investment, sources of alternative project funding, and project financing and ownership alternatives.

## TAX CREDITS

### Types of Tax Credits

Companies that install cogeneration equipment are eligible for the 10 percent investment credit and the accelerated depreciation allowance permitted by state and federal tax laws. Cogeneration applications using primary fuels other than oil or natural gas qualify for an additional 10 percent energy tax credit under COWPTA. If up to one-fifth of the fuel used is oil or gas (used as a backup or to stabilize the flame in the boiler), a facility can still qualify. This credit is available only for cogeneration equipment installed between January 1, 1980, and December 31, 1982. An additional 10 percent tax credit is available under the Energy Tax Act of 1978 for cogeneration facilities that use waste heat recovery boilers.

Investment tax credit provisions of the Internal Revenue Code can reduce the cost of certain types of business property by as much as 25 percent by allowing a credit against tax liability for a portion of the cost. This credit is available as a regular investment tax credit and as an energy tax credit.

The regular investment tax credit is equal to a percentage of the cost of qualifying property, which increases as the useful life of the property increases. The maximum regular investment tax credit of 10 percent applies to qualifying property with a useful life of 7 years or more.

The energy tax credit is a percentage of the cost of certain types of property deemed likely to result in conservation of energy or production or utilization of energy from sources other than oil or gas. The credit ranges from 10 to 15 percent, depending on the type of property, and for qualifying cogeneration equipment is 10 percent of the cost.

Both these credits offset tax liability up to the first $25,000 and 80 percent of the balance for 1981 (90 percent for 1982 and thereafter). Credits in excess of these limits may be carried back 3 years and forward 7 years.

In computing the limitation based on tax liability, the regular investment credit is applied first. If property giving rise to a credit is disposed of or ceases to be used in the manner which originally entitled the owner to a credit, some or all of the credit may be recaptured as an addition to tax liability.

## Regular Investment Tax Credit

Property which qualifies for the regular investment tax credit includes tangible personal property, other than an air-conditioning or heating unit, and any other tangible property, except a building or its structural components, but only if such property is used as an integral part of a manufacturing, production, or extraction operation or for furnishing transportation, communications, electric energy, gas, water, or sewage disposal services. Such property must be depreciable and have a useful life of 3 years or more. The percentage of the cost of qualifying property which may be claimed as a credit varies with the useful life of the property. If the useful life is 3 or 4 years, the credit will be $3\frac{1}{3}$ percent of the property's cost; 5 or 6 years, $6\frac{2}{3}$ percent; and 7 years or more, 10 percent. To qualify, the property must be new in the hands of the taxpayer.

Except in limited circumstances, boilers primarily fueled by petroleum or petroleum products (including natural gas) do not qualify for the regular investment tax credit. The only exception which may be applicable here is for a boiler at a location where federal or state air pollution regulations prohibit the use of coal as fuel.

## Energy Tax Credit

Investment in certain types of property, called energy property, the use of which is seen as furthering the government's energy policy, entitles the owner to a credit of 10 to 15 percent of the cost of such property. To qualify, equipment must be depreciable, be new in the hands of the owner, and have a useful life of at least 3 years. Cogeneration equipment is classified as energy property and, if placed in service within prescribed time periods (as discussed below), entitles the owner to a credit of up to 10 percent of the cost. Cogeneration equipment is defined as property which is an integral part of a system that uses the same fuel to produce both qualifying energy and electricity at an industrial or commercial facility at which, as of January 1, 1980, electricity or qualifying energy was produced. Cogeneration equipment does not qualify for the energy tax credit, however, if it is added to a system that uses oil, natural gas, or a product of oil or natural gas as a fuel other than for the purpose of start-up, flame control, or backup. Where oil, natural gas, or the product of either is used for any of these purposes, it may not constitute more than 20 percent of the total fuel (measured on a Btu basis) consumed by the system during any taxable year. Cogeneration equipment includes new equipment added to an existing system to begin generating a secondary energy product or to expand an existing cogeneration capacity. If the latter, it qualifies for the credit only to the extent that it increases the capacity of the system to produce the secondary energy product.

The types of equipment embraced by the term "cogeneration equipment" have been illustrated as follows. In the case of an energy-using system where the primary energy product is steam, heat, or other useful energy (such as shaft power) for process or space heating purposes, qualifying cogeneration equipment includes a turbine and generator to produce electricity, and any other equipment up to the electrical transmission stage. Where electricity is the primary product, qualifying equipment includes that necessary to recover and distribute, but not to use, excess energy after the electricity is generated.

Property must be placed in service within a specified period to qualify for the energy credit. In the case of cogeneration equipment, the period generally is from January 1, 1980, through December 31, 1982. Property whose construction, reconstruction, or erection requires 2 years or more to complete qualifies for the credit, however, if it is placed in service before the end of 1990 and both of the following conditions are met.

1. Before 1983, engineering studies that identify the site, process, and equipment required have been completed and all necessary construction and environmental permits have been applied for.
2. Before 1986, binding contracts have been entered into for the acquisition or construction of at least 50 percent of all equipment specially designed for the project (measured by costs as of December 31, 1985).

Property which does not qualify as cogeneration equipment may qualify for the energy tax credit as another type of energy property, called alternate-energy property. The credit for alternate-energy property is 10 percent.

To be considered alternate-energy property, however, a significant percentage of the fuel used by such property must be other than oil or gas. Alternate-energy property includes boilers and burners which use primarily (more than 50 percent on a Btu basis) a fuel other than oil or gas or any of their products. Alternate-energy property also includes equipment for modifying existing equipment which uses oil or natural gas so that it will use either a substance other than oil or natural gas, or oil mixed with a substance other than oil and natural gas where the other substance is at least 25 percent of the fuel.

## OPERATING AND MAINTENANCE COSTS

Operating and maintenance costs for cogeneration projects range from $0.003 to $0.008 per kilowatthour depending on the major equipment and cycle selection. A typical 10-year schedule for combustion turbine generator maintenance is given in Table 8-1.

## INSTALLATION COSTS AND PAYBACK PERIOD

Total recovery of the investment in a cogeneration installation can generally be realized in several years through savings in electricity and fuel bills. Payback periods of between 2 to 5 years are common.

Two prime factors affect the economics of any cogeneration application: the plant load and the fuel cost relative to the cost of purchased energy. Therefore, the ideal application is one having a high load factor for electric power usage and a favorable steam or heat utilization factor along with a relatively low fuel cost as compared to that of purchased electric power. One factor being very favorable can compensate for the other being less than favorable. Plants operating 7 days a week at 24 hours per day are the most desirable.

A simplified profit basis is established by calculating the payback period as shown in Figure 8-1. This is the total installation cost for each application adjusted for specific financial criteria for the industry such as taxes, insurance, rate of return on investment, and applicable tax credits for a qualified cogenerator, divided by the difference between the predicted annual electric utility power usage and the demand charges before and after the cogeneration installation (with the reduced peak demand charge considered), minus the estimated total operating cost consisting of the annual predicted fuel cost plus the annual estimated maintenance, overhaul, and labor costs prorated over the life of the equipment.

Typical financial criteria include the following:

- Federal income tax rate
- State income tax rate
- Economic life and rate of return (depreciation period and method, that is, declining balance or double declining balance)
- Investment tax credit
- Annual property tax and insurance

The payback period for return on the investment determines the economic feasibility of a cogeneration facility. Payback is determined from the total installed cost, revenues, and operating and maintenance costs. The total installed cost includes costs for equipment, buildings, materials, and labor. Revenues are obtained from the electricity, steam, and exhaust heat sold or from the avoided costs of purchasing power from an outside vendor. Operating and maintenance costs

**Table 8-1 Typical 10-Year Maintenance Schedule for a Combustion Turbine Generator Set[a]**

| Maintenance period, hr | Job | Shutdown hours required | Work hours required |
|---|---|---|---|
| 750 | Turbine operational maintenance | 0 | 2 |
| 1,500 | Turbine standard maintenance | 7 | 7 |
| 2,250 | Turbine operational maintenance | 0 | 2 |
| 3,000 | Turbine standard maintenance | 7 | 7 |
| 3,750 | Turbine operational maintenance | 0 | 2 |
| 4,500 | Turbine standard maintenance | 7 | 7 |
| 5,250 | Turbine operational maintenance | 0 | 2 |
| 6,000 | Turbine intermediate maintenance | 15 | 30 |
| 6,750 | Turbine operational maintenance | 0 | 2 |
| 7,500 | Turbine standard maintenance | 7 | 7 |
| 8,250 | Turbine major maintenance | 12 | 22 |
| | Total—first year | 55 | 90 |
| 16,500 | | | |
| | Total—second year[b] | 55 | 90 |
| | Total—third year[c] | 205 | 390 |
| | Total—fourth year[d] | 55 | 90 |
| | Total—fifth year[e] | 55 | 90 |
| | Total—sixth year[f] | 205 | 390 |
| | Total—seventh year[e] | 55 | 90 |
| | Total—eighth year[d] | 55 | 90 |
| | Total—ninth year[g] | 205 | 390 |
| | Total—tenth year[e] | 55 | 90 |
| | Total—after 10 years | 1000 | 1800 |

[a] For turbines, other maintenance and minor breakdowns may require another 200 hours of downtime per year. On this basis theoretical highest availability factor is, for a 10-year period, 96.5 percent

[b] Turbine second year same as first year.

[c] Turbine third year same as first year except for major disassembly and overhaul at 25,000 hours; add 150 shutdown hours and 300 work hours.

[d] Same as second year.

[e] Same as first year.

[f] Turbine requires overhaul during sixth year.

[g] Same as third year.

include the incremental cost of fuel, maintenance, income taxes (federal, state, and local), property taxes, and insurance. The payback period is equal to the total installed cost divided by the revenues minus the operating costs.

## ALTERNATIVE PROJECT FUNDING SOURCES

A project may qualify for state and federal financing programs if a cogenerator desires to supplement the traditional sources of capital investment financing.

I. Annual savings = present costs without cogeneration

− future costs with cogeneration

= revenues

− operating costs

= [sale of electric power or savings of cost avoided or displaced by power generated

+ value of reduced electric power capacity or demand charge

+ heat the combustion turbine exhaust displaces (drying, hot water, steam)

+ cooling effect displaced]

− [combustion turbine fuel costs

+ staffing and operating costs

+ taxes

+ insurance

+ maintenance and overhaul costs]

II. Payback period = project total installed cost*

÷ annual savings

*Adjusted for taxes, insurance, rate of return on investment, amortization, and applications tax credits

**Fig. 8-1 Economic payback analysis.**

Capital financing is available for certain gas utilities which will fund up to a maximum of $50,000 for the feasibility study and design of a qualified cogeneration project. For instance, the Southern California Gas Company may invest up to $50,000 in any one project. The California Energy Commission also finances feasibility and design studies primarily for municipal cogeneration projects.

In some cases funding can be obtained for a cogeneration project study from both a gas utility and state energy commission. Some states have programs offering low-interest loans financed by revenue bonds repayable from the revenues earned by a cogeneration facility.

Pollution control devices may be financed by money borrowed from the state. Some states loan money to agricultural and industrial businesses at rates approximately two to five percentage points lower than commercial rates. Money may be raised by selling revenue bonds, based on the credit of the contracting business and other collateral as required.

## PROJECT FINANCING AND OWNERSHIP

### Types of Financing

Internal and external approaches to the financing of cogeneration facilities by corporations, investors, the public, and joint venture arrangements are discussed in this section. Each method carries with it a level of financial risk, responsibilities, benefits, and liabilities. An optimum balance between the risks and the rate of return on the investment is a desirable goal.

The financing and ownership of cogeneration projects must be decided on early in the development of a facility. Financing and ownership are directly related to tax benefits, the depreciation period, selection of financing alternatives, cost of capital formation, participant sharing of goals and risks, return on investment, and degree of control over fuel, steam, heat, energy, and electric power costs. A financial analysis will determine the economic attractiveness of the project. These goals and constraints will impact the financial alternatives available.

Project financing involves securing funding, and available methods of financing are borrowing and leasing. Financing includes the formation of capital, operating, and maintenance funds for a cogeneration project.

Funding risks are reduced by obtaining contracts and agreements involving permits, engineering, construction, operation, maintenance, electric power sales, and thermal and/or mechanical energy sales. The earning power of the income will dictate the financial viability of the project.

Contracts for electric power sales and thermal and mechanical energy sales, in addition to purchase agreements for fuel supply and water rights, are most important. These contracts should be made with reliable and reputable firms of proven financial status and business experience that are expected to continue competing in their product market area while fulfilling their long-term contractual obligations, thus minimizing the funding risks associated with the cogeneration facility. Cogeneration financing by utilities includes put bonds, tax-exempt commercial paper, floating-rate bonds, municipal bond insurance, and other financial guarantees such as letters of credit.

Private investment cogeneration financing by state, municipal, county, city, and industry is growing as economic conditions change in the financial markets. The level of investments from the financial market are tied to areas such as volatility of the capital markets, unprecedented high interest rates, investor perceptions of project viability, capitalized interest requirements, fuel cost increases and fuel availability, and energy (steam, hot water, electricity) sale contractual agreements.

A third-party financier provides legal contractual obligations under take-or-pay agreements involving energy, operational costs, maintenance costs, and debt service payments on bonds issued to finance the project.

Issuers who finance new technologies through the municipal bond market may use a variety of unusual financing techniques to make these bonds more attractive to investors.

Cogeneration financing will continue to attract new and innovative methods of both conventional and unconventional financing. The financial advantages of cogeneration over conventional power-generating facilities result from shorter lead times, flexibility in planning for smaller projects, for example 5 megawatts as compared to 500 megawatts, reduced capitalized interest requirements, a reduced impact of fuel cost increases, and availability of fuel on a priority basis.

A methodology for determining whether a cogeneration investment is worthwhile should maximize profits, utilize tax and depreciation benefits associated with the facility, result in the lowest cost of financing and owning the facility, reduce unexpected environmental risks and equipment costs, avoid regulations as a public utility, minimize the impact on corporate future capital formation, and maintain a high corporate financial rating.

A corporation should maintain a technical and financial staff to implement a long-term corporate financing and energy policy, thus avoiding the need to accept alternatives not compatible financially or technically with expansion objectives.

Ownership ranges from direct to third-party-type arrangements each having different goals, risks, and rewards. Ownership is often a prime factor in selecting equity accounting to remove a financial obligation from the balance sheet and income statement for purposes of future debt issuance and financial ratings. Cogeneration project ownership may involve corporations, individuals, partnerships, joint ventures, banks, leasing companies, insurance companies, investor-owned utilities, public utilities, and municipal, county, and city governments.

### Corporate Bonds

Some corporate bonds are secured by the mortgage of a plant, equipment, or other property as security for a promise to pay off the bonds.

Other corporate bond financial alternatives such as debentures are bonds not backed by mortgaged property but only by a promise to pay and the company's general credit rating.

### Tax-Exempt Bonds

State, county, city, and local governments can issue tax-exempt municipal bonds to finance cogeneration facilities. The investor income from municipal bonds is exempt from federal income tax and generally from state income tax in the state issued. Bond issuers are expected to seek alternatives with the lowest risk in order to minimize liability exposure upon default. Therefore revenue bonds are in most cases preferred by the issuing group over general obligation bonds.

Revenue bonds can be issued by state and local governments to finance a facility producing revenues from long-term contract agreements for the sale of electric power, thermal heat, steam, and mechanical energy. The risk to the issuing group is limited to the revenues generated once the facility is in operation.

This risk reduction for the issuing group results in increased risk for the investor. Thus revenue bonds carry a higher interest rate to compensate the investor and provide the necessary investment incentive.

Industrial development or revenue bonds can be issued by state and local governments to finance a cogeneration facility for lease or sale to a private entity. The private entity is then entitled to the benefits of investment tax credits, depreciation, and reduced overall project capital costs resulting from a tax-exempt bond interest rate less than that required for corporate bond issues. These bonds have restrictions involving the size and type of facility and amount of capital.

General obligation bonds issued by state and local governments are another alternative for financing cogeneration facilities that can then be leased to industry. General obligation bonds are backed by the reputation and credit rating of the issuing group. The good faith, credit, and taxing power of the issuing group is irrevocably pledged to the punctual payment of the principal and interest on the bonds. Banks underwrite, issue, and deal in general obligation bonds.

## Leasing

Another form of financing available for cogeneration facilities is leasing. The lessor leases the equipment and facilities to the lessee. The lessor takes the financial risk by investing his or her capital, owns the facility, and takes advantage of the tax benefits and allowance for depreciation of the equipment at the facility. Leasing may provide more borrowing leverage relative to equity since it is not recognized as debt.

The lessee takes possession of the facility and equipment with minimum investment of his or her own capital and can deduct the lease payments as business expenses. The lessee thus conserves capital for current uses while creating no fixed debt.

The lease payments over the contracted duration generally cover the expenses of the lessor and amortize debt without appearing on the balance sheet of the lessee as a liability or asset. The period of the lease is mutually negotiated to satisfy the financial needs of the lessor and the lessee.

The lessor can transfer the tax benefits to the lessee through a conditional sales lease arrangement. This arrangement consists of a short-term (approximately 12-month) lease with an option to purchase at the end of this period. The cost of this alternative is higher than that of conventional leases in order to offset the risk assumed by the lessor in negotiating a new lease.

A bond-issuing agency or the owner of the facility may lease the facility to a lessee along with the associated tax benefits. The lease may be arranged as a sale with the lessee as owner for tax benefit reasons. The lessee may then sublease the facility to another operator. Payments from the operator can be applied to amortization of the bonds.

## Safe Harbor Leasing

The safe harbor leasing provisions of the Economic Recovery Act of 1981 provide for corporations not having sufficient tax liability but unusable tax credits or depreciation deductions to sell them to others. Thus corporations without tax liability can benefit from the same federal tax incentives as those having sufficient tax liability.

Tax law revisions during the last quarter of 1982 require cogeneration-related buying and selling tax credit transactions be completed before July 2, 1983. The restrictions imposed on this mode of leasing have significantly reduced the tax benefits for both the buyer and seller. The investment credit must now be spread over 5 years instead of 1 year. The new tax law prohibits buyers of tax benefits from reducing their current annual tax liability by more than 50 percent through October 1, 1985. Sellers can lease only 40 percent of their eligible property each year until 1986, and benefits cannot be used to reduce taxes from past years. These restrictions basically apply until 1986.

## Leverage Leasing

A tax leverage lease provides maximum utilization of tax benefits while sometimes involving a lender, a lessor, and a lessee. The lender and the lessor provide capital and equity to finance the facility. A typical lessor may provide 25 to 40 percent of the equity with the lender financing the remaining portion on a nonrecourse basis. The lender secures the loan risk by a leveraged lease for the facility and equipment.

Benefits to the lessor are a decrease in the amount of capital needed in order to leverage its financing of the project and secure a lower interest rate than that available for conventional financing.

Ownership provides the lessor with tax advantages from depreciation and investment tax credits to offset other income while planning for sufficient after-tax return on the capital equity investment in the facility. Tax credit benefits are not available to state, county, city, and local governments. These benefits can be purchased by the lessor with payments made to the lender.

## Lease Purchase

In a lease purchase agreement, the lease payments can serve as loan payments toward the purchase of the equipment. At the expiration of the lease, the lessee can purchase the cogeneration facility for a minimum percentage of its original cost.

This arrangement should appeal to companies seeking ownership at a later date while immediately benefiting from the revenues and more efficient energy utilization associated with a cogeneration facility.

## Equity Lease

An equity lease is a financing option in which the lessor pays for the facility and then leases it to the lessee. The lease payment is a percentage of the profits from the system, allowing the lessor to profit from his or her investment. The lessor incurs the risk and receives the rewards associated with the facility.

## Purchase Buy-Back Agreements

This alternative financing option provides for risk sharing between corporations. Once a project meets agreed-on criteria, then one proponent is obligated to buy back the other participant's holdings. This money can then be used by the participants for other projects. This alternative allows many corporations to continue with projects during times of high interest rates with minimum equity available for debt coverage.

## Sale and Leaseback

A conventional sale and leaseback alternative occurs when a company arranges for a lessor to purchase cogeneration equipment and facilities and then leases them back to the company at a reduced rate relative to the tax benefits. The lease payments should completely amortize the investment and also provide a return to the lessor. Ownership of the property can be structured for recapture by the lessee at the end of the lease period. This method is sometimes used by the initial owner or company to avoid a shortage of funds. This contractual arrangement allows the initial owner to convert fixed assets to capital to meet liquidity goals.

In another type of arrangement a buyer purchases some equipment on its balance sheet and then leases it to an operator who sells energy back to the lessor. This is an approach used in working with some tax-exempt organizations.

## Credit

Credit primarily consists of a line of credit or a letter of credit from a lender with which a borrower can obtain funds from another lender for financing a cogeneration facility. While the issuer of a letter of credit carries the ultimate credit risk, the amount of the line of credit is generally negotiated on the basis of the borrower's equity and established securities. However, the line of credit is sometimes unsecured and bears interest tied to the base or prime rate for loans.

A line of credit is revolving in nature, meaning that funds can be drawn upon, paid, and borrowed in a continuous cycle for a defined duration. A short-term line of credit could typically be for less than 2 or 3 years. Since a short-term line of credit is relatively expensive, its use may be limited to providing funds on a temporary basis until long-term, lower-cost financing can be arranged.

A line of credit can also be issued to support the offering of notes. This financial alternative provides liquidity to the borrower to retire short-term debt. The lender does not expect the borrower to exercise the extended credit but to benefit from the lender's support to reduce the interest rate on offering securities to investors.

A letter of credit issued by a bank substitutes the generally higher credit rating of the bank for the assumed lower credit rating of the bond issuer. The letter of credit promising payment to the lender is irrevocable and represents the bank's commitment of support and assumption of the burden of future financial obligation. The letter of credit places the risk of nonpayment on the bank. This financial alternative therefore is more expensive, which may be offset by improved rating and enhanced desirability, thus reducing the cost of the bonds or notes issued.

### Internal Capital Funding

Annual budgets and internal capital constraints indicate the amount of capital available for various product line expansions and improvements. Cogeneration projects for most industrial applications compete for limited funding available in the overall corporate budget. The merits of a cogeneration project depend on the cash flow, source of financing, schedule for project construction, and operation.

### Loans, Grants, and Other Incentives

Federal loan and grant programs are available for certain types of energy projects. This approach can supplement the financial arrangements worked out to fund a project. Many states have developed programs designed to help finance energy projects. Some state public utility commissions have authorized investor-owned utilities to use rate-base funds to subsidize energy conservation projects, and other states are considering similar moves.

Some states have created financing authorities specifically charged with assisting in energy conservation and efficiency improvement. In addition, some states also offer property tax incentives.

### Third-Party Financing, Ownership, and Operation

A third party may finance, own, engineer, design, purchase equipment for, construct, and operate the cogeneration facility without using the electricity or the heat generated.

It can then lease the facility back to others to maintain the facility, use the heat or steam, and use or sell the electricity. Any combination of the above is also possible. Third-party ownership may include profit sharing from energy and electricity sales agreements with the user. Ownership benefits include depreciation and investment tax credits. Technical and economic risks involving completion, operation, and financing are assumed by the third-party owner.

Each financial backer has individual legal requirements, limitations, financial capabilities, and motivations which will affect the support of cogeneration projects. Cooperation among investors and participants should be encouraged where appropriate, to reduce the cost of capital investment and to facilitate completion of the project.

Third-party owners and partners should be carefully qualified for financial stability, credibility, long-term binding contractual commitments, and options for future ownership transfer of the facility.

# POWER PURCHASE AND SALES AGREEMENTS

Section 210 of PURPA mandates that utilities offer to purchase electricity from qualifying cogeneration facilities. The two types of possible contracts are standard offers by utilities to purchase power from qualified cogenerators and agreements resulting from negotiations between utilities and qualified cogenerators. Some cogenerators seek to negotiate contracts containing terms and conditions more closely tailored to their particular circumstances and utilities and cogenerators are free to negotiate purchase agreements under terms and conditions differing from those of standard offers.

PURPA, FERC, and PUC rule-making policies establish standards governing the prices, terms, and conditions for utility purchases of electric power from cogeneration facilities. PURPA does not require that electric utilities make contracts with qualified cogenerators.

Power sales agreements are intended to be economically equitable for all parties including the cogenerator, electric utility and its rate payers, and stock and bond shareholders. Specific risks can be dealt with in a power sales agreement. The use of a long-term power contract to sell any electric power output in excess of in-house electrical demand to the electric utility mitigates the marketing risks as well as the financial risks. Marketing risks include those associated with the rate of escalation of operating and maintenance costs for production of the electricity being sold.

Power sales agreements reduce the financial risks of a project to a lender by providing debt guarantee in contracts between cogenerators and private investors or public electric utilities. The utility assumes this risk to the extent of the avoided cost agreement.

A power sales agreement can be based on the avoided cost at the time energy is delivered or at the time the obligation is entered into. The avoided cost at the time energy is delivered can be immediately determined by comparing the current price of oil, gas, or other alternate fuel with the incremental heat rates of the utility.

Minor changes in the power sales agreement may require costly changes and impact on engineering, design, financing, permit and regulation compliance, construction, and operation schedules. The power sales agreement negotiations with the utility should be considered when preparing the project schedule.

The power sales or purchase agreement negotiated between a cogenerator and an electric utility is of primary concern for any cogeneration project that will sell excess power. Power sales provisions are negotiable, since few standard power sales contracts are satisfactory to the cogenerator and utility without modifications.

A mutually acceptable power sales agreement with an electric utility must conform to corporate financial policy and the economic criteria considered when financing a cogeneration project producing excess power. The agreement indicates the price the utility will pay for power received from the cogenerator and clarifies the regulatory, technical, ownership, operational, maintenance risks, and liability to be assumed by the utility or the cogenerator. The power sales agreement requires the cogenerator to supply the contracted amount of power to the electric utility, except for the minimum amount of time stipulated for system emergencies or low-load periods. A power sales agreement may be negotiated for a simultaneous buy and sell agreement which gives the cogenerator the option to sell all the power it produces at the avoided cost rate while simultaneously buying all on-site power required from the utility at its standard rate.

The state PUC may require the approval of nonstandard power purchase contracts entered into by the electric utility. The nonstandard contracts reviewed by the PUC are generally unique or unusual and generally place a significant risk to or a substantial financial burden on rate payers, the utility, or the cogenerator. The possibility that the PUC will not approve the contract constitutes an inherent risk. The process by which the PUC grants its approval may affect the overall project schedule.

The most important risk in the contract is that it frees the qualifying facility from future public utility commission adjustments. The burden of a reasonable nonstandard contract price is placed on the utility, which must administer avoided cost and protect the rate payer. Advance approval of nonstandard contracts by the PUC assures the utility that it will not be exposing its shareholders and rate payers to unnecessary risks, and also protects the cogenerator from contract adjustments. The utility and subsequently the cogenerator may both be at risk should the PUC not allow a passthrough of costs to a cogenerator based on a nonstandard power purchase contract. Future federal and state PUC decisions may modify power purchase agreements and the use of contracts and contract rates and terms. The electric utility may be reluctant to take these risks and expose its shareholders to this uncertainty. Utilities seek assurance that payments made for power purchased under contracts resulting from nonstandard offers will be recovered in rates. This burden is removed from the utility when the PUC approves a pricing agreement permitting the utility to recover power purchase costs from rate payers.

Power sales agreements can be take or pay, take and pay, or take as required. The take-or-pay contract is often referred to as a "hell-or-high-water" contract whereby the utility purchases a specified amount of electric power even if it never receives power from the cogenerator. This revenue thus guarantees debt service and is, therefore, the strongest argument for financing.

A take-and-pay contract is the most common type. The utility purchases all the power produced by the cogenerator except in certain specific instances. Contracts limit periods of nonpurchase of power to a specific minimum number of hours per year during outages for emergency and maintenance. During this period, negotiated contract terms with the utility can be based on the average kilowatthours sold on the days immediately before the outage.

Under the terms of a take-as-required contract, the utility dispatches power on the basis of economy and air quality within its electric service territory. A utility having lower-cost power available to its system could disconnect from the cogenerator and take advantage of the lower-cost power produced. The utility is obligated to pay only for energy it accepts. The California PUC Order Instituting Rulemaking No. 2 (OIR 2) Decision 82-01-103 of January 21, 1982, ordered major California electric utilities to file standard offers for power purchases based on avoided cost. These offers are available to all qualified cogeneration facilities in California. The utilities are assured that purchases of energy under these four standard offers are reasonable and will be passed to rate payers via the utility energy cost adjustment clause proceedings. These four standard offers are based on cogeneration facilities providing as-available electricity payments, firm capacity payments, a 5-year contract with payment for energy tied to either an as-available or a firm capacity option, and a long-term contract with energy and capacity payments based on the utility's long-run avoided cost.

A more comprehensive review by potential cogenerators, utilities, and regulators structuring cogeneration programs is given in the four contracts for Pacific Gas and Electric Company in Appendix K.

# 10

## BENEFITS OF COGENERATION

The benefits of cogeneration listed below are derived from improved power cycle efficiency and an associated reduction in fuel consumption as compared to the less efficient conventional power plant utilizing fuel solely for generating electric energy. These economic advantages and other incentives are available from cogeneration facilities.

- Higher efficiency can be achieved by utilizing the same fuel to provide electricity and heat:

    Reduced fuel consumption

    Reduced fuel cost

    Reduced electric utility bills

    Reduced production or operating costs

    An economic competitive advantage through a maximized return on investment capital

    The most efficient use of capital investment

- The amount of useful energy produced can be increased through the recovery of otherwise wasted heat.
- State and federal regulations require electric utilities to purchase electricity from qualified cogeneration facilities.
- The electric utility must pay the cogenerator its avoided cost: the utility cost for generating the same amount of electricity itself or of purchasing it from another utility.
- A lesser effect on the environment because of efficient fuel use:

    Reduced air emissions (sulfur dioxide, nitrogen oxides, particulates)

    Reduced thermal pollution

- A reliable source of power and process steam or heat: availability of utility backup.
- Energy independence:

  Production of process steam or heat for internal use

  Selling to a neighboring user

  Generation of electricity for internal use and selling of excess to a public utility
- Operating flexibility whereby the total electrical load can be provided, a portion of the load can be provided, or excess power and steam or heat can be sold to others:

  Total independence from utility power

  Total independence from utility backup

  Partial independence from utility power
- On-site generation can eliminate energy losses (5 to 10 percent) due to transmission and distribution systems.
- Cogeneration plants can be placed in operation in 1 to 4 years compared to 8 to 10 years for conventional power plants.
- Proven cogeneration technology and air emission controls.
- Tax credits.
- Depreciation allowance.
- Energy conservation:

  Improve energy efficiency by implementing an energy conservation plan

  Convert any existing standby power generator set to a continuous profit producer

  Recover heat energy from cooling and exhaust systems to add profits to the operation

  Minimize energy demand

  Utilize available energy for internal process applications and displace other less efficient sources

The following specific conditions indicate good potential cogeneration applications.

- Existence of a large thermal load:

  Constant daily and annual thermal loads

  Cost-effective cogeneration investment

- Redundancy for standby purposes is not required.
- Cogenerator purchases standby electricity from the utility when its electricity demands exceed production.

Utilities generally compute demand charges based on the highest peak loads recorded during a billing period. Utility bills consist of energy charges for Kilowatthours used and Peak demand (for the highest peak load). In many cases a peak load month can determine the demand charge for the entire year. If emergency or standby power is available, then the peak shaving benefit may be obtained with little additional investment and a profit can be made from the savings associated with peak shaving. Peak shaving is the practice of reducing the peak load demand billing charge. An existing emergency standby power generator can become a profit producer, in addition to providing on-going tests of the equipment's operability while not significantly increasing equipment maintenance costs.

# 11

# TECHNICAL AND ECONOMIC FEASIBILITY

The objective of the hypothetical project described in this chapter is to provide a cogeneration system having

- The least complexity
- The highest reliability
- Simplicity of operation and interconnection with building heating and cooling systems
- Minimum maintenance
- Redundancy or backup features
- Low capital installed costs
- Ease of construction

This facility is for the cogeneration of electricity and thermal energy for heating and cooling systems. The cogenerator sells all electric power produced to a local electric utility and buys back all site power. Revenues are generated from the sale of electricity, heat, and cooling thermal energy. Heating and cooling revenues are received from district heating consumers.

The financial, design, and operating criteria for this cogeneration project must provide

- Maximum efficiency and energy conservation features
- Satisfaction of environmental and regulatory requirements
- The least environmental impact
- The highest return on the investment at the least possible risk

This project is thermally matched to the system demand to optimize efficiency. The unit loading tracks the thermal demand for process steam and hot water while generating electricity, thus requiring the least fuel use. The project is evaluated on the basis of owner financing for economic comparisons.

Equipment and sizing selection criteria include heat rates and exhaust temperatures for achieving the greatest operating efficiency. This sizing approach fully utilizes the combustion turbine generator exhaust heat and meets the FERC topping-cycle efficiency standard of 42.5 percent, thus permitting the facility to benefit from the regulatory and tax advantages available to qualifying facilities.

The economic implication is that this thermally optimized unit has the operating flexibility to satisfy thermal load energy requirements—peak, seasonal, and daily load variations—at an optimum capacity factor, resulting in attractive overall operating revenues.

The design life of this plant is expected to be in excess of 20 years. This facility has been specified, engineered, and designed to meet best available control technology (BACT) standards and to satisfy emission control requirements.

This example consists of two alternative equipment and system arrangements. Alternate 1 is shown in Figure 4-1 and consists of one combustion turbine generator exhausting to one waste heat recovery boiler. Alternate 2 consists of two combustion turbine generators each separately exhausting to one of two waste heat recovery boilers.

Each combustion turbine generator has a rating of 3000 kilowatts net output, a lower heating valve heat rate of 12,970 Btu per kilowatthour, and a heat input of $40.2 \times 10^6$ Btu per hour with an exhaust temperature of 930°F.

The waste heat recovery boiler converts the combustion turbine generator exhaust gas mass flow rate of 126,000 pounds per hour at 930°F to hot water for heating the building. High-temperature hot water is pumped and distributed to heating units to supply the heating load.

Absorption chillers convert the combustion turbine exhaust heat to chilled water to supply the building cooling load, and chilled water is pumped through the cooling units. Mechanical draft cooling towers provide cooling water for the heat exchangers. Hot- and cold-water storage heat sinks are provided. The overall project schedule is 24 months.

An economic comparison of the engineering aspects of the two alternatives includes estimated equipment, materials, engineering, and construction costs for the 24-month construction period. At the top of the next page is a summary of the costs incurred.

Alternative 1 is based on one 3000-kilowatt combustion turbine generator and one waste-heat recovery steam generator (WHRSG). An economic comparison and cash-flow tabulation are given in Table 11-1. Alternative 2 is based on two 3000-kilowatt combustion turbine generators and two WHRSGs. An economic comparison and cash-flow tabulation are given in Table 11-2.

An interest rate of 11.5 percent is assumed during the 24 months of construction.

| | Alternative 1 | Alternative 2 |
|---|---|---|
| Equipment and materials | $4,600,000 | $ 5,800,000 |
| Engineering and construction | 3,100,000 | 3,950,000 |
| Total installed cost before construction | $7,700,000 | $ 9,750,000 |
| Interest during construction | 886,000 | 1,122,000 |
| Total cost | $8,586,000 | $10,870,000 |

## FINANCIAL AND CASH-FLOW ANALYSIS

The project is financed with 30 percent equity and 70 percent debt. Construction costs are financed 30 percent from equity and 70 percent from short-term debt. This short-term debt is refinanced at the end of the 24-month construction period with an 8-year, 16 percent loan to be repaid in eight equal annual payments.

### Operating Expenses

Operating expenses consist of fuel costs, operating and maintenance costs, property taxes and insurance, and interest.

***Fuel Costs*** Fuel costs are based on the plant operating at 75 percent of its capacity and utilizing natural gas, with an annual base fuel consumption of 264,114 $\times 10^6$ Btu for alternate 1. In the first year of operation the cost per therm (100,000 Btu per therm) is $0.6441. The cost of natural gas fuel escalates at 13.55 percent per year. (See Table 11-3.)

Annual fuel consumption = (combustion turbine fuel consumption, Btu/hr) (hr/yr) (average capacity factor)

Alternative 1 annual fuel consumption = $(40.2 \times 10^6$ Btu/hr) (8760 hr/yr) (75% capacity factor)

= $264{,}114 \times 10^6$ Btu/yr

Alternative 2 annual fuel consumption = $(80.4 \times 10^6$ Btu/hr) (8760 hr/yr) (75% capacity factor)

= $528{,}228 \times 10^6$ Btu/yr

The fuel gas cost assumes 13.55 percent escalation per year after the base year of initial operation.

The annual fuel gas cost, \$/yr = (annual fuel consumption, Btu/yr) (fuel gas cost, $\$/10^6$ Btu)

Fuel gas cost, $\$/10^6$ Btu = (fuel gas cost, \$/therm) ÷ (100,000 Btu/therm)

**Table 11-1 Alternative 1**
Economic Comparison and Cash Flow

| | Year | | | | | | | | | |
|---|---|---|---|---|---|---|---|---|---|---|
| | 1 | 2 | 3 | 4 | 5 | 6 | 7 | 8 | 9 | 10 |
| **Total plant cost** | — | — | 8,586,000 | — | — | — | — | — | — | — |
| 30% Equity | — | — | 2,575,800 | — | — | — | — | — | — | — |
| 70% Long-term debt | — | — | 6,010,200 | — | — | — | — | — | — | — |
| Principal | — | — | 751,275 | 751,275 | 751,275 | 751,275 | 751,275 | 751,275 | 751,275 | 751,275 |
| **Operating expenses** | | | | | | | | | | |
| Fuel costs | — | — | 1,701,158 | 1,931,730 | 2,193,467 | 2,490,595 | 2,828,133 | 3,211,362 | 3,646,358 | 4,140,515 |
| Operating and maintenance costs | — | — | 270,000 | 288,900 | 309,129 | 330,762 | 353,915 | 378,689 | 405,197 | 433,561 |
| Property taxes and insurance | — | — | 92,000 | 92,000 | 92,000 | 92,000 | 92,000 | 92,000 | 92,000 | 92,000 |
| Interest | — | — | 961,632 | 841,428 | 721,224 | 601,020 | 480,816 | 360,612 | 240,408 | 120,204 |
| **Total operating expenses** | — | — | 3,024,790 | 3,154,058 | 3,315,820 | 3,514,377 | 3,754,864 | 4,042,663 | 4,383,963 | 4,786,280 |
| **Revenues** | | | | | | | | | | |
| Electricity | — | — | 1,675,155 | 1,825,062 | 1,981,461 | 2,200,932 | 2,359,647 | 2,574,363 | 2,819,568 | 3,089,583 |
| Thermal heating | — | — | 763,360 | 872,130 | 951,480 | 1,075,770 | 1,191,100 | 1,317,090 | 1,473,500 | 1,643,670 |
| Depreciation, federal | — | — | 1,287,900 | 1,888,920 | 1,803,060 | 1,803,060 | 1,803,060 | — | — | — |
| Depreciation, state | — | — | 1,717,200 | 1,717,200 | 1,717,200 | 1,717,200 | 1,717,200 | — | — | — |
| Investment tax credit | — | — | 772,740 | — | — | — | — | — | — | — |
| **Total revenues** | — | — | 6,216,355 | 6,303,312 | 6,453,201 | 6,796,962 | 7,071,007 | 3,891,453 | 4,293,068 | 4,733,253 |
| **Cash-flow summary** | | | | | | | | | | |
| Total payment to principal | — | — | 751,275 | 751,275 | 751,275 | 751,275 | 751,275 | 751,275 | 751,275 | 751,275 |
| Total operating expenses | — | — | 3,024,790 | 3,154,058 | 3,315,820 | 3,514,377 | 3,754,864 | 4,042,663 | 4,383,963 | 4,786,280 |
| Total revenues | — | — | 6,216,355 | 6,303,312 | 6,453,201 | 6,796,962 | 7,071,007 | 3,891,453 | 4,293,068 | 4,733,253 |
| Net income | — | — | 2,440,290 | 2,397,979 | 2,386,106 | 2,531,310 | 2,564,868 | (902,485) | (842,170) | (804,302) |

**Table 11-2 Alternative 2**
Economic Comparison and Cash Flow

| | Year | | | | | | | | | |
|---|---|---|---|---|---|---|---|---|---|---|
| | 1 | 2 | 3 | 4 | 5 | 6 | 7 | 8 | 9 | 10 |
| **Total plant cost** | — | — | 10,870,000 | — | — | — | — | — | — | — |
| 30% Equity | — | — | 3,261,000 | — | — | — | — | — | — | — |
| 70% Long-term debt | — | — | 7,609,000 | — | — | — | — | — | — | — |
| Principal | — | — | 951,125 | 951,125 | 951,125 | 951,125 | 951,125 | 951,125 | 951,125 | 951,125 |
| **Operating expenses** | | | | | | | | | | |
| Fuel costs | — | — | 3,402,317 | 3,863,460 | 4,386,334 | 4,981,190 | 5,656,265 | 6,422,724 | 7,292,716 | 8,281,030 |
| Operating and maintenance costs | — | — | 180,000 | 192,600 | 206,082 | 220,508 | 235,943 | 252,459 | 270,131 | 289,041 |
| Property taxes and insurance | — | — | 116,000 | 116,000 | 116,000 | 116,000 | 116,000 | 116,000 | 116,000 | 116,000 |
| Interest | — | — | 1,217,440 | 1,065,260 | 913,080 | 760,900 | 608,720 | 456,540 | 304,360 | 152,180 |
| Total operating expenses | — | — | 4,915,757 | 5,237,320 | 5,621,496 | 6,078,598 | 6,616,928 | 7,247,723 | 7,983,207 | 8,838,251 |
| **Revenues** | | | | | | | | | | |
| Electricity | — | — | 3,350,310 | 3,650,310 | 3,962,222 | 4,401,864 | 4,719,294 | 5,148,726 | 5,639,136 | 6,179,166 |
| Thermal heating | — | — | 1,526,720 | 1,744,260 | 1,902,960 | 2,151,540 | 2,382,220 | 2,634,180 | 2,947,000 | 3,287,340 |
| Depreciation, federal | — | — | 1,630,500 | 2,391,400 | 2,282,700 | 2,282,700 | 2,282,700 | — | — | — |
| Depreciation, state | — | — | 2,174,000 | 2,174,000 | 2,174,000 | 2,174,000 | 2,174,000 | — | — | — |
| Investment tax credit | — | — | 978,300 | — | — | — | — | — | — | — |
| Total | — | — | 9,659,830 | 9,959,784 | 10,322,582 | 11,010,104 | 11,558,214 | 7,782,906 | 8,586,136 | 9,466,506 |
| **Cash-flow summary** | | | | | | | | | | |
| Total payment to principal | — | — | 951,125 | 951,125 | 951,125 | 951,125 | 951,125 | 951,125 | 951,125 | 951,125 |
| Total operating expenses | — | — | 4,915,757 | 5,237,320 | 5,621,496 | 6,078,598 | 6,616,928 | 7,247,723 | 7,983,207 | 8,838,251 |
| Total revenues | — | — | 9,659,830 | 9,959,784 | 10,322,582 | 11,010,104 | 11,558,214 | 7,782,906 | 8,586,136 | 9,466,506 |
| Net income | — | — | 3,792,948 | 3,771,339 | 3,749,961 | 3,980,381 | 3,990,161 | (415,942) | (348,196) | (322,870) |

Table 11-3 Fuel Costs

| Year | Fuel gas cost | | Annual fuel cost | |
|---|---|---|---|---|
| | Dollars per therm | Dollars per $10^6$ Btu | Alternative 1, $ | Alternative 2, $ |
| 1 | — | — | — | — |
| 2 | — | — | — | — |
| 3 | 0.6441 | 6.441 | 1,701,158 | 3,402,317 |
| 4 | 0.7314 | 7.314 | 1,931,730 | 3,863,460 |
| 5 | 0.8305 | 8.305 | 2,193,467 | 4,386,334 |
| 6 | 0.9430 | 9.430 | 2,490,595 | 4,981,190 |
| 7 | 1.0708 | 10.708 | 2,828,133 | 5,656,265 |
| 8 | 1.2159 | 12.159 | 3,211,362 | 6,422,724 |
| 9 | 1.3806 | 13.806 | 3,646,358 | 7,292,716 |
| 10 | 1.5677 | 15.677 | 4,140,515 | 8,281,030 |

## OPERATING AND MAINTENANCE COSTS

The operating staff provides supervision, administration, and technical support for the plant and is also involved in billing and collecting revenues. Technical maintenance includes repair and overhaul for the facility. Other maintenance costs include those for materials used in the maintenance of nontechnical items. The work is performed by the operating staff.

Operating staff and maintenance costs (Table 11-4) are assumed to be $45 per kilowatt times the net plant electricity output for alternative 2, or ($45 per kilowatt) (6000 kilowatts) = $270,000, based on two combustion turbine generators and two WHRSGs.

For alternative 1 $0.003 per kilowatthour is deducted for one combustion turbine generator and $25,000 per year for the controls and one WHRSG. The deduction for alternative 1 is therefore

| | |
|---|---|
| (3000 kW) (8760 hr/yr) (85% capacity) ($0.003/kWhr) | = $67,014/yr |
| Deduction for controls and WHRSG | = $25,000/yr |
| Combined deduction for combustion turbine generator and WHRSG | = $92,014/yr |
| Deduction approximately | = $90,000/yr |

Costs escalate at 7 percent per year after the first year of operation.

***Property Taxes and Insurance*** Property taxes and insurance are assumed to be 2 percent of the equipment and materials capital cost per year. The sale of steam, heat, or electricity by a qualifying cogeneration facility is exempt from state sales taxes.

For alternative 1 the annual property taxes and insurance are equal to 2 percent

Table 11-4 Annual Operating, Staffing, and Maintenance Costs

| Year | Alternative 1, $ | Alternative 2, $ |
|---|---|---|
| 1 | — | — |
| 2 | — | — |
| 3 | 180,000 | 270,000 |
| 4 | 192,600 | 288,900 |
| 5 | 206,082 | 309,139 |
| 6 | 220,508 | 330,762 |
| 7 | 235,943 | 353,915 |
| 8 | 252,459 | 378,689 |
| 9 | 270,131 | 405,197 |
| 10 | 289,041 | 433,561 |

of the $4,600,000 equipment and materials capital cost, or $92,000. For alternative 2 the annual property taxes and insurance are equal to 2 percent of the $5,800,000 equipment and materials capital cost, or $116,000.

***Interest*** Interest on the long-term 70 percent 8-year debt is 16 percent (Table 11-5). The annual principal amount is equal to the long-term debt divided by the 8-year period.

For alternative 1 this is

$$(\$6{,}010{,}200) \div (8 \text{ yr}) = \$751{,}275/\text{yr}$$

The interest for year 10 is

$$(16\%)(\text{principal}) = (0.16)\ (\$751{,}275) = \$120{,}204$$

For alternative 2 this is

$$(\$7{,}609{,}000) \div (8 \text{ yr}) = \$951{,}125/\text{yr}$$

The interest for year 10 is

$$(16\%)(\text{principal}) = (0.16)\ (\$951{,}125) = \$152{,}180$$

## REVENUES

Revenues consist of electricity revenues, thermal revenues, depreciation, and energy investment tax credits.

### Electricity Revenue

The electricity revenue consists of a capacity price and an energy price (Table 11-6). The capacity price is based on a power purchase contract price from the electric utility of $112 per kilowatt on firm capacity for an 8-year contract life. The firm

**Table 11-5 Interest on Debt**

| Year | Alternative 1, $ | Alternative 2, $ |
|---|---|---|
| 1 | — | — |
| 2 | — | — |
| 3 | 961,632 | 1,217,440 |
| 4 | 841,428 | 1,065,260 |
| 5 | 721,224 | 913,080 |
| 6 | 601,020 | 760,900 |
| 7 | 480,816 | 608,720 |
| 8 | 360,612 | 456,540 |
| 9 | 240,408 | 304,360 |
| 10 | 120,204 | 152,180 |

capacity is 85 percent of the total capacity. The energy price is based on the average guaranteed rate for the first 8 years of operation. The following energy and capacity price projections assume that the cogeneration facility will operate at an average of 75 percent of its capacity. The guaranteed rate is the minimum price. The electric revenues estimated for these examples are therefore conservative and may be higher should the cogenerator negotiate a higher price from the electric utility.

The capacity price for alternative 1 is

$$(\$112/\text{kw yr})\ (3000\ \text{kW})\ (85\%\ \text{firm capacity}) = \$285{,}600$$

The energy price for alternative 1 is

$$(\$0.0705/\text{kW yr})\ (8760\ \text{hr/yr})\ (75\%\ \text{capacity factor})\ (3000\ \text{kW}) = \$1{,}389{,}555$$

***Thermal Revenue*** The thermal revenue for heating demand is assumed to be sold to the customer at a cost equivalent to producing heat in a natural-gas-fired boiler having an efficiency of 65 percent.

Alternative 1 is based on producing $40 \times 10^9$ Btu of heat, whereas alternative 2 is based on producing $80 \times 10^9$ Btu of heat. The heating revenues for alternatives 1 and 2 are computed in Table 11-7.

The thermal revenue for a cooling demand of $125{,}257.1 \times 10^6$ Btu for alternative 2 is converted to kilowatthours by dividing the cooling demand of $125{,}257.1 \times 10^6$ Btu by 3413 Btu per kilowatthour to obtain $36.7 \times 10^6$ kilowatthours. The thermal revenue is based on the peak period electric power cost adjusted for a coefficient of performance equal to 4.3. The cooling revenues for alternatives 1 and 2 are computed in Table 11-8 and the combined thermal revenues are shown in Table 11-9.

***Federal Depreciation Allowances*** Should the cogenerator be able to take advantage of accelerated depreciation benefits, then it may choose a 5-year term, except for buildings which may have a 15-year term.

**Table 11-6 Electricity Revenue**

| Year | Energy price, $/kWh | Capacity price, $/kW yr | Capacity price at 85% capacity factor | | Price of energy | | Total electricity revenue | |
|---|---|---|---|---|---|---|---|---|
| | | | Alternative 1, $ | Alternative 2, $ | Alternative 1, $ | Alternative 2, $ | Alternative 1, $ | Alternative 2, $ |
| 1 | — | — | — | — | — | — | — | — |
| 2 | — | — | — | — | — | — | — | — |
| 3 | 0.0705 | 112 | 285,600 | 571,200 | 1,389,555 | 2,779,110 | 1,675,155 | 3,350,310 |
| 4 | 0.0772 | 119 | 303,450 | 606,900 | 1,521,612 | 3,043,224 | 1,825,062 | 3,650,124 |
| 5 | 0.0841 | 127 | 323,850 | 647,700 | 1,657,611 | 3,315,222 | 1,981,461 | 3,962,922 |
| 6 | 0.0942 | 135 | 344,250 | 688,500 | 1,856,682 | 3,713,364 | 2,200,932 | 4,401,864 |
| 7 | 0.1007 | 147 | 374,850 | 749,700 | 1,984,797 | 3,969,594 | 2,359,647 | 4,719,294 |
| 8 | 0.1103 | 157 | 400,350 | 800,700 | 2,174,013 | 4,348,026 | 2,574,363 | 5,148,726 |
| 9 | 0.1208 | 172 | 438,600 | 877,200 | 2,380,968 | 4,761,936 | 2,819,568 | 5,639,136 |
| 10 | 0.1323 | 189 | 481,950 | 963,900 | 2,607,633 | 5,215,266 | 3,089,583 | 6,179,166 |

**Table 11-7. Thermal Revenues for Heating**

| Year | Fuel cost, \$/$10^6$ Btu | Adjusted fuel cost for 65% efficient boiler, \$/$10^6$ Btu | Annual heating demand | | Heating revenue | |
|---|---|---|---|---|---|---|
| | | | Alternative 1, $\times 10^6$ Btu | Alternative 2, $\times 10^6$ Btu | Alternative 1, \$/yr | Alternative 2, \$/yr |
| 1 | — | — | — | — | — | — |
| 2 | — | — | — | — | — | — |
| 3 | 6.441 | 9.909 | 40,000 | 80,000 | 396,360 | 792,720 |
| 4 | 7.314 | 11.252 | 40,000 | 80,000 | 450,080 | 900,160 |
| 5 | 8.305 | 12.777 | 40,000 | 80,000 | 511,080 | 1,022,160 |
| 6 | 9.430 | 14.508 | 40,000 | 80,000 | 580,320 | 1,160,640 |
| 7 | 10.708 | 16.474 | 40,000 | 80,000 | 658,960 | 1,317,920 |
| 8 | 12.159 | 18.706 | 40,000 | 80,000 | 748,240 | 1,496,480 |
| 9 | 13.806 | 21.240 | 40,000 | 80,000 | 849,600 | 1,699,200 |
| 10 | 15.677 | 24.118 | 40,000 | 80,000 | 964,720 | 1,929,440 |

**Table 11-8 Thermal Revenues for Cooling**

| Year | Peak period electricity cost, \$/kWh | Adjusted electricity cost, for chiller COP* = 4.3 is \$/kWh ÷ COP \$/kWh | Annual cooling demand | | Cooling revenue | |
|---|---|---|---|---|---|---|
| | | | Alternative 1, $\times 10^6$ kWh | Alternative 2, $\times 10^6$ kWh | Alternative 1, \$/yr | Alternative 2, \$/yr |
| 1 | — | — | — | — | — | — |
| 2 | — | — | — | — | — | — |
| 3 | 0.087 | 0.020 | 18.35 | 36.7 | 367,000 | 734,000 |
| 4 | 0.098 | 0.023 | 18.35 | 36.7 | 422,050 | 844,100 |
| 5 | 0.104 | 0.024 | 18.35 | 36.7 | 440,400 | 880,800 |
| 6 | 0.117 | 0.027 | 18.35 | 36.7 | 495,450 | 990,900 |
| 7 | 0.123 | 0.029 | 18.35 | 36.7 | 532,150 | 1,064,300 |
| 8 | 0.134 | 0.031 | 18.35 | 36.7 | 568,850 | 1,137,700 |
| 9 | 0.146 | 0.034 | 18.35 | 36.7 | 623,900 | 1,247,800 |
| 10 | 0.159 | 0.037 | 18.35 | 36.7 | 678,950 | 1,357,900 |

*Coefficient of performance.

**Table 11-9 Combined Revenues**

| | 3000 kilowatts | | | 6000 kilowatts | | |
|---|---|---|---|---|---|---|
| Year | Heating revenue, \$/yr | Cooling revenue, \$/yr | Combined thermal revenue, \$/yr | Heating revenue, \$/yr | Cooling revenue, \$/yr | Combined thermal revenue, \$/yr |
| 1 | — | — | — | — | — | — |
| 2 | — | — | — | — | — | — |
| 3 | 396,360 | 367,000 | 763,360 | 792,720 | 734,000 | 1,526,720 |
| 4 | 450,080 | 422,050 | 872,130 | 900,160 | 844,100 | 1,744,260 |
| 5 | 511,080 | 440,400 | 951,480 | 1,022,160 | 880,800 | 1,902,960 |
| 6 | 580,320 | 495,450 | 1,075,770 | 1,160,640 | 990,900 | 2,151,540 |
| 7 | 658,960 | 532,150 | 1,191,110 | 1,317,920 | 1,064,300 | 2,382,220 |
| 8 | 748,240 | 568,850 | 1,317,090 | 1,496,480 | 1,137,700 | 2,634,180 |
| 9 | 849,600 | 623,900 | 1,473,500 | 1,699,200 | 1,247,800 | 2,947,000 |
| 10 | 964,720 | 678,950 | 1,643,670 | 1,929,440 | 1,357,900 | 3,287,340 |

Depreciation for this example is based on the accelerated cost recovery system (ACRS) of the Economic Recovery Tax Act of 1981. The recovery period for depreciable cogeneration property is 5 years. Depreciation for year 1 is 15 percent, for year 2 it is 22 percent, and for years 3 through 5 it is 21 percent per year for federal tax purposes.

Depreciation is based on the total plant cost of \$8,586,000 for alternative 1 and \$10,870,000 for alternative 2 (Table 11-10).

***State Depreciation Allowances*** The state is assumed to have not adopted the federal ACRS depreciation provision. This issue is dependent on state policies. For state tax purposes the straight-line method and a 60-month term are assumed. The straight-line method of depreciation yields an annual amount obtained by dividing the total plant cost by the 60-month amortization period.

**Table 11-10 Depreciation Rates**

| Year | Depreciation rate, % | Alternative 1, \$ | Alternative 2, \$ |
|---|---|---|---|
| 1 | 15 | 1,287,900 | 1,630,500 |
| 2 | 22 | 1,888,920 | 2,391,400 |
| 3 | 21 | 1,803,060 | 2,282,700 |
| 4 | 21 | 1,803,060 | 2,282,700 |
| 5 | 21 | 1,803,060 | 2,282,700 |
| Total | 100 | 8,586,000 | 10,870,000 |

The straight-line depreciation annual amount for alternative 1 is

$$\$8,586,000 \div 5 \text{ yr} = \$1,717,200$$

The straight-line depreciation annual amount for alternative 2 is

$$\$10,870,000 \div 5 \text{ yr} = \$2,174,000$$

***Energy Investment Tax Credit*** The project will use the same fuel source to produce both qualifying thermal energy and electric power. It is estimated that 90 percent of the project's cost will qualify for the 10 percent energy investment tax credit for cogeneration equipment. The remaining 10 percent of the project cost is for buildings which do not qualify for the regular investment credit.

For alternative 1 the investment tax credit is

$$(90\%)(\$8,586,000)(10\%) = \$772,740$$

For alternative 2 the investment tax credit is

$$(90\%)(\$10,870,000)(10\%) = \$978,300$$

# APPENDIX A

## GLOSSARY

**Aggregate value** The value of electricity production to the entire utility system as opposed to the closest plant to the cogenerator (which may be more or less efficient than the system average); for example, avoided transmission and distribution costs may be aggregated over the utility's entire system and paid to the qualifying facility (QF) on a prorated basis without regard to its distance from the load center.

**As-available energy or capacity** Electricity provided by a QF to a utility as it becomes available, rather than at prearranged times and in prearranged quantities.

**Auxiliary power source (APS)** Electricity-generating facilities designed to be used in the event of an outage at the local utility grid.

**Average-cost pricing** The pricing of electric service designed to recover the total costs of a system in order to make total revenues (including rate of return) equal to total costs. Total costs are based on costs recorded in books of account and forecasted to be recorded in such accounts.

**Avoided costs** The incremental costs to an electric utility of electric energy or capacity or both which, but for purchase from the QF or QFs, such utility would generate itself or purchase from its source.

**Back-up power** Electric energy or capacity supplied by an electric utility to replace energy ordinarily generated by a facility's own generation equipment during an unscheduled outage at the facility.

**Base load** The minimum continuous load on a power system over a given period of time.

**Biomass** Any organic material not derived from fossil fuels.

**Biomass conversion** The process by which any organic material not derived from fossil fuels (such as wood waste, rice hulls, and walnut shells) is changed into electricity or energy.

**Bottoming-cycle cogeneration facility** A cogeneration facility in which first the energy input to the system is applied to a useful thermal energy process, and then the reject heating emerging from the process is used for power production.

**Capability** The maximum load a generator, turbine, transmission circuit, apparatus, station, or system can supply under specified conditions for a given time interval without exceeding approved temperature and stress limits.

**Capacity** The load for which a generator, turbine, transformer, transmission circuit, apparatus, station, or system is rated. Capacity is also used synonymously with capability.

**Capacity costs** Costs associated with capital investments in electricity production and delivery.

**Capacity factor** The ratio of average load on a generating resource to its capacity rating during a specified period of time expressed as a percentage.

**Capacity payments** Payments which reflect the value capacity received.

**Cogeneration** The sequential production of electricity and heat, steam, or useful work from the same fuel source.

**Cogeneration facility** Equipment used to produce electric energy and forms of useful thermal energy (such as heat or steam), used for industrial or commercial heating or cooling purposes, through the sequential use of energy.

**Combined cycle** The use of waste heat from a gas turbine topping cycle for the generation of electricity in a steam turbine generator system, thereby increasing the efficiency of heat utilization.

**Curtailment** A period during which a utility declines to purchase available electricity from a QF, generally because of low demand or system maintenance requirements.

**Debt service** Periodic payments due on loans made to a QF.

**Dispatchability** A condition of the QF whereby, through engineering design, installed equipment, operating conditions, and procedures, the electric utility has the ability to dispatch the facility for operation at any time in a manner agreed upon by the parties concerned.

**Energy cost adjustment clause (ECAC)** Periodic adjustments of utility rates to reflect fuel and related costs.

**Energy costs** Costs associated with fuel use in electricity production.

**Escalated payments** A payment commitment for future years determined from forecast rates of avoided costs.

**Firm capacity payments** Payments for electricity provided in predetermined quantities and at predetermined times, which may be based on avoided costs at the time of delivery or at the time the obligation is incurred.

**Firm power** Power available at all times during the period covered by the commitment, except for forced outages and scheduled maintenance. Firm power is provided with sufficient legally enforceable guarantees of deliverability to permit the purchasing electric utility to avoid the need to construct a generating unit, to build a smaller and less expensive plant, or to purchase less firm power from another facility.

**Forecast** Figure values determined by a mathematical model.

**Heat rate** A measure of generating station thermal efficiency generally expressed in Btu per net kilowatthour. The average heat rate is computed by dividing the total Btu content of the fuel burned by the resulting net kilowatthours generated. The marginal heat rate is calculated as the additional (saved) Btu's needed to produce or (not produce) the next kilowatthour.

**Increase in supply** An economic model of electrical production by QFs whereby such production is viewed as increasing the aggregate supply of energy, implying that QFs should be subject to operating and performance standards comparable to those of a utility's own generating plants for purposes of determining pricing.

**Interconnection** The physical system of electrical transmission between a QF and a utility.

**Interconnection costs** Reasonable costs of connection, switching, metering, transmission, distribution safety provisions, and administration incurred by the electric utility directly related to the installation and maintenance of the physical facilities necessary to permit interconnected operations with a QF, to the extent such costs are in excess of the corresponding costs the electric utility would have incurred if it had not engaged in interconnected operations but instead generated an equivalent amount of electric energy itself or purchased an equivalent amount of electric energy or capacity from other sources. Interconnection costs do not include costs included in the calculation of avoided costs.

**Interruptible power** Electric energy or capacity supplied by an electric utility subject to interruption by the electric utility under specified conditions.

**Kilowatt (kW)** An electrical unit of power equal to 1000 watts.

**Kilowatthour (kWh)** A basic unit of electric energy equal to the use of 1 kilowatt for a period of 1 hour.

**Levelization** A financial arrangement whereby payments are constant over a specified period and are based on forecasted values and the value of money over time.

**Line losses** Losses in electricity which occur during its transmission and distribution.

**Load** The amount of electric power delivered to a given point on a system, or the total amount of demand on the system.

**Load factor** The ratio of average to peak use during a specified period of time, expressed as a percentage.

**Loan or bond guarantees** A utility which guarantees the repayment of a bond or loan on behalf of a QF in the event that it is unable to make timely payments.

**Maintenance power** Electric energy or capacity supplied by an electric utility during scheduled outages of the QF.

**Marginal cost pricing** The pricing of electrical service designed to equate the rates for electrical service with the marginal costs of that service.

**Marginal cost** The change in total cost caused by a change in output. Marginal cost can also be understood as the additional cost to produce an additional unit of output, or the savings from producing one unit less of output (i.e., avoided cost).

**Monopoly** A market structure in which there are many buyers but only one seller.

**Monopsony** A market structure in which there are many sellers and only one buyer.

**Natural gas** Unmixed natural gas or any mixture of natural gas and artificial gas.

**Nonfirm power** Electric power available as surplus only, which is supplied by the power producer at the producer's option and can be interrupted by the power producer at will.

**Nonstandard contracts** Negotiated contracts between a utility and a QF which do not conform to standard offer guidelines previously approved by the PUC.

**Oil** Crude oil, residual fuel oil, liquid natural gas, or any refined petroleum product.

**Peak load** The maximum electric load consumed or produced in a stated period of time. It may also be characterized as the minimum instantaneous load within a designated interval of a stated period of time.

**Purchase** The buying of electric energy or capacity or both from a QF by an electric utility.

**Qualifying facility** A cogeneration facility or a small power production facility which is a QF under 18 CFR, Chapter I, Part 292, Subpart B of the FERC regulations.

**Rate** Any price, rate, charge, or classification made, demanded, observed, or received with respect to the sale or purchase of electrical energy or capacity, or any rule, regulation, or practice respecting any such rate, charge, or clarification, and any contract pertaining to the sale or purchase of electrical energy or capacity.*

**Rate base** The utility investment on which a utility is allowed to earn a rate of return.

**Reduction in demand** An economic model of electrical production by QFs whereby such production is viewed as reducing aggregate demand for energy, implying that QFs need not be subject to utility operating and performance standards for purposes of pricing determinations.

**Refusal to purchase** See Curtailment.

**Refused-derived fuels** Fuels derived from municipal waste used as fuel for electrical energy production or low-Btu gases from sewage treatment plants for use in turbines.

**Reliability** The conformance of a QF to a specified set of standards of the electric utility and the system.

**Reserve margins** Extra capacity available to (1) meet anticipated demands for power or (2) serve load in the event of a loss of generation resulting from an unscheduled outage. The reserve margin is the ratio of excess capacity to anticipated peak load expressed as a percentage.

*Rates are defined in the California Public Utility Code to include rates, fares, and charges (§210). §451 of the code identifies rates as all charges demanded or received by any public utility for a product or commodity furnished. We recognize the FERC definition of rate to include any price for purchase and use the term "rate" herein to refer also to prices. However, the term "price" is also used to allow for separate identification, in some cases, of payments for purchases from the more traditional use of the term rates as referring to charges demanded or received under filed tariffs.

**Sale** The sale of electrical energy or capacity or both by an electric utility to a qualifying facility.

**Simultaneous purchase and sale** A regulatory convention that allows a QF to simultaneously sell its own generation to the utility while purchasing its requirements from the utility; an exchange of electrical flow does not necessarily occur—the difference is cash flow.

**Small power production** Any unregulated electricity production facility as defined primarily in the FERC rules, including hydroelectric, geothermal, biomass refuse-derived fuel, and wind facilities.

**Social costs** Tangible but hard to quantify costs to society of an economic or technological activity.

**Spinning reserves** Reserves operated at less than the rated capacity to relieve imbalance on the system.

**Standard offer** A utility offer to purchase electricity from a QF that is formed within guidelines previously adopted by the California PUC.

**Supplementary firing** An energy input to the cogeneration facility used only in the thermal process of a topping-cycle cogeneration facility or only in the electricity-generating process of a bottoming-cycle cogeneration facility.

**Supplementary power** Electric energy or capacity supplied by an electric utility, regularly used by a qualifying facility in addition to that which the facility generates itself.

**System emergency** A condition of a utility's system likely to result in imminent significant disruption of service to customers or imminently linked to endanger life or property.

**System power values (SPV)** The Pacific Gas and Electric Company model of the marginal costs of additional capacity and energy, based on Application 58545, OII 26, and PG&E price offers on a combined cycle plant as the marginal plant.

**Time-differentiated payments** Payments made according to time-of-day or time-of-year delivery periods.

**Topping-cycle cogeneration facility** A cogeneration facility in which the energy input to the facility is first used to produce useful power output, and the reject heat from power production is then used to provide useful thermal energy.

**Total energy input** The total energy in all forms supplied by external sources other than supplementary firing to the facility.

**Total energy output of a topping-cycle cogeneration facility** The sum of the useful power output and useful thermal energy output.

**Up-front payments** Initial large outlays provided to QFs by utilities to assist in financing large capital expenses incurred for construction.

**Useful power output of a cogeneration facility** The electrical or mechanical energy made available for use, exclusive of any such energy used in the power production process.

**Useful thermal energy output of a topping-cycle cogeneration facility** The thermal energy made available for use in any industrial or commercial process or used in any heating or cooling application.

**Values to perpetuity** The costs (values) of owning and operating a generating plant for an infinite number of years, assuming plant replacement at the end of its useful life. Values are then levelized to the initial year.

**Waste** By-product materials other than biomass.

**Wheeling** The use of transmission facilities of one utility system to transmit power to another utility system or between customer facilities within a single utility system or between utility systems.

# APPENDIX B

## REPRESENTATIVE EQUIPMENT AND LAYOUTS

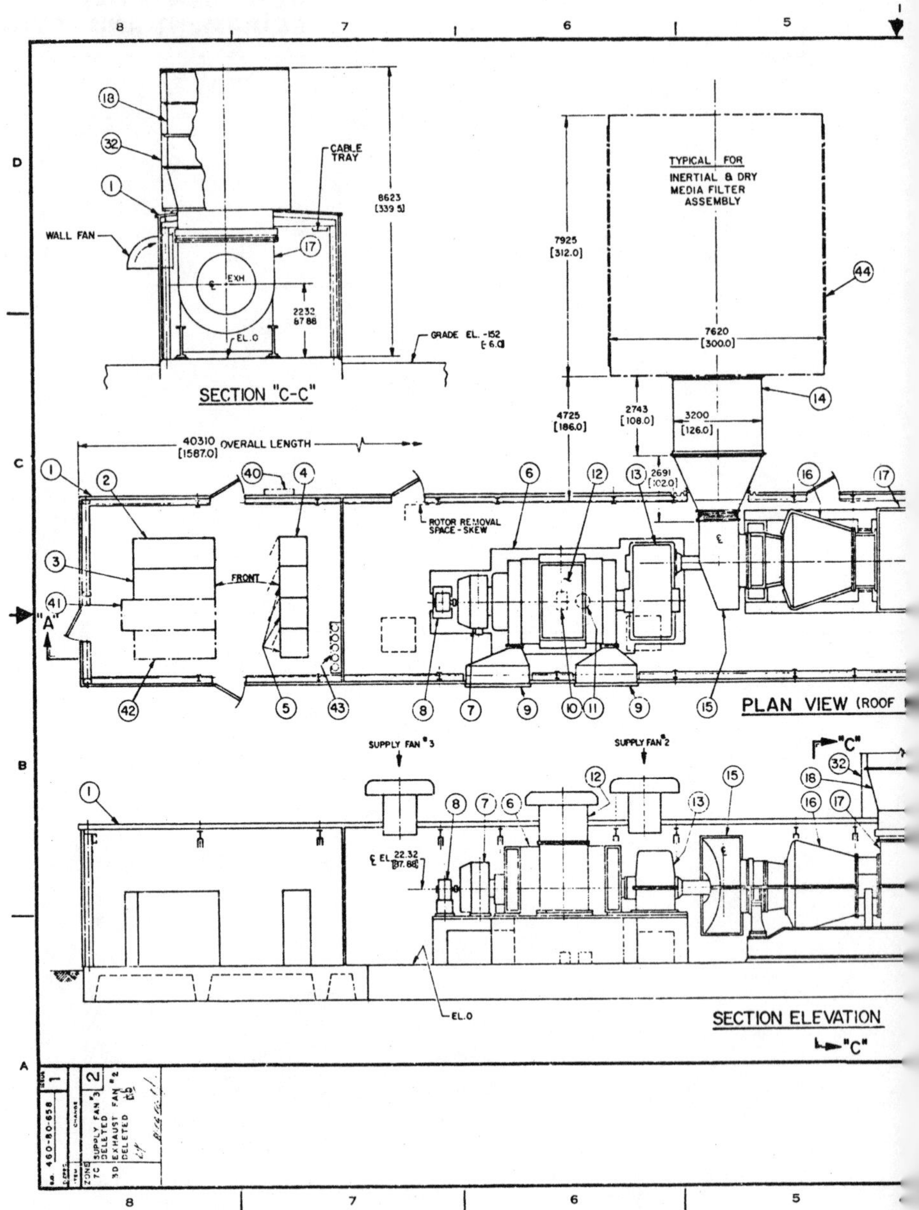


Figure B-1 Gas turbine generator general arrangement. (*Westinghouse Standard Proposal for the W191 Econopac Gas Turbine Power Plant*)

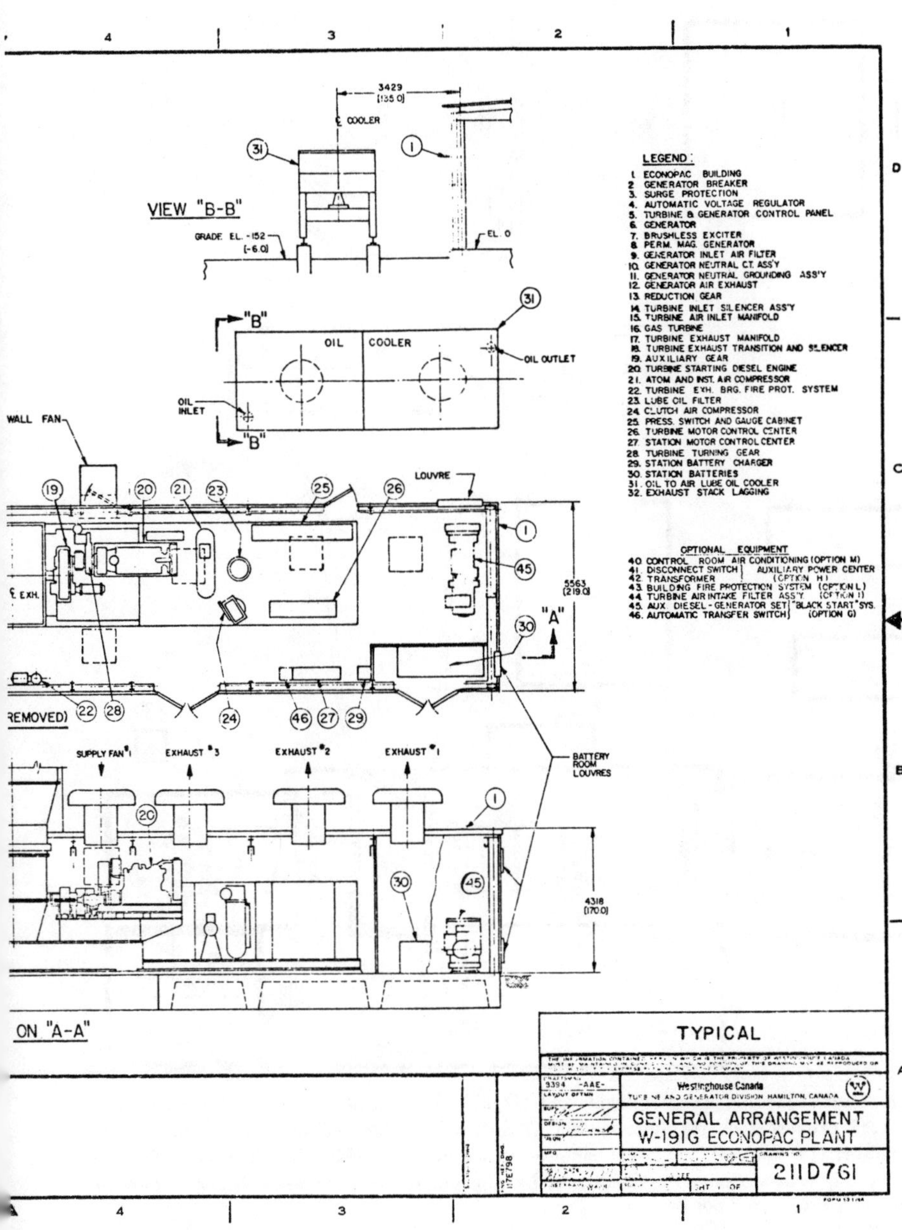

VIEW "B-B"
3429
(135.0)
℄ COOLER
GRADE EL. -152
(-6.0)
EL. 0
"B"
OIL
COOLER
OIL OUTLET
OIL INLET
"B"
WALL FAN
LOUVRE
℄ EXH.
5563
(219.0)
"A"
(REMOVED)
SUPPLY FAN #1
EXHAUST #3
EXHAUST #2
EXHAUST #1
BATTERY ROOM LOUVRES
4318
(170.0)
ON "A-A"
LEGEND:
1. ECONOPAC BUILDING
2. GENERATOR BREAKER
3. SURGE PROTECTION
4. AUTOMATIC VOLTAGE REGULATOR
5. TURBINE & GENERATOR CONTROL PANEL
6. GENERATOR
7. BRUSHLESS EXCITER
8. PERM. MAG. GENERATOR
9. GENERATOR INLET AIR FILTER
10. GENERATOR NEUTRAL C.T. ASS'Y
11. GENERATOR NEUTRAL GROUNDING ASS'Y
12. GENERATOR AIR EXHAUST
13. REDUCTION GEAR
14. TURBINE INLET SILENCER ASS'Y
15. TURBINE AIR INLET MANIFOLD
16. GAS TURBINE
17. TURBINE EXHAUST MANIFOLD
18. TURBINE EXHAUST TRANSITION AND SILENCER
19. AUXILIARY GEAR
20. TURBINE STARTING DIESEL ENGINE
21. ATOM AND INST. AIR COMPRESSOR
22. TURBINE EXH. BRG. FIRE PROT. SYSTEM
23. LUBE OIL FILTER
24. CLUTCH AIR COMPRESSOR
25. PRESS. SWITCH AND GAUGE CABINET
26. TURBINE MOTOR CONTROL CENTER
27. STATION MOTOR CONTROL CENTER
28. TURBINE TURNING GEAR
29. STATION BATTERY CHARGER
30. STATION BATTERIES
31. OIL TO AIR LUBE OIL COOLER
32. EXHAUST STACK LAGGING
OPTIONAL EQUIPMENT
40. CONTROL ROOM AIR CONDITIONING (OPTION M)
41. DISCONNECT SWITCH } AUXILIARY POWER CENTER
42. TRANSFORMER } (OPTION H)
43. BUILDING FIRE PROTECTION SYSTEM (OPTION L)
44. TURBINE AIR INTAKE FILTER ASS'Y (OPTION I)
45. AUX. DIESEL-GENERATOR SET } "BLACK START" SYS.
46. AUTOMATIC TRANSFER SWITCH } (OPTION G)
TYPICAL
Westinghouse Canada
TURBINE AND GENERATOR DIVISION HAMILTON, CANADA
GENERAL ARRANGEMENT
W-191G ECONOPAC PLANT
211D761
117E798

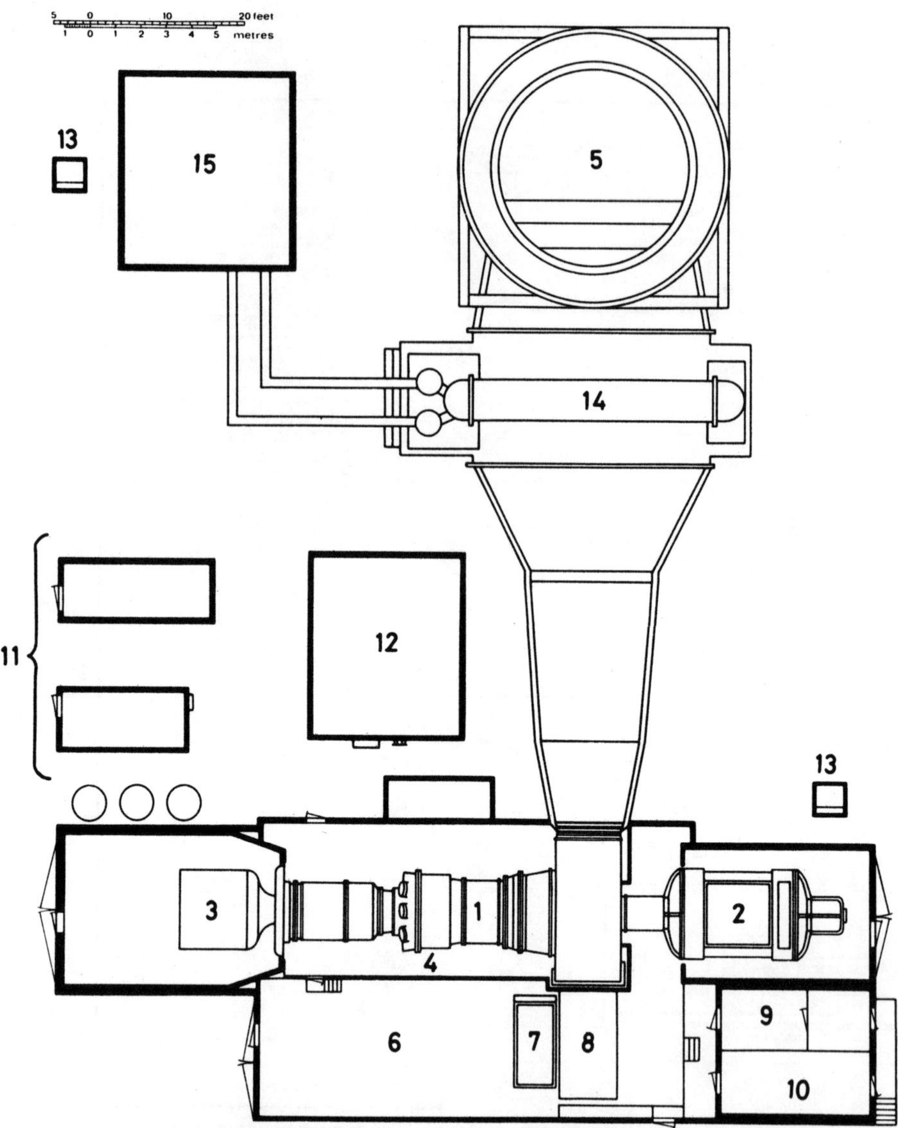


**Figure B-2** **District heating power station layout: (1) gas turbine, (2) alternator, (3) starting unit, (4) pipe gallery, (5) exhaust stack, (6) unloading bay, (7) governing oil and fuel unit, (8) lube oil unit, (9) auxiliary power, (10) control room, (11) compressed air modules, (12) lube oil cooler, (13) unit transformer, (14) heat exchanger, (15) district heating pumps, etc. (*Stal-Laval, Inc., "Turbine Power," Bulletin 572E9.79.4.000*)**

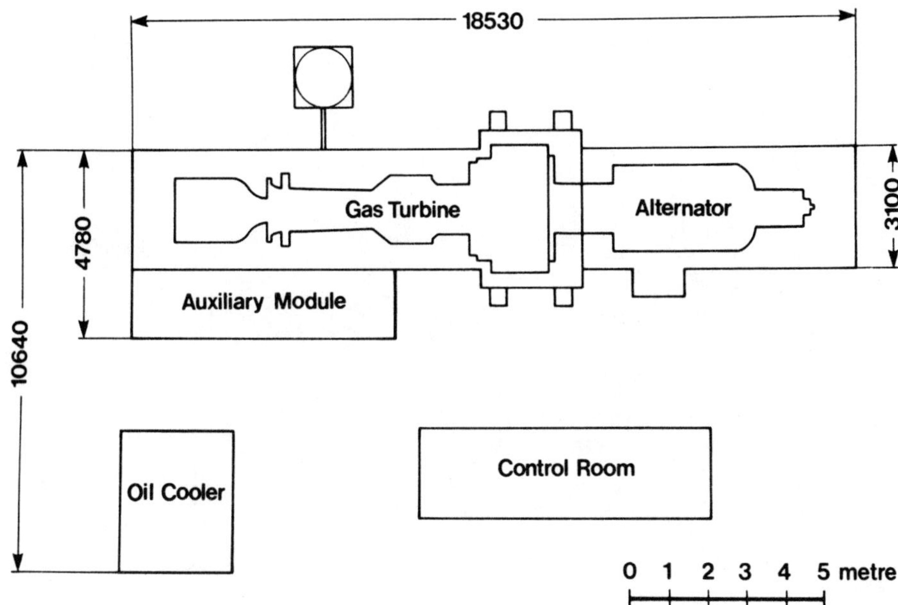


**Figure B-3** Combustion turbine generator basic arrangement. (*Stal-Laval, Inc., "The GT 35 Gas Turbine 10-15MW," Bulletin 588E02.801.000*)

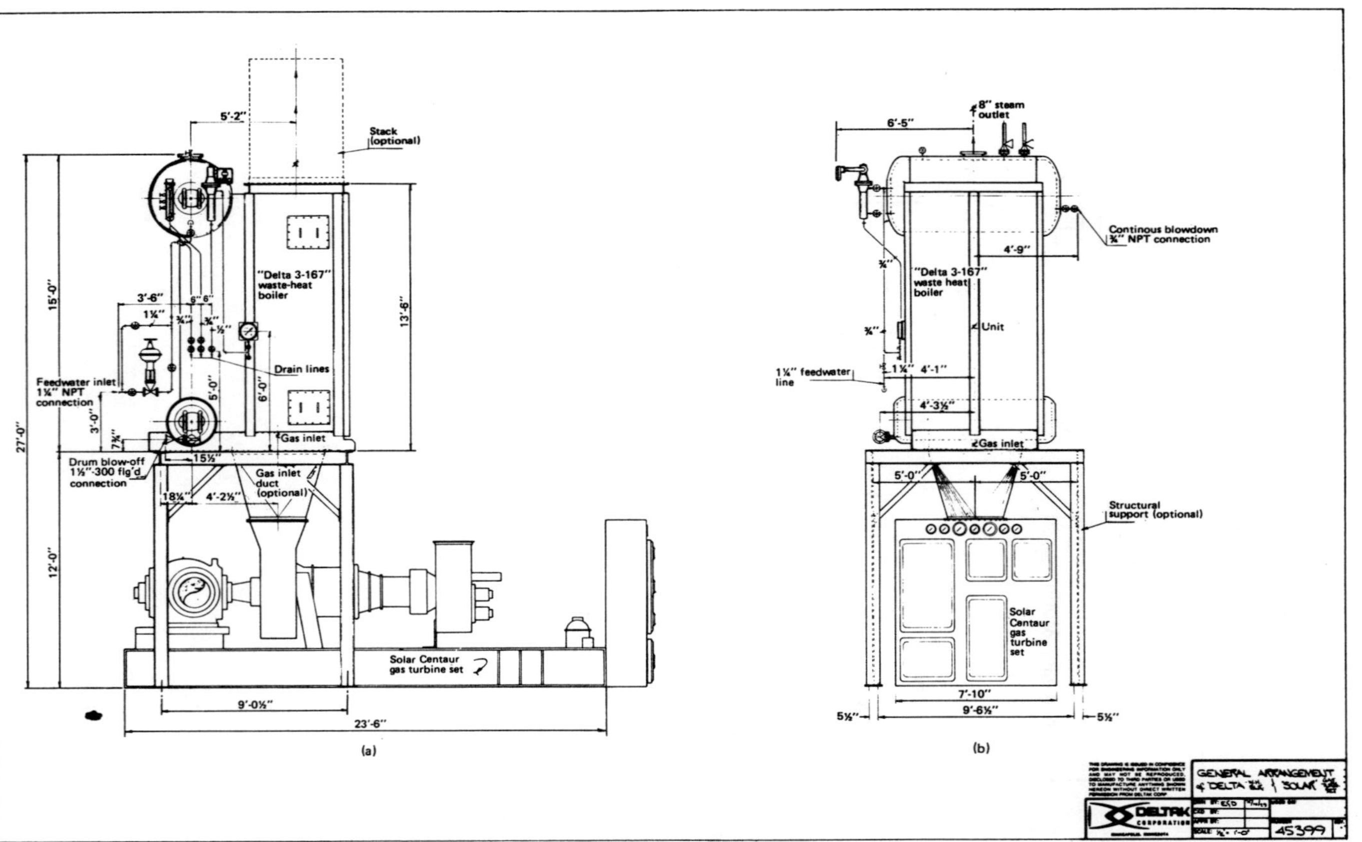


**Figure B-4** Waste heat recovery steam generator and combustion turbine generator general arrangement. (*a*) side elevation, (*b*) end elevation. (*Deltak Corporation.*)

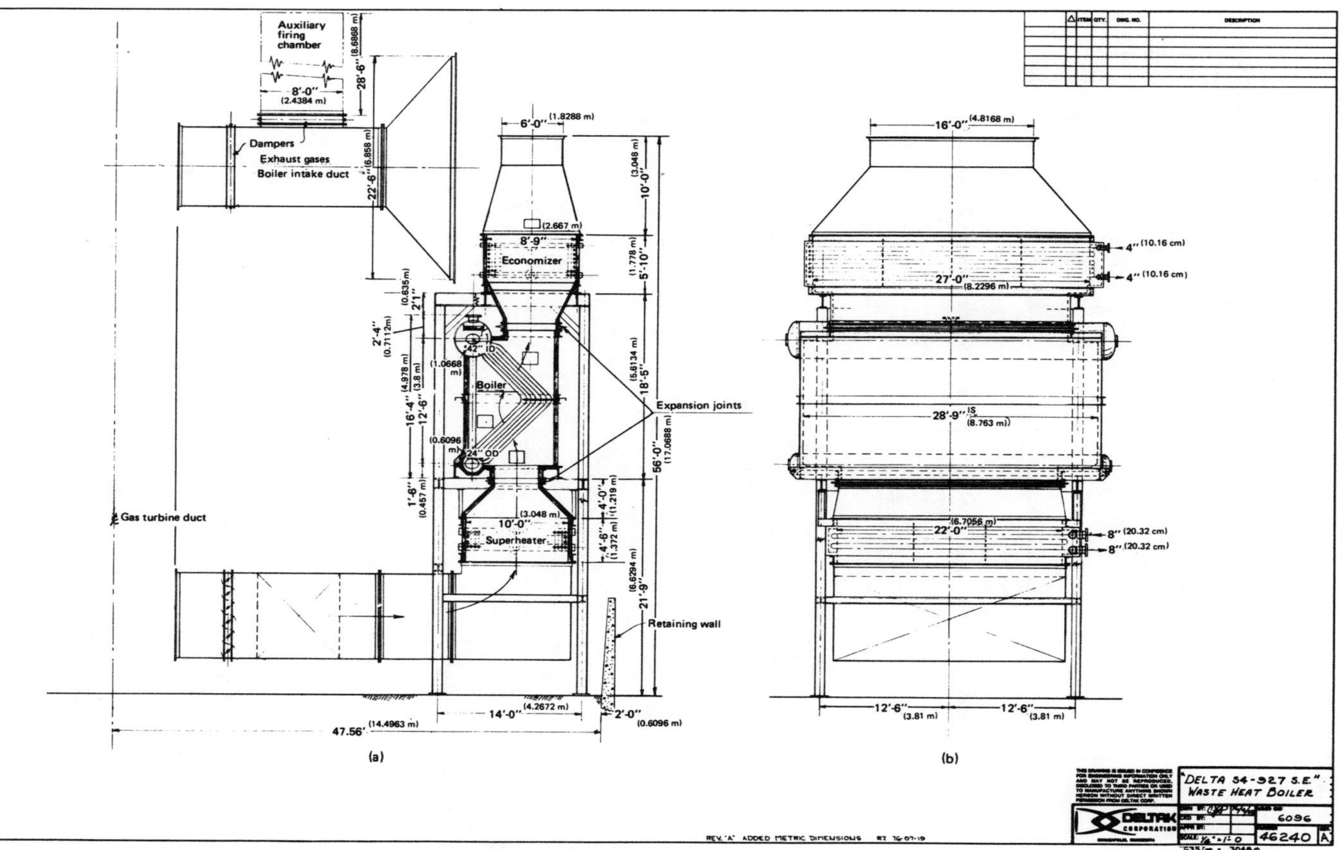


**Figure B-5** Waste heat recovery steam generator layout: (*a*) side elevation, (*b*) end elevation. (*Deltak Corporation*)

**Dimensional Data:**

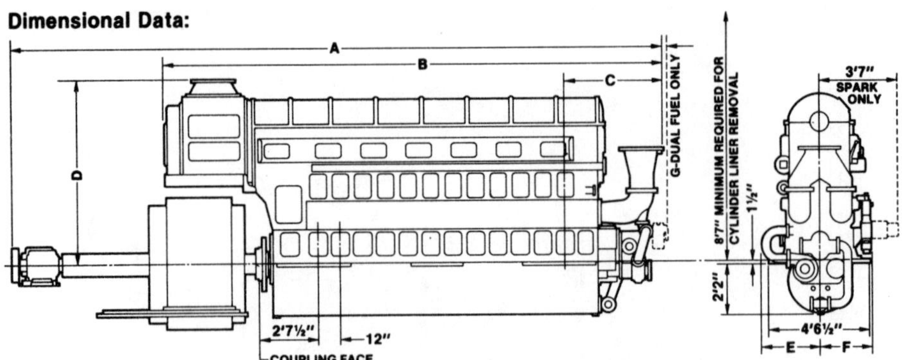


**Models 38DD8-⅛ Dual Fuel and 38DS8-⅛ Spark Ignition Blower Scavenged Engines**

| CYL. | A | B | C | D | E | F | G |
|---|---|---|---|---|---|---|---|
| # 4 | 17'11" | 12'6⅞" | 3'10½" | 7'5" | 2'5" | 2'4⅞" | 4⅜" |
| # 5 | 19'1" | 13'6⅞" | 3'10½" | 7'5" | 2'5" | 2'4⅞" | 4⅜" |
| 6 | 20'1" | 14'2⅛" | 3'10½" | 7'5" | 2'5" | 2'4⅞" | 4⅜" |
| 8 | 22'4¼" | 16'10⅝" | 4'0¾" | 7'5" | 2'5" | 2'4⅞" | 2⅛" |
| # 9 | 23'4¼" | 18'1⅜" | 4'0¾" | 7'5" | 2'5" | 2'4⅞" | 2⅛" |
| 10 | 24'10" | *19'1⅜" | 4'0¾" | †7'5" | 2'5" | 2'4⅞" | 2⅛" |
| 10 | 27'8⅞" | 22'2⅜" | 4'4¾" | 8'0" | 2'5" | 2'4⅞" | 2⅛" |

* Spark ignition dimension "B" is 19'10⅜". † Spark ignition dimension "D" is 8'0".
# These dimensions are for Dual Fuel Engines only.

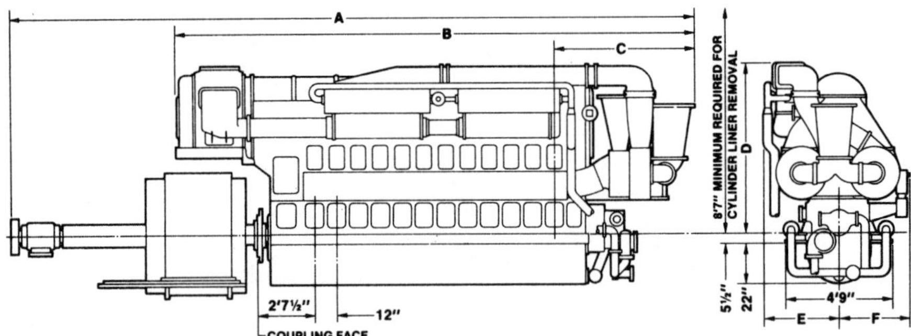


**Model 38TDD8-⅛ Turbocharged Dual Fuel Engine**

| CYL. | A | B | C | D | E | F |
|---|---|---|---|---|---|---|
| 6 | 22'6" | 15'2⅛" | 4'10½" | 7'6" | 3'7⅜" | 3'2¾" |
| 9 | 26'0" | 19'0⅝" | 5'8¾" | 7'7⅝" | 3'5⅛" | 3'7⅞" |
| 12 | 30'3" | 23'8½" | 6'6⅝" | 7'9⅜" | 3'5⅛" | 3'6⅜" |

**Model 38TDS8-⅛ Turbocharged Spark Ignition Engines**

| CYL. | A | B | C | D | E | F |
|---|---|---|---|---|---|---|
| 6 | 23'6" | 16'2" | 5'7¾" | 7'10¼" | 2'5½" | 2'1½" |
| 8 | 26'4" | 19'4¼" | 6'2¾" | 7'10¼" | 2'5½" | 2'1½" |
| 10 | 28'4" | 21'4¼" | 6'2¾" | 8'2" | 2'6" | 2'1½" |
| 12 | 31'8" | 25'1¾" | 7'11¾" | 7'9¼" | 2'11½" | 2'11½" |

All drawings are for illustrative purposes only. For installations obtain certified prints.
All ratings subject to factory approved application.

(a)

**Figure B-6** Internal-combustion engine dimensional data. (*Colt Industries, Fairbanks Morse Engine Division*)

COLT-PIELSTICK PC2 ENGINES

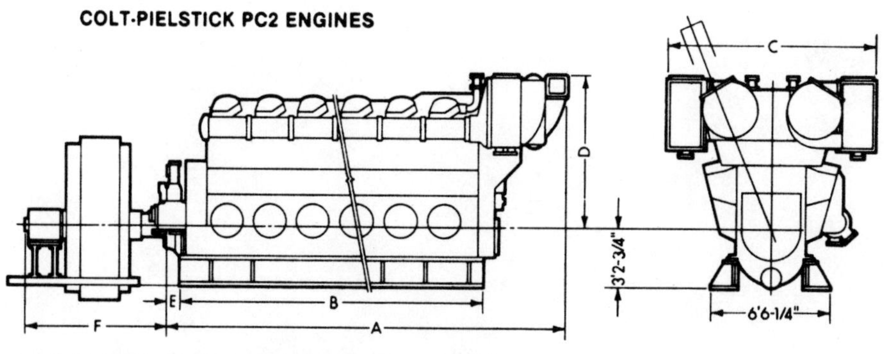


| Cyl. | A | B | C | D | E | F |
|---|---|---|---|---|---|---|
| 12 | 23'3" | 16'11" | 11'5¾" | 8'5" | 18½" | 8'6¾" |
| 14 | 25'8" | 19'4¼" | 11'5¾" | 8'5" | 18½" | 10'0¼" |
| 16 | 28'7" | 21'9½" | 12'1¾" | 9'8¼" | 18½" | 10'10½" |
| 18 | 31'0" | 24'2½" | 12'1¾" | 9'8¼" | 18½" | 10'10½" |

Drawings are for illustration only. For installations obtain certified prints. Ratings subject to factory approved application.

(b)

## DESIGN INFORMATION CHECKLIST

### PURPOSE

The questionnaire below can be used to gather essential data for a specific site. The data can then be used to integrate the cogeneration expansion into the existing operating facilities and systems. In filling out the questionnaire, it is important to include all information pertinent to the project power cycle, start-up procedures, operating procedures, peak loads, system support, and shutdown procedures for the existing facilities. Site and facility characteristics should be identified, as well as all existing engineering data. Engineering data may include performance and emissions tests, water quality analyses, engineering and plant arrangement drawings, operations manuals, test procedures, and foundation soil analyses. It is important that unique operating characteristics associated with any drying process—such as air and exhaust balances and humidity requirements—be identified in this questionnaire.

### GENERAL INFORMATION

Customer: ______________________________

Facility location

Street address: ______________________________

City, state, zip code: ______________________________

Facilty contacts

Name: ______________________________

Position: ______________________________

Telephone: ______________________________

Date of meeting: ______________________________

Meeting attendees

| Corporation | Name |
|---|---|
| ____________________ | ____________________ |
| ____________________ | ____________________ |
| ____________________ | ____________________ |
| ____________________ | ____________________ |
| ____________________ | ____________________ |
| ____________________ | ____________________ |

General description of the facilities and location:

Access Major highway: ____________________

Railroad spur: Yes ____ No ____

Size Facilities occupy ____________ acres

Temperatures, wet bulb or dry bulb

| | Maximum, °F | Minimum, °F | Design, °F |
|---|---|---|---|
| Outdoor | | | |
| Indoor | | | |

Approximate periods of *normal* facility operation

Daily: ____________________ hours

Weekly: ____________________ days

Annual: ____________________ weeks

General description of facility's existing boiler plant:

Equipment Manufacture: ____________________

Year installed: ____________________

Type: ____________________

Location: ____ Indoor

____ Outdoor

____ Enclosures (specify) ____________

Access: ____ Easy

____ Difficult

Building ____ Old ____ New

Type of structure: ____________________

Fuels consumed
- Gas ____
- No. 2 oil ____
- Residual ____
- Coal ____
- Lignite ____
- Coke ____
- Biomass ____

Description of any HVAC modifications to the buildings or enclosure that would involve heat production or consumption:

## STEAM AND EXHAUST HEAT PROCESS USES

### Present Plant Operating Characteristics

Describe the steam and exhaust heat demands during normal operations and during start-up and shutdown of the process. Identify standby equipment. Identify hours of operation and transient conditions.

Description of steam and exhaust heat uses:

**MAJOR EQUIPMENT REQUIRING STEAM**

| | Steam demand | | |
|---|---|---|---|
| | **Flow, lb/hr** | **Temp., °F** | **Press, psig** |
| Maximum | | | |
| Average | | | |
| Minimum | | | |

## MAJOR EQUIPMENT REQUIRING EXHAUST HEAT

| | Exhaust demand | |
|---|---|---|
| | Heat, Btu/hr | Temp., °F |
| Maximum | | |
| Average | | |
| Minimum | | |

Description of normal operations including steam demand, steam quality, and hours of operation:

Description of transients during normal operations:

Description of start-up operations, including steam demand, steam quality, and hours of operation:

Description of shutdown operations, including steam demand, steam quality, and hours of operation:

**SUMMARY OF NORMAL DEMAND CYCLE FOR 24 HOURS:**

| | Steam demand | | |
|---|---|---|---|
| Time, hr | lb/hr | °F | psig |
| 0–2400 | | | |

Description of any plant changes that will affect the present operating conditions:

Probable sources of process low-pressure extraction points where steam can be taken for deaerator use:

## DESIGN INFORMATION FOR EXISTING EQUIPMENT

## Mechanical

---

**BOILER DATA**

Manufacturer: ____________________

____Balanced draft ____Pressurized

____Natural circulation ____Controlled circulation

____Subcritical

Design flows:

| | Pounds per hour | psig | Temperature, °F |
|---|---|---|---|
| Outlet | | | |

**TANKS**

Blowdown tanks: ____ Yes ____ No

No. of tanks: ____________________

Dimensions: ________ Height ________ Diameter ________ Conn. size

Design pressure: ________ psig

Flash tanks ____ Yes ____ No

No. of tanks: ________

Dimensions: ________Height ________ Diameter ________ Conn. size

Design pressure: ________ psig

**FUELS AVAILABLE**

| | No. 2 oil | | Gas |
|---|---|---|---|
| | Pressure, psig | Temperature, °F | Pressure, psig |
| | | | |

Start-up

Low load

Normal

Emergency

---

**FANS**

| | FD | ID |
|---|---|---|
| Number | | |
| 1–100% capacity | | |
| 2–50% capacity | | |
| Other | | |
| Type: centrifugal flow | | |

Describe probable interconnection point on the existing steam supply line to process.

**CLOSED FEEDWATER HEATERS**

____ Yes ____ No

Manufacturer: ____________________

____ Number of stages of low-pressure feedwater heating

____ Number of stages of high-pressure feedwater heating

**OPEN FEEDWATER HEATER**

____ Yes ____ No

Manufacturer: ____________________

Design conditions:

Steam ________ lb/hr ________ psig ________ °F

Feedwater ________ lb/hr ________ psig ________ °F

**BOILER FEED PUMPS**

Main boiler feed pumps

Manufacturer: ____________________

Number: ____________________

Type of driver: ____________________

________ hp ________ V ____ θ

Capacity: ____% total system flow

Rating: ____ gal/min at ____ TDH

**AIR COMPRESSORS**

Separate instrument air compressor ____ Yes ____ No

Instrument

No. of compressors: ____

Rating: ____

Pressure: ____

Dryers: ____

Dew point: ____

Service air

No. of compressors: ____

Rating: ____

Pressure: ____

**STANDBY EQUIPMENT (description)**

Manufacturer: ____________________

Type: ____________________

Steam generation: ____________________

Electric Power: ____________________

**WATER AND WASTE TREATING**

| | | Type | Quantity Used | Quantity Available | Pressure, psig |
|---|---|---|---|---|---|
| Water source: city main | Treated water | | | | |
| | General service | | | | |

Treatment:

____ Flash evaporator

____ Demineralizer

____ Other (specify) ____________________

Condensate demineralization system:

____ Deep bed type

____ Precoat type

____ None

____ Other (specify) ____________________

Outlet water quality: ____________________

Capacity: ____________________

Manufacturer: ____________________

Date installed: ____________________

Condition: ____________________

Sanitary waste disposal:

____ Independent sewage treatment plant

____ Municipal sewage treatment plant tie-in

Outlet water quality: ____________________

Capacity: ____________________

Manufacturer: ____________________

Date installed: ____________________

Condition: ____________________

Industrial wastewater disposal:

____ Impoundment

____ Other (specify) ____________________

Regularity restrictions (special); ____________________

____________________

Outlet water quality: ____________________

Capacity manufacturer: ____________________

Date installed: ____________________

Condition: ____________________

Fuel supply:

Natural gas pressure (psig)

Current ____

Optional future ____ (min.) ____ (max.)

Optional future ____ (min.) ____ (max.)

# Electrical

---

## UTILITY INTERFACE AND RESPONSIBILITIES

Review of client single line and protective relaying:

Guidelines on minimum requirements:

Design and installation of interconnection equipment; description including parameters for client single-line calculation:

Transformer impedance window:

System nominal voltage ±:

Transformer tops availability:

System impedance:

## DESIGN INFORMATION FOR EXISTING EQUIPMENT

Description of client preliminary investigation on electrical equipment modification

Existing breakers to be removed: ______

Location: ______

New outdoor switchgear: ______

Modification to existing indoor cubicle bus: ______

Existing switchgear to be retained

Provide manufacturer, MVA rating, interruption, and momentary current rating:

Original shop order number of manufacturer: ______

Description of existing feeders

Complete information on electrical equipment; motor horsepower, MCC rating, etc.:

Motors

Provide complete parameters for existing machines. Motor manufacturer, original shop order number, motor data sheet, etc., require $\times''d$, $\times''d$, $\times'd$, $\times$ applicable time contents for transient study:

**NEW DESIGN**

Description of special features required by client

Automatic synchronization capability:

Manual synchronization capability:

Degree of automation sequential synchronization:

---

## Instrumentation

---

Existing burner control system

| | | | |
|---|---|---|---|
| Type of system | ___ Solid state | | ___ Relay |
| Type of operation | ___ Local | ___ Remote | ___ Auto |
| Power supply | ___ Single | ___ Redundant | |

Manufacturer: ____________________

Date installed: ____________________

Code compliance: ____________________

Condition: ____________________

Preferred burner control system

Type of system: ____________________

Type of operation: ____________________

Power supply: ____________________

Existing boiler control system

Analog portion ____ Solid state

____ Pneumatic

____ Other (specify) ____________________

Digital portion ____ Solid state

____ Relay

Type of operation ____ Local ____ Remote

Degree of automation: ____________________

Date installed: ____________________

Condition: ____________________

Manufacturer: ____________________

Preferred boiler control system

Type of system: ____________________

Type of operation: ____________________

Location of central controls: ____________________

Other instrumentation

**HEATING AND VENTILATION**

List heating and ventilation conditions of existing buildings:

| Building | Heating | Air conditioning | Ventilation | None |
|---|---|---|---|---|

## Civil and Structural Factors

### AVAILABILITY OF REQUIRED FACILITIES

| | Existing at site | | Tie into existing | | New required | |
|---|---|---|---|---|---|---|
| | Yes | No | Yes | No | Yes | No |
| Storm sewer | | | | | | |
| Sanitary sewer | | | | | | |
| Sanitary treatment | | | | | | |
| Industrial waste treatment | | | | | | |
| Gas | | | | | | |
| Plant road | | | | | | |
| Railroad | | | | | | |
| Fire protection loop system | | | | | | |

## EXISTING OPERATING, MAINTENANCE, AND ENERGY CONSERVATION PROCEDURES AND PREFERENCES

### DESCRIPTION OF STANDBY PHILOSOPHY

Redundancy, excess capacity, back-up systems:

### DESCRIPTION OF OPERATING PHILOSOPHY

Local control, remote, equipment sequencing:

### CURRENT PLANT MAINTENANCE SCHEDULE

Number of plant operators: ______________________________

Maintenance personnel (normal operations): ______________________

Current estimated maintenance work hours per year: ________________

### ENERGY CONSERVATION

Energy conservation systems, processes, and equipment design conditions.

---

## SUMMARY OF ENGINEERING DRAWINGS AND DOCUMENTS REQUIRED

---

### DRAWINGS REQUIRED

____ Mechanical P&I (flow) diagrams

____ Plot plans

____ Equipment arrangements

____ Piping drawings

____ Piping isometrics and BM

____ Piping line lists

____ Material handling drawings

____ Architectural drawings

____ HVAC drawings

____ Plumbing drawings

____ Concrete drawings

____ Structural drawings

____ Civil drawings

____ Instrumentation plans and layouts

____ Instrumentation schematics

____ Instrumentation logic diagrams

____ Electrical diagrams, plans, and layouts

____ Electrical control elementary diagrams

____ Electrical interconnection wiring diagrams

____ Electrical cable, conduit and cable tray schedules

**EQUIPMENT LISTS**

____ Mechanical equipment

____ Electrical motors

____ Instrument index

____ Valve index

**OTHER ENGINEERING DOCUMENTS**

____ Air and exhaust flow diagrams

____ Performance tests

____ Emission tests

____ Boiler operating manual

____ Meterological data

____ Soils information

____ Other

**SURVEY (BOUNDARY AND CONTOUR)**

____ Aerial photograph

____ Tie-in points (gas, sanitation, sewer, storm sewer, contaminated waste, water, oil, electric, steam, exhaust heat, fire water)

# APPENDIX D

## REPRESENTATIVE COGENERATION PROJECT SCHEDULES

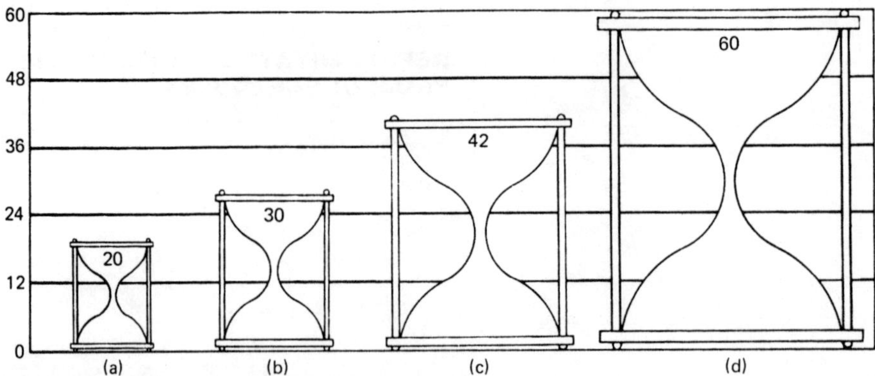


**Figure D-1** **Comparative lead times for various power cycle alternatives times are given in months from order to operation, exclusive of permit cycle times. Lower equipment costs and shorter lead times for combustion turbine systems combine to give lower installed costs. (*a*) Combustion turbine (simple cycle), (*b*) combined cycle, (*c*) oil- or gas-fired steam plant, (*d*) coal-fired steam plant. (*Westinghouse Combustion Turbines, "The Ready Source of Power Worldwide," Bulletin SA 11152*)**

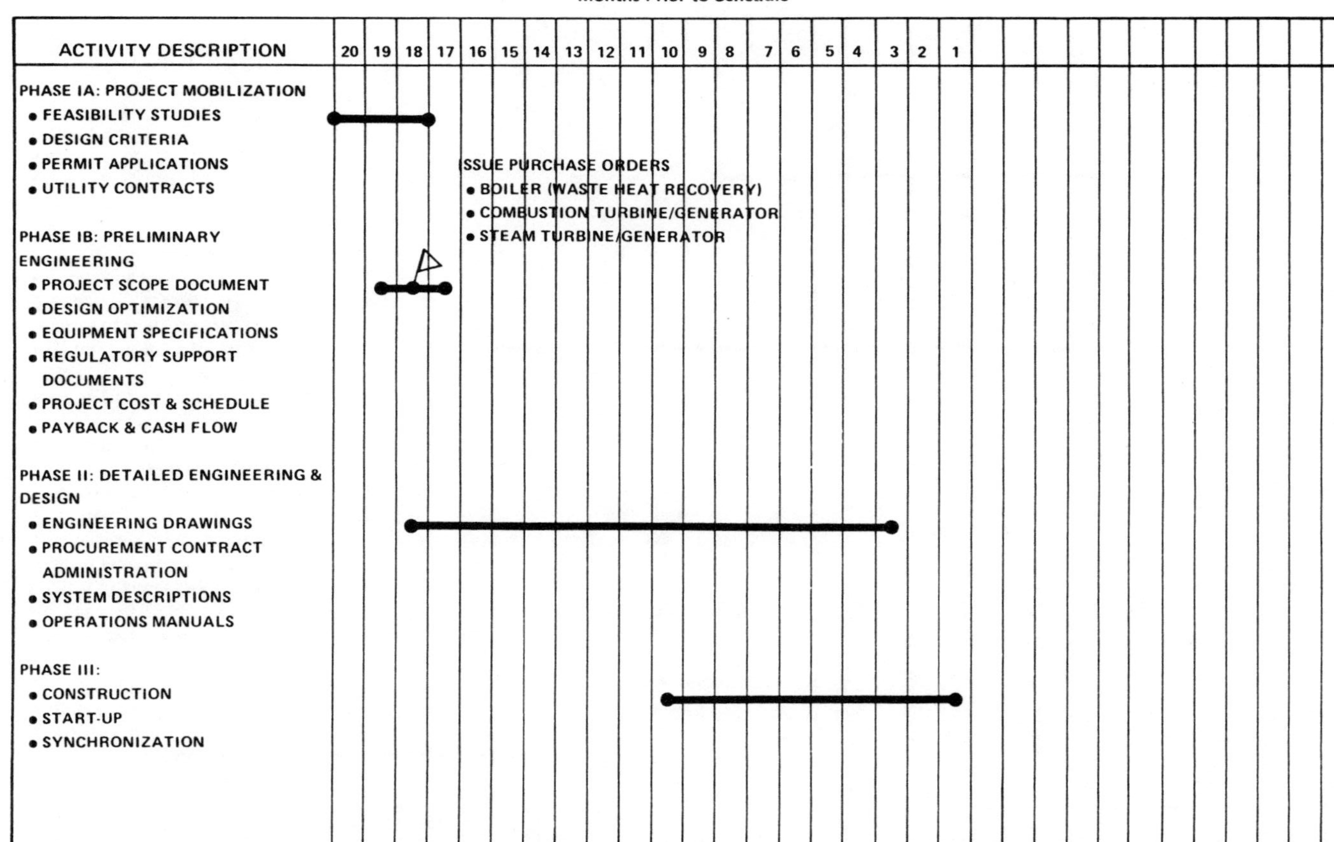


**Figure D-2** Representative cogeneration project schedule (fast track).

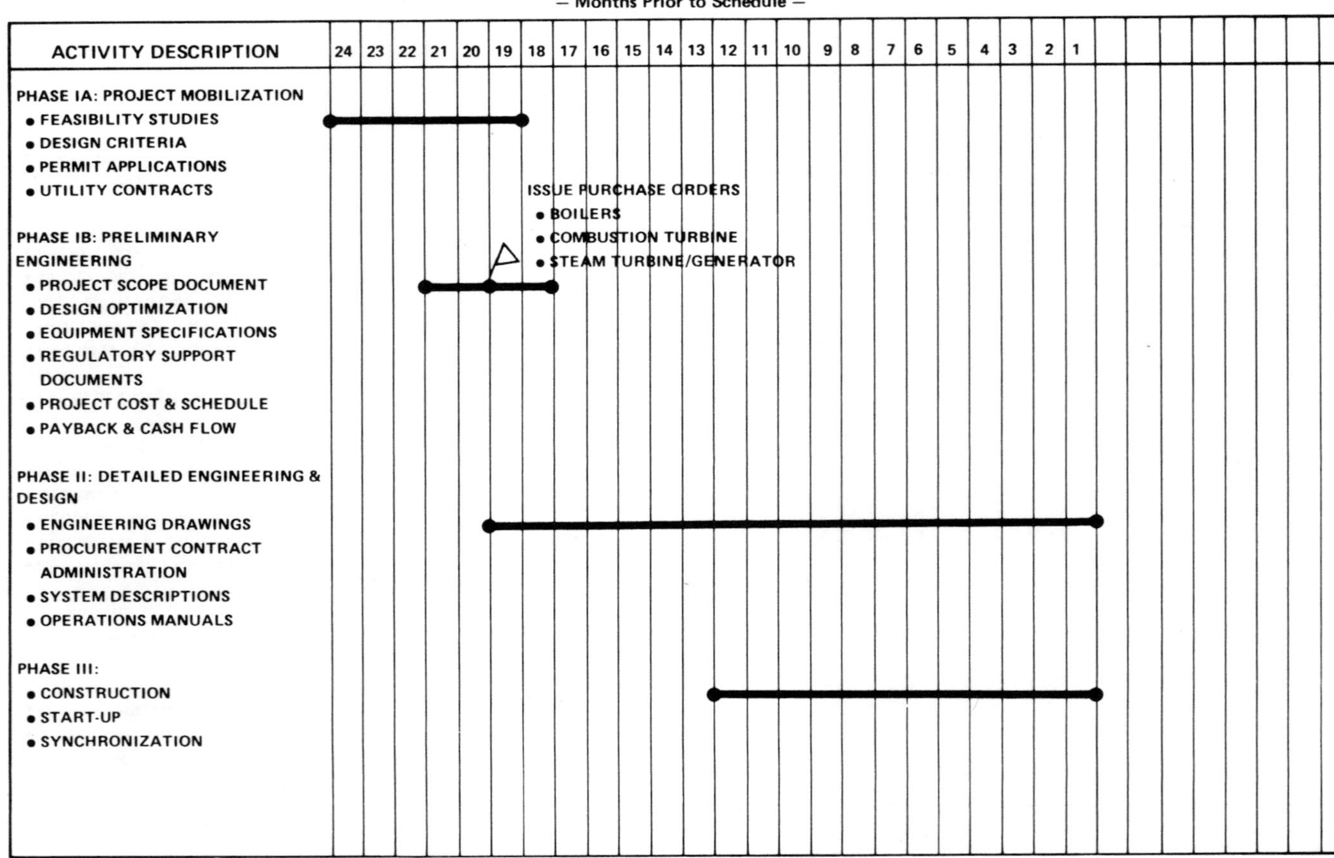


**Figure D-3** Representative cogeneration project schedule (nominal).

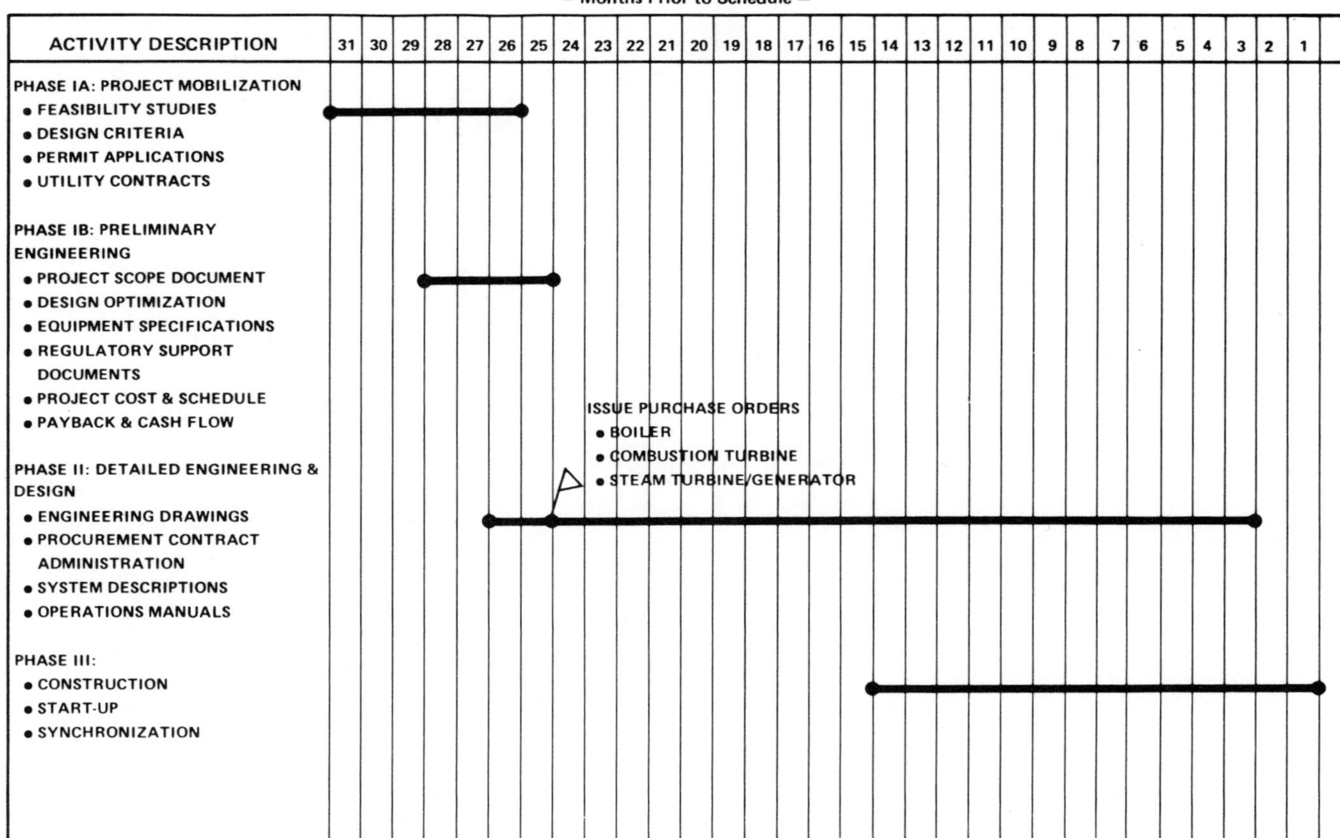


Figure D-4 Representative cogeneration project schedule (large).

# APPENDIX E

## SUPPLIERS OF COMBUSTION TURBINE GENERATOR SETS

Below is a listing of the addresses of some equipment representatives and suppliers of combustion turbine generator sets.

ACEC
Gas Turbine Systems Division
BP4, 6000 Charleroi, Belgium

Aeg-Kanis Turbinenfabrik
Marketing Services
Altendorfer Str. 39-85
D-4300 Essen, West Germany

Alsthom-Atlantique
38 Avenue Kleber
75 795 Paris Cedex 16, France

Avco Industrial Engine Operation
17220 Park Row
Houston, TX 77084

BBC Brown Boveri & Co., Ltd.
P.O. Box 85
Ch 5401 Baden, Switzerland

Brown Boveri Turbomachinery, Inc.
711 Anderson Avenue N
St. Cloud, MN 56301

Centrax, Ltd.
Gas Turbine Division
Shaldon Road, Newton Abbot
Devon, TQ12 4SQ England

Commonwealth Aircraft Corporation, Ltd.
304 Lorimer St.,
Melbourne
Victoria 3207, Australia

Cooper Energy Services
North Sandusky Street
Mount Vernon, OH 43050

Creusot Loire
15 Rue Pasquier
75008 Paris, France

Curtiss-Wright Corporation
Power Systems Group
1 Passaic Street
Woodridge, NJ 07075

Detroit Diesel Allison
Division of General Motors Company
P.O. Box 894
Indianapolis, IN 46206

Dresser Clark
Division of Dresser Industries, Inc.
P.O. Box 560
Olean, NY 14760

Energy Transformation Company
Swinehart and Rick Roads
Boyertown, PA 19512

Fiat TTG S.p.A
Via Cuneo 20
Torino 10152, Italy

Garrett Corporation
Industrial Engines & Prod.
P.O. Box 5217
Phoenix, AZ 85010

GEC Gas Turbines Ltd.
Cambridge Road
Whetstone
Leicester LE8 3LH, England

General Electric Company
Gas Turbine Division
1 River Road
Schenectady, NY 12345

Hawker Siddeley Canada Ltd.
Orenda Division
Box 6001 Toronto Amf
Toronto, Ontario L5P 1B3, Canada

Hispano Suiza, Inc.
10633 Shadow Wood Drive
Houston, TX 77043

Hitachi, Ltd.
Thermal Plant Dept.
2-6-2 Otemachi
Tokyo, Japan

Ingersoll-Rand Company
Turbo Products Division
Phillipsburg, NJ 08865

Ishikawajima-Harima Heavy Industries Co., Ltd.
Aero-Engine & Space Operations
3-5 Mukodai-Cho
Tanashi-shi
Tokyo 188, Japan

John Brown Engineering Ltd.
Clydebank G81 1YA
Dunbartonshire, Scotland

Kawasaki Heavy Industries, Ltd.
Jet Engine Division
1-1 Kawasaki-Cho
Akashi-shi Hyogo, Japan

Klockner-Humboldt-Deutz
Hochhaus 11
Deutz-Kalker Str. 18-26
D500 Koln, West Germany

Kongsberg Vapenfabrikk
Gas Turbines & Power Systems
P.O. Box 25 N-3601
Kongsberg, Norway

Kraftwerk Union AG
Hamnmerbacherstrasse 12-14
8520 Erlangen, West Germany

Kvaerner Brug A/S
Steam & Gas Turbine Division
P.O. Box 3610
Oslo, Norway

Lucas Aerospace, Ltd.
Shirley Solihull
West Midlands, England

M.A.N. Unternehmensbereich
GHH Sterkrade
P.O. Box 110240 D-4200
Oberhausen 11, West Germany

Mitsubishi Heavy Industries, Ltd.
5-1 Marunouchi 2-Chome
Chiyoda-ku
Tokyo 100, Japan

Mitsui Engineering & Shipbuilding Co., Ltd.
6-4 Tsukiji 5-Chome
Chuo-ku
Tokyo, Japan

MTU Motoren-u Turbinen
Union Muchen Gmbh.
Dachauer Str. 665
8000 Munchen 50, West Germany

Noel Penny Turbines Ltd.
Siskin Drive
Toll Bar End
Coventry CV3 4FE, England

North American Turbine Corporation
P.O. Box 40510
Houston, TX 77040

Nuovo Pignone S.p.A.
Via Felice Matteucci 2
50100 Firenze, Italy

Pratt & Whitney Aircraft of Canada
400 Main Street
East Hartford, CT 06108

Prvni Brnenska Strojirna
First Brno Eng Works
Olomouchka 7/9
656 66 Brno, Czechoslovakia

Roland Marine, Inc.
1 State Street Plaza
New York, NY 10004

Rolls-Royce, Ltd.
Industrial and Marine Division
P.O. Box 72
Ansty, Coventry CV79JR, England

RSV Thomassen Holland
P.O. Box 70 6990 Ab Rheden
Netherlands, Holland

Ruston Gas Turbines, Ltd.
P.O. Box 1
Lincoln, England

Snecma Soc National Detude &
Construction De Moteurs
2 Boulevard Victor, Cdx 15
75724 Paris, France

Solar Turbines International
2200 Pacific Highway
P.O. Box 80966
San Diego, CA 92138

Stal-Laval Turbin Ab
S-612 20
Finspong, Sweden

Sulzer Brothers, Ltd.
Dept. Thermal Turbomachinery
CH-8401 Winterthur, Switzerland

Teledyne Cae Turbine Engines
1330 Laskey Road
Toledo, OH 43612

Toshiba Corporation
Heavy Apparatus Division
Toshiba Mita Building
13-12 Mita 3 Chome
Minatoku, Japan

Turbine Industrie Services Commerciaux
1, Rue de la Fonderie
B.P. 1210
68054 Mulhouse Cedex, France

Turbine Hispano Ogasco
10633 Shadow Wood Drive
Houston, TX 77043

Turbomeca
1 Rue Beaujon
64320 Bordes Bizanos
Paris 8, France

United Technologies Corporation
Power Systems Division
1690 New Britain Avenue
Farmington, CT 06032

U.S.S.R. Energomachexport
35 Mosfilmovskaya
Moscow V-330, U.S.S.R.

Westinghouse Canada Ltd.
Turbine & Generator Division
P.O. Box 510
Hamilton, Ontario L8N 3K2, Canada

Williams Research Corporation
2280 W. Maple Road
Walled Lake, MI 48088

# APPENDIX F

## SUPPLIERS OF STEAM TURBINE GENERATOR SETS

Below is a list of the names and addresses of some equipment representatives and suppliers of steam turbine generator sets.

ACEC-Stg Division
Ateliers de Constructions Electriques de Charleroi
P.O. Box 4
Charleroi, Belgium

Aeg-Kanis Turbinenfabrik
Marketing Services
Altendorfer Str 39-85
D-4300 Essen, West Germany

Alsthom-Atlantique
Dem Division
38 Avenu Kleber Cedex 16
75784 Paris, France

American Man Corporation
1114 Avenue of the Americas
New York, NY 10036

APE-Allen, Limited
P.O. Box 43
Queens Engineering Works
Bedford, England MK40 4JB

BBC Brown Boveri & Co., Ltd.
P.O. Box CH-5401
Baden, Switzerland

Blohm & Voss AG
Turbine Sales Department
Postfach 10 07 20
2000 Hamburg 1, Germany

Borsig GMBH
Gruppe Deutsche Babcock
Postfach 1000
Berlin 27, West Germany

Carling Turbine Blower
8 Nebraska Street
Worcester, MA 01613

Coppus Engineering Corporation
344 Park Avenue
Worcester, MA 01610

Creusot-Loire
Department Compresseurs et Turbines Industrielle
15 Rue Pasquier
75383 Paris Cedex 08, France

Crouse Power Turbine
Generator Division
Box 247
North Stonington, CT 06359

Delaval Turbine, Inc.
Turbine Compressor Division
P.O. Box 251
Trenton, NJ 08602

Elliott Company
Division of Carrier Corporation
North Fourth Street
Jeannette, PA 15644

GEC Turbine Generators Ltd.
Industrial & Marine Steam Turbine Division
Box 131
Manchester, M60 1AE, England

General Electric Company
Medium Steam Turbine Department
1100 Western Avenue
Lynn, MA 01910

Hayward Tyler Ltd.
24 Crawley Green Road
P.O. Box 2
Luton, England LU1 3LW

Ishikawajima-Harima Heavy Industries Co., Ltd.
Otemachi 2-2-1
Chiyoda-ku
Tokyo 100, Japan

Kawasaki Heavy Industries, Ltd.
Machinery Export Department
World Trade Center Building
4-1, 2 Chome Hamamatsu-Cho
Minatoku, Tokyo, Japan

Kraftwerk Union AG
Technik Mulheim
Wiesenstrasse 35
Postfach 0114 20
4330 Mulheim (Ruhr), West Germany

M.A.N. Unternehmensbereich
GHH Sterkrade
P.O. Box 110240
D-4200 Oberhausen 11, West Germany

Mitsubishi Heavy Industries
5-1Marunouchi 2-Chome
Chiyoda-Ku
Tokyo 100, Japan

Mitsui Engineering & Shipbuilding Co. Ltd.
6-4 Tsukiji 5-Chome, Chuo-ku
Tokyo 104, Japan

Murray Turbines
Process Division
The Trane Company
3600 Pammel Creek Road
La Crosse, WI 54601

Nuovo Pignone
2, F. Matteucci
Box 414
Florence, Italy

Peter Brotherhood Ltd.
Lincoln Road
Peterborough PE4 6AB, England

Prvni Brnenska Strojirna
Olomouchka 7/9,
656 66 Brno, Czechoslovakia

Siemens AG
Industrial Turbines & Power Plants
Dept. E25
Werner von Siemens Str. 50
8520 Erlangen, West Germany

Skinner Engine Company
337 W. 12th Street
Erie, PA 16512

Stal-Laval Turbin AB
S-612 20 Finspong, Sweden

Terry Steam Turbine Company
P.O. Box 555
Windsor, CT 06095

Thermo Electron Corporation
101 First Avenue
Waltham, MA 02154

Thomassen Holland BV
Rhine-Schelde-Verolme
P.O. Box 3
De Steeg, Netherlands

Toshiba Corporation
International Operations
1-6 Uchisaiwaicho 1 Chome
Tokyo 100, Japan

Trane Murry
3600 Pammel Creek Road
La Crosse, WI 54601

Transamerica Delaval Inc.
Turbine & Compressor Division
P.O. Box 8788
Trenton, NJ 08650

Turbodyne Division
McGraw-Edison Company
Coats Street
Wellsville, NY 14895

Turbonetic Energy, Inc.
968 Albany-Shaker Road
Latham, NY 12110

United Technologies Corporation
Power Systems Division
1690 New Britain Avenue
Farmington, CT 06032

Utility Power Corporation
1135 South 70th Street
West Allis, WI 53214

Veb Stromungsmachinen
83 Pirna Sonnestein
Ddr, West Germany

Westinghouse Canada, Inc.
Turbine & Generator Division
P.O. Box 510
Hamilton, Ontario L8N 3K2, Canada

Westinghouse Electric Corporation
Lester Branch
P.O. Box 9175
Philadelphia, PA 19113

# APPENDIX G

## SUPPLIERS OF WASTE HEAT RECOVERY EQUIPMENT

Below is a list of the names and addresses of some representatives and suppliers of waste heat recover steam generators, heat exchangers, and heat recovery equipment.

Abco Industries, Inc.
2675 East Highway 80
P.O. Box 268
Abilene, TX 79604

Alfa-Laval Inc.
Thermal Division
2115 Linwood Avenue
Ft. Lee, NJ 07024

American Standard
P.O. Box 1102
Buffalo, NY 14240

APV Spiro-Gills, Ltd.
Pulborough
London Road
West Sussex, England

B V Machinefabriek Breda
Boilers and Process Equipment
Speelhuslaan 173/Pob 3260
Noord-Brabant, Netherlands

Bigelow Company
P.O. Box 706
New Haven, CT 06503

Borsig Gmbh
Gruppe Duetsche Babcock
Postfach 1000
Berlin 27, West Germany

Busch Company
4907 Penn Avenue
Pittsburgh, PA 15224

Coaltech Farrier, Inc.
P.O. Box 1602
York, PA 17405

Cockerill SA
Mechanical Division
Post Code 4100
Seraing, Belgium

Conesco, Inc.
611 North Road
Medford, WI 54451

Covrad
Heat Transfer Division
Sr. Henry Parkes Road W. Mdld.
Coventry CV5 6BN, England

Deltak Corporation
P.O. Box 9496
Minneapolis, MN 55440

Emerson Electric Company
Edwin L. Wiegand Division

Four Allegheny Center
Pittsburgh, PA 15212

Engineering Controls Division
Pott Industries, Inc.
611 East Marceau Street
St. Louis, MO 63111

Escoa Fintube Corporation
P.O. Box 399
Pryor, OK 74361

Esex, Inc.
2959 West 21st Street Pob 9516
Tulsa, OK 74107

Farrier Products
Division of Combustion Turbine Power
P.O. Box 1602
York, PA 17405

Fiat TTG S.p.A.
Via Cuneo 20
Torino 10152
Italy

Foster Wheeler Power Products
Hampstead Road
Greater London House
London NW 1, England

GEA Kohlturmbau Und
Luftkondensation Gmbh
Pob 100807
Bochum 4630, West Germany

GTE Sylvania Inc.
Chemical & Metallurgical
Hawes Street
Towanda, PA 18848

General Motors Corporation
Harrison Radiator Division
Lockport, NY 14094

Harris Thermal Trans Prod.
19830 SW 102d Street
Tualatin, OR 97062

Hartford Steam Boiler
Inspection, Technical Services
56 Prospect Street
Hartford, CT 06102

Hayden, Inc.
Industrial Division
1531 Pomona Road
Corona, CA 91720

Heg/Nem Maatschapy Bv.
P.O. Box 6
2300 Aa Leiden
Holland

Henry Vogt Machine Company
P.O. Box 1918
Louisville, KY 40201

Hudson & West Pty Ltd. NSW 2014
2-6 Mentmore Avenue Rosebery
Sydney 2018, Australia

International Boiler Works
P.O. Box 498
E. Stroudsberg, PA 18301

International Power Machinery Co.
834 Terminal Tower Building
Cleveland, OH 44113

John Zink Company
4401 S. Peoria
Tulsa, OK 74105

Killebrew Engineering Corporation
2501 Adie Road
Maryland Heights, MO 63043

Lytron, Inc.
Dragon Court
Woburn, MA 01801

MAN Steam Turbines
Department Tdtp
Katzwanger Strasse 101
Nurnberg 115, West Germany

Mitsubishi-Ce Heavy Industries
5-1 Marunouchi 2 Chome
Chiyoda-Ku
Tokyo, Japan

Modine Manufacturing Company
1500 Dekoven Avenue
Racine, WI 53401

Nebraska Boiler Company
70th Cornhusker Highway
Lincoln, NE 68501

Northern Engineering Industries
Power Plant Division
Saltmeadows Road
Gateshead NE8 1YZ, England

Norton Company
1 New Bond Street
Worcester, MA 01606

Peter Brotherhood, Ltd.
Marketing Manager—Turbines
Lincoln Road Walton
Peterborough PE4 6AB, England

Riley Stoker Corporation
Division of the Riley Company
P.O. Box 547
Worcester, MA 01613

Societe Alsacienne de Constructions Mecaniques
1 Rue de Fonderie Bp 1210
68054 Mulhse, France

Stewart-Warner Corporation
South Wind Division
1514 Drover Street
Indianapolis, IN 46221

Struthers Wells Corporation
P.O. Box 8
Warren, PA 16365

Sundstrand Energy Systems
4747 Harrison Avenue
Rockford, IL 61101

York Division of Borg-Warner
Box 1592
York, PA 17405

Young Radiator Company
2825 Four Mile Road
Racine, WI 53404

Zurn Industries, Inc.
Energy Division
1422 East Avenue
Erie, PA 16503

# APPENDIX H

## SUPPLIERS OF HVAC ABSORPTION AND CENTRIFUGAL CHILLER EQUIPMENT

Below is a list of the names and addresses of some equipment representatives and suppliers of building heating, ventilating, and air conditioning (HVAC) absorption, refrigeration, and centrifugal chiller equipment.

Carrier Air Conditioning
Division of Carrier Corporation
Carrier Parkway
P.O. Box 4808
Syracuse, NY 13221

Gas Energy, Inc.
(Hitachi Equipment Representative)
195 Montague Street
Brooklyn, NY 11201

Trane Company
Commercial Air Conditioning Division
3600 Pammel Creek Road
La Crosse, WI 54601

York Division
Borg-Warner Corporation
P.O. Box 1592
York, Pennsylvania 17405

# APPENDIX I

## SUPPLIERS OF RECIPROCATING INTERNAL-COMBUSTION ENGINES

Below is a list of the names and addresses of some equipment representatives and suppliers of reciprocating internal-combustion engines.

Alco Power, Inc.
100 Orchard Street
Auburn, NY 13021

Allis-Chalmers Corporation
P.O. Box 512
Milwaukee, WI 53201

Caterpillar Tractor Company
Engine Division
100 N.E. Adams
Peoria, IL 61629

Cummins Engine Company
Box 3005
Columbus, IN 47201

Detroit Diesel Allison
Division of General Motors Co.
P.O. Box 894
Indianapolis, IN 46206

Fairbanks Morse
Engine Division
701 Lawton Avenue
Beloit, WI 53511

International Harvester
Engine Division
401 N. Michigan Avenue
Chicago, IL 60611

Lister Diesels, Inc.
Division of Hawker Siddeley, Inc.
555 E. 56 Highway
Olathe, KS 66061

Onan Corporation
A Subsidiary of McGraw-Edison Company
1400 73d Avenue, N.E.
Minneapolis, MN 55432

Sullair Corporation
3700 E. Michigan Blvd.
Michigan City, IN 46360

Sulzer Brothers, Inc.
200 Park Avenue, 22d Floor
New York, NY 10166

Wabash Power Equipment Company
444 Carpenter Avenue
Wheeling, IL 60090

Waukesha Engine Division
Dresser Industries, Inc.
1000 West St. Paul Avenue
Waukesha, WI 53187

# APPENDIX J

## SUPPLIERS OF MAINTENANCE AND OVERHAUL SERVICES

The following is a listing of the names and addresses of some representatives of maintenance and overhaul service.

AAR Technical Service Center
Energy Services Division
25 Buena Vista Avenue
Lawrence, Long Island, NY 11559

Airco Temescal
2850 Seventh Street
Berkeley, CA 94710

Aircraft Turbine Service
1100 Shames Drive
Westbury, NY 11590

Airesearch Aviation Company
17250 Chanute Road
Houston, TX 77060

Airwork Corporation
Municipal Airport
Millville, NJ 08332

Ames Industrial Corporation
55 Orville Drive
Bohemia, NY 11716

Aviation Power Supply
3111 Kenwood Street
Burbank, CA 91505

BBC Brown Boveri & Company, Ltd.
P.O. Box CH 5401
Baden, Switzerland

Besco
2525 W. Edgewood Avenue
Jacksonville, FL 32209

Bethlehem Steel Corporation
Ship Repair Sales Office
1 State Street Plaza
New York, NY 10004

Brown Boveri & Cie AG
Postfach 351
D-6800 Mannheim, West Germany

Brown Boveri Turbomachinery, Inc.
711 Anderson Avenue N
St. Cloud, MN 56301

Caprice Engineering Company, Inc.
509 S. Hindry Avenue
Inglewood, CA 90301

Chromalloy American Corporation
Turbine Support Division
4430 Director Drive
San Antonio, TX 78219

Chromalloy
Research & Technology Division
Blaisdell Road
Orangeburg, NY 10962

Coatings Technology Corporation
Box 223
Branford, CT 06405

Combustion Engineering
C-E Power Systems
1000 Prospect Hill Road
Windsor, CT 06095

Cooper Airmotive
Engine Division
P.O. Box 7086
Dallas, TX 75209

Creole Production Service, Inc.
Box 1181
Houston, TX 77001

Crouse Power Turbine
Generator Division
Box 247
N. Stonington, CT 06359

Curtiss-Wright Corporation
Overhaul Department
1 Passaic Street
Woodridge, NJ 07075

D.R. Michael Company, Inc.
787 Granite Street
Braintree, MA 02184

Dresser Machinery Repair
Dresser Industries
Dresser Tower
601 Jefferson
Houston, TX 77002

Durblade Corporation
1022 Wirt Road
Houston, TX 77055

Elbar Bv
Industrieterrein Spikwein
5944 ZG Arcen, Netherlands

Eldim Bv
Industrieterrein Spkweien
P.O. Box 4315
5944 Zg Arcen, Netherlands

Elliot Service Company
North Fourth Street
Jeanette, PA 15644

Energy Maintenance Corporation
The Exchange
Farmington, CT 06032

Exline Inc.
P.O. Box 1267
Salina, KS 67401

Fiat TTG S.p.A.
Via Cuneo 20
Torino 10152, Italy

Finnair
Technical Center, SF 01530
Helsinki-Ventaa-Lento, Finland

Gas Turbine Corporation
P.O. Box G
East Granby, CT 06026

General Electric Company
Apparatus Service Division
Building 6, Room 213
1 River Road
Schenectady, NY 12345

George Engine Company
Box 8
Harvey, LA 70059

Gulf Engineering Company
1000 S. Peters Street
New Orleans, LA 70130

Hahn & Clay Machine & Boiler
5100 Clinton Drive
Houston, TX 77020

Hawker Siddeley Canada, Ltd.
Orenda Division
Box 6001 Toronto Amf
Toronto, Ontario L5P 1B3, Canada

Hong Kong Aircraft & Engineering
Kai Tak Airport Kowloon
P.O. Box 5728
Hong Kong, Japan

Howmet Turbine Company, Corporation
500 Terrace Plaza
Muskegon, MI 49443

I. W. Hickham Company, Inc.
11518 Old La Porte Road
La Porte, TX 77571

John Brown Engineering Ltd.
Clydebank G81 1YA
Dunbartonshire, Scotland

Jones & Rickard Ltd.
P.O. Box 198, Waterloo
New South Wales 17, Australia

Klockner-Humboldt-Deutz
Deutz-Kalker Str. 18-26
D500 Kohn, West Germany

Mechanical Technology
968 Albany-Shaker Road
Latham, NY 12110

Metal Improvement Company
Curtiss-Wright Corporation
1485 Teaneck Road
Teaneck, NJ 07666

Mobile Inspection Service
9110 Dice Road
Santa Fe Spring, CA 90670

Murray Turbines
1106 Washington Street
Burlington, IA 52601

Power Services Caribe
Suite 302, General Computer Building
Ave Ponce de Leon 1590
Rio Piedras, Puerto Rico 00926

Powmat Limited
Northway 10 Exec. Park
Ballstond Lake, NY 12019

Preco Turbine Services
5571 Maudlin
Houston, TX 77087

Rolls-Royce Canada Ltd.
P.O. Box 1000
Montreal AMF
Montreal, Quebec H4Y 1B7, Canada

Sargent Industries
Sweeney Division
6300 Stapleton Drive
Denver, CO 80216

Shackelford-Wattner
Division of Texas Custom Builders
5824 Waltrip Street
Houston, TX 77087

Solar Turbines International
International Harvester
2200 Pacific, Highway
San Diego, CA 92138

Spar Aerospace
825 Caledonia Road
Toronto M6B 3X8, Canada

Stewart & Stevenson Services, Inc.
Gas Turbine Engine Division
P.O. Box 1637
Houston, TX 77001

Sulzer Brothers Ltd.
Thermal Turbomachinery
P.O. Box 8023
Zurich, Switzerland

TK International
P.O. Box 45587
Tulsa, OK 74145

Torque Systems, Inc.
1719 E. 28th Street
Long Beach, CA 90806

Torvac Midland Ltd.
Brindley Road, N. Exhall
Coventry CB4 4HE, England

Transamerica Delaval Inc.
Deltex
5710 Ransom Street
Houston, TX 77087

Turbine Associates, Inc.
P.O. Box 53648
Lafayette, LA 70505

Turbine Components
1 Commercial Street
Branford, CT 06405

Turbine Engine Services
P.O. Box 116
East Granby, CT 06026

Turbine Specialties Inc.
Box 207, W. State Street Road
Salina, KS 67401

Turbine Technologies, Inc.
Kettle Cove Industrial Park
Magnolia, MA 01930

Turbo Spec & Machine Company
4301 W. Country Road
Odessa, TX 79760

Twombly, Inc.
172 Broadway
Woodcliff Lake, NJ 07675

Union Carbide Corporation
Coatings Service Department
1500 Polco Street
Indianapolis, IN 46224

Vac-Aero International
1371 Speers Road
Oakville, Ontario L6L 2X5, Canada

Vac-Hyd
182 New Boston Park
Woburn, MA 01801

Vacuum Heatreat
300 Taft Street
Hollywood, FL 33021

Volvo Flygmotor Ab
Trollhattan 461 81, Sweden

W. D. Machine Works
8139 Tidwell Pob 23342
Houston, TX 77028

Werkspoor-Sneek
21 Kamerlingh Onnesstraat
8606 Jn Sneek, Holland

Westinghouse Canada Ltd.
Turbine & Generator Division
P.O. Box 510
Hamilton, Ontario L8N 3K2, Canada

Westingthouse Electric Corporation
Lester Branch
P.O. Box 9175
Philadelphia, PA 19113

Worthington Service Corporation
1009 Executive Parkway
St. Louis, MO 63141

# APPENDIX K

## POWER PURCHASE AGREEMENTS: PACIFIC GAS AND ELECTRIC COMPANY

PACIFIC GAS AND ELECTRIC COMPANY

STANDARD OFFER #1

POWER PURCHASE AGREEMENT

FOR

AS-DELIVERED CAPACITY AND ENERGY

STANDARD OFFER #1:

AS-DELIVERED CAPACITY AND ENERGY

POWER PURCHASE AGREEMENT

CONTENTS

**AS-DELIVERED CAPACITY AND ENERGY**
**POWER PURCHASE AGREEMENT**
**BETWEEN**
________________________________
**AND**
**PACIFIC GAS AND ELECTRIC COMPANY**

________________________________________ ("Seller"), and PACIFIC GAS AND ELECTRIC COMPANY ("PGandE"), referred to collectively as "Parties" and individually as "Party", agree as follows:

ARTICLE 1 QUALIFYING STATUS

Seller warrants that, at the date of first power deliveries from Seller's Facility[1] and during the term of agreement, its Facility shall meet the qualifying facility requirements established as of the effective date of this Agreement by the Federal Energy Regulatory Commission's rules (18 Code of Federal Regulations 292) implementing the Public Utility Regulatory Policies Act of 1978 (16 U.S.C.A. 796, et seq.).

ARTICLE 2 PURCHASE OF POWER

(a) Seller shall sell and deliver and PGandE shall purchase and accept from the Facility having a nameplate rating of _______ kW located at ________________________________________

________________________________________________________________

the as-delivered capacity and energy at the voltage level of ______ kV. Seller has chosen ________________________[2] as its energy sale option. Seller may convert its energy sale option as provided in Section A-3 of Appendix A.

(b) The scheduled operation date when Seller estimates first delivery of electric energy from the Facility to PGandE is ________________________ (Date). At the end of each calendar quarter Seller shall give to PGandE written notice of any change in the scheduled operation date.

(c) Seller shall limit the Facility's actual rate of delivery into the PGandE system to _______ kW.

(d) The primary energy source for the Facility is ________________________________________.

ARTICLE 3 PURCHASE PRICE

PGandE shall pay Seller for as-delivered capacity at prices authorized from time to time by the CPUC and for energy at prices equal to PGandE's full short run avoided operating costs as approved by the CPUC.

---

1 Underlining identifies those terms which are defined in Section A-1 of Appendix A.

2 Insert either "net energy output" or "surplus energy output" to show the energy sale option selected by Seller.

### ARTICLE 4 NOTICES

All written notices shall be directed as follows:

to PGandE: Pacific Gas and Electric Company
Attention: Vice President -
Electric Operations
77 Beale Street
San Francisco, CA 94106

to Seller: ____________________
____________________
____________________
____________________
____________________

### ARTICLE 5 DESIGNATED SWITCHING CENTER

The designated PGandE switching center shall be unless changed by PGandE:

____________________
(Name)

____________________
(Location)

____________________
(Phone number)

### ARTICLE 6 TERMS AND CONDITIONS

This Agreement includes the following appendices which are attached and incorporated by reference:

Appendix A - GENERAL TERMS AND CONDITIONS
Appendix B - INSURANCE

### ARTICLE 7 TERM OF AGREEMENT

This Agreement shall be binding upon execution and remain in effect for _____ years from the date of execution; provided, however, that it shall terminate if Seller does not commence deliveries of as-available capacity and energy within five years of this Agreement's effective date.

IN WITNESS WHEREOF, the Parties hereto have caused this Agreement to be executed by their duly authorized representatives and effective as of the last date set forth below.

| ____________________ (SELLER) | PACIFIC GAS AND ELECTRIC COMPANY |
|---|---|
| BY: ____________________ | BY: ____________________ |
| ____________________ (Type Name) | ____________________ (Type Name) |
| TITLE: ____________________ | TITLE: ____________________ |
| DATE SIGNED: ____________________ | DATE SIGNED: ____________________ |

S.O. #1

# APPENDIX A

## GENERAL TERMS AND CONDITIONS

CONTENTS

# APPENDIX A

## GENERAL TERMS AND CONDITIONS

A-1. DEFINITIONS

Whenever used in this Agreement, appendices, and attachments hereto, the following terms shall have the following meanings:

CPUC - The Public Utilities Commission of the State of California.

Designated PGandE switching center - That switching center or other PGandE installation identified in Article 5.

Facility - That generation apparatus described in Article 2 and all associated equipment owned, maintained, and operated by Seller.

Interconnection facilities - All means required and apparatus installed to interconnect and deliver power from the Facility to the PGandE system including, but not limited to, connection, transformation, switching, metering, communications, and safety equipment, such as equipment required to protect (1) the PGandE system and its customers from faults occurring at the Facility, and (2) the Facility from faults occurring on the PGandE system or on the systems of others to which the PGandE system is directly or indirectly connected. Interconnection facilities also include any necessary additions and reinforcements by PGandE to the PGandE system required as a result of the interconnection of the Facility to the PGandE system.

Net energy output - The Facility's gross output in kilowatt-hours less station use and transformation and transmission losses to the point of delivery into the PGandE system.

Prudent electrical practices - Those practices, methods, and equipment, as changed from time to time, that are commonly used in prudent electrical engineering and operations to design and operate electric equipment lawfully and with safety, dependability, efficiency, and economy.

Special facilities - Those parts of the interconnection facilities furnished by PGandE at Seller's request, metering and data processing equipment, and those additions and reinforcements to the PGandE system which are needed to accommodate the maximum delivery of energy and capacity from the Facility as provided in this Agreement. All special facilities shall be furnished at Seller's expense pursuant to PGandE's electric Rule No. 21 by a separate agreement.

Station use - Energy used to operate the Facility's auxiliary equipment. The auxiliary equipment includes, but is not limited to, forced and induced draft fans, cooling towers, boiler feed pumps, lubricating oil systems, plant lighting, fuel handling systems, control systems, and sump pumps.

Surplus energy output - The Facility's gross output, in kilowatt-hours, less station use, and any other use by the Seller, and transformation and transmission losses to the point of delivery into the PGandE system.

Term of agreement - The period of time during which this Agreement will be in effect, as provided in Article 7.

Voltage level - The voltage at which the Facility interconnects with the PGandE system, measured at the point of delivery.

A-2. CONSTRUCTION

A-2.1 Land Rights

Seller hereby grants to PGandE all necessary rights of way and easements to install, operate, maintain, replace, and remove the special facilities, including adequate and continuing access rights on property of Seller. Seller agrees to execute such other grants, deeds, or documents as PGandE may require to enable it to record such rights of way and easements. If any part of PGandE's equipment is to be installed on property owned by other than Seller, Seller shall, at its own cost and expense, obtain from the owners thereof all necessary rights of way and easements, in a form satisfactory to PGandE, for the construction, operation, maintenance, and replacement of PGandE's equipment upon such property. If Seller is unable to obtain these rights of way and easements, Seller shall reimburse PGandE for all costs incurred by PGandE in obtaining them. PGandE shall at all times have the right of ingress to and egress from the Facility at all reasonable hours for any purposes reasonably connected with this Agreement or the exercise of any and all rights secured to PGandE by law or its tariff schedules.

A-2.2 Design, Construction, Ownership, and Maintenance

(a) Seller shall design, construct, install, own, operate, and maintain all interconnection facilities, except special facilities, to the point of interconnection with the PGandE system as required for PGandE to receive as-delivered capacity and energy from the Facility. The Facility and interconnection facilities shall meet all requirements of applicable codes and all standards of prudent electrical practices and shall be managed in a safe and prudent manner.

(b) Seller shall submit all specifications necessary for the design of the interconnection facilities to PGandE for review and written acceptance prior to their release for construction purposes. PGandE's review and acceptance of these specifications shall not be construed as confirming or endorsing the design or as warranting their safety, durability, or reliability. PGandE shall not, by reason of such review or lack of review, be responsible for strength, details of design, adequacy, or capacity of equipment built pursuant to such specifications, nor shall PGandE's acceptance be deemed to be an endorsement of any of such equipment. Seller shall change the interconnection facilities as may be reasonably required by PGandE to meet changing requirements of the PGandE system.

(c) In the event it is necessary for PGandE to install interconnection facilities for the purposes of this Agreement, they shall be installed as special facilities.

A-2.3 Meter Installation

(a) PGandE shall specify, provide, install, own, operate, and maintain as special facilities all metering and data processing equipment for the registration and recording of energy and other related parameters which are required for the reporting of data to PGandE and for computing the payment due Seller from PGandE.

(b) Seller shall provide, construct, install, own, and maintain at Seller's expense all that is required to accommodate the metering and data processing equipment, such as, but not limited to, metal-clad switchgear, switchboards, cubicles, metering panels, enclosures, conduits, rack structures, and equipment mounting pads.

A-3. ENERGY SALE OPTIONS

A-3.1 General

Seller has two energy sale options, net energy output or surplus energy output. Seller has made its initial selection in Article 2(a).

A-3.2 Energy Sale Conversion

(a) Seller is entitled to convert from one option to the other 12 months after execution of this Agreement, and thereafter at least 12 months after the effective date of the most recent conversion, subject to the following conditions:

(1) Seller shall provide PGandE with a written request to convert its energy sale option.

(2) Seller shall comply with all applicable tariffs on file with the CPUC and contracts in effect between the Parties at the time of conversion covering the existing and proposed (i) facilities used to serve Seller's premises and (ii) interconnection facilities.

(3) Seller shall install and operate equipment required by PGandE to prevent PGandE from serving any part of Seller's load which is served by the Facility and not under contract for PGandE standby service. At Seller's request, PGandE shall provide this equipment as special facilities.

(b) PGandE shall not be required to remove or reserve capacity of facilities made idle by Seller's energy sale conversion and may dedicate such facilities at any time to serve other customers or to interconnect with other electric power sources.

(c) PGandE shall process requests for conversion in the order received. The effective date of conversion shall depend on the completion of the changes required to accommodate Seller's energy sale conversion.

A-4. OPERATION

A-4.1 Inspection and Approval

Seller shall not operate the Facility in parallel with PGandE's system until an authorized PGandE representative has inspected the interconnection facilities, and PGandE has given written approval to begin parallel operation.

A-4.2 Facility Operation and Maintenance

Seller shall operate and maintain its Facility according to prudent electrical practices, applicable laws, orders, rules, and tariffs and shall provide such reactive power support as may be reasonably required by PGandE to maintain system voltage level and power factor. Seller shall operate the Facility at the power factors or voltage levels prescribed by PGandE's system dispatcher or designated representative. If Seller fails to provide reactive power support, PGandE may do so at Seller's expense.

A-4.3 Point of Delivery

Seller shall deliver the energy at the point where Seller's electrical conductors contact PGandE's existing system or at such other point as the Parties may agree in writing.

A-4.4 Operating Communications

(a) Seller shall maintain operating communications with the designated PGandE switching center. The operating communications shall include, but not be limited to, system paralleling or separation, scheduled and unscheduled shutdowns, equipment clearances, levels of operating voltage or power factors, and daily capacity and generation reports.

(b) Seller shall keep a daily operations log for each generating unit which shall include information on unit availability, maintenance outages, circuit breaker trip operations requiring a manual reset, and any significant events related to the operation of the Facility.

(c) If Seller makes deliveries greater than one megawatt, Seller shall measure and register on a graphic recording device power in kW and voltage in kV at a location within the Facility agreed to by both Parties.

(d) If Seller makes deliveries greater than one and up to and including ten megawatts, Seller shall report to the designated PGandE switching center, twice a day at agreed upon times for the current day's operation, the hourly readings in kW of capacity delivered and the energy in kWh delivered since the last report.

(e) If Seller makes deliveries of greater than ten megawatts, Seller shall telemeter the delivered capacity and energy information, including real power in kW, reactive power in kVAR, and energy in kWh to a switching center selected by PGandE. PGandE may also require Seller to telemeter transmission kW, kVAR, and kV data depending on the number of generators and transmission configuration. Seller shall provide and maintain the data circuits required for telemetering. When telemetering is inoperative, Seller shall report daily the capacity delivered each hour and the energy delivered each day to the designated PGandE switching center.

A-4.5 Meter Testing and Inspection

(a) All meters used to provide data for the computation of the payments due Seller from PGandE shall be sealed, and the seals shall be broken only by PGandE when the meters are to be inspected, tested, or adjusted.

(b) PGandE shall inspect and test all meters upon their installation and annually thereafter. At Seller's request and expense, PGandE shall inspect or test a meter more frequently. PGandE shall give reasonable notice to Seller of the time when any inspection or test shall take place, and Seller may have representatives present at the test or inspection. If a meter is found to be inaccurate or defective, PGandE shall adjust, repair, or replace it at its expense in order to provide accurate metering.

A-4.6 Adjustments to Meter Measurements

If a meter fails to register, or if the measurement made by a meter during a test varies by more than two percent from the measurement made by the standard meter used in the test, an adjustment shall be made correcting all measurements made by the inaccurate meter for (1) the actual period during which inaccurate measurements were made, if the period can be determined, or if not, (2) the period immediately preceding the test of the meter equal to one-half the time from the date of the last previous test of the meter, provided that the period covered by the correction shall not exceed six months.

A-5. PAYMENT

PGandE shall mail to Seller not later than 30 days after the end of each monthly billing period (1) a statement showing the kilowatt-hours delivered to PGandE during on-peak, partial-peak, and off-peak periods during the monthly billing period, (2) PGandE's computation of the payment due Seller, and (3) PGandE's check in payment of said amount. If within 30 days of receipt of the statement Seller does not make a report in writing to PGandE of an error, Seller shall be deemed to have waived any error in PGandE's statement, computation, and payment, and they shall be considered correct and complete.

A-6. ADJUSTMENTS OF PAYMENTS

(a) In the event adjustments to payments are required as a result of inaccurate meters, PGandE shall use the corrected measurements described in Section A-4.6 to recompute the amount due from PGandE to Seller for the as-delivered capacity and energy delivered under this Agreement during the period of inaccuracy.

(b) The additional payment to Seller or refund to PGandE shall be made within 30 days of notification of the owing Party of the amount due.

A-7 ACCESS TO RECORDS

Each Party, after reasonable written notice to the other Party, shall have the right of access to all metering and related records including the operations logs of the Facility.

A-8. CURTAILMENT OF DELIVERIES AND HYDRO SPILL CONDITIONS

(a) PGandE shall not be obligated to accept or pay for and may require Seller to interrupt or reduce deliveries of as-delivered capacity and energy (1) when necessary in order to construct, install, maintain, repair, replace, remove, investigate, or inspect any of its equipment or any part of its system, or (2) if it determines that interruption or reduction is necessary because of emergencies, forced outages, force majeure, or compliance with prudent electrical practices.

(b) In anticipation of a period of hydro spill conditions, as defined by the CPUC, PGandE may notify Seller that any purchases of energy from Seller during such period shall be at hydro savings prices quoted by PGandE. If Seller delivers energy to PGandE during any such period, Seller shall be paid hydro savings prices for those deliveries in lieu of prices which would otherwise be applicable. The hydro savings prices shall be calculated by PGandE using the following formula:

$$\frac{AQF - S}{AQF} \times PP$$

where: AQF = Energy, in kWh, projected to be available during hydro spill conditions from all qualifying facilities under agreements containing hydro savings price provisions.

S = Potential energy, in kWh, from PGandE hydro facilities which will be spilled if all AQF is delivered to PGandE.

PP = Prices published by PGandE for purchases during other than hydro spill conditions.

(c) PGandE shall not be obligated to accept or pay for and may require Seller with a Facility with a nameplate rating of one megawatt or greater to interrupt or reduce deliveries of as-delivered capacity and energy during periods when purchases under this Agreement would result in costs greater than those which PGandE would incur if it did not make such purchases but instead generated an equivalent amount of energy itself.

(d) Whenever possible, PGandE shall give Seller reasonable notice of the possibility that interruption or reduction of deliveries under subsections (a) or (c), above, may be required. PGandE shall give Seller notice of general periods when hydro spill conditions are anticipated, and shall give Seller as much advance notice as practical of any specific hydro spill period and the hydro savings price which will be applicable during such period. Before interrupting or reducing deliveries under subsection (c), above, and before invoking hydro savings prices under subsection (b) above, PGandE shall take reasonable steps to make economy sales of the surplus energy giving rise to the condition. If such economy sales are made, while the surplus energy condition exists Seller shall be paid at the economy sales price obtained by PGandE in lieu of the otherwise applicable prices.

(e) If Seller is selling net energy output to PGandE and simultaneously purchasing its electrical needs from PGandE, energy curtailed pursuant to subsections (b) or (c) above shall not be used by Seller to meet its electrical needs. When Seller elects not to sell energy to PGandE at the hydro savings price pursuant to subsection (b) or when PGandE curtails deliveries of energy pursuant to subsection (c), Seller shall continue to purchase all its electrical needs from PGandE. If Seller is selling surplus energy output to PGandE, subsections (b) or (c) shall only apply to the surplus energy output being delivered to PGandE, and Seller can continue to internally use that generation it has retained for its own use.

A-9. FORCE MAJEURE

(a) The term force majeure as used herein means unforeseeable causes beyond the reasonable control of and without the fault or negligence of the Party claiming force majeure, including but not limited to, acts of God, labor disputes, sudden actions of the elements, and actions by federal, state, municipal, or any other government agency.

(b) If either Party because of force majeure is rendered wholly or partly unable to perform its obligations under this Agreement, that Party shall be excused from whatever performance is affected by the force majeure to the extent so affected provided that:

(1) The non-performing Party, within two weeks after the occurrence of the force majeure, gives the other Party written notice describing the particulars of the occurrence,

(2) the suspension of performance is of no greater scope and of no longer duration than is required by the force majeure,

(3) the non-performing Party uses its best efforts to remedy its inability to perform (this subsection shall not require the settlement of any strike, walkout, lockout, or other labor dispute on terms which, in the sole judgment of the Party involved in the dispute, are contrary to its interest. It is understood and agreed that the settlement of strikes, walkouts, lockouts, or other labor disputes shall be at the sole discretion of the Party having the difficulty), and

(4) when the non-performing Party is able to resume performance of its obligations under this Agreement, that Party shall give other Party written notice to that effect.

A-10. INDEMNITY

Each Party as indemnitor shall save harmless and indemnify the other Party and the directors, officers, and employees of such other Party against and from any and all loss and liability for injuries to persons including employees of either Party, and property damages, including property of either Party, resulting from or arising out of (1) the engineering, design, construction, maintenance, or operation of or (2) the making of replacements, additions, or betterments to, the indemnitor's facilities. This indemnity and save harmless provision shall apply notwithstanding the active or passive negligence of the indemnitee. Neither Party shall be indemnified hereunder for liability or loss resulting from its sole negligence or willful misconduct. The indemnitor shall, on the other Party's request, defend any suit asserting a claim covered by this indemnity and shall pay all costs, including reasonable attorney fees, that may be incurred by the other Party in enforcing this indemnity.

A-11. LIABILITY; DEDICATION

(a) Nothing in this Agreement shall create any duty to, any standard of care with reference to, or any liability to any person not a Party to it. Neither Party shall be liable to the other Party for consequential damages.

(b) Each Party shall be responsible for protecting its facilities from possible damage by reason of electrical disturbances or faults caused by the operation, faulty operation, or nonoperation of the other Party's facilities, and such other Party shall not be liable for any such damages so caused.

(c) No undertaking by one Party to the other under any provision of this Agreement shall constitute the dedication of that Party's system or any portion thereof to the other Party or to the public nor affect the status of PGandE as an independent public utility corporation or Seller as an independent individual or entity and not a public utility.

A-12. SEVERAL OBLIGATIONS

Except where specifically stated in this Agreement to be otherwise, the duties, obligations, and liabilities of the Parties are intended to be several and not joint or collective. Nothing contained in this Agreement shall ever be construed to create an association, trust, partnership, or joint venture or impose a trust or partnership duty, obligation, or liability on or with regard to either Party. Each Party shall be liable individually and severally for its own obligations under this Agreement.

A-13. NON-WAIVER

Failure to enforce any right or obligation by either Party with respect to any matter arising in connection with this Agreement shall not constitute a waiver as to that matter or any other matter.

A-14. ASSIGNMENT

Neither Party shall voluntarily assign its rights nor delegate its duties under this Agreement, or any part of such rights or duties, without the written consent of the other Party, except in connection with the sale or merger of a substantial portion of its properties. Any such assignment or delegation made without such written consent shall be null and void. Consent for assignment will not be withheld unreasonably. Such assignment shall include, unless otherwise specified therein, all of Seller's rights to any refunds which might become due under this Agreement.

**A-15. CAPTIONS**

All indexes, titles, subject headings, section titles, and similar items are provided for the purpose of reference and convenience and are not intended to affect the meaning of the contents or scope of this Agreement.

**A-16. CHOICE OF LAWS**

This Agreement shall be interpreted in accordance with the laws of the State of California, excluding any choice of law rules which may direct the application of the laws of another jurisdiction.

**A-17. GOVERNMENTAL JURISDICTION AND AUTHORIZATION**

Seller warrants that the Facility and interconnection facilities, except special facilities, shall at all times conform to all applicable laws and regulations, and Seller shall obtain any governmental authorizations and permits required for the construction and operation thereof. If at any time Seller does not hold such authorizations and permits, PGandE may refuse to accept deliveries of power hereunder.

**A-18. NOTICES**

Any notice, demand, or request required or permitted to be given by either Party to the other, and any instrument required or permitted to be tendered or delivered by either Party to the other, shall be in writing and so given, tendered, or delivered, as the case may be, by depositing the same in any United States Post Office with postage prepaid for transmission by certified mail, return receipt requested, addressed to the Party, or personally delivered to the Party, at the address in Article 4 of this Agreement. Changes in such designation may be made by notice similarly given.

**APPENDIX B**

**INSURANCE**

**B-1. WORKERS' COMPENSATION**

Seller shall furnish PGandE a certificate of workers' compensation or self-insurance indicating compliance with the Labor Code of California, including Employer's Liability insurance, with a minimum of $2,000,000 for injury or death of any one person. This certificate shall provide for 30-day's written notice to PGandE prior to cancellation, termination, alteration, or material change of such insurance.

**B-2. COMPREHENSIVE GENERAL AND COMPREHENSIVE AUTOMOBILE LIABILITY COVERAGE**

(a) Seller shall maintain during the performance hereof, Comprehensive General Liability and Comprehensive Automobile Liability of not less than $3,000,000 combined single limit or equivalent for bodily injury, personal injury, and property damage as the result of any one occurrence.

(b) Comprehensive General Liability shall include coverage for Premises-Operations, Owners and Contractors Protective, Products/Completed Operations Hazard, Explosion, Collapse, Underground, Contractual Liability, and Broad Form Property Damage including Completed Operations. Comprehensive Automobile Liability shall include coverage for Owned, Hired, and Non-Owned automobiles.

(c) Such insurance, by endorsement to the policy(ies), shall include PGandE as an additional insured insofar as work performed by Seller for PGandE is concerned, shall contain a severability of interest clause, shall provide that PGandE shall not by reason of its inclusion as an additional insured incur liability to the insurance carrier for payment of premium for such insurance, and shall provide for 30-day's written notice to PGandE prior to cancellation, termination, alteration, or material change of such insurance.

**B-3. ADDITIONAL INSURANCE PROVISIONS**

(a) Evidence of coverage described above in Sections B-1 and B-2 shall state that coverage provided is primary and is not excess to or contributing with any insurance or self-insurance maintained by PGandE.

(b) PGandE shall have the right to inspect or obtain a copy of the original policy(ies) of insurance.

(c) Seller shall furnish the required certificates and endorsements to PGandE prior to the date construction of the <u>Facility</u> begins or the effective date of this Agreement, whichever is later.

(d) All insurance certificates, endorsements, cancellations, terminations, alterations, and material changes of such insurance shall be issued and submitted to the following:

PACIFIC GAS AND ELECTRIC COMPANY<br>
Attention: Manager - Insurance Department<br>
77 Beale Street<br>
San Francisco, CA 94106

PACIFIC GAS AND ELECTRIC COMPANY

STANDARD OFFER #2

POWER PURCHASE AGREEMENT

FOR

FIRM CAPACITY AND ENERGY

**STANDARD OFFER #2:**

**FIRM CAPACITY AND ENERGY**

**POWER PURCHASE AGREEMENT**

CONTENTS

FIRM CAPACITY AND ENERGY
POWER PURCHASE AGREEMENT
BETWEEN

______________________________
AND
PACIFIC GAS AND ELECTRIC COMPANY

______________________________ ("Seller"), and PACIFIC GAS AND ELECTRIC COMPANY ("PGandE"), referred to collectively as "Parties" and individually as "Party", agree as follows:

ARTICLE 1 QUALIFYING STATUS

Seller warrants that, at the date of first power deliveries from Seller's Facility[1] and during the term of agreement, its Facility shall meet the qualifying facility requirements established as of the effective date of this Agreement by the Federal Energy Regulatory Commission's rules (18 Code of Federal Regulations 292) implementing the Public Utility Regulatory Policies Act of 1978 (16 U.S.C.A. 796, et seq.).

ARTICLE 2 PURCHASE OF POWER

(a) Seller shall sell and deliver and PGandE shall purchase and accept delivery of firm capacity and energy at the voltage level of _____ kV as indicated below--

1. Contract capacity - _____ kW; and

2. Energy - ______________________.[2]

Seller may convert its energy sale option as provided in Section A-3 of Appendix A.

(b) Seller shall provide the firm capacity and energy set forth above from its ______________________ kW
[Nameplate rating of generators(s)]

Facility located at ____________________________________________

____________________________________________________________.

(c) The scheduled operation date of the Facility is ________________ [Date]. At the end of each calendar quarter Seller shall give written notice to PGandE of any change in the scheduled operation date.

(d) Seller shall limit the Facility's actual rate of delivery into the PGandE system to _______ kW.

(e) The primary energy source for the Facility is ______________________________________.

ARTICLE 3 PURCHASE PRICE

(a) PGandE shall pay Seller for firm capacity at the contract capacity price under Option _____ set forth in Section C-5 of Appendix C and for energy at prices equal to PGandE's full short run avoided operating costs as approved by the CPUC. PGandE's obligation to pay for the contract capacity shall begin on the actual operation date. The contract capacity price shall be subject to adjustment as provided for in Appendix D.

(b) The contract capacity prices in Table C of Appendix C are applicable for deliveries of capacity beginning after December 30, 1982.

---

1 Underlining identifies those terms which are defined in Section A-1 of Appendix A.

2 Insert either "net energy output" or "surplus energy output" to show the energy sale option selected by Seller.

ARTICLE 4 NOTICES

All written notices shall be directed as follows:

to PGandE: Pacific Gas and Electric Company
Attention: Vice President -
Electric Operations
77 Beale Street
San Francisco, CA 94106

to Seller: ____________________
____________________
____________________
____________________
____________________

ARTICLE 5 DESIGNATED SWITCHING CENTER

The designated PGandE switching center shall be unless changed by PGandE:

____________________
(Name)

____________________
(Location)

____________________
(Phone number)

ARTICLE 6 TERMS AND CONDITIONS

This Agreement includes the following appendices which are attached and incorporated by reference:

Appendix A - GENERAL TERMS AND CONDITIONS
Appendix B - INSURANCE
Appendix C - FIRM CAPACITY PRICE SCHEDULE
Appendix D - ADJUSTMENT OF CAPACITY PAYMENTS IN THE EVENT OF TERMINATION OR REDUCTION

ARTICLE 7 TERM OF AGREEMENT

This Agreement shall be binding upon execution and remain in effect for ________years from the actual operation date; provided, however, that it shall terminate if the actual operation date does not occur within five years of the effective date.

IN WITNESS WHEREOF, the Parties hereto have caused this Agreement to be executed by their duly authorized representatives and effective as of the last date set forth below.

| ____________________ (SELLER) | PACIFIC GAS AND ELECTRIC COMPANY |
|---|---|
| BY: ____________________ | BY: ____________________ |
| ____________________ (Type Name) | ____________________ (Type Name) |
| TITLE: ____________________ | TITLE: ____________________ |
| DATE SIGNED: ____________________ | DATE SIGNED: ____________________ |

S.O. #2

# APPENDIX A

## GENERAL TERMS AND CONDITIONS

CONTENTS

APPENDIX A

GENERAL TERMS AND CONDITIONS

A-1. DEFINITIONS

Whenever used in this Agreement, appendices, and attachments hereto, the following terms shall have the following meanings:

Actual operation date - The day following the day during which all features and equipment of the Facility are demonstrated to PGandE's satisfaction to be capable of operating simultaneously to deliver power continuously into PGandE's system as provided in this Agreement.

Adjusted capacity price - The $/kW-year purchase price from Table C, Appendix C for the period of Seller's actual performance.

Capacity sale reduction - A reduction in the amount of capacity provided or to be provided under this Agreement, other than a temporary reduction during probationary periods under Section C-5.

Contract capacity - That capacity identified in Article 2(a) except as otherwise changed as provided herein.

Contract capacity price - The capacity price applicable for the period from the actual operation date through the term of agreement from either the firm capacity price schedule, Table C of Appendix C, or the successor to Table C in effect on the actual operation date, whichever is higher.

Contract termination - The early termination of this Agreement.

CPUC - The Public Utilities Commission of the State of California.

Current firm capacity price - The $/kW-year capacity price from the firm capacity price schedule published by PGandE at the time notice of termination or reduction of contract capacity is given, for a term equal to the period from the date of termination or reduction to the end of of the term of agreement.

Designated PGandE switching center - That switching center or other PGandE installation identified in Article 5.

Dispatchable - The Facility is operable and can be called upon at any time to increase its deliveries of capacity to any level up to the contract capacity.

Facility - That generation apparatus described in Article 2 and all associated equipment owned, maintained, and operated by Seller.

Firm capacity price schedule - The periodically published schedule of the $/kW-year prices that PGandE offers to pay for capacity. See Table C, Appendix C.

Forced outage - Any outage resulting from a design defect, inadequate construction, operator error or a breakdown of the mechanical or electrical equipment that fully or partially curtails the electrical output of the Facility.

Interconnection facilities - All means required and apparatus installed to interconnect and deliver power from the Facility to the PGandE system including, but not limited to, connection, transformation, switching, metering, communications, and safety equipment, such as equipment required to protect (1) the PGandE system and its customers from faults occurring at the Facility, and (2) the Facility from faults occurring on the PGandE system or on the systems of others to which the PGandE system is directly or indirectly connected. Interconnection facilities also include any necessary additions and reinforcements by PGandE to the PGandE system required as a result of the interconnection of the Facility to the PGandE system.

Net energy output - The Facility's gross output in kilowatt-hours less station use and transformation and transmission losses to the point of delivery into the PGandE system.

Prudent electrical practices - Those practices, methods, and equipment, as changed from time to time, that are commonly used in prudent electrical engineering and operations to design and operate electric equipment lawfully and with safety, dependability, efficiency, and economy.

Scheduled operation date - The day specified in Article 2(c) when the Facility is, by Seller's estimate, expected to produce energy and capacity that will be available for delivery to PGandE.

Special facilities - Those parts of the interconnection facilities furnished by PGandE at Seller's request, metering and data processing equipment, and those additions and reinforcements to the

PGandE system which are needed to accommodate the maximum delivery of energy and capacity from the Facility as provided in this Agreement. All special facilities shall be furnished at Seller's expense pursuant to PGandE's electric Rule No. 21 by a separate agreement.

Station use - Energy used to operate the Facility's auxiliary equipment. The auxiliary equipment includes, but is not limited to, forced and induced draft fans, cooling towers, boiler feed pumps, lubricating oil systems, plant lighting, fuel handling systems, control systems, and sump pumps.

Surplus energy output - The Facility's gross output, in kilowatt-hours, less station use, and any other use by Seller, and transformation and transmission losses to the point of delivery into the PGandE system.

Term of agreement - The period of time during which this Agreement will be in effect, as provided in Article 7.

Voltage level - The voltage at which the Facility interconnects with the PGandE system, measured at the point of delivery.

A-2. CONSTRUCTION

A-2.1 Land Rights

Seller hereby grants to PGandE all necessary rights of way and easements, including adequate and continuing access rights on property of Seller, to install, operate, maintain, replace, and remove the special facilities. Seller agrees to execute such other grants, deeds, or documents as PGandE may require to enable it to record such rights of way and easements. If any part of PGandE's equipment is to be installed on property owned by other than Seller, Seller shall, at its own cost and expense, obtain from the owners thereof all necessary rights of way and easements, in a form satisfactory to PGandE, for the construction, operation, maintenance, and replacement of PGandE's equipment upon such property. If Seller is unable to obtain such rights of way and easements, Seller shall reimburse PGandE for all costs incurred by PGandE in obtaining them. PGandE shall at all times have the right of ingress to and egress from the Facility at all reasonable hours for any purposes reasonably connected with this Agreement or the exercise of any and all rights secured to PGandE by law or its tariff schedules.

A-2.2 Design, Construction, Ownership, and Maintenance

(a) Seller shall design, construct, install, own, operate, and maintain all interconnection facilities, except special facilities, to the point of interconnection with the PGandE system as required for PGandE to receive firm capacity and energy from the Facility. The Facility and interconnection facilities shall meet all requirements of applicable codes and all standards of prudent electrical practices and shall be managed in a safe and prudent manner.

(b) Seller shall submit to PGandE all specifications for the interconnection facilities and, at PGandE's option, the Facility, for review and written acceptance prior to their release for construction purposes. PGandE's review and acceptance of these specifications shall not be construed as confirming or endorsing the design or as warranting their safety, durability, or reliability. PGandE shall not, by reason of such review or lack of review, be responsible for strength, details of design, adequacy, or capacity of equipment built pursuant to such specifications, nor shall PGandE's acceptance be deemed to be an endorsement of any of such equipment. Seller shall change the interconnection facilities as may be reasonably required by PGandE to meet changing requirements of the PGandE system.

(c) In the event it is necessary for PGandE to install interconnection facilities for the purposes of this Agreement, they shall be installed as special facilities.

A-2.3 Meter Installation

(a) PGandE shall specify, provide, install, own, operate, and maintain as special facilities all metering and data processing equipment for the registration and recording of energy and other related parameters which are required for the reporting of data to PGandE and for computing the payment due Seller from PGandE.

(b) Seller shall provide, construct, install, own, and maintain at Seller's expense all that is required to accommodate the metering and data processing equipment, such as, but not limited to, metal-clad switchgear, switchboards, cubicles, metering panels, enclosures, conduits, rack structures, and equipment mounting pads.

A-3. ENERGY SALE OPTIONS

A-3.1 General

Seller has two energy sale options, net energy output or surplus energy output. Seller has made its initial selection in Article 2(a).

A-3.2 Energy Sale Conversion

(a) Seller is entitled to convert from one option to the other 12 months after execution of this Agreement, and thereafter at least 12 months after the effective date of the most recent conversion, subject to the following conditions:

(1) Seller shall provide PGandE with a written request to convert its energy sale option.

(2) Seller shall comply with all applicable tariffs on file with the CPUC and contracts in effect between the Parties at the time of conversion covering the existing and proposed (i) facilities used to serve Seller's premises and (ii) interconnection facilities.

(3) Seller shall install and operate equipment required by PGandE to prevent PGandE from serving any part of Seller's load which is served by the Facility and not under contract for PGandE standby service. At Seller's request PGandE shall provide this equipment as special facilities.

(4) If the energy sale conversion results in a capacity sale reduction, the provisions in Appendix D shall apply.

(b) PGandE shall not be required to remove or reserve capacity of facilities made idle by Seller's energy sale conversion and may dedicate such facilities at any time to serve other customers or to interconnect with other electric power sources.

(c) PGandE shall process requests for conversion in the order received. The effective date of conversion shall depend on the completion of the changes required to accommodate Seller's energy sale conversion.

A-4. OPERATION

A-4.1 Inspection and Approval

Seller shall not operate the Facility in parallel with PGandE's system until an authorized PGandE representative has inspected the interconnection facilities, and PGandE has given written approval to begin parallel operation.

A-4.2 Facility Operation and Maintenance

Seller shall operate and maintain its Facility according to prudent electrical practices, applicable laws, orders, rules, and tariffs and shall provide such reactive power support as may be reasonably required by PGandE to maintain system voltage level and power factor. Seller shall operate the Facility at the power factors or voltage levels prescribed by PGandE's system dispatcher or designated representative. If Seller fails to provide reactive power support, PGandE may do so at Seller's expense.

A-4.3 Point of Delivery

Seller shall deliver the energy at the point where Seller's electrical conductors contact PGandE's existing system or at such other point as the Parties may agree in writing.

A-4.4 Operating Communications

(a) Seller shall maintain operating communications with the designated PGandE switching center. The operating communications shall include, but not be limited to, system paralleling or separation, scheduled and unscheduled shutdowns, equipment clearances, levels of operating voltage or power factors and daily capacity and generation reports.

(b) Seller shall keep a daily operations log for each generating unit which shall include information on unit availability, maintenance outages, circuit breaker trip operations requiring a manual reset, and any significant events related to the operation of the Facility.

(c) If Seller makes deliveries greater than one megawatt, Seller shall measure and register on a graphic recording device power in kW and voltage in kV at a location within the Facility agreed to by both Parties.

(d) If Seller makes deliveries greater than one and up to and including ten megawatts, Seller shall report to the designated PGandE switching center, twice a day at agreed upon times for the current day's operation, the hourly readings in kW of capacity delivered and the energy in kWh delivered since the last report.

(e) If Seller makes deliveries of greater than ten megawatts, Seller shall telemeter the delivered capacity and energy information, including real power in kW, reactive power in kVAR, and energy in kWh to a switching center selected by PGandE. PGandE may also require Seller to telemeter transmission kW, kVAR, and kV data depending on the number of generators and transmission configuration. Seller shall provide and maintain the data circuits required for telemetering. When telemetering is inoperative, Seller shall report daily the capacity delivered each hour and the energy delivered each day to the designated PGandE switching center.

(f) If Seller provides dispatchable capacity greater than ten megawatts pursuant to Option 1 in Section C-5 of Appendix C, Seller may be required by PGandE to provide telemetering and control equipment to allow the Facility to respond to system load frequency requirements on digital control from PGandE.

A-4.5 Meter Testing and Inspection

(a) All meters used to provide data for the computation of the payments due Seller from PGandE shall be sealed, and the seals shall be broken only by PGandE when the meters are to be inspected, tested, or adjusted.

(b) PGandE shall inspect and test all meters upon their installation and annually thereafter. At Seller's request and expense, PGandE shall inspect or test a meter more frequently. PGandE shall give reasonable notice to Seller of the time when any inspection or test shall take place, and Seller may have representatives present at the test or inspection. If a meter is found to be inaccurate or defective, PGandE shall adjust, repair, or replace it at its expense in order to provide accurate metering.

A-4.6 Adjustments to Meter Measurements

If a meter fails to register, or if the measurement made by a meter during a test varies by more than two percent from the measurement made by the standard meter used in the test, an adjustment shall be made correcting all measurements made by the inaccurate meter for -- (1) the actual period during which inaccurate measurements were made, if the period can be determined, or if not, (2) the period immediately preceding the test of the meter equal to one-half the time from the date of the last previous test of the meter, provided that the period covered by the correction shall not exceed six months.

A-5. PAYMENT

PGandE shall mail to Seller not later than 30 days after the end of each monthly billing period (1) a statement showing the capacity and energy delivered to PGandE during on-peak, partial-peak, and off-peak periods during the monthly billing period, (2) PGandE's computation of the amount due Seller, and (3) PGandE's check in payment of said amount. If within 30 days of receipt of the statement Seller does not make a report in writing to PGandE of an error, Seller shall be deemed to have waived any error in PGandE's statement, computation, and payment, and they shall be considered correct and complete.

A-6. ADJUSTMENTS OF PAYMENTS

(a) In the event adjustments to payments are required as a result of inaccurate meters, PGandE shall use the corrected measurements described in Section A-4.6 to recompute the amount due from PGandE to Seller for the firm capacity and energy delivered under this Agreement during the period of inaccuracy.

(b) The additional payment to Seller or refund to PGandE shall be made within 30 days of notification of the owing Party of the amount due.

A-7. ACCESS TO RECORDS

Each Party, after giving reasonable written notice to the other Party, shall have the right of access to all metering and related records including operations logs of the Facility.

A-8. CURTAILMENT OF DELIVERIES AND HYDRO SPILL CONDITIONS

(a) PGandE shall not be obligated to accept or pay for and may require Seller to interrupt or reduce deliveries of energy (1) when necessary in order to construct, install, maintain, repair, replace, remove, investigate, or inspect any of its equipment or any part of its system, or (2) if it determines that interruption or reduction is necessary because of emergencies, forced outages, force majeure, or compliance with prudent electrical practices.

(b) In anticipation of a period of hydro spill conditions, as defined by the CPUC, PGandE may notify Seller that any purchases of energy from Seller during such period shall be at hydro savings prices quoted by PGandE. If Seller delivers energy to PGandE during any such period, Seller shall be paid hydro savings prices for those deliveries in lieu of prices which would otherwise be applicable. The hydro savings prices shall be calculated by PGandE using the following formula:

$$\frac{AQF - S}{AQF} \times PP$$

where: AQF = Energy, in kWh, projected to be available during hydro spill conditions from all qualifying facilities under agreements containing hydro savings price provisions.

S = Potential energy, in kWh, from PGandE hydro facilities which will be spilled if all AQF is delivered to PGandE.

PP = Prices published by PGandE for purchases during other than hydro spill conditions.

(c) PGandE shall not be obligated to accept or pay for and may require Seller with a Facility with a nameplate rating of one megawatt or greater to interrupt or reduce deliveries of energy during periods when purchases under this Agreement would result in costs greater than those which PGandE would incur if it did not make such purchases but instead generated an equivalent amount of energy itself.

(d) Whenever possible, PGandE shall give Seller reasonable notice of the possibility that interruption or reduction of deliveries under subsections (a) or (c), above, may be required. PGandE shall give Seller notice of general periods when hydro spill conditions are anticipated, and shall give Seller as much advance notice as practical of any specific hydro spill period and the hydro savings price which will be applicable during such period. Before interrupting or reducing deliveries under subsection (c), above, and before invoking hydro savings prices under subsection (b), above, PGandE shall take reasonable steps to make economy sales of the surplus energy giving rise to the condition. If such economy sales are made, while the surplus energy condition exists Seller shall be paid at the economy sales price obtained by PGandE in lieu of the otherwise applicable prices.

(e) If Seller is selling net energy output to PGandE and simultaneously purchasing its electrical needs from PGandE, energy curtailed pursuant to subsections (b) or (c) above shall not be used by Seller to meet its electrical needs. When Seller elects not to sell energy to PGandE at the hydro savings price pursuant to subsection (b) or when PGandE curtails deliveries of energy pursuant to subsection (c), Seller shall continue to purchase all its electrical needs from PGandE. If Seller is selling surplus energy output to PGandE, subsections (b) or (c) shall only apply to the surplus energy output being delivered to PGandE, and Seller can continue to internally use that generation it has retained for its own use.

A-9. FORCE MAJEURE

(a) The term force majeure as used herein means unforeseeable causes, other than forced outages, beyond the reasonable control of and without the fault or negligence of the Party claiming force majeure, including but not limited to, acts of God, labor disputes, sudden actions of the elements, and actions by federal, state, municipal, or any other government agency.

(b) If either Party because of force majeure is rendered wholly or partly unable to perform its obligations under this Agreement, that Party shall be excused from whatever performance is affected by the force majeure to the extent so affected provided that:

(1) the non-performing Party, within two weeks after the occurrence of the force majeure, gives the other Party written notice describing the particulars of the occurrence,

(2) the suspension of performance is of no greater scope and of no longer duration than is required by the force majeure,

(3) the non-performing Party uses its best efforts to remedy its inability to perform (this subsection shall not require the settlement of any strike, walkout, lockout or other labor dispute on terms which, in the sole judgment of the Party involved in the dispute, are contrary to its interest. It is understood and agreed that the settlement of strikes, walkouts, lockouts or other labor disputes shall be at the sole discretion of the Party having the difficulty),

(4) when the non-performing Party is able to resume performance of its obligations under this Agreement, that Party shall give the other Party written notice to that effect, and

(5) capacity payments during such periods of force majeure on Seller's part shall be governed by Section C-2(c) of Appendix C.

A-10. INDEMNITY

Each Party as indemnitor shall save harmless and indemnify the other Party and the directors, officers, and employees of such other Party against and from any and all loss and liability for injuries to persons including employees of either Party, and property damages including property of either Party resulting from or arising out of (1) the engineering, design, construction, maintenance, or operation of, or (2) the making of replacements, additions, or betterments to, the indemnitor's facilities. This indemnity and save harmless provision shall apply notwithstanding the active or passive negligence of the indemnitee. Neither Party shall be indemnified hereunder for its liability or loss resulting from its sole negligence or willful misconduct. The indemnitor shall, on the other Party's request, defend any suit asserting a claim covered by this indemnity and shall pay all costs, including reasonable attorney fees, that may be incurred by the other Party in enforcing this indemnity.

A-11. LIABILITY; DEDICATION

(a) Nothing in this Agreement shall create any duty to, any standard of care with reference to, or any liability to any person not a Party to it. Neither Party shall be liable to the other Party for consequential damages.

(b) Each Party shall be responsible for protecting its facilities from possible damage by reason of electrical disturbances or faults caused by the operation, faulty operation, or nonoperation of the other Party's facilities, and such other Party shall not be liable for any such damages so caused.

(c) No undertaking by one Party to the other under any provision of this Agreement shall constitute the dedication of that Party's system or any portion thereof to the other Party or to the public or affect the status of PGandE as an independent public utility corporation or Seller as an independent individual or entity and not a public utility.

A-12. SEVERAL OBLIGATIONS

Except where specifically stated in this Agreement to be otherwise, the duties, obligations, and liabilities of the Parties are intended to be several and not joint or collective. Nothing contained in this Agreement shall ever be construed to create an association, trust, partnership, or joint venture or impose a trust or partnership duty, obligation, or liability on or with regard to either Party. Each Party shall be liable individually and severally for its own obligations under this Agreement.

A-13. NON-WAIVER

Failure to enforce any right or obligation by either Party with respect to any matter arising in connection with this Agreement shall not constitute a waiver as to that matter or any other matter.

A-14. ASSIGNMENT

Neither Party shall voluntarily assign its rights nor delegate its duties under this Agreement, or any part of such rights or duties, without the written consent of the other Party, except in connection with the sale or merger of a substantial portion of its properties. Any such assignment or delegation made without such written consent shall be null and void. Consent for assignment shall not be withheld unreasonably. Such assignment shall include, unless otherwise specified therein, all of Seller's rights to any refunds which might become due under this Agreement.

A-15. CAPTIONS

All indexes, titles, subject headings, section titles, and similar items are provided for the purpose of reference and convenience and are not intended to affect the meaning of the contents or scope of this Agreement.

A-16. CHOICE OF LAWS

This Agreement shall be interpreted in accordance with the laws of the State of California, excluding any choice of law rules which may direct the application of the laws of another jurisdiction.

A-17. GOVERNMENTAL JURISDICTION AND AUTHORIZATION

Seller warrants that the Facility and interconnection facilities, except special facilities, shall at all times conform to all applicable laws and regulations, and Seller shall obtain any governmental authorizations and permits required for the construction and operation thereof. If at any time Seller does not hold such authorizations and permits PGandE may refuse to accept deliveries of power hereunder.

A-18. NOTICES

Any notice, demand, or request required or permitted to be given by either Party to the other, and any instrument required or permitted to be tendered or delivered by either Party to the other, shall be in writing and so given, tendered, or delivered, as the case may be, by depositing the same in any United States Post Office with postage prepaid for transmission by certified mail, return receipt requested, addressed to the Party, or personally delivered to the Party, at the address in Article 4 of this Agreement. Changes in such designation may be made by notice similarly given.

## APPENDIX B

## INSURANCE

B-1. WORKERS' COMPENSATION

Seller shall furnish PGandE a certificate of workers' compensation or self-insurance indicating compliance with the Labor Code of California, including Employer's Liability insurance with a minimum of $2,000,000 for injury or death of any one person. This certificate shall provide for 30-day's written notice to PGandE prior to cancellation, termination, alteration, or material change of such insurance.

B-2. COMPREHENSIVE GENERAL AND COMPREHENSIVE AUTOMOBILE LIABILITY COVERAGE

(a) Seller shall maintain during the performance hereof, Comprehensive General Liability and Comprehensive Automobile Liability of not less than $3,000,000 combined single limit or equivalent for bodily injury, personal injury, and property damage as the result of any one occurrence.

(b) Comprehensive General Liability shall include coverage for Premises-Operations, Owners and Contractors Protective, Products/Completed Operations Hazard, Explosion, Collapse, Underground, Contractual Liability, and Broad Form Property Damage including Completed Operations. Comprehensive Automobile Liability shall include coverage for Owned, Hired, and Non-Owned automobiles.

(c) Such insurance, by endorsement to the policy(ies), shall include PGandE as an additional insured insofar as work performed by Seller for PGandE is concerned, shall contain a severability of interest clause, shall provide that PGandE shall not by reason of its inclusion as an additional insured incur liability to the insurance carrier for payment of premium for such insurance, and shall provide for 30-day's written notice to PGandE prior to cancellation, termination, alteration, or material change of such insurance.

B-3. ADDITIONAL INSURANCE PROVISIONS

(a) Evidence of coverage described above in Sections B-1 and B-2 shall state that coverage provided is primary and is not excess to or contributing with any insurance or self-insurance maintained by PGandE.

(b) PGandE shall have the right to inspect or obtain a copy of the original policy(ies) of insurance.

(c) Seller shall furnish the required certificates and endorsements to PGandE prior to the date construction of the Facility begins or the effective date of this Agreement, whichever is later.

(d) All insurance certificates, endorsements, cancellations, terminations, alterations, and material changes of such insurance shall be issued and submitted to the following:

PACIFIC GAS AND ELECTRIC COMPANY
Attention: Manager - Insurance Department
77 Beale Street
San Francisco, CA 94106

APPENDIX C

FIRM CAPACITY PRICE SCHEDULE

CONTENTS

## APPENDIX C

## FIRM CAPACITY PRICE SCHEDULE

### C-1 GENERAL

This Appendix C establishes conditions and prices under which PGandE shall pay for firm capacity.

### C-2 MINIMUM PERFORMANCE REQUIREMENTS

(a) To receive full capacity payments the Facility must meet the following requirements:

(1) The contract capacity shall be available[1] for all of the on-peak hours[2] in the peak months on the PGandE system, which are presently the months of June, July, and August, subject to a 20 percent allowance for forced outages in any month. Compliance with this provision shall be based on the Facility's total on-peak availability[1] for each of the peak months and shall exclude any energy associated with generation levels greater than the contract capacity.

(2) If Seller selects Option 1, the contract capacity shall be dispatchable throughout the year, subject to (i) a monthly allowance for forced outages of 20% of the hours Seller is called upon to deliver power to PGandE and (ii) the allowances for scheduled maintenance outages. Except during the peak months on the PGandE system, Seller may accumulate and apply the 20 percent allowance for forced outages for any consecutive three month period. Seller shall demonstrate that the Facility is fueled by a reliable fuel supply and adequate fuel storage is available to deliver power as requested by PGandE's system dispatcher.

(b) If Seller is prevented from meeting the minimum performance requirements because of a forced outage on the PGandE system or a condition set forth in Section A-8, PGandE shall continue capacity payments. Under Option 2, capacity payments will be calculated in the same manner used for scheduled maintenance outages.

(c) If Seller is prevented from meeting the minimum performance requirements because of force majeure, PGandE shall continue capacity payments for ninety days from the occurrence of the force majeure. Thereafter, Seller shall be deemed to have failed to have met the minimum performance requirements. Under Option 2, capacity payments will be calculated in the same manner used for scheduled maintenance outages.

(d) If Seller is prevented from meeting the minimum performance requirements because of exteme dry year conditions, PGandE shall continue capacity payments. Extreme dry year conditions are drier than those used to establish contract capacity pursuant to Section C-8. Seller shall warrant to PGandE that the Facility is a hydroelectric facility and that such conditions are the sole cause of Seller's inability to meet its contract capacity obligations. Under Option 1, starting with the month in which Seller cannot provide its contract capacity, payments shall be made under Option 2 for a one-year period, and if at the end of this one-year period Seller is not able to resume the contract capacity due solely to continued extreme dry year conditions, Seller shall continue to receive payments under Option 2 for additional one-year periods as long as such conditions continue to exist.

(e) If Seller is prevented from meeting the minimum performance requirements for reasons other than those described above in Sections C-2(b), (c), or (d):

(1) Seller shall receive the reduced capacity payments as provided in Section C-5 for a probationary period not to exceed 15 months, or as otherwise agreed to by the Parties.

(2) If, at the end of the probationary period Seller has not demonstrated that the Facility can meet the minimum performance requirements, PGandE may derate the contract capacity pursuant to Section C-4(b).

---

[1] For purposes of Option 1, "available" means either dispatchable by PGandE or actually delivered to PGandE. For purposes of Option 2, "available" means actually delivered to PGandE.

[2] On-peak, partial-peak, and off-peak hours are defined in Table A.

C-3 SCHEDULED MAINTENANCE

Outage periods for scheduled maintenance shall not exceed 840 hours (35 days) in any 12-month period. This allowance may be used in increments of an hour or longer on a consecutive or nonconsecutive basis. Seller may accumulate unused maintenance hours from one 12-month period to another up to a maximum of 1,080 hours (45 days). This accrued time must be used consecutively and only for major overhauls. Seller shall provide PGandE with the following advance notices: 24 hours for scheduled outages less than one day, one week for a scheduled outage of one day or more (except for major overhauls), and six months for a major overhaul. Seller shall not schedule major overhauls during the peak months (presently June, July and August). Seller shall make reasonable efforts to schedule or reschedule routine maintenance outside the peak months, and in no event shall outages for scheduled maintenance exceed 30 peak hours during the peak months. Seller shall confirm in writing to PGandE pursuant to Article 4, within 24 hours of the original notice, all notices Seller gives for scheduled maintenance.

C-4 ADJUSTMENTS TO CONTRACT CAPACITY

(a) Seller may increase the contract capacity with the approval of PGandE and receive payment for the additional capacity thereafter in accordance with the applicable capacity purchase price published by PGandE at the time the increase is first delivered to PGandE.

(b) Seller may reduce the contract capacity at any time by giving notice thereof to PGandE, subject to the provisions of Appendix D if the reduction occurs after the actual operation date. PGandE may reduce the contract capacity as a result of appropriate tests, studies, or prior performance. The amount by which the contract capacity is reduced by PGandE shall be deemed a capacity sale reduction without notice as provided in Section D-3 of Appendix D.

(c) Either Party may request, when it reasonably appears that the capacity of the Facility may have changed for any reason, that a new contract capacity be determined.

C-5 PAYMENT OPTIONS

Seller has two options for calculation of capacity payments and Seller has made its selection in Article 3(a). As used below in this section, month refers to a calendar month. The two options are as follows:

Option 1

When Seller meets the requirements of Section C-2 the monthly payment for capacity will be one-twelfth of the product of the contract capacity price, the contract capacity, the appropriate capacity loss adjustment factor from Table B based on the Facility's interconnection voltage, and the appropriate performance bonus factor, if any, from Table D. Capacity payments will continue during scheduled maintenance outages provided that the provisions of Section C-3 are met.

During a probationary period Seller's monthly payment for capacity shall be determined by substituting for the contract capacity, the capacity at which Seller would have met the minimum performance requirements. In any month during the probationary period that Seller does not meet the minimum performance requirements at whatever capacity was determined for the previous month, Seller's monthly payment for capacity shall be determined by substituting the capacity at which Seller would have met the minimum performance requirements. The performance bonus factor shall not be applied during a probationary period.

Option 2

The monthly payment for capacity will be the product of the Period Price Factor (PPF), the Monthly Delivered Capacity (MDC), the appropriate capacity loss adjustment factor from Table B based on the Facility's interconnection voltage, and the appropriate performance bonus factor, if any, from Table D, plus any allowable payment for outages due to scheduled maintenance. Firm capacity prices shall be applied to meter readings taken during the separate times and periods as illustrated in Table A of this Appendix.

The PPF is determined by multiplying the contract capacity price by the following Option 2 Allocation Factors[1]:

| | Option 2 Allocation Factor | x | Contract Capacity Price | = | PPF ($/kW-month) |
|---|---|---|---|---|---|
| Seasonal Period A | .16479 | | ____ | | ____ |
| Seasonal Period B | .02515 | | ____ | | ____ |

The MDC is determined in the following manner:

(1) Determine the Performance Factor (P), which is defined as the lesser of 1.0 or the following quantity:

$$P = \frac{A}{C \times (B-S) \times (0.8^2)} \qquad (\leq 1.0)$$

Where:

A = Total kilowatt-hours delivered during all on-peak and partial-peak hours excluding any energy associated with generation levels greater than the contract capacity.

C = Contract capacity in kilowatts.

B = Total on-peak and partial-peak hours during the month.

S = Total on-peak and partial-peak hours during the month Facility is out of service on scheduled maintenance.

(2) Determine the Monthly Capacity Factor (MCF), which is computed using the following expression:

$$MCF = P \times (1.0 - \frac{M}{D})$$

Where:

M = The number of hours during the month Facility is out of service on scheduled maintenance.

D = The number of hours in the month.

(3) Determine the MDC by multiplying the MCF by C:

MDC (kilowatts) = MCF x C

The monthly payment for capacity is then determined by multiplying the PPF by the MDC, by the appropriate capacity loss adjustment factor presented from Table B, and by the appropriate performance bonus factor, if any, from Table D.

monthly payment for capacity = PPF x MDC x capacity loss adjustment factor x performance bonus factor

Furthermore, the payment for a month in which there is an outage for scheduled maintenance shall also include an amount equal to the product of the average hourly capacity payment[3] for the most recent month in the same type of Seasonal Period (i.e., Seasonal Period A or Seasonal Period B) during which deliveries were made times the number of hours of outage for scheduled maintenance in the current month. Capacity payments will continue during the outage periods for scheduled maintenance provided that the provisions of Section C-3 are met.

During a probationary period Seller's monthly payment for capacity shall be determined by substituting for the contract capacity, the capacity at which Seller would have met the minimum performance requirements. In the event that during the probationary period Seller does not meet the minimum performance requirements at whatever capacity was established for the previous month, Seller's monthly payment for capacity shall be determined by substituting the capacity at which Seller would have met the minimum performance requirements. The performance bonus factor shall not be applied during probationary periods.

---

[1] All allocation factors are subject to change by PGandE based on PGandE's marginal capacity cost allocation, as determined in general rate case proceedings before the CPUC. Seasonal Periods A and B are defined in Table A of this Appendix.

[2] 0.8 reflects a 20% allowance for forced outage.

[3] Total monthly payment divided by the total number of hours in the monthly billing period.

TABLE A[1]

| | Monday through Friday[2] | Saturdays[2] | Sundays and Holidays |
|---|---|---|---|
| **Seasonal Period A (May 1 through September 30)** | | | |
| On-Peak | 12:30 p.m. to 6:30 p.m. | | |
| Partial-Peak | 8:30 a.m. to 12:30 p.m. | 8:30 a.m. to 10:30 p.m. | |
| | 6:30 p.m. to 10:30 p.m. | | |
| Off-Peak | 10:30 p.m. to 8:30 a.m. | 10:30 p.m. to 8:30 a.m. | All Day |
| **Seasonal Period B (October 1 through April 30)** | | | |
| On-Peak | 4:30 p.m. to 8:30 p.m. | | |
| Partial-Peak | 8:30 p.m. to 10:30 p.m. | 8:30 a.m. to 10:30 p.m. | |
| | 8:30 a.m. to 4:30 p.m. | | |
| Off-Peak | 10:30 p.m. to 8:30 a.m. | 10:30 p.m. to 8:30 a.m. | All Day |

TABLE B

If the Facility is non-remote[3] the capacity loss adjustment factors are as follows:

| Interconnection Voltage | Capacity Loss Adjustment Factor |
|---|---|
| Transmission | .989 |
| Distribution | .991 |

If the Facility is remote the capacity loss adjustment factor is ____________[4].

---

1 This table is subject to change to accord with the on-peak, partial-peak, and off-peak periods as defined in PGandE's own rate schedules for the sale of electricity to its large industrial customers.

2 Except the following holidays: New Year's Day, Washington's Birthday, Memorial Day, Independence Day, Labor Day, Veteran's Day, Thanksgiving, and Christmas, as said days are specified in Public Law 90-363 (5 U.S.C.A. Section 6103(a)).

3 As defined by the CPUC.

4 Determined individually.

S.O. #2

**TABLE C**

**Firm Capacity Price Schedule**

**(Levelized $/kW-year)**

| Actual Operation Date (Year) | Life of Contract 1 | 2 | 3 | 4 | 5 | 6 | 7 | 8 | 9 | 10 | 11 | 12 | 13 | 14 | 15 | 20 | 25 | 30 |
|---|---|---|---|---|---|---|---|---|---|---|---|---|---|---|---|---|---|---|
| 1982 | 65 | 68 | 70 | 72 | 75 | 77 | 79 | 81 | 84 | 86 | 88 | 90 | 91 | 93 | 95 | 103 | 109 | 113 |
| 1983 | 70 | 73 | 75 | 78 | 80 | 83 | 85 | 88 | 90 | 92 | 94 | 96 | 98 | 100 | 102 | 110 | 117 | 122 |
| 1984 | 76 | 78 | 81 | 84 | 86 | 89 | 92 | 94 | 97 | 99 | 101 | 103 | 106 | 108 | 110 | 118 | 125 | 130 |
| 1985 | 81 | 84 | 87 | 90 | 93 | 96 | 99 | 101 | 104 | 106 | 109 | 111 | 113 | 115 | 118 | 127 | 134 | 140 |
| 1986 | 88 | 91 | 94 | 97 | 100 | 103 | 106 | 109 | 112 | 114 | 117 | 119 | 122 | 124 | 126 | 136 | 144 | 150 |
| 1987 | 95 | 98 | 101 | 105 | 108 | 111 | 114 | 117 | 120 | 123 | 125 | 128 | 130 | 133 | 135 | 146 | 154 | 160 |

**TABLE D**

**Performance Bonus Factor**

The following shall be the performance bonus factors applicable to the calculation of the monthly payments for capacity delivered by the Facility after it has demonstrated a capacity factor in excess of 85%.

| DEMONSTRATED CAPACITY FACTOR (%) | PERFORMANCE BONUS FACTOR |
|---|---|
| 85 | 1.000 |
| 90 | 1.059 |
| 95 | 1.118 |
| 100 | 1.176 |

After the Facility has delivered power during the span of all of the peak months on the PGandE system (presently June, July, and August) in any year (span),

(i) the capacity factor for each such month shall be calculated in the following manner:

$$\text{CAPACITY FACTOR (\%)} = \frac{A}{B \times C} \times 100$$

Where:

A = Total kilowatt-hours delivered by Seller in any peak month during all on-peak hours excluding any energy associated with generation levels greater than the contract capacity.

C = Contract capacity in kilowatts.

B = Total on-peak hours during the month.

(ii) the arithmetic average of the above capacity factors shall be determined for that span,

(iii) the average of the above arithmetic average capacity factors for the most recent span(s), not to exceed 5, shall be calculated and shall become the Demonstrated Capacity Factor.

To calculate the performance bonus factor for a Demonstrated Capacity Factor not shown in Table D use the following formula:

$$\text{Performance Bonus Factor} = \frac{\text{Demonstrated Capacity Factor (\%)}}{85\%}$$

THE FOLLOWING SECTIONS SHALL APPLY ONLY TO HYDROELECTRIC PROJECTS

C-6 DETERMINATION OF NATURAL FLOW DATA

Natural flow data shall be based on a period of record of at least 50 years and which includes historic critically dry periods. In the event Seller demonstrates that a natural flow data base of at least 50 years would be unreasonably burdensome, PGandE shall accept a shorter period of record with a corresponding reduction in the averaging basis set forth in Section C-8. Seller shall determine the natural flow data by month by using one of the following methods:

Method 1

If stream flow records are available from a recognized gauging station on the water course being developed in the general vicinity of the project, Seller may use the data from them directly.

Method 2

If directly applicable flow records are not available, Seller may develop theoretical natural flows based on correlation with available flow data for the closest adjacent and similar area which has a recognized gauging station using generally accepted hydrologic estimating methods.

C-7 THEORETICAL OPERATION STUDY

Based on the monthly natural flow data developed under Section C-6 a theoretical operation study shall be prepared by Seller. Such a study shall identify the monthly capacity rating in kW and the monthly energy production in kWh for each month of each year. The study shall take into account all relevant operating constraints, limitations, and requirements including but not limited to --

(1) Release requirements for support of fish life and any other operating constraints imposed on the project;

(2) Operating characteristics of the proposed equipment of the Facility such as efficiencies, minimum and maximum operating levels, project control procedures, etc.;

(3) The design characteristics of project facilities such as head losses in penstocks, valves, tailwater elevation levels, etc.; and

(4) Release requirements for purposes other than power generation such as irrigation, domestic water supply, etc.

The theoretical operation study for each month shall assume an even distribution of generation throughout the month unless Seller can demonstrate that the Facility has water storage characteristics. For the study to show monthly capacity ratings, the Facility shall be capable of operating during all on-peak hours in the peak months on the PGandE system, which are presently the months of June, July, and August. If the project does not have this capability throughout each such month, the capacity rating in that month of that year shall be set at zero for purposes of this theoretical operation study.

C-8 DETERMINATION OF AVERAGE DRY YEAR CAPACITY RATINGS

Based on the results of the theoretical operation study developed under Section C-7, the average dry year capacity rating shall be established for each month. The average dry year shall be based on the average of the five years of the lowest annual generation as shown in the theoretical operation study. Once such years of lowest annual generation are identified, the monthly capacity rating is determined for each month by averaging the capacity ratings from each month of those years. The contract capacity shown in Article 2(a) shall not exceed the lowest average dry year monthly capacity ratings for the peak months on the PGandE system, which are presently the months of June, July, and August.

C-9 INFORMATION REQUIREMENTS

Seller shall provide the following information to PGandE for its review:

(1) A summary of the average dry year capacity ratings based on the theoretical operation study as provided in Table E;

(2) A topographic project map which shows the location of all aspects of the Facility and locations of stream gauging stations used to determine natural flow data;

(3) A discussion of all major factors relevant to project operation;

(4) A discussion of the methods and procedures used to establish the natural flow data. This discussion shall be in sufficient detail for PGandE to determine that the methods are consistent with those outlined in Section C-6 and are consistent with generally accepted engineering practices; and

(5) Upon specific written request by PGandE, Seller's theoretical operation study.

C-10 ILLUSTRATIVE EXAMPLE

(1) Determine natural flows - These flows are developed based on historic stream gauging records and are compiled by month, for a long-term period (normally at least 50 years or more) which covers dry periods which historically occurred in the 1920's and 30's and more recently in 1976 and 77. In all but unusual situations this will require application of hydrological engineering methods to records that are available, primarily from the USGS publication "Water Resources Data for California".

(2) Perform Theoretical Operation Study - Using the natural flow data compiled under (1) above a theoretical operation study is prepared which determines, for each month of each year, energy generation (kWh) and capacity rating (kW). This study is performed based on the Facility's design, operating capabilities, constraints, etc., and should take into account all factors relevant to project operation. Generally such a study is done by computer which routes the natural flows through project features, considering additions and withdrawals from storage, spill past the project, releases for support of fish life, etc., to determine flow available for generation. Then the generation and capacity amounts are computed based on equipment performance, efficiencies, etc.

(3) Determine Average Dry Year Capacity Ratings - After the theoretical project operation study is complete the five years in which the annual generation (kWh) would have been the lowest are identified. Then for each month, the capacity rating (kW) is averaged for the five years to arrive at a monthly average capacity rating. The contract capacity is then set by the Seller based on the monthly average dry year capacity ratings and the minimum performance requirements of Appendix C. An example project is shown in the attached completed Table E.

EXAMPLE

TABLE E

Summary of Theoretical Operation Study

Project: New Creek 1 Dispatchable: Yes ___ No X

Water Source: West Fork New Creek

Mode of Operation: Run of the river

Type of Turbine: Francis Design Flow: 100 cfs Design Head: 150 feet

Operating Characteristics:[1]

| | Flow (cfs) | Head (feet) Gross | Head (feet) Net | Output (kW) | Efficiency (%) Turbine | Efficiency (%) Generator |
|---|---|---|---|---|---|---|
| Normal Operation | 100 | 160 | 150 | 1,120 | 90 | 98 |
| Maximum Operation | 110 | 160 | 148 | 1,150 | 85 | 98 |
| Minimum Operation | 30 | 160 | 155 | 290 | 75 | 98 |

Average Dry Year Operation - Based on the average of the following lowest generation years: 1930, 1932, 1934, 1949, 1977.

| Month | Energy Generation (kWh) | Capacity Output (kW) | Percent of Total Hours Operated[2] |
|---|---|---|---|
| January | 855,000 | 1,150 | 100 |
| February | 753,000 | 1,120 | 100 |
| March | 818,000 | 1,100 | 100 |
| April | 727,000 | 1,010 | 100 |
| May | 699,000 | 940 | 100 |
| June | 612,000 | 850 | 100 |
| July | 484,000 | 650 | 100 |
| August | 305,000 | 410 | 100 |
| September | 245,000 | 340 | 100 |
| October | 148,800 | 200 | 100 |
| November | 468,000 | 650 | 100 |
| December | 595,000 | 800 | 100 |

Maximum Contract Capacity: 410 kW

1 If Facility has a variable head, operating curves should be provided.

2 For this to be less than 100%, Facility must be dispatchable.

# APPENDIX D

## ADJUSTMENT OF CAPACITY PAYMENTS IN THE EVENT OF TERMINATION OR REDUCTION

CONTENTS

## APPENDIX D

## ADJUSTMENT OF CAPACITY PAYMENTS IN THE EVENT OF TERMINATION OR REDUCTION

D-1 GENERAL PROVISIONS

(a) This Appendix shall be applicable in the event there is a contract termination or a capacity sale reduction (each sometimes referred to as "termination" in this Appendix D).

(b) The Parties agree that the amount which PGandE pays Seller for the capacity which Seller makes available to PGandE is based on the agreed value to PGandE of Seller's performance of capacity obligations during the full period of the term of agreement. The Parties further agree that in the event PGandE does not receive such full performance by reason of a termination:

(1) PGandE shall be deemed damaged by reason thereof,

(2) it would be impracticable or extremely difficult to fix the actual damages to PGandE resulting therefrom,

(3) the refunds and payments as provided in Sections D-2 and D-3, as applicable, are in the nature of adjustments in capacity prices and liquidated damages, and not a penalty, and are fair and reasonable, and

(4) such refunds and payments represent a reasonable endeavor by the Parties to estimate a fair compensation for the reasonable losses that would result from such termination or reduction.

(c) In the event of a capacity sale reduction, the quantity by which the contract capacity is reduced shall be used to calculate the payments due PGandE in accordance with Sections D-2 and D-3, as applicable.

(d) Seller shall be invoiced by PGandE for all refunds and payments due under this Appendix D and the special facilities agreement. From the date of the notice of termination or the date of termination, whichever is earlier, Seller shall pay interest, compounded monthly, on all overdue amounts, at the published Federal Reserve Board three months Prime Commercial Paper rate.

(e) If Seller does not make payments pursuant to Section D-1(d), PGandE shall have the right to offset any amounts due it against any present or future payments due Seller.

(f) Notices of termination shall be made in accordance with Section A-18 of Appendix A.

D-2 TERMINATION WITH PRESCRIBED NOTICE

In the event Seller terminates this entire Agreement, or all or part of the contract capacity thereof, with the following prescribed written notice:

| Amount of Contract Capacity Terminated | Length of Notice Required |
|---|---|
| 1,000 kW or under | 3 months |
| over 1,000 kW through 10,000 kW | 9 months |
| over 10,000 kW through 25,000 kW | 12 months |
| over 25,000 kW through 50,000 kW | 36 months |
| over 50,000 kW through 100,000 kW | 48 months |
| over 100,000 kW | 60 months |

Then the following provisions shall apply:

(1) With respect to the amount by which the contract capacity is reduced, Seller shall refund to PGandE an amount equal to the difference between (a) the capacity payments already paid by PGandE, based on the original term of agreement and (b) the total capacity payments which PGandE would have paid based on the period of Seller's actual performance using the adjusted capacity price. Additionally, Seller shall pay interest, compounded monthly, on all overpayments, at the published Federal Reserve Board three months' Prime Commercial Paper rate.

(2) From the date PGandE receives the termination notice to the date of actual termination, PGandE shall make capacity payments based on the adjusted capacity price for the amount of contract capacity being terminated.

(3) From the date PGandE receives the termination notice, PGandE shall continue to pay for the amount of contract capacity not being terminated, if any, at the original contract capacity price.

D-3 TERMINATION WITHOUT PRESCRIBED NOTICE

(a) If Seller terminates this Agreement, or all or a part of the contract capacity thereof, without the notice prescribed in Section D-2, the provisions prescribed in Section D-2 will all apply. Additionally:

(b) Seller shall pay PGandE a sum equal to the amount by which the contract capacity is being terminated times the difference between the current firm capacity price on the date of termination for a term equal to the balance of the term of agreement and the contract capacity price, pro-rated for the length of notice given by multiplying by the difference between the prescribed length of notice and the actual notice given, with the difference divided by 12. In the event that the current firm capacity price is less than the contract capacity price, no payment under this Section D-3 shall be due either Party.

This additional payment shall be computed using the following formula:

$$G = CC \times (T - CCP) \times \frac{J - H}{12}$$

where $G \geqq 0$

and where:

| | | |
|---|---|---|
| G | = | additional payment. |
| CC | = | the amount by which the contract capacity is being terminated. |
| T | = | the current firm capacity price. |
| CCP | = | the contract capacity price. |
| H | = | the actual number of months notice given. |
| J | = | the prescribed length of notice. |

D-4. TERMINATION EXAMPLES

These examples demonstrate how to calculate capacity payment adjustments when capacity sales are terminated.

(a) Termination with Prescribed Notice

(1) Example Based on Option 1

Assumptions

i. Term of agreement is 15 years;

ii. Actual operation date is July 1, 1984;

iii. Prescribed notice is given on July 1, 1985;

iv. Contract capacity to be reduced by 10,000 kW on July 1, 1986; actual performance to be from July 1, 1984 through July 1, 1986[1]; and

v. The applicable capacity loss adjustment factor is .989.

vi. No performance bonus for capacity has been earned.

---

1 The capacity payment is adjusted upon receiving notice, so no refund is necessary for the last month of the first twelve months of operation and all of the second twelve months (June 1, 1985 to July 1, 1986). Seller performed for eleven months prior to payment adjustment. (Note that due to the 30-day interval between delivery and payment, performance in the twelfth month (June 1985) can be paid for at the adjusted capacity price.

The amount of overpayment (E) made by PGandE to Seller during each monthly billing period is calculated as follows:

E = (A - B) x C x L x U

Where:

A = contract capacity price per month for the actual operation date (July 1, 1984) and the term of agreement which is 15 years = $110/kW-yr ÷ 12 mo/yr = $9.17/kW-mo.

B = adjusted capacity price per month for the actual operation date (July 1, 1984) and a two-year agreement term = $78/kW-yr ÷ 12 mo/yr = $6.50/kW-mo.

C = amount by which the contract capacity is being reduced = 10,000 kW

L = capacity loss adjustment factor = .989

U = performance bonus factor; when seller does not qualify for a performance bonus factor, as in this example, U is removed from the above calculation of E.

Therefore:

E = ($9.17/kW-mo - $6.50/kW-mo) x 10,000 kW x .989 = $26,406 per month

Table A shows a step-by-step derivation of the refund Seller owes PGandE for the early termination outlined above. The $305,963 that Seller owes PGandE appears at the lower right-hand corner of the table. All other figures of this table represent intermediate calculation steps.

TABLE A

| (a) Monthly Billing Period[1] | (b) Date of Payment[2] | (c) Amount of Overpayment[3] | (d) Accumulated Overpayment[4] | (e) Interest Rate[5] | (f) Interest Charge[6] on Accumulated Overpayment (f)=(d)x(e) | (g) Balance[7] (g)=(c)+(d)+(f) |
|---|---|---|---|---|---|---|
| | | $ | $ | % | $ | $ |
| 7/84 | 8/30/84 | 26,406 | 0 | 1.2 | 0 | 26,406 |
| 8/84 | 9/30/84 | 26,406 | 26,406 | 1.0 | 264 | 53,076 |
| 9/84 | 10/30/84 | 26,406 | 53,076 | 0.9 | 478 | 79,960 |
| 10/84 | 11/30/84 | 26,406 | 79,960 | 0.8 | 640 | 107,006 |
| 11/84 | 12/30/84 | 26,406 | 107,006 | 0.7 | 749 | 134,161 |
| 12/84 | 1/30/85 | 26,406 | 134,161 | 0.8 | 1,073 | 161,640 |
| 1/85 | 3/2/85 | 26,406 | 161,640 | 0.9 | 1,455 | 189,501 |
| 2/85 | 3/30/85 | 26,406 | 189,501 | 1.0 | 1,895 | 217,802 |
| 3/85 | 4/30/85 | 26,406 | 217,802 | 1.1 | 2,396 | 246,604 |
| 4/85 | 5/30/85 | 26,406 | 246,604 | 1.2 | 2,959 | 275,969 |
| 5/85 | 6/30/85 | 26,406 | 275,969 | 1.3 | 3,588 | 305,963 |

[1] The month in which power deliveries were made. For purposes of simplification, the monthly billing period will coincide exactly with each calendar month.

[2] The date on which payment for the monthly billing period stated in column (a) is made.

[3] The amount of overpayment made by PGandE to Seller during each monthly billing period.

[4] The amount of overpayment accumulated up through last month's date of payment.

[5] The interest rate for the period between the date of payment for the previous monthly billing period and the date of payment for this monthly billing period. These interest rates are arbitrarily chosen for use in this example.

[6] The amount of interest charge accrued between the date of payment for the previous monthly billing period and the date of payment for this monthly billing period on the accumulated overpayment balance existing as of the previous monthly billing period's date of payment.

[7] The amount Seller owes PGandE at this stage of the calculation. The balance (g) for a given monthly billing period equals the accumulated overpayment (d) for the monthly billing period immediately following.

S.O. #2

(a) **Termination with Prescribed Notice (cont'd)**

(2) **Example Based on Option 2**

**Assumptions:**

i. Term of agreement is 15 years;

ii. Actual operation date is July 1, 1984;

iii. Prescribed notice is given on July 1, 1985;

iv. Contract capacity is reduced by 10,000 kW on July 1, 1986; actual performance is from July 1, 1984 through July 1, 1986[1];

v. Scheduled outage for maintenance: 18 days in November = 432 hours;

vi. The applicable capacity loss adjustment factor is .989.

vii. Listed below is Seller's Performance Factor (P) and where applicable, the performance bonus factor (U) earned for each of the monthly billing periods[2] prior to the time capacity payment is adjusted. Also listed below are the number of hours the Facility was out of service for scheduled maintenance (M) and the number of hours in the month (D) for each of these months.

| Monthly Billing Period | | P | U | M | D |
|---|---|---|---|---|---|
| July | 1984 | 1.00 | - | 0 | 744 |
| August | 1984 | 0.90 | - | 0 | 744 |
| September | 1984 | 1.00 | - | 0 | 720 |
| October | 1984 | 0.96 | - | 0 | 744 |
| November | 1984 | 0.98 | - | 432 | 720 |
| December | 1984 | 1.00 | - | 0 | 744 |
| January | 1985 | 1.00 | 1.117 | 0 | 744 |
| February | 1985 | 0.92 | - | 0 | 672 |
| March | 1985 | 0.85 | - | 0 | 744 |
| April | 1985 | 0.78 | - | 0 | 720 |
| May | 1985 | 1.00 | 1.059 | 0 | 744 |

The amount of overpayment (E) made by PGandE to Seller during each monthly billing period is calculated as follows:

$$E = [P \times (1 - \frac{M}{D}) \times K \times L \times U \times (A - B) \times C] + [\frac{M}{D} \times R]$$

Where:

P = performance factor.

M = number of hours of scheduled maintenance for that monthly billing period.

D = number of hours in that monthly billing period.

K = allocation factor from Section C-5.

L = capacity loss adjustment factor = .989.

U = performance bonus factor; when seller does not qualify for a performance bonus factor, U is removed from the above calculation of E.

A = contract capacity price for the actual operation date (July 1, 1984) and term of agreement which is 15 years = $110/kW-yr.

---

1 The capacity payment is adjusted upon receiving notice, so no refund is necessary for the last month of the first twelve months of operation and all of the second twelve months (June 1, 1985 to July 1, 1986). Seller performed for eleven months prior to payment adjustment. [Note that due to the 30-day interval between delivery and payment, performance in the twelfth month (June 1985) can be paid for at the adjusted capacity purchase price.]

2 For purposes of simplification, the monthly billing period will coincide exactly with each calendar month.

B = adjusted capacity price for the actual operation date and a two-year agreement term = $78/kW-yr.

C = amount by which the contract capacity is being reduced = 10,000 kW.

R = amount of overpayment for the most recent monthly billing period in the same Seasonal Period (i.e., Seasonal Period A or Seasonal Period B).

The results of the calculations are:

| Monthly Billing Period | | Amount of Overpayment (E) |
|---|---|---|
| July | 1984 | $52,153 |
| August | 1984 | 46,937 |
| September | 1984 | 52,153 |
| October | 1984 | 7,641 |
| November | 1984 | 7,705 |
| December | 1984 | 7,959 |
| January | 1985 | 8,891 |
| February | 1985 | 7,323 |
| March | 1985 | 6,766 |
| April | 1985 | 6,208 |
| May | 1985 | 55,230 |

Table B shows a step-by-step derivation of the refund Seller owes PGandE for the early termination outlined above. The $275,136 that Seller owes PGandE appears at the lower right-hand corner of the table. All other figures of this table represent intermediate calculation steps.

TABLE B

| (a) Monthly Billing Period[1] | (b) Date of Payment[2] | (c) Amount of Overpayment[3] | (d) Accumulated Overpayment[4] | (e) Interest Rate[5] | (f) Interest Charge[6] on Accumulated Overpayment (f)=(d)x(e) | (g) Balance[7] (g)=(c)+(d)+(f) |
|---|---|---|---|---|---|---|
| | | $ | $ | % | $ | $ |
| 7/84 | 8/30/84 | 52,153 | 0 | 1.2 | 0 | 52,153 |
| 8/84 | 9/30/84 | 46,937 | 52,153 | 1.0 | 522 | 99,612 |
| 9/84 | 10/30/84 | 52,153 | 99,612 | 0.9 | 897 | 152,662 |
| 10/84 | 11/30/84 | 7,641 | 152,662 | 0.8 | 1,221 | 161,524 |
| 11/84 | 12/30/84 | 7,705 | 161,524 | 0.7 | 1,131 | 170,360 |
| 12/84 | 1/30/85 | 7,959 | 170,360 | 0.8 | 1,363 | 179,682 |
| 1/85 | 3/2/85 | 8,891 | 179,682 | 0.9 | 1,617 | 190,190 |
| 2/85 | 3/30/85 | 7,323 | 190,190 | 1.0 | 1,902 | 199,415 |
| 3/85 | 4/30/85 | 6,766 | 199,415 | 1.1 | 2,194 | 208,375 |
| 4/85 | 5/30/85 | 6,208 | 208,375 | 1.2 | 2,501 | 217,084 |
| 5/85 | 6/30/85 | 55,230 | 217,084 | 1.3 | 2,822 | 275,136 |

[1] The month in which power deliveries were made. For purposes of simplification, the monthly billing period will coincide exactly with each calendar month.

[2] The date on which payment for the monthly billing period stated in column (a) is made.

[3] The amount of overpayment made by PGandE to Seller during each monthly billing period.

[4] The amount of overpayment accumulated up through last month's date of payment.

[5] The interest rate for the period between the date of payment for the previous monthly billing period and the date of payment for this monthly billing period. These interest rates are arbitrarily chosen for use in this example.

[6] The amount of interest charge accrued between the date of payment for the previous monthly billing period and the date of payment for this monthly billing period on the accumulated overpayment balance existing as of the previous monthly billing period's date of payment.

[7] The amount Seller owes PGandE at this stage of the calculation. The balance (g) for a given monthly billing period equals the accumulated overpayment (d) for the monthly billing period immediately following.

(b) Termination without Prescribed Notice

If Seller terminates without prescribed notice, Seller will owe PGandE a refund [the calculation of which is described in Sections D-4(a)(1) and D-4(a)(2) of this example] and payment (G). This example demonstrates how the payment (G) is calculated.

**Assumptions**

i. Term of agreement is 15 years;

ii. Actual operation date is July 1, 1984;

iii. Notice is given on January 1, 1989; and

iv. Contract capacity is to be reduced by 10,000 kW on July 1, 1989; actual performance is from July 1, 1984 through July 1, 1989.

The payment (G) is calculated as follows:

$$G = CC \times (T - CCP) \times \left(\frac{J-H}{12}\right) \qquad G \geq 0$$

Where:

CC = the amount of contract capacity being terminated = 10,000 kW.

T = the current firm capacity price $140/kW-yr is arbitrarily chosen for use in this example for a July 1, 1989 Operation Date and 10-year agreement term.

CCP = the contract capacity price = $110/kW-yr

H = the actual number of months given = six months

J = the prescribed notice = twelve months

The sample calculation is:

$$G = CC \times (T - CCP) \times \left(\frac{J-H}{12}\right)$$

$$G = 10{,}000 \text{ kW} \times (\$140/\text{kW-yr} - \$110/\text{kW-yr}) \times \left(\frac{12 \text{ mos.} - 6 \text{ mos.}}{12 \text{ mos./yr}}\right)$$

$$G = \underline{\underline{\$150{,}000}}$$

PACIFIC GAS AND ELECTRIC COMPANY

STANDARD OFFER #3

POWER PURCHASE AGREEMENT

FOR

AS-DELIVERED CAPACITY AND ENERGY

FROM

FACILITIES OF 100 KILOWATTS OR LESS

**AS-DELIVERED CAPACITY AND ENERGY**
**FROM**
**FACILITIES OF 100 KILOWATTS OR LESS**

**POWER PURCHASE AGREEMENT**
**BETWEEN**
**_______________________**
**AND**
**PACIFIC GAS AND ELECTRIC COMPANY**

_______________________ ("Seller"), and PACIFIC GAS AND ELECTRIC COMPANY ("PGandE"), referred to collectively as "Parties" and individually as "Party", agree as follows:

1. Seller states that its facility located at _______________________

_______________________

and described as Make _______________, Model __________, Serial No. __________,

fuel or energy source _______________, and having a nameplate output rating of _____kW,

___volts, _____phase, 60 hertz will be ready to deliver power for sale on or about _______________.

Seller has chosen to sell PGandE _______________.*

2. Seller has two options for payment for as-delivered capacity and energy delivered to PGandE. They are:

Option No. 1: Seller elects to have the value of the purchased power credited to its monthly bill from PGandE for electric service in the next billing period (assumes electric service account).

Option No. 2: Seller elects to receive payment check within approximately 30 days of the meter reading date if the value of the purchased power is at least $50, and if less, to have the value of the purchased power credited to its monthly bill from PGandE for electric service in the next billing period.

Seller has selected Option No. ________.

3. On and after the date PGandE gives its written approval for parallel operation, PGandE shall pay Seller for as-delivered capacity at prices authorized from time to time by the California Public Utilities Commission (CPUC) and for energy at prices equal to PGandE's full short run avoided operating costs as approved by the CPUC.

4. Seller shall pay for designing, installing, operating, and maintaining the facility in accordance with all applicable laws and regulations and shall comply with PGandE's electric Rule No. 21.

5. Seller shall deliver the as-delivered capacity and energy at the agreed point of interconnection with PGandE's system as shown in Appendix A.

6. PGandE shall, at its expense, furnish and install a standard watt-hour meter (and current transformers if required) in a meter socket and enclosure equipment provided and installed by Seller at or near the point of interconnection. At Seller's option and expense, PGandE shall furnish and install a time-of-delivery meter pursuant to a separate special facilities agreement referred to in PGandE's electric Rule No. 21.

7. Seller shall (a) maintain the facility and interconnection facilities, except facilities installed by PGandE, in conformance with all applicable laws and regulations, (b) obtain any governmental authorization and permits required for the construction and operation thereof, and (c) manage the facility and interconnection facilities in a safe and prudent manner. If at any time Seller does not hold such authorizations and permits, PGandE may refuse to accept deliveries of power hereunder.

---

* Insert either "net energy output" or "surplus energy output" to show the energy sale option selected by Seller.

8. (a) PGandE may enter Seller's premises (1) to inspect at any reasonable times Seller's protective devices and read or test meters, and (2) to disconnect, without notice, the interconnection facilities if, in PGandE's opinion, a hazardous condition exists and such immediate action is necessary to protect persons, or PGandE's facilities, or other customers' facilities from damage or interference caused by Seller's facility, or lack of properly operating protective devices.

   (b) PGandE shall not be obligated to accept or pay for and may require Seller to interrupt or reduce deliveries of as-delivered capacity and energy (1) when necessary in order to construct, install, maintain, repair, replace, remove, investigate, or inspect any of its equipment or part of its system, or (2) if it determines that curtailment, interruption, or reduction is necessary because of emergencies, forced outages, force majeure, or compliance with prudent electrical practices.

   (c) Whenever possible, PGandE shall give Seller reasonable notice of the possibility that interruption or reduction of deliveries may be required.

9. Each Party as indemnitor shall save harmless and indemnify the other Party and the directors, officers, and employees of such other Party against and from any and all loss and liability for injuries to persons including employees of either Party, and damages, including property of either Party, resulting from or arising out of (a) the engineering, design, construction, maintenance, or operation of or (b) the making of replacements, additions, or betterments to the indemnitor's facilities. This indemnity and save harmless provision shall apply notwithstanding the active or passive negligence of the indemnitee. Neither Party shall be indemnified for liability or loss resulting from its sole negligence or willful misconduct. The indemnitor shall, on the other Party's request, defend any suit asserting a claim covered by this indemnity and shall pay all costs, including reasonable attorney fees, that may be incurred by the other Party in enforcing this indemnity.

10. Nothing in this Agreement shall create any duty to, any standard of care with reference to, or any liability to any person not a Party to it. Neither Party shall be liable to the other Party for consequential damages.

11. Each Party shall be responsible for protecting its facilities from possible damage by reason of the electrical disturbances or faults caused by the operation, faulty operation, or nonoperation of the other Party's facilities, and such other Party shall not be liable for any such damages so caused.

12. This Agreement shall be in effect when signed by the Seller and PGandE for an initial term of one year and shall remain in effect thereafter month to month unless terminated by either Party on 30 (thirty) days advance written notice to the other.

13. Any notice required under this Agreement shall be in writing and mailed at any United States Post Office with postage prepaid for transmission by certified mail, return receipt requested, addressed to the Party, or personally delivered to the Party, at the address below. Changes in such designation may be made by notice similarly given.

All written notice shall be directed as follows:

to PGandE: Pacific Gas and Electric Company
Attention: Manager, Commercial Department
77 Beale Street
San Francisco, CA 94106

to Seller: ______________________
______________________
______________________
______________________
______________________

S.O. #3

14. This Agreement includes the following appendix which is attached and incorporated by reference:

Appendix A - POINT OF INTERCONNECTION LOCATION SKETCH

IN WITNESS WHEREOF, the Parties hereto have caused this Agreement to be executed by their duly authorized representatives as of the last date set forth below.

| ______________________ (SELLER) | PACIFIC GAS AND ELECTRIC COMPANY |
|---|---|
| BY: ______________________ | BY: ______________________ |
| ______________________ (Type Name) | ______________________ (Type Name) |
| TITLE: ______________________ | TITLE: ______________________ |
| DATE SIGNED: ______________________ | DATE SIGNED: ______________________ |

Mailing Address:

______________________

______________________

S.O. #3

APPENDIX A

POINT OF INTERCONNECTION LOCATION SKETCH

S.O. #3

PACIFIC GAS AND ELECTRIC COMPANY

STANDARD OFFER #4

POWER PURCHASE AGREEMENT

FOR

LONG-TERM ENERGY AND CAPACITY

# STANDARD OFFER #4:
# LONG-TERM ENERGY AND CAPACITY
# POWER PURCHASE AGREEMENT

CONTENTS

# LONG-TERM ENERGY AND CAPACITY
# POWER PURCHASE AGREEMENT
# BETWEEN

______________________________

# AND

# PACIFIC GAS AND ELECTRIC COMPANY

________________________________________ ("Seller"), and PACIFIC GAS AND ELECTRIC COMPANY ("PGandE"), referred to collectively as "Parties" and individually as "Party", agree as follows:

## ARTICLE 1 QUALIFYING STATUS

Seller warrants that, at the date of first power deliveries from Seller's Facility[1] and during the term of agreement, its Facility shall meet the qualifying facility requirements established as of the effective date of this Agreement by the Federal Energy Regulatory Commission's rules (18 Code of Federal Regulations 292) implementing the Public Utility Regulatory Policies Act of 1978 (16 U.S.C.A. 796, et seq.).

---

1 Underlining identifies those terms which are defined in Section A-1 of Appendix A.

## ARTICLE 2 COMMITMENT OF PARTIES

The prices to be paid Seller for energy and/or capacity delivered pursuant to this Agreement have wholly or partly been fixed at the time of execution. Actual avoided costs at the time of energy and/or capacity deliveries may be substantially above or below the prices fixed in this Agreement. Therefore, the Parties expressly commit to the prices fixed in this Agreement for the applicable period of performance and shall not seek to or have a right to renegotiate such prices for any reason. As part of its consideration for the benefit of fixing part or all of the energy and/or capacity prices under this Agreement, Seller waives any and all rights to judicial or other relief from its obligations and/or prices set forth in Appendices B, D and E, or modification of any other term or provision for any reasons whatsoever.

This Agreement contains certain provisions which set forth methods of calculating damages to be paid to PGandE in the event Seller fails to fulfill certain performance obligations. The inclusion of such provisions is not intended to create any express or implied right in Seller to terminate this Agreement prior to the expiration of the term of agreement. Termination of this Agreement by Seller prior to its expiration date shall constitute a breach of this Agreement and the damages expressly set forth in this

S.O. #4

Agreement shall not constitute PGandE's sole remedy for such breach.

ARTICLE 3 PURCHASE OF POWER

(a) Seller shall sell and deliver and PGandE shall purchase and accept delivery of capacity and energy at the voltage level of _____ kV.

(b) Seller shall provide capacity and energy from its

______________________________________________ kW
[Nameplate rating of generator(s)]
Facility located at ________________________________

______________________________________________.

(c) The scheduled operation date of the Facility is

__________________. At the end of each calendar quarter
[Date]
Seller shall give written notice to PGandE of any change in the scheduled operation date.

(d) Seller shall limit the Facility's actual rate of delivery into the PGandE system to ______ kW.

(e) The primary energy source for the Facility is

______________________________________________.

## ARTICLE 4 ENERGY PRICE

PGandE shall pay Seller for its ________________[1] under the energy payment option checked below:[2]

______ Energy Payment Option 1 - Forecasted Energy Prices

During the fixed price period, Seller shall be paid for energy delivered at prices equal to _______[3] percent of the prices set forth in Table B-1, Appendix B, plus _______[4] percent of PGandE's full short-run avoided operating costs.

For the remaining years of the term of agreement, Seller shall be paid for energy delivered at prices equal to PGandE's full short-run avoided operating costs.

If Seller's Facility is not an oil or gas-fired cogeneration facility, Seller may convert from Energy Payment Option 1 to Energy Payment Option 2 and be

---

1 Insert either "net energy output" or "surplus energy output" to show the energy sale option selected by Seller.

2 Option 2 is not available to oil or gas-fired cogenerators.

3 Insert either 0, 20, 40, 60, 80, or 100, at Seller's option. If Seller's Facility is an oil or gas-fired cogeneration facility, either 0 or 20 must be inserted.

4 Insert the difference between 100 and the percentage selected under footnote 3 above.

subject to the conditions therein, provided that Seller shall not change the percentage of energy prices to be based on PGandE's full short-run avoided operating costs. Such conversion must be made at least 90 days prior to the date of initial energy deliveries and must be made by written notice in accordance with Section A-17, Appendix A.

______ Energy Payment Option 2 - Levelized Energy Prices

During the fixed price period, Seller shall be paid for energy delivered at prices equal to ________[1] percent of the levelized energy prices set forth in Table B-2, Appendix B for the year in which energy deliveries begin and term of agreement, plus _______[2] percent of PGandE's full short-run avoided operating costs. During the fixed price period, Seller shall be subject to the conditions and terms set forth in Appendix B, Energy Payment Option 2.

For the remaining years of the term of agreement, Seller shall be paid for energy delivered at prices equal to PGandE's full short-run avoided operating costs.

---

1 Insert either 20, 40, 60, 80, or 100, at Seller's option.

2 Insert the difference between 100 and the percentage selected under footnote 1 above.

Seller may convert from Energy Payment Option 2 to Energy Payment Option 1, provided that Seller shall not change the percentage of energy prices to be based on PGandE's full short-run avoided operating costs. Such conversion must be made at least 90 days prior to the date of initial energy deliveries and must be made by written notice in accordance with Section A-17, Appendix A.

______ Energy Payment Option 3 - Incremental Energy Rate

Beginning with the date of initial energy deliveries and continuing until ______________,[1] Seller shall be paid monthly for energy delivered at prices equal to PGandE's full short-run avoided operating costs, provided that adjustments shall be made annually to the extent set forth in Appendix B, Energy Payment Option 3.

The Incremental Energy Rate Band Widths specified by Seller in Table I below shall be used in determining the annual adjustment, if any.

---

1 Specified by Seller. Must be December 31, 1998 or prior.

Table I

| Year | Incremental Energy Rate Band Widths (must be multiples of 100 or zero) |
|---|---|
| 1984 | ________ |
| 1985 | ________ |
| 1986 | ________ |
| 1987 | ________ |
| 1988 | ________ |
| 1989 | ________ |
| 1990 | ________ |
| 1991 | ________ |
| 1992 | ________ |
| 1993 | ________ |
| 1994 | ________ |
| 1995 | ________ |
| 1996 | ________ |
| 1997 | ________ |
| 1998 | ________ |

After ______________, Seller shall be paid for energy delivered at prices equal to PGandE's full short-run avoided operating costs.

ARTICLE 5 . CAPACITY ELECTION AND CAPACITY PRICE

Seller may elect to deliver either firm capacity or as-delivered capacity, and Seller's election is indicated below:

______ Firm capacity - ______ kW for____ years from the firm capacity availability date with payment determined in accordance with Appendix E. Except for hydroelectric facilities, PGandE shall pay Seller for capacity delivered in excess of firm capacity on an as-delivered capacity basis in accordance with As-Delivered Capacity Payment Option _______ set forth in Appendix D.

OR

______ As-delivered capacity with payment determined in accordance with As-Delivered Capacity Payment Option ______ set forth in Appendix D.

ARTICLE 6 LOSS ADJUSTMENT FACTORS

Capacity Loss Adjustment Factors shall be as shown in Appendix D and Appendix E, dependent upon Seller's capacity election set forth in Article 5 of this Agreement.

Energy Loss Adjustment Factors shall be considered as unity for all energy payments related to Energy Payment Options 1 and 2 set forth in Appendix B for the entire fixed price period of this Agreement, except for the percentage of payments that Seller elected in Article 4 to have calculated based on PGandE's full short-run avoided operating costs. Energy Loss Adjustment Factors for all payments related to PGandE's full short-term avoided operating costs are subject to CPUC rulings for the entire term of agreement.

ARTICLE 7 CURTAILMENT

Seller has two options regarding possible curtailment

by PGandE of Seller's deliveries, and Seller's selection is indicated below:

____ Curtailment Option A - Hydro Spill and Negative Avoided Cost

____ Curtailment Option B - Adjusted Price Period

The two options are described in Appendix C.

ARTICLE 8 RETROACTIVE APPLICATION OF CPUC ORDERS

Pursuant to Ordering Paragraph 1(f) of CPUC Decision No. 83-09-054 (September 7, 1983), after the effective date of the CPUC's Application 82-03-26 decision relating to 1) line loss factors; 2) interconnection provisions involving future lines and system upgrades; and 3) insurance, Seller has the option to retain the relevant terms of this Agreement or have the results of that decision incorporated into this Agreement. To retain the terms herein, Seller shall provide written notice to PGandE within 30 days after the effective date of the relevant CPUC decision on Application 82-03-26. Failure to provide such notice will result in the amendment of this Agreement to comply with that decision.

As soon as practicable following the issuance of a decision in Application 82-03-26, PGandE shall notify Seller of the effective date thereof and its results.

S.O. #4

## ARTICLE 9 NOTICES

All written notices shall be directed as follows:

To PGandE: Pacific Gas and Electric Company
Attention: Vice President - Electric Operations
77 Beale Street
San Francisco, CA 94106

To Seller: ______________________________

______________________________

______________________________

______________________________

______________________________

## ARTICLE 10 DESIGNATED SWITCHING CENTER

The designated PGandE switching center shall be, unless changed by PGandE:

______________________________
(Name)

______________________________
(Location)

______________________________
(Phone number)

## ARTICLE 11 TERMS AND CONDITIONS

This Agreement includes the following appendices which are attached and incorporated by reference:

Appendix A - GENERAL TERMS AND CONDITIONS

Appendix B - ENERGY PAYMENT OPTIONS

S.O. #4

Appendix C - CURTAILMENT OPTIONS

Appendix D - AS-DELIVERED CAPACITY

Appendix E - FIRM CAPACITY

## ARTICLE 12 TERM OF AGREEMENT

This Agreement shall be binding upon execution and remain in effect for ________ years[1] from the date of initial energy deliveries; provided, however, that it shall terminate if energy deliveries do not start within five years of the effective date.

IN WITNESS WHEREOF, the Parties hereto have caused this Agreement to be executed by their duly authorized representatives and it is effective as of the last date set forth below.

| | |
|---|---|
| ______________________ (SELLER) | PACIFIC GAS AND ELECTRIC COMPANY |
| BY: ______________________ | BY: ______________________ |
| ______________________ (Type Name) | ______________________ (Type Name) |
| TITLE: ______________________ | TITLE: ______________________ |
| DATE SIGNED: ______________ | DATE SIGNED: ______________ |

[1] The minimum contract term is 15 years and the maximum contract term is 30 years.

# APPENDIX A

## GENERAL TERMS AND CONDITIONS

CONTENTS

# APPENDIX A

## GENERAL TERMS AND CONDITIONS

### A-1 DEFINITIONS

Whenever used in this Agreement, appendices, and attachments hereto, the following terms shall have the following meanings:

Adjusted firm capacity price - The $/kW-year purchase price for firm capacity from Table E-2, Appendix E for the period of Seller's actual performance.

As-delivered capacity - Capacity delivered to PGandE in excess of firm capacity or in lieu of a firm capacity commitment.

CPUC - The Public Utilities Commission of the State of California.

Current firm capacity price - The $/kW-year capacity price from PGandE's firm capacity price schedule effective at the time PGandE derates the firm capacity pursuant to Section E-4(b), Appendix E or Seller terminates performance under this Agreement, for a term equal to the period from the date of deration or termination to the end of the term of agreement.

Designated PGandE switching center - That switching center or other PGandE installation identified in Article 10.

Facility - That generation apparatus described in Article 3 and all associated equipment owned, maintained, and operated by Seller.

Firm capacity - That capacity, if any, identified as firm in Article 5 except as otherwise changed as provided herein.

Firm capacity availability date - The day following the day during which all features and equipment of the Facility are demonstrated to PGandE's satisfaction to be capable of operating simultaneously to deliver firm capacity continuously into PGandE's system as provided in this Agreement.

Firm capacity price - The price for firm capacity applicable for the firm capacity availability date and the number of years of firm capacity delivery from either the firm capacity price schedule, Table E-2, Appendix E, or the successor to Table E-2 in effect on the <u>firm capacity availability date</u>, whichever is higher.

<u>Firm capacity price schedule</u> - The periodically published schedule of the $/kW-year prices that PGandE offers to pay for <u>firm capacity</u>. See Table E-2, Appendix E.

<u>Fixed price period</u> - The period during which forecasted or levelized energy prices, and/or forecasted <u>as-delivered capacity</u> prices, are in effect; defined as the first five years of the <u>term of agreement</u> if the <u>term of agreement</u> is 15 or 16 years; the first six years of the <u>term of agreement</u> if the <u>term of agreement</u> is 17, 18, or 19 years; or the first ten years of the <u>term of agreement</u> if the <u>term of agreement</u> is anywhere from 20 through 30 years.

<u>Forced outage</u> - Any outage resulting from a design defect, inadequate construction, operator error or a breakdown of the mechanical or electrical equipment that fully or partially curtails the electrical output of the <u>Facility</u>.

<u>Full short-run avoided operating costs</u> - <u>CPUC</u>-approved costs which are the basis of PGandE's published energy prices. PGandE's current energy price calculation is shown in Table B-5, Appendix B. PGandE's

published off-peak hours' prices shall be adjusted, as appropriate, if Seller has selected Curtailment Option B.

Interconnection facilities - All means required and apparatus installed to interconnect and deliver power from the Facility to the PGandE system including, but not limited to, connection, transformation, switching, metering, communications, and safety equipment, such as equipment required to protect (1) the PGandE system and its customers from faults occurring at the Facility, and (2) the Facility from faults occurring on the PGandE system or on the systems of others to which the PGandE system is directly or indirectly connected. Interconnection facilities also include any necessary additions and reinforcements by PGandE to the PGandE system required as a result of the interconnection of the Facility to the PGandE system.

Net energy output - The Facility's gross output in kilowatt-hours less station use and transformation and transmission losses to the point of delivery into the PGandE system.

Prudent electrical practices - Those practices, methods, and equipment, as changed from time to time, that are commonly used in prudent electrical engineering and operations to design and operate electric equipment lawfully and with safety, dependability, efficiency, and economy.

S.O. #4

Scheduled operation date - The day specified in Article 3(c) when the Facility is, by Seller's estimate, expected to produce energy that will be available for delivery to PGandE.

Special facilities - Those parts of the interconnection facilities furnished by PGandE at Seller's request, metering and data processing equipment, and those additions and reinforcements to the PGandE system which are needed to accommodate the maximum delivery of energy and capacity from the Facility as provided in this Agreement. All special facilities shall be furnished at Seller's expense pursuant to PGandE's electric Rule No. 21 by a separate agreement.

Station use - Energy used to operate the Facility's auxiliary equipment. The auxiliary equipment includes, but is not limited to, forced and induced draft fans, cooling towers, boiler feed pumps, lubricating oil systems, plant lighting, fuel handling systems, control systems, and sump pumps.

Surplus energy output - The Facility's gross output, in kilowatt-hours, less station use, and any other use by Seller, and transformation and transmission losses to the point of delivery into the PGandE system.

Term of agreement - The number of years this Agreement will remain in effect from the date of initial energy deliveries, as provided in Article 12.

Voltage level - The voltage at which the Facility interconnects with the PGandE system, measured at the point of delivery.

A-2 CONSTRUCTION

A-2.1 Land Rights

Seller hereby grants to PGandE all necessary rights of way and easements, including adequate and continuing access rights on property of Seller, to install, operate, maintain, replace, and remove the special facilities. Seller agrees to execute such other grants, deeds, or documents as PGandE may require to enable it to record such rights of way and easements. If any part of PGandE's equipment is to be installed on property owned by other than Seller, Seller shall, at its own cost and expense, obtain from the owners thereof all necessary rights of way and easements, in a form satisfactory to PGandE, for the construction, operation, maintenance, and replacement of PGandE's equipment upon such property. If Seller is unable to obtain such rights of way and easements, Seller shall reimburse PGandE for all costs incurred by PGandE in obtaining them. PGandE shall at all times have the right of ingress to and egress from the Facility at all reasonable hours for any purposes reasonably connected with this Agreement or the exercise of any and all rights secured to PGandE by law or its tariff schedules.

A-2.2 Design, Construction, Ownership, and Maintenance

(a) Seller shall design, construct, install, own, operate, and maintain all interconnection facilities, except special facilities, to the point of interconnection with the PGandE system as required for PGandE to receive capacity and energy from the Facility. The Facility and interconnection facilities shall meet all requirements of applicable codes and all standards of prudent electrical practices and shall be managed in a safe and prudent manner.

(b) Seller shall submit to PGandE all specifications for the interconnection facilities (except special facilities) and, at PGandE's option, the Facility, for review and written acceptance prior to their release for construction purposes. PGandE's review and acceptance of these specifications shall not be construed as confirming or endorsing the design or as warranting their safety, durability, or reliability. PGandE shall not, by reason of such review or lack of review, be responsible for strength, details of design, adequacy, or capacity of equipment built pursuant to such specifications, nor shall PGandE's acceptance be deemed to be an endorsement of any of such equipment. Seller shall change the interconnection facilities as may be reasonably required by PGandE to meet changing requirements of the PGandE system.

(c) In the event it is necessary for PGandE to install interconnection facilities for the purposes of this Agreement, they shall be installed as special facilities.

A-2.3 Meter Installation

(a) PGandE shall specify, provide, install, own, operate, and maintain as special facilities all metering and data processing equipment for the registration and recording of energy and other related parameters which are required for the reporting of data to PGandE and for computing the payment due Seller from PGandE.

(b) Seller shall provide, construct, install, own, and maintain at Seller's expense all that is required to accommodate the metering and data processing equipment, such as, but not limited to, metal-clad switchgear, switchboards, cubicles, metering panels, enclosures, conduits, rack structures, and equipment mounting pads.

## A-3 OPERATION

### A-3.1 Inspection and Approval

Seller shall not operate the Facility in parallel with PGandE's system until an authorized PGandE representative has inspected the interconnection facilities, and PGandE has given written approval to begin parallel operation.

### A-3.2 Facility Operation and Maintenance

Seller shall operate and maintain its Facility according to prudent electrical practices, applicable laws, orders, rules, and tariffs and shall provide such reactive power support as may be reasonably required by PGandE to maintain system voltage level and power factor. Seller shall operate the Facility at the power factors or voltage levels prescribed by PGandE's system dispatcher or designated representative. If Seller fails to provide reactive power support, PGandE may do so at Seller's expense.

### A-3.3 Point of Delivery

Seller shall deliver the energy at the point where Seller's electrical conductors contact PGandE's existing system or at such other point as the Parties may agree in writing.

S.O. #4

A-3.4 Operating Communications

(a) Seller shall maintain operating communications with the designated PGandE switching center. The operating communications shall include, but not be limited to, system paralleling or separation, scheduled and unscheduled shutdowns, equipment clearances, levels of operating voltage or power factors and daily capacity and generation reports.

(b) Seller shall keep a daily operations log for each generating unit which shall include information on unit availability, maintenance outages, circuit breaker trip operations requiring a manual reset, and any significant events related to the operation of the Facility.

(c) If Seller makes deliveries greater than one megawatt, Seller shall measure and register on a graphic recording device power in kW and voltage in kV at a location within the Facility agreed to by both Parties.

(d) If Seller makes deliveries greater than one and up to and including ten megawatts, Seller shall report to the designated PGandE switching center, twice a day at agreed upon times for the current day's operation, the hourly readings in kW of capacity delivered and the energy in kWh delivered since the last report.

(e) If Seller makes deliveries of greater than ten megawatts, Seller shall telemeter the delivered capacity and energy information, including real power in kW, reactive power in kVAR, and energy in kWh to a switching center selected by PGandE. PGandE may also require Seller to telemeter transmission kW, kVAR, and kV data depending on the number of generators and transmission configuration. Seller shall provide and maintain the data circuits required for telemetering. When telemetering is inoperative, Seller shall report daily the capacity delivered each hour and the energy delivered each day to the designated PGandE switching center.

A-3.5 Meter Testing and Inspection

(a) All meters used to provide data for the computation of the payments due Seller from PGandE shall be sealed, and the seals shall be broken only by PGandE when the meters are to be inspected, tested, or adjusted.

(b) PGandE shall inspect and test all meters upon their installation and annually thereafter. At Seller's request and expense, PGandE shall inspect or test a meter more frequently. PGandE shall give reasonable notice to Seller of the time when any inspection or test shall take place, and Seller may have representatives present at the test or inspection. If a meter is found to be inaccurate or defective, PGandE shall adjust, repair, or replace it at its expense in order to provide accurate metering.

A-3.6 Adjustments to Meter Measurements

If a meter fails to register, or if the measurement made by a meter during a test varies by more than two percent from the measurement made by the standard meter used in the test, an adjustment shall be made correcting all measurements made by the inaccurate meter for -- (1) the actual period during which inaccurate measurements were made, if the period can be determined, or if not, (2) the period immediately preceding the test of the meter equal to one-half the time from the date of the last previous test of the meter, provided that the period covered by the correction shall not exceed six months.

A-4 PAYMENT

PGandE shall mail to Seller not later than 30 days after the end of each monthly billing period (1) a statement showing the energy and capacity delivered to PGandE during on-peak, partial-peak, and off-peak periods during the monthly billing period, (2) PGandE's computation of the amount due Seller, and (3) PGandE's check in payment of said amount. If within 30 days of receipt of the statement Seller does not make a report in writing to PGandE of an error, Seller shall be deemed to have waived any error in

PGandE's statement, computation, and payment, and they shall be considered correct and complete.

A-5 ADJUSTMENTS OF PAYMENTS

(a) In the event adjustments to payments are required as a result of inaccurate meters, PGandE shall use the corrected measurements described in Section A-3.6 to recompute the amount due from PGandE to Seller for the capacity and energy delivered under this Agreement during the period of inaccuracy.

(b) The additional payment to Seller or refund to PGandE shall be made within 30 days of notification of the owing Party of the amount due.

A-6 ACCESS TO RECORDS

Each Party, after giving reasonable written notice to the other Party, shall have the right of access to all metering and related records including operations logs of the <u>Facility</u>.

A-7 INTERRUPTION OF DELIVERIES

PGandE shall not be obligated to accept or pay for and may require Seller to interrupt or reduce deliveries of energy (1) when necessary in order to construct, install, maintain, repair, replace, remove, investigate, or inspect any of its equipment or any part of its system, or (2) if it determines that interruption or reduction is necessary because of PGandE system emergencies, forced outages, force majeure, or compliance with <u>prudent</u> <u>electrical</u> <u>practices</u>; provided that PGandE shall not interrupt deliveries pursuant to this section in order to take advantage, or make purchases, of less expensive energy elsewhere. Whenever possible, PGandE shall give Seller reasonable notice of the possibility that interruption or reduction of deliveries may be required.

A-8 FORCE MAJEURE

(a) The term force majeure as used herein means unforeseeable causes, other than <u>forced</u> <u>outages</u>, beyond the reasonable control of and without the fault or negligence of the Party claiming force majeure, including but not limited to, acts of God, labor disputes, sudden actions of the elements, and actions by federal, state, municipal, or any other government agency.

(b) If either Party because of force majeure is rendered wholly or partly unable to perform its obligations under this Agreement, that Party shall be excused from whatever performance is affected by the force majeure to the extent so affected provided that:

(1) the non-performing Party, within two weeks after the occurrence of the force majeure, gives the other Party written notice describing the particulars of the occurrence,

(2) the suspension of performance is of no greater scope and of no longer duration than is required by the force majeure,

(3) the non-performing Party uses its best efforts to remedy its inability to perform (this subsection shall not require the settlement of any strike, walkout, lockout or other labor dispute on terms which, in the sole judgment of the Party involved in the dispute, are contrary to its interest. It is understood and agreed that the settlement of strikes, walkouts, lockouts or other labor disputes shall be at the sole discretion of the Party having the difficulty),

(4) when the non-performing Party is able to resume performance of its obligations under this Agreement, that Party shall give the other Party written notice to that effect, and

(5) capacity payments during such periods of force majeure on Seller's part shall be governed by Section E-2(c), Appendix E.

A-9 INDEMNITY

Each Party as indemnitor shall save harmless and indemnify the other Party and the directors, officers, and employees of such other Party against and from any and all loss and liability for injuries to persons including employees of either Party, and property damages including property of either Party resulting from or arising out of (1) the engineering, design, construction, maintenance, or operation of, or (2) the making of replacements, additions, or betterments to, the indemnitor's facilities. This indemnity and save harmless provision shall apply notwithstanding the active or passive negligence of the indemnitee. Neither Party shall be indemnified hereunder for its liability or loss resulting from its sole negligence or willful misconduct. The indemnitor shall, on the other Party's request, defend any suit asserting a claim covered by this indemnity and shall pay all costs, including reasonable attorney fees, that may be incurred by the other Party in enforcing this indemnity.

A-10 LIABILITY; DEDICATION

(a) Nothing in this Agreement shall create any duty to, any standard of care with reference to, or any liability to any person not a Party to it. Neither Party shall be liable to the other Party for consequential damages.

S.O. #4

(b) Each Party shall be responsible for protecting its facilities from possible damage by reason of electrical disturbances or faults caused by the operation, faulty operation, or nonoperation of the other Party's facilities, and such other Party shall not be liable for any such damages so caused.

(c) No undertaking by one Party to the other under any provision of this Agreement shall constitute the dedication of that Party's system or any portion thereof to the other Party or to the public or affect the status of PGandE as an independent public utility corporation or Seller as an independent individual or entity and not a public utility.

A-11 SEVERAL OBLIGATIONS

Except where specifically stated in this Agreement to be otherwise, the duties, obligations, and liabilities of the Parties are intended to be several and not joint or collective. Nothing contained in this Agreement shall ever be construed to create an association, trust, partnership, or joint venture or impose a trust or partnership duty, obligation, or liability on or with regard to either Party. Each Party shall be liable individually and severally for its own obligations under this Agreement.

A-12 NON-WAIVER

Failure to enforce any right or obligation by either Party with respect to any matter arising in connection with this Agreement shall not constitute a waiver as to that matter or any other matter.

A-13 ASSIGNMENT

Neither Party shall voluntarily assign its rights nor delegate its duties under this Agreement, or any part of such rights or duties, without the written consent of the other Party, except in connection with the sale or merger of a substantial portion of its properties. Any such assignment or delegation made without such written consent shall be null and void. Consent for assignment shall not be withheld unreasonably. Such assignment shall include, unless otherwise specified therein, all of Seller's rights to any refunds which might become due under this Agreement.

A-14 CAPTIONS

All indexes, titles, subject headings, section titles, and similar items are provided for the purpose of reference and convenience and are not intended to affect the meaning of the contents or scope of this Agreement.

A-15 CHOICE OF LAWS

This Agreement shall be interpreted in accordance with the laws of the State of California, excluding any choice of law rules which may direct the application of the laws of another jurisdiction.

A-16 GOVERNMENTAL JURISDICTION AND AUTHORIZATION

Seller warrants that the <u>Facility</u> and <u>interconnection facilities</u>, except <u>special</u> <u>facilities</u>, shall at all times conform to all applicable laws and regulations, and Seller shall obtain any governmental authorizations and permits required for the construction and operation thereof. If at any time Seller does not hold such authorizations and permits PGandE may refuse to accept deliveries of power hereunder.

A-17 NOTICES

Any notice, demand, or request required or permitted to be given by either Party to the other, and any instrument required or permitted to be tendered or delivered by either Party to the other, shall be in writing (except as provided in Section E-3) and so given, tendered, or delivered, as the case may be, by depositing the same in any United States Post Office with postage prepaid for transmission by certified mail, return receipt requested, addressed to the

Party, or personally delivered to the Party, at the address in Article 9 of this Agreement. Changes in such designation may be made by notice similarly given.

A-18 INSURANCE

A-18.1 Workers' Compensation

Seller shall furnish PGandE a certificate of workers' compensation or self-insurance indicating compliance with the Labor Code of California, including Employer's Liability insurance with a minimum of $2,000,000 for injury or death of any one person. This certificate shall provide for 30-days' written notice to PGandE prior to cancellation, termination, alteration, or material change of such insurance.

A-18.2 Comprehensive General and Comprehensive Automobile Liability Coverage

(a) Seller shall maintain during the performance hereof, Comprehensive General Liability and Comprehensive Automobile Liability of not less than $3,000,000 combined single limit or equivalent for bodily injury, personal injury, and property damage as the result of any one occurrence.

(b) Comprehensive General Liability shall include coverage for Premises-Operations, Owners and Contractors Protective, Products/Completed Operations Hazard, Explosion, Collapse, Underground, Contractual Liability, and Broad Form Property Damage including Completed Operations. Comprehensive Automobile Liability shall include coverage for Owned, Hired, and Non-Owned automobiles.

(c) Such insurance, by endorsement to the policy(ies), shall include PGandE as an additional insured insofar as work performed by Seller for PGandE is concerned, shall contain a severability of interest clause, shall provide that PGandE shall not by reason of its inclusion as an additional insured incur liability to the insurance carrier for payment of premium for such insurance, and shall provide for 30-days' written notice to PGandE prior to cancellation, termination, alteration, or material change of such insurance.

A-18.3 Additional Insurance Provisions

(a) Evidence of coverage described above in Sections A-18.1 and A-18.2 shall state that coverage provided is primary and is not excess to or contributing with any insurance or self-insurance maintained by PGandE.

(b) PGandE shall have the right to inspect or obtain a copy of the original policy(ies) of insurance.

S.O. #4

(c) Seller shall furnish the required certificates and endorsements to PGandE prior to the date construction of the Facility begins or the effective date of this Agreement, whichever is later.

(d) All insurance certificates, endorsements, cancellations, terminations, alterations, and material changes of such insurance shall be issued and submitted to the following:

PACIFIC GAS AND ELECTRIC COMPANY
Attention: Manager - Insurance Department
77 Beale Street
San Francisco, CA 94106

S.O. #4

# APPENDIX B

# ENERGY PRICES

## Energy Payment Option 1 - Forecasted Energy Prices

Pursuant to Article 4, the energy payment calculation for Seller's energy deliveries during each year of the fixed price period shall include the appropriate prices for such year in Table B-1, multiplied by the percentage Seller has specified in Article 4. If Seller has selected Curtailment Option B in Article 7, the forecasted off-peak hours' energy prices listed in Table B-1 shall be adjusted upward by 7.7% for Period A and 9.6% for Period B.

TABLE B-1

Forecasted Energy Price Schedule

| Year of Energy Deliveries | Forecasted Energy Prices*, ¢/kWh | | | | | | Weighted Annual Average |
|---|---|---|---|---|---|---|---|
| | Period A | | | Period B | | | |
| | On-Peak | Partial-Peak | Off-Peak | On-Peak | Partial-Peak | Off-Peak | |
| 1983 | 5.36 | 5.12 | 4.94 | 5.44 | 5.31 | 5.19 | 5.18 |
| 1984 | 5.66 | 5.40 | 5.22 | 5.74 | 5.61 | 5.48 | 5.47 |
| 1985 | 5.75 | 5.48 | 5.30 | 5.83 | 5.69 | 5.56 | 5.55 |
| 1986 | 5.99 | 5.72 | 5.52 | 6.08 | 5.94 | 5.80 | 5.79 |
| 1987 | 6.38 | 6.08 | 5.88 | 6.47 | 6.32 | 6.17 | 6.16 |
| 1988 | 6.94 | 6.62 | 6.39 | 7.03 | 6.87 | 6.71 | 6.70 |
| 1989 | 7.60 | 7.25 | 7.00 | 7.70 | 7.53 | 7.35 | 7.34 |
| 1990 | 8.12 | 7.74 | 7.48 | 8.23 | 8.04 | 7.85 | 7.84 |
| 1991 | 8.64 | 8.24 | 7.96 | 8.75 | 8.56 | 8.35 | 8.34 |
| 1992 | 9.33 | 8.90 | 8.60 | 9.46 | 9.24 | 9.02 | 9.01 |
| 1993 | 10.10 | 9.63 | 9.30 | 10.23 | 10.00 | 9.76 | 9.75 |
| 1994 | 10.91 | 10.41 | 10.06 | 11.06 | 10.81 | 10.55 | 10.54 |
| 1995 | 11.79 | 11.25 | 10.87 | 11.96 | 11.68 | 11.40 | 11.39 |
| 1996 | 12.67 | 12.09 | 11.68 | 12.85 | 12.56 | 12.25 | 12.24 |
| 1997 | 13.61 | 12.98 | 12.54 | 13.79 | 13.48 | 13.15 | 13.14 |

* These prices are differentiated by the time periods as defined in Table B-4.

S.O. #4

## Energy Payment Option 2 - Levelized Energy Prices

Pursuant to Article 4, the energy payment calculation for Seller's energy deliveries during the fixed price period shall include the appropriate prices set forth in Table B-2 for the year in which energy deliveries begin and term of agreement, multiplied by the percentage Seller has specified in Article 4. If Seller has selected Curtailment Option B in Article 7, the levelized off-peak hours' energy prices listed in Table B-2 shall be adjusted upward by 7.7% for Period A and 9.6% for Period B. The discount specified in (c)(vi) below, if applicable, will be applied to the energy payments during the fixed price period.

During the fixed price period, Seller shall be subject to the following conditions and terms:

(a) Minimum Damages

The Parties agree that the levelized energy prices which PGandE pays Seller for the energy which Seller delivers to PGandE is based on the agreed value to PGandE of Seller's energy deliveries during the entire fixed price period. In the event PGandE does not receive such full performance by reason of a termination, Seller shall pay PGandE an amount based on the difference between the net present values, at the time of termination, of the payments Seller would receive at the forecasted energy prices in Table B-1 and the payments Seller would receive at the levelized energy prices, for the remaining years of the <u>fixed price period</u>. This amount shall be calculated by assuming that Seller continued to generate for the remaining years of the <u>fixed price period</u> at a level equal to the average annual energy generation during the period of performance, and by applying the weighted annual average levelized price applicable to Seller's <u>Facility</u> and the weighted annual average forecasted energy prices in Table B-1 for the remaining years of the <u>fixed price period</u>. The following formula shall be used to make this calculation:

$$P = \sum_{n=1}^{Y} \frac{(F_n)(A)(W)}{(1.15)^n} - \sum_{n=1}^{Y} \frac{(L)(A)(W)}{(1.15)^n}$$

where:

$P$ = amount due PGandE

$Y$ = number of years remaining in the <u>fixed price period</u>

$F_n$ = weighted annual average forecasted energy price in the $n^{th}$ year after the breach, failure to perform, or expiration of security, as shown in Table B-1 for the corresponding calendar year

S.O. #4

L = weighted annual average levelized energy price applicable to Seller's Facility

A = average annual energy generation by Seller during the period of performance

n = summation index; refers to the $n^{th}$ year following termination

W = percent of Seller's energy payments based on the levelized energy prices, as specified in Article 4

(b) Performance Requirements

Seller shall operate and maintain the Facility in accordance with prudent electrical practices in order to maximize the likelihood that the Facility's output as delivered to PGandE during the part of the fixed price period when the levelized price is below the forecasted price ("last part") shall equal or exceed 70% of the Facility's output during the part of the fixed price period when the levelized price is above the forecasted price ("first part"). In the event that the Facility's output during any year or series of years in the last part of the fixed price period is less than 70% of the average annual production during the first part of the fixed price period, PGandE may, at its discretion (taking into consideration events occurring during such year or series of years such as curtailment by PGandE, Seller's choice not to operate during adjusted price periods, or scheduled maintenance including major overhauls, and the probability that Seller's future performance will be adequate), either request payment from Seller or immediately draw on the security posted, up to the amount equal to

$P \times \frac{A-B}{A}$, where:

P and A are as defined in Section (a) above

B = Seller's average annual energy generation during the year or series of years in which the 70% performance requirement was not met.

PGandE shall not request payment from Seller or draw on the security posted if the Facility's output during the last part of the fixed price period falls below 70% of the average annual energy generation during the first part of the fixed price period solely because of force majeure as defined in Section A-8, Appendix A or a lack of or limited availability of the primary energy resource of the Facility, if such energy resource is wind, water, or sunlight.

(c) Security

(1) As security for amounts which Seller may be obligated to pay PGandE pursuant to Sections (a) and (b) above, Seller shall provide and maintain one or more of the following in an amount as

described in Section (c)(2) below.

(i) An irrevocable bank letter of credit delivered to and in favor of PGandE with terms acceptable to PGandE.

(ii) A payment bond providing for payment to PGandE in the event of any failure to meet the performance requirements set forth in Section (b) above or breach of this Agreement by Seller. Such bond shall be issued by a surety company acceptable to PGandE and shall have terms acceptable to PGandE.

(iii) Fully paid up, noncancellable Project Failure Insurance made payable to PGandE with terms of such policy(ies) acceptable to PGandE.

(iv) A performance bond providing for payment to PGandE in the event of any failure to meet the performance requirements set forth in Section (b) above or breach of this Agreement by Seller. Such bond shall be issued by a surety company acceptable to PGandE and shall have terms acceptable to PGandE.

(v) A corporate guarantee of payment to PGandE which PGandE deems, in its sole discretion, to provide at least the same quality of security as subsections (i) through (iv) above.

(vi) Other forms of security which PGandE does not deem to be equivalent security to those listed in subsections (i) through (v) above, and which PGandE, in its sole discretion, deems adequate. Such other forms of security may include, for example, a corporate guarantee or a lien, mortgage or deed of trust on the Facility or land upon which it is located. A 1.5% discount will be applied against the levelized energy price portion of PGandE's payments to Seller during the fixed price period if this type of security is provided.

(2) (i) Commencing 90 days prior to the scheduled operation date and continuing until December 1 of the following calendar year, security as described in Section (c)(1) above shall be in place in an amount calculated in accordance with the formula set forth in Section (a) above, assuming Seller delivered energy through the end of the following calendar year and then terminated this Agreement. For purposes of determining the

required amount of security, it shall be assumed that Seller's deliveries through the end of the following calendar year would equal R x C x H, where:

R = nameplate rating, in kWh, of the Facility

C = estimated capacity factor of the Facility, which shall be established by mutual agreement of the Parties at the time of execution of this Agreement.

H = number of hours from the scheduled operation date through the end of the following calendar year.

(ii) In the second calendar year of operation and each year thereafter until the end of the fixed price period, from December 1 through December 1 of the following year, security shall be in place in an amount calculated by the formula set forth in Section (a) above assuming Seller continued to deliver energy in each month through the end of the following calendar year, at a level equal to the average monthly energy deliveries to date, and then terminated this Agreement.

(3) Security must be maintained throughout the fixed price period as specified above. Any security with a fixed expiration date must be renewed by Seller prior to that date. If such security is not renewed at least 30 days prior to its expiration, PGandE may, at its discretion, either request payment from Seller or immediately draw on the security posted, up to the amount calculated in accordance with the formula set forth in Section (a) above.

(4) If, at any time during the fixed price period, PGandE believes Seller is in material breach of this Agreement, PGandE shall so notify Seller in writing and Seller must remedy such breach within a reasonable period of time. If Seller does not so remedy, PGandE may, at its discretion, either request payment from Seller or immediately draw upon the security posted, up to the amount calculated in accordance with the formula set forth in Section (a) above, provided that if during Seller's period to remedy, Seller disputes PGandE's conclusion that Seller is in material breach, and PGandE elects to draw upon the security, the amount drawn upon by PGandE shall be deposited in an interest earning escrow account and held in such account until the dispute is resolved in accordance with Section (c)(5) below.

S.O. #4

(5) Upon the written request of either Party, any controversy or dispute between the Parties concerning Section (c)(4) above shall be subject to arbitration in accordance with the provisions of the California Arbitration Act, Sections 1280-1294.2 of the California Code of Civil Procedure except as provided otherwise in this section. Either Party may demand arbitration by first giving written notice of the existence of a dispute and then within 30 days of such notice giving a second written notice of the demand for arbitration.

Within ten days after receipt of the demand for arbitration, each Party shall appoint one person, who shall not be an employee of either Party, to hear and determine the dispute. After both arbitrators have been appointed, they shall within five (5) days select a third arbitrator.

The arbitration hearing shall take place in San Francisco, California, within 30 days of the appointment of the arbitrators, at such time and place as they select. The arbitrators shall give written notice of the time of the hearing to both Parties at least ten days prior to the hearing. The arbitrators shall not be authorized to alter, extend, or modify the terms of this Agreement. At the hearing, each Party shall submit a proposed written decision, and any relevant evidence may be presented. The decision of the arbitrators must consist of selection of one of the two proposed decisions, in its entirety.

The decision of any two arbitrators shall be binding and conclusive as to disputes relating to Section (c)(4) only. Upon determining the matter, the arbitrators shall promptly execute and acknowledge their decision and deliver a copy to each Party. A judgment confirming the award may be rendered by any superior court having jurisdiction. Each Party shall bear its own arbitration costs and expenses, including the cost of the arbitrator it selected, and the costs and expenses of the third arbitrator shall be divided equally between both Parties, except as provided otherwise elsewhere in this Agreement.

Pending resolution of any controversy or dispute hereunder, performance by each Party shall continue so as to maintain the status quo prior to notice of such controversy or dispute. Resolution of the controversy or dispute shall include payment of any interest accrued in the escrow account.

S.O. #4

TABLE B-2
Levelized Energy Price Schedule

For a term of agreement of 15-16 years:

| Year in Which Energy Deliveries Begin | Levelized Energy Prices*, ¢/kWh | | | | | | Weighted Annual Average |
|---|---|---|---|---|---|---|---|
| | Period A | | | Period B | | | |
| | On-Peak | Partial-Peak | Off-Peak | On-Peak | Partial-Peak | Off-Peak | |
| 1983 | 5.76 | 5.50 | 5.31 | 5.85 | 5.71 | 5.58 | 5.57 |
| 1984 | 6.06 | 5.78 | 5.58 | 6.14 | 6.00 | 5.86 | 5.85 |
| 1985 | 6.41 | 6.11 | 5.91 | 6.50 | 6.35 | 6.20 | 6.19 |
| 1986 | 6.85 | 6.54 | 6.32 | 6.95 | 6.79 | 6.63 | 6.62 |
| 1987 | 7.37 | 7.03 | 6.79 | 7.47 | 7.30 | 7.13 | 7.12 |
| 1988 | 7.96 | 7.60 | 7.34 | 8.07 | 7.89 | 7.70 | 7.69 |

For a term of agreement of 17-19 years:

| Year in Which Energy Deliveries Begin | Levelized Energy Prices*, ¢/kWh | | | | | | Weighted Annual Average |
|---|---|---|---|---|---|---|---|
| | Period A | | | Period B | | | |
| | On-Peak | Partial-Peak | Off-Peak | On-Peak | Partial-Peak | Off-Peak | |
| 1983 | 5.90 | 5.63 | 5.44 | 5.98 | 5.84 | 5.71 | 5.70 |
| 1984 | 6.23 | 5.95 | 5.74 | 6.32 | 6.18 | 6.03 | 6.02 |
| 1985 | 6.60 | 6.30 | 6.08 | 6.69 | 6.53 | 6.38 | 6.37 |
| 1986 | 7.06 | 6.73 | 6.51 | 7.16 | 7.00 | 6.83 | 6.82 |
| 1987 | 7.60 | 7.25 | 7.00 | 7.70 | 7.53 | 7.35 | 7.34 |
| 1988 | 8.21 | 7.83 | 7.57 | 8.32 | 8.13 | 7.94 | 7.93 |

For a term of agreement of 20-30 years:

| Year in Which Energy Deliveries Begin | Levelized Energy Prices*, ¢/kWh | | | | | | Weighted Annual Average |
|---|---|---|---|---|---|---|---|
| | Period A | | | Period B | | | |
| | On-Peak | Partial-Peak | Off-Peak | On-Peak | Partial-Peak | Off-Peak | |
| 1983 | 6.49 | 6.20 | 5.98 | 6.58 | 6.43 | 6.28 | 6.27 |
| 1984 | 6.90 | 6.58 | 6.35 | 6.99 | 6.83 | 6.67 | 6.66 |
| 1985 | 7.34 | 7.00 | 6.76 | 7.44 | 7.27 | 7.10 | 7.09 |
| 1986 | 7.88 | 7.51 | 7.26 | 7.99 | 7.81 | 7.62 | 7.61 |
| 1987 | 8.49 | 8.10 | 7.82 | 8.61 | 8.41 | 8.21 | 8.20 |
| 1988 | 9.16 | 8.74 | 8.44 | 9.29 | 9.08 | 8.86 | 8.85 |

* These prices are differentiated by the time periods as defined in Table B-4.

S.O. #4

**Energy Payment Option 3 - Incremental Energy Rate**

During the period specified in Article 4, annual adjustments to Seller's energy payments shall be made as described below.

At the end of each calendar year, the Derived Incremental Energy Rate (with units expressed in Btu/kWh) will be calculated as follows:

$$\text{Derived Incremental Energy Rate (DIER)} = \frac{B}{A \times C}$$

where:

A = the total kWh delivered by Seller during the calendar year, excluding any kWh delivered when Seller was asked to curtail deliveries under Curtailment Option A or when Seller was asked to take adjusted prices under Curtailment Option B;

B = the total dollars paid for the energy described for A above;

C = the weighted average price paid during the calendar year by PGandE's Electric Department for oil and natural gas for PGandE's fossil steam plants, expressed in $/Btu on a gas Btu basis.

S.O. #4

If the DIER is between the upper and lower Incremental Energy Rate Bounds specified for that year in Table B-3 for the curtailment option selected by Seller, no additional payment is due either Party.

If the DIER is below the lower Incremental Energy Rate Bound, PGandE shall pay Seller an amount calculated as follows:

$$P_S = (\text{Lower Incremental Energy Rate Bound} - \text{DIER})(A)(C)$$

where:

$P_S$ = additional payment due Seller

DIER = Derived Incremental Energy Rate

PGandE shall add this payment to the first payment made to Seller following the calculation.

If the DIER is above the upper Incremental Energy Rate Bound, Seller shall pay PGandE an amount calculated as follows:

$$P_B = (\text{DIER} - \text{Upper Incremental Energy Rate Bound})(A)(C)$$

where:

$P_B$ = amount due PGandE

DIER = Derived Incremental Energy Rate

This amount shall be deducted from the first payment made to Seller following the calculation. If there is any remaining amount due PGandE, PGandE may, at its option, invoice Seller with such payment due within 30 days or deduct this amount from future payments due Seller.

Table B-3

Forecasted Incremental Energy Rates and Incremental Energy Rate Bounds

Curtailment Option A:

| Year | Forecasted Incremental Energy Rates, Btu/kWh (a) | Incremental Energy Rate Band Width from Article 4, Btu/kWh (b) | Upper Incremental Energy Rate Bound, Btu/kWh [column (a) plus column (b)] | Lower Incremental Energy Rate Bound, Btu/kWh [column (a) minus column(b)] |
|---|---|---|---|---|
| 1984 | 9,000 | ______ | ______ | ______ |
| 1985 | 9,050 | ______ | ______ | ______ |
| 1986 | 8,840 | ______ | ______ | ______ |
| 1987 | 8,850 | ______ | ______ | ______ |
| 1988 | 8,960 | ______ | ______ | ______ |
| 1989 | 8,820 | ______ | ______ | ______ |
| 1990 | 8,540 | ______ | ______ | ______ |
| 1991 | 8,540 | ______ | ______ | ______ |
| 1992 | 8,540 | ______ | ______ | ______ |
| 1993 | 8,540 | ______ | ______ | ______ |
| 1994 | 8,540 | ______ | ______ | ______ |
| 1995 | 8,540 | ______ | ______ | ______ |
| 1996 | 8,540 | ______ | ______ | ______ |
| 1997 | 8,540 | ______ | ______ | ______ |
| 1998 | 8,540 | ______ | ______ | ______ |

S.O. #4

Curtailment Option B:

| Year | Forecasted Incremental Energy Rates, Btu/kWh (a) | Incremental Energy Rate Band Width from Article 4, Btu/kWh (b) | Upper Incremental Energy Rate Bound, Btu/kWh [column (a) plus column (b)] | Lower Incremental Energy Rate Bound, Btu/kWh [column (a) minus column(b)] |
|---|---|---|---|---|
| 1984 | 9,440 | ______ | ______ | ______ |
| 1985 | 9,500 | ______ | ______ | ______ |
| 1986 | 9,280 | ______ | ______ | ______ |
| 1987 | 9,290 | ______ | ______ | ______ |
| 1988 | 9,400 | ______ | ______ | ______ |
| 1989 | 9,270 | ______ | ______ | ______ |
| 1990 | 8,970 | ______ | ______ | ______ |
| 1991 | 8,970 | ______ | ______ | ______ |
| 1992 | 8,970 | ______ | ______ | ______ |
| 1993 | 8,970 | ______ | ______ | ______ |
| 1994 | 8,970 | ______ | ______ | ______ |
| 1995 | 8,970 | ______ | ______ | ______ |
| 1996 | 8,970 | ______ | ______ | ______ |
| 1997 | 8,970 | ______ | ______ | ______ |
| 1998 | 8,970 | ______ | ______ | ______ |

S.O. #4

TABLE B-4[1]

| | Monday through Friday[2] | Saturdays[2] | Sundays and Holidays |
|---|---|---|---|
| Seasonal Period A (May 1 through September 30) | | | |
| On-Peak | 12:30 p.m. to 6:30 p.m. | | |
| Partial-Peak | 8:30 a.m. to 12:30 p.m.<br>6:30 p.m. to 10:30 p.m. | 8:30 a.m. to 10:30 p.m. | |
| Off-Peak | 10:30 p.m. to 8:30 a.m. | 10:30 p.m. to 8:30 a.m. | All Day |
| Seasonal Period B (October 1 through April 30) | | | |
| On-Peak | 4:30 p.m. to 8:30 p.m. | | |
| Partial-Peak | 8:30 p.m. to 10:30 p.m.<br>8:30 a.m. to 4:30 p.m. | 8:30 a.m. to 10:30 p.m. | |
| Off-Peak | 10:30 p.m. to 8:30 a.m. | 10:30 p.m. to 8:30 a.m. | All Day |

---

1 This table is subject to change to accord with the on-peak, partial-peak, and off-peak periods as defined in PGandE's own rate schedules for the sale of electricity to its large industrial customers.

2 Except the following holidays: New Year's Day, Washington's Birthday, Memorial Day, Independence Day, Labor Day, Veteran's Day, Thanksgiving, and Christmas, as said days are specified in Public Law 90-363 (5 U.S.C.A. Section 6103(a)).

S.O. #4

**TABLE B-5**
**Energy Prices Effective August 1 - October 31, 1983**

The energy purchase price calculations which will apply to energy deliveries determined from meter readings taken during August, September, and October 1983 are as follows:

| Time Period | Incremental Heat Rate[1] (a) | Cost of Energy[2] (b) | Energy Purchase Price[3,4] (c) = (a) x (b) |
|---|---|---|---|
| May 1-Sept. 30 (Period A) | | | |
| Time of Delivery Basis: | | | |
| On-Peak | 11,447 Btu/kWh | $5.3948/10^6 Btu | $0.06175/kWh |
| Partial-Peak | 10,919 | 5.3948 | 0.05891 |
| Off-Peak | 10,549 | 5.3948 | 0.05691 |
| Seasonal Average (Period A) | 10,821 | 5.3948 | 0.05838 |
| Oct. 1-April 30 (Period B) | | | |
| Time of Delivery Basis: | | | |
| On-Peak | 11,605 Btu/kWh | $5.3948/10^6 Btu | $0.06261/kWh |
| Partial-Peak | 11,341 | 5.3948 | 0.06118 |
| Off-Peak | 11,067 | 5.3948 | 0.05970 |
| Seasonal Average (Period B) | 11,231 | 5.3948 | 0.06059 |

1 Incremental heat rates are derived from marginal energy costs (including variable operating and maintenance expense) adopted by CPUC Decision No. 93887. They are adjusted to reflect the use of natural gas as the incremental fuel.

2 Cost of natural gas under PGandE Gas Schedule No. G-55 effective 7/1/83 per Advice No. 1225-G (Supplemental), Resolution No. 2545.

3 Energy Purchase Price = Incremental Heat Rate x Cost of Energy.

4 This figure excludes the applicable energy line loss adjustment factors. However, as ordered by Ordering Paragraph No. 12(j) of Decision No. 82-12-120, this figure is currently 1.0 for transmission and primary distribution loss adjustments and is equal to marginal cost line loss adjustment factors for the secondary distribution voltage level. These factors may be changed by the CPUC in the future.

S.O. #4

# APPENDIX C

## CURTAILMENT OF DELIVERIES

Seller has two options regarding curtailment of energy deliveries and Seller has made its selection in Article 6. The two options are as follows:

### CURTAILMENT OPTION A - HYDRO SPILL AND NEGATIVE AVOIDED COST

(a) In anticipation of a period of hydro spill conditions, as defined by the CPUC, PGandE may notify Seller that any purchases of energy from Seller during such period shall be at hydro savings prices quoted by PGandE. If Seller delivers energy to PGandE during any such period, Seller shall be paid hydro savings prices for those deliveries in lieu of prices which would otherwise be applicable. The hydro savings prices shall be calculated by PGandE using the following formula:

$$\frac{AQF - S}{AQF} \times PP \qquad (\geq 0)$$

where:

AQF = Energy, in kWh, projected to be available during hydro spill conditions from all qualifying facilities under agreements containing hydro savings price provisions.

S.O. #4

S = Potential energy, in kWh, from PGandE hydro facilities which will be spilled if all AQF is delivered to PGandE.

PP = Prices published by PGandE for purchases during other than hydro spill conditions.

PGandE shall give Seller notice of general periods when hydro spill conditions are anticipated, and shall give Seller as much advance notice as practical of any specific hydro spill period and the hydro savings price which will be applicable during such period.

(b) PGandE shall not be obligated to accept or pay for and may require Seller with a Facility with a nameplate rating of one megawatt or greater to interrupt or reduce deliveries of energy during periods when PGandE would incur negative avoided costs (as defined by the CPUC) due to continued acceptance of energy deliveries under this Agreement.

Whenever possible, PGandE shall give Seller reasonable notice of the possibility that interruption or reduction of deliveries may be required.

(c) Before interrupting or reducing deliveries under subsection (b), above, and before invoking hydro savings prices under subsection (a), above, PGandE shall take reasonable steps to make economy sales of the surplus energy giving rise to the condition. If such economy sales are made, while the surplus energy condition exists Seller shall be paid at the economy sales price obtained by PGandE in lieu of the otherwise applicable prices.

(d) If Seller is selling net energy output to PGandE and simultaneously purchasing its electrical needs from PGandE and Seller elects not to sell energy to PGandE at the hydro savings price pursuant to subsection (a) or when PGandE curtails deliveries of energy pursuant to subsection (b), Seller shall not use such energy to meet its electrical needs but shall continue to purchase all its electrical needs from PGandE. If Seller is selling surplus energy output to PGandE, subsections (a) or (b) shall only apply to the surplus energy output being delivered to PGandE, and Seller can continue to internally use that generation it has retained for its own use.

CURTAILMENT OPTION B - ADJUSTED PRICE PERIOD

(a) In each calendar year, the price which PGandE is obligated to pay Seller for energy deliveries during 1,000 off-peak hours (as defined in Table B-4, Appendix B) may be adjusted to a price equal to, but not in excess of, PGandE's available alternative source. This adjusted price shall be effective under any of the following conditions:

S.O. #4

(i) when PGandE's energy source at the margin is not a PGandE oil- or gas-fueled plant, and PGandE can replace Seller's energy with energy from this source at a cost less than the price paid to Seller;

(ii) when PGandE would incur negative avoided costs (as defined by the CPUC) due to continued acceptance of energy deliveries under this Agreement; or

(iii) when PGandE is experiencing minimum system operations.

During any of the conditions described above the adjusted price may be zero.

(b) Whenever possible, PGandE shall give Seller reasonable notice of any price adjustment for energy deliveries and its probable duration.

(c) If Seller is selling net energy output to PGandE and simultaneously purchasing its electrical needs from PGandE and Seller elects not to sell energy to PGandE at the adjusted price, Seller shall not use such energy to meet its electrical needs but shall continue to purchase all its electrical needs from PGandE.

(d) After Seller receives notice of the probable duration of the period during which the adjusted price will be paid, Seller may elect to perform maintenance during such period and so inform the PGandE employee in charge at the designated PGandE switching center prior to the time when the adjusted price period is expected to begin. If Seller makes such election, the number of off-peak hours of probable duration quoted in PGandE's notice to Seller shall be applied to the 1,000 hour calendar year limitation set forth in this section. After an election to do maintenance, if Seller makes any deliveries of energy during the quoted probable duration period, Seller shall be paid the adjusted price quoted in its notice from PGandE without regard to any subsequent changes on the PGandE system which may alter the adjusted price or shorten the actual duration of the condition.

## APPENDIX D

## AS-DELIVERED CAPACITY PRICES

### D-1 AS-DELIVERED CAPACITY PAYMENT OPTIONS

Seller has two options for as-delivered capacity payments and Seller has made its selection in Article 5. The two options are as follows:

#### AS-DELIVERED CAPACITY PAYMENT OPTION 1

PGandE shall pay Seller for as-delivered capacity at prices authorized from time to time by the CPUC. The as-delivered capacity prices in effect on the date of execution are calculated as shown in Exhibit D-1, with a shortage cost of $70 per kilowatt-year.

#### AS-DELIVERED CAPACITY PAYMENT OPTION 2

During the fixed price period, the as-delivered capacity prices will be calculated in accordance with Exhibit D-1 and the forecasted shortage costs in Table D-2.

For the remaining years of the term of agreement, PGandE shall pay Seller for as-delivered capacity at the higher of:

(i) prices authorized from time to time by the CPUC;

(ii) the as-delivered capacity prices that were paid Seller in the last year of the fixed price period; or

(iii) the as-delivered capacity prices in effect in the first year following the end of the fixed price period, provided that the annualized shortage cost from which these prices are derived does not exceed the annualized value of a gas turbine.

D-2 AS-DELIVERED CAPACITY IN EXCESS OF FIRM CAPACITY

The amount of capacity delivered in excess of firm capacity will be considered as-delivered capacity. This as-delivered capacity is based on the total kilowatt-hours delivered each month during all on-peak, partial-peak and off-peak hours excluding any energy associated with generation levels equal to or less than the firm capacity.

Seller has the two options listed in Section D-1 for payment for such as-delivered capacity. Seller has made its selection in Article 5.

EXHIBIT D-1

The as-delivered capacity price (in cents per kW-hr) for power delivered by the Facility is the product of three factors:

(a) The shortage cost in each year the Facility is operating.

(b) A capacity loss adjustment factor which provides for the effect of the deliveries on PGandE's transmission and distribution losses based on the Seller's interconnection voltage level. The applicable capacity loss adjustment factors for non-remote[1] Facilities are presented in Table D-1(a). Capacity loss adjustment factors for remote Facilities shall be calculated individually.

(c) An allocation factor which accounts for the different values of as-delivered capacity in different time periods and converts dollars per kW-year to cents per kWh. The current allocation factors are presented in Table D-1(b). The time periods to which they apply are shown in Table B-4, Appendix B. The allocation factors are subject to change from time to time.

---

1 As defined by the CPUC.

TABLE D-1(a)

Capacity Loss Adjustment Factors
For Non-Remote[1] Facilities

| Voltage Level | Loss Adjustment Factor |
|---|---|
| Transmission | .989 |
| Distribution | .991 |

If the Facility is remote, the capacity loss adjustment factor is ______________[2].

TABLE D-1(b)

Allocation Factors
For As-Delivered Capacity[3]

| | Peak (¢-yr/$-hr) | Partial Peak (¢-yr/$-hr) | Off-Peak (¢-yr/$-hr) |
|---|---|---|---|
| Seasonal Period A | .09982 | .01635 | .00000 |
| Seasonal Period B | .02023 | .00306 | .00001 |

---

1 As defined by the CPUC. The capacity loss adjustment factors for remote Facilities are determined individually.

2 Determined individually.

3 The units for the allocation factor, ¢-yr/$-hr, are derived from the conversion of $/kW-yr into ¢/kWh as follows:

$$\frac{¢/\text{kWh}}{\$/\text{kW-yr}} = \frac{¢/\text{kW-hr}}{\$/\text{kW-yr}} = \frac{¢\text{-yr}}{\$\text{-hr}}$$

The allocation factors are subject to change from time to time.

TABLE D-2

Forecasted Shortage Cost Schedule

| Year | Forecast Shortage Cost, $/kW-Yr |
|---|---|
| 1983 | 70 |
| 1984 | 76 |
| 1985 | 81 |
| 1986 | 88 |
| 1987 | 95 |
| 1988 | 102 |
| 1989 | 110 |
| 1990 | 118 |
| 1991 | 126 |
| 1992 | 135 |
| 1993 | 144 |
| 1994 | 154 |
| 1995 | 164 |
| 1996 | 176 |
| 1997 | 188 |

S.O. #4

# APPENDIX E

## FIRM CAPACITY

CONTENTS

# APPENDIX E

# FIRM CAPACITY

## E-1 GENERAL

This Appendix E establishes conditions and prices under which PGandE shall pay for firm capacity.

PGandE's obligation to pay for firm capacity shall begin on the firm capacity availability date. The firm capacity price shall be subject to adjustment as provided for in this Appendix E.

The firm capacity prices in Table E-2 are applicable for deliveries of firm capacity beginning after December 30, 1982.

## E-2 MINIMUM PERFORMANCE REQUIREMENTS

(a) To receive full capacity payments, the firm capacity shall be delivered for all of the on-peak hours[1] in the peak months on the PGandE system, which are presently the months of June, July, and August, subject to a 20 percent allowance for forced outages in any month.

---

[1] On-peak, partial-peak, and off-peak hours are defined in Table B-4, Appendix B.

Compliance with this provision shall be based on the Facility's total on-peak deliveries for each of the peak months and shall exclude any energy associated with generation levels greater than the firm capacity.

(b) If Seller is prevented from meeting the minimum performance requirements because of a forced outage on the PGandE system, a PGandE curtailment of Seller's deliveries, or a condition set forth in Section A-7, Appendix A, PGandE shall continue capacity payments. Firm capacity payments will be calculated in the same manner used for scheduled maintenance outages.

(c) If Seller is prevented from meeting the minimum performance requirements because of force majeure, PGandE shall continue capacity payments for ninety days from the occurrence of the force majeure. Thereafter, Seller shall be deemed to have failed to have met the minimum performance requirements. Firm capacity payments will be calculated in the same manner used for scheduled maintenance outages.

(d) If Seller is prevented from meeting the minimum performance requirements because of exteme dry year conditions, PGandE shall continue capacity payments. Extreme dry year conditions are drier than those used to establish firm capacity pursuant to Section E-8. Seller shall warrant to PGandE that the Facility is a hydroelectric facility and that such conditions are the sole cause of Seller's inability to meet its firm capacity obligations.

(e) If Seller is prevented from meeting the minimum performance requirements for reasons other than those described above in Sections E-2(b), (c), or (d):

(1) Seller shall receive the reduced firm capacity payments as provided in Section E-5 for a probationary period not to exceed 15 months, or as otherwise agreed to by the Parties.

(2) If, at the end of the probationary period Seller has not demonstrated that the Facility can meet the minimum performance requirements, PGandE may derate the firm capacity pursuant to Section E-4(b).

E-3 SCHEDULED MAINTENANCE

Outage periods for scheduled maintenance shall not exceed 840 hours (35 days) in any 12-month period. This allowance may be used in increments of an hour or longer on a consecutive or nonconsecutive basis. Seller may accumulate unused maintenance hours from one 12-month period to another up to a maximum of 1,080 hours (45 days). This accrued time must be used consecutively and only for major overhauls. Seller shall provide PGandE with the following advance notices: 24 hours for scheduled outages less than one day, one week for a scheduled outage of one day or more (except for major overhauls), and six months for a major overhaul. Seller shall not schedule major overhauls during the peak months (presently June, July and August). Seller shall make reasonable efforts to schedule or reschedule routine maintenance outside the peak months, and in no event shall outages for scheduled maintenance exceed 30 peak hours during the peak months. Seller shall confirm in writing to PGandE pursuant to Article 9, within 24 hours of the original notice, all notices Seller gives personally or by telephone for scheduled maintenance.

If Seller has selected Curtailment Option B, off-peak hours of maintenance performed pursuant to Section (d) of Curtailment Option B, Appendix C shall not be deducted from Seller's scheduled maintenance allowances set forth above.

E-4 ADJUSTMENTS TO FIRM CAPACITY

(a) Seller may increase the firm capacity with the approval of PGandE and receive payment for the additional capacity thereafter in accordance with the applicable capacity purchase price published by PGandE at the time the increase is first delivered to PGandE.

(b) Seller may reduce the firm capacity at any time prior to the firm capacity availability date by giving written notice thereof to PGandE. PGandE may derate the firm capacity as a result of appropriate tests, studies, or prior performance.

E-5 FIRM CAPACITY PAYMENTS

The method for calculation of firm capacity payments is shown below. As used below in this section, month refers to a calendar month.

The monthly payment for firm capacity will be the product of the Period Price Factor (PPF), the Monthly Delivered Capacity (MDC), the appropriate capacity loss adjustment factor from Table E-1 based on the Facility's interconnection voltage, and the appropriate performance bonus factor, if any, from Table E-3, plus any allowable payment for outages due to scheduled maintenance. The firm capacity price shall be applied to meter readings taken during the separate times and periods as illustrated in Table B-4, Appendix B.

The PPF is determined by multiplying the firm capacity price by the following Allocation Factors[1]:

| | Allocation Factor | x | Firm Capacity Price | = | PPF ($/kW-month) |
|---|---|---|---|---|---|
| Seasonal Period A | .16479 | | ______ | | ______ |
| Seasonal Period B | .02515 | | ______ | | ______ |

[1] All allocation factors are subject to change by PGandE based on PGandE's marginal capacity cost allocation, as determined in general rate case proceedings before the CPUC. Seasonal Periods A and B are defined in Table B-4, Appendix B.

S.O. #4

The MDC is determined in the following manner:

(1) Determine the Performance Factor (P), which is defined as the lesser of 1.0 or the following quantity:

$$P = \frac{A}{C \times (B-S) \times (0.8^{*})} \qquad (\leq 1.0)$$

Where:

A = Total kilowatt-hours delivered during all on-peak and partial-peak hours excluding any energy associated with generation levels greater than the firm capacity.

C = Firm capacity in kilowatts.

B = Total on-peak and partial-peak hours during the month.

S = Total on-peak and partial-peak hours during the month Facility is out of service on scheduled maintenance.

(2) Determine the Monthly Capacity Factor (MCF), which is computed using the following expression:

$$MCF = P \times (1.0 - \frac{M}{D})$$

Where:

M = The number of hours during the month Facility is out of service on scheduled maintenance.

D = The number of hours in the month.

---

* 0.8 reflects a 20% allowance for forced outage.

S.O. #4

(3) Determine the MDC by multiplying the MCF by C:

MDC (kilowatts) = MCF x C

The monthly payment for firm capacity is then determined by multiplying the PPF by the MDC, by the appropriate capacity loss adjustment factor presented from Table E-1, and by the appropriate performance bonus factor, if any, from Table E-3.

monthly payment for firm capacity = PPF x MDC x capacity loss adjustment factor x performance bonus factor

Furthermore, the payment for a month in which there is an outage for scheduled maintenance shall also include an amount equal to the product of the average hourly firm capacity payment[1] for the most recent month in the same type of Seasonal Period (i.e., Seasonal Period A or Seasonal Period B) during which deliveries were made times the number of hours of outage for scheduled maintenance in the current month. Firm capacity payments will continue during the outage periods for scheduled maintenance provided that the provisions of Section E-3 are met.

During a probationary period Seller's monthly payment for firm capacity shall be determined by substituting for the firm capacity, the capacity at which

---

1 Total monthly payment divided by the total number of hours in the monthly billing period.

Seller would have met the minimum performance requirements. In the event that during the probationary period Seller does not meet the minimum performance requirements at whatever firm capacity was established for the previous month, Seller's monthly payment for firm capacity shall be determined by substituting the firm capacity at which Seller would have met the minimum performance requirements. The performance bonus factor shall not be applied during probationary periods.

TABLE E-1

If the Facility is non-remote[1] the firm capacity loss adjustment factors are as follows:

| Interconnection Voltage | Capacity Loss Adjustment Factor |
|---|---|
| Transmission | .989 |
| Distribution | .991 |

If the Facility is remote the firm capacity loss adjustment factor is _____________[2].

1 As defined by the CPUC.

2 Determined individually.

TABLE E-2

## Firm Capacity Price Schedule

(Levelized $/kW-year)

| Firm Capacity Avail-ability Date | Number of Years of Firm Capacity Delivery | | | | | | | | | | | | | | | | | |
|---|---|---|---|---|---|---|---|---|---|---|---|---|---|---|---|---|---|---|
| (Year) | 1 | 2 | 3 | 4 | 5 | 6 | 7 | 8 | 9 | 10 | 11 | 12 | 13 | 14 | 15 | 20 | 25 | 30 |
| 1982 | 65 | 68 | 70 | 72 | 75 | 77 | 79 | 81 | 84 | 86 | 88 | 90 | 91 | 93 | 95 | 103 | 109 | 113 |
| 1983 | 70 | 73 | 75 | 78 | 80 | 83 | 85 | 88 | 90 | 92 | 94 | 96 | 98 | 100 | 102 | 110 | 117 | 122 |
| 1984 | 76 | 78 | 81 | 84 | 86 | 89 | 92 | 94 | 97 | 99 | 101 | 103 | 106 | 108 | 110 | 118 | 125 | 130 |
| 1985 | 81 | 84 | 87 | 90 | 93 | 96 | 99 | 101 | 104 | 106 | 109 | 111 | 113 | 115 | 118 | 127 | 134 | 140 |
| 1986 | 88 | 91 | 94 | 97 | 100 | 103 | 106 | 109 | 112 | 114 | 117 | 119 | 122 | 124 | 126 | 136 | 144 | 150 |
| 1987 | 95 | 98 | 101 | 105 | 108 | 111 | 114 | 117 | 120 | 123 | 125 | 128 | 130 | 133 | 135 | 146 | 154 | 160 |

S.O. #4

TABLE E-3

Performance Bonus Factor

The following shall be the performance bonus factors applicable to the calculation of the monthly payments for firm capacity delivered by the Facility after it has demonstrated a firm capacity factor in excess of 85%.

| DEMONSTRATED FIRM CAPACITY FACTOR (%) | PERFORMANCE BONUS FACTOR |
|---|---|
| 85 | 1.000 |
| 90 | 1.059 |
| 95 | 1.118 |
| 100 | 1.176 |

After the Facility has delivered power during the span of all of the peak months on the PGandE system (presently June, July, and August) in any year (span),

(i) the firm capacity factor for each such month shall be calculated in the following manner:

$$\text{FIRM CAPACITY FACTOR (\%)} = \frac{A}{B \times C} \times 100$$

Where:

A = Total kilowatt-hours delivered by Seller in any peak month during all on-peak hours excluding any energy associated with generation levels greater than the firm capacity.

B = Total on-peak hours during the month.

C = Firm capacity in kilowatts.

(ii) the arithmetic average of the above firm capacity factors shall be determined for that span,

(iii) the average of the above arithmetic average firm capacity factors for the most recent span(s), not to exceed 5, shall be calculated and shall become the Demonstrated Firm Capacity Factor.

To calculate the performance bonus factor for a Demonstrated Firm Capacity Factor not shown in Table E-3 use the following formula:

$$\text{Performance Bonus Factor} = \frac{\text{Demonstrated Firm Capacity Factor (\%)}}{85\%}$$

SECTIONS E-6 THROUGH E-10 SHALL APPLY ONLY TO HYDROELECTRIC PROJECTS

E-6 DETERMINATION OF NATURAL FLOW DATA

Natural flow data shall be based on a period of record of at least 50 years and which includes historic critically dry periods. In the event Seller demonstrates that a natural flow data base of at least 50 years would be

S.O. #4

unreasonably burdensome, PGandE shall accept a shorter period of record with a corresponding reduction in the averaging basis set forth in Section E-8. Seller shall determine the natural flow data by month by using one of the following methods:

Method 1

If stream flow records are available from a recognized gauging station on the water course being developed in the general vicinity of the project, Seller may use the data from them directly.

Method 2

If directly applicable flow records are not available, Seller may develop theoretical natural flows based on correlation with available flow data for the closest adjacent and similar area which has a recognized gauging station using generally accepted hydrologic estimating methods.

E-7 THEORETICAL OPERATION STUDY

Based on the monthly natural flow data developed under Section E-6 a theoretical operation study shall be prepared by Seller. Such a study shall identify the monthly capacity rating in kW and the monthly energy production in kWh for each month of each year. The study shall take into account all relevant operating constraints, limitations, and requirements including but not limited to --

(1) Release requirements for support of fish life and any other operating constraints imposed on the project;

(2) Operating characteristics of the proposed equipment of the Facility such as efficiencies, minimum and maximum operating levels, project control procedures, etc.;

(3) The design characteristics of project facilities such as head losses in penstocks, valves, tailwater elevation levels, etc.; and

(4) Release requirements for purposes other than power generation such as irrigation, domestic water supply, etc.

The theoretical operation study for each month shall assume an even distribution of generation throughout the month unless Seller can demonstrate that the Facility has water storage characteristics. For the study to show monthly capacity ratings, the Facility shall be capable of operating during all on-peak hours in the peak months on the PGandE system, which are presently the months of June, July, and August. If the project does not have this capability throughout each such month, the capacity rating in that month of that year shall be set at zero for purposes of this theoretical operation study.

E-8 DETERMINATION OF AVERAGE DRY YEAR CAPACITY RATINGS

Based on the results of the theoretical operation study developed under Section E-7, the average dry year capacity rating shall be established for each month. The average dry year shall be based on the average of the five years of the lowest annual generation as shown in the theoretical operation study. Once such years of lowest annual generation are identified, the monthly capacity rating is determined for each month by averaging the capacity ratings from each month of those years. The firm capacity shown in Article 5 shall not exceed the lowest average dry year monthly capacity ratings for the peak months on the PGandE system, which are presently the months of June, July, and August.

E-9 INFORMATION REQUIREMENTS

Seller shall provide the following information to PGandE for its review:

(1) A summary of the average dry year capacity ratings based on the theoretical operation study as provided in Table E-4;

(2) A topographic project map which shows the location of all aspects of the Facility and locations of stream gauging stations used to determine natural flow data;

(3) A discussion of all major factors relevant to project operation;

(4) A discussion of the methods and procedures used to establish the natural flow data. This discussion shall be in sufficient detail for PGandE to determine that the methods are consistent with those outlined in Section E-6 and are consistent with generally accepted engineering practices; and

(5) Upon specific written request by PGandE, Seller's theoretical operation study.

E-10 ILLUSTRATIVE EXAMPLE

(1) Determine natural flows - These flows are developed based on historic stream gauging records and are compiled by month, for a long-term period (normally at least 50 years or more) which covers dry periods which historically occurred in the 1920's and 30's and more recently in 1976 and 77. In all but unusual situations this will require application of hydrological engineering methods to records that are available, primarily from the USGS publication "Water Resources Data for California".

(2) Perform theoretical operation study - Using the natural flow data compiled under (1) above a theoretical operation study is prepared which determines, for each month of each year, energy generation (kWh) and capacity rating (kW). This study is performed based on the <u>Facility's</u> design, operating capabilities, constraints, etc., and should take into account all factors relevant to project

operation. Generally such a study is done by computer which routes the natural flows through project features, considering additions and withdrawals from storage, spill past the project, releases for support of fish life, etc., to determine flow available for generation. Then the generation and capacity amounts are computed based on equipment performance, efficiencies, etc.

(3) Determine average dry year capacity ratings - After the theoretical project operation study is complete the five years in which the annual generation (kWh) would have been the lowest are identified. Then for each month, the capacity rating (kW) is averaged for the five years to arrive at a monthly average capacity rating. The <u>firm capacity</u> is then set by the Seller based on the monthly average dry year capacity ratings and the minimum performance requirements of this appendix. An example project is shown in the attached completed Table E-4.

S.O. #4

EXAMPLE

TABLE E-4

Summary of Theoretical Operation Study

Project: New Creek 1

Water Source: West Fork New Creek

Mode of Operation: Run of the river

Type of Turbine: Francis Design Flow: 100 cfs, Design Head: 150 feet

Operating Characteristics:[1]

| | Flow (cfs) | Head (feet) Gross | Net | Output (kW) | Efficiency (%) Turbine | Generator |
|---|---|---|---|---|---|---|
| Normal Operation | 100 | 160 | 150 | 1,120 | 90 | 98 |
| Maximum Operation | 110 | 160 | 148 | 1,150 | 85 | 98 |
| Minimum Operation | 30 | 160 | 155 | 290 | 75 | 98 |

Average Dry Year Operation - Based on the average of the following lowest generation years: 1930, 1932, 1934, 1949, 1977.

| Month | Energy Generation (kWh) | Capacity Output (kW) | Percent of Total Hours Operated |
|---|---|---|---|
| January | 855,000 | 1,150 | 100 |
| February | 753,000 | 1,120 | 100 |
| March | 818,000 | 1,100 | 100 |
| April | 727,000 | 1,010 | 100 |
| May | 699,000 | 940 | 100 |
| June | 612,000 | 850 | 100 |
| July | 484,000 | 650 | 100 |
| August | 305,000 | 410 | 100 |
| September | 245,000 | 340 | 100 |
| October | 148,800 | 200 | 100 |
| November | 468,000 | 650 | 100 |
| December | 595,000 | 800 | 100 |

Maximum firm capacity: 410 kW

---

1 If Facility has a variable head, operating curves should be provided.

S.O. #4

E-11 MINIMUM DAMAGES

(a) In the event the firm capacity is derated or Seller terminates this Agreement, the quantity by which the firm capacity is derated or the firm capacity shall be used to calculate the payments due PGandE in accordance with Section (d).

(b) Seller shall be invoiced by PGandE for all amounts due under this section. Payment shall be due within 30 days of the date of invoice.

(c) If Seller does not make payments pursuant to Section (b), PGandE shall have the right to offset any amounts due it against any present or future payments due Seller.

(d) Seller shall pay to PGandE:

(i) an amount equal to the difference between (a) the firm capacity payments already paid by PGandE, based on the original term of agreement and (b) the total firm capacity payments which PGandE would have paid based on the period of Seller's actual performance using the adjusted firm capacity price. Additionally, Seller shall pay interest, compounded monthly from the date the excess capacity payment was made until the date

Seller repays PGandE, on all overpayments, at the published Federal Reserve Board three months' Prime Commercial Paper rate; plus

(ii) a sum equal to the amount by which the firm capacity is being terminated or derated times the difference between the current firm capacity price on the date of termination or deration for a term equal to the balance of the term of agreement and the firm capacity price, multiplied by the appropriate factor shown in Table E-5 below. In the event that the current firm capacity price is less than the firm capacity price, no payment under this subsection (ii) shall be due either Party.

TABLE E-5

| Amount of Firm Capacity Terminated or Derated | Factor |
|---|---|
| 1,000 kW or under | 0.25 |
| over 1,000 kW through 10,000 kW | 0.75 |
| over 10,000 kW through 25,000 kW | 1.00 |
| over 25,000 kW through 50,000 kW | 3.00 |
| over 50,000 kW through 100,000 kW | 4.00 |
| over 100,000 kW | 5.00 |

# APPENDIX L

## GAS TURBINE GENERATOR AND STEAM TURBINE GENERATOR SPECIFICATIONS

# 5 Turbomachinery Specifications

## TURBOMACHINERY SPECIFICATIONS

| Manufacturer | Gas Turbine | Gas Turbine Generating Set | Turbo Compressor | Turbo Expander | Hydro Turbine | Steam Turbine |
|---|---|---|---|---|---|---|
| ACEC | 326 | 362 | | | | 403 |
| AEG-KANIS Turbinenfabrik | 326 | | | | | 403 |
| AiResearch Manufacturing | 326 | | | | | |
| Allis-Chalmers | | | 372 | | 400 | |
| Alsthom-Atlantique | 328 | | 372 | 392 | | 403 |
| Alturdyne | | 362 | 372 | | | |
| Ateliers Bouvier | | | | | 400 | |
| Atlas-Copco Turbonetics | | | 372 | 392 | | |
| Avco Industrial Engine | | 362 | | | | |
| Avco Lycoming | 328 | | | | | |
| BBC Brown Boveri | 328 | | | | | 403 |
| Bet-Shemesh Engines | 328 | | | | | |
| Blohm & Voss | | | | | | 403 |
| Borsig GmbH | | | 374 | 392 | | 403 |
| Bouvier Hydropower | | | | | 400 | |
| Brown Boveri Turbomachinery | | | | | | 403 |
| Cantieri Navali Ruiniti | | | | | | |
| Carling Turbine Blower | | | | | | 404 |
| CE/N Hydro Power | | | | | 400 | |
| Centrax Ltd. | 330 | 362 | | | | |
| Commonwealth Aircraft | 330 | | | | | |
| Cooper Energy Services | | | 374 | | | |
| Cooper Rolls | 330 | | | | | |
| Coppus Engineering | | | | | | 404 |
| Creusot-Loire | 330 | 362 | 376 | | | 404 |
| Curtiss-Wright | 330 | | | | | |
| Detroit Diesel Allison | 332 | 364 | | | | |
| Dresser Clark | 332 | | 376 | 392 | | |
| Elliott Company | | | 376 | 392 | | 404 |
| Escher Wyss Ltd. Hydraulics | | | | | 401 | |
| Fabrique Nationale Herstal | 332 | | | | | |
| Fiat TTG | 332 | | | | | |
| First Brno Engineering | | | | 392 | | |
| Garrett Turbine Engine | 332 | | | | | |
| Ganz-Mavag | | | | | 401 | |
| GEC Gas Turbines | 334 | 364 | | | | |
| GEC Turbine Generators | | | | | | 404 |
| GE Co.-Marine & Industrial Engine Division | 336 | | | | | |
| GE Co.-Mechanical Drive | | | 378 | | | |
| GE Co.-Steam Turbine Dept. | | | | | | 404 |
| GE Co.-Gas Turbine Division | 334 | | | | | |
| George Engine Company | | 364 | | | | |
| Hants & Sussex Aviation | 338 | | | | | |
| Hayward Tyler | | | | | 401 | 404 |
| Hispano Suiza | 338 | 364 | 378 | | | |
| Hitachi Ltd. | 338 | | 380 | | | |
| I.A.M. Rinaldo Piaggio | 338 | | | | | |
| Ingersoll-Rand | 338 | | 380 | 394 | | |
| Ishikawajima-Harima Heavy Industries (IHI, Inc.) | 340 | 366 | 382 | 394 | | 404 |
| John Brown Engineering | 340 | | | | | |
| Joy Manufacturing | | | 382 | | | |
| Kawasaki Heavy Industries | 340 | 366 | 382 | | | 404 |

## MASTER INDEX



Turbomachinery International Handbook

Turbomachinery Specifications

# CONVERSIONS AND EQUIVALENTS

## Power Conversions

| | W | ft.pdl/s | cal/s | kcal/h | Btu/s | Btu/h | ft.lbf/s | Chu/s | Hp |
|---|---|---|---|---|---|---|---|---|---|
| 1 W | 1 | 23.7 | 0.239 | 0.860 | 9.48 − 04 | 3.41 | 0.738 | 5.27 − 04 | 1.341 − 03 |
| 1 ft.pdl/s | 4.21 − 02 | 1 | 1.007 − 02 | 3.63 − 02 | 3.99 − 05 | 0.1438 | 3.11 − 02 | 2.22 − 05 | 5.64 − 05 |
| 1 cal/s | 4.18 | 99.3 | 1 | 3.6 | 3.97 − 03 | 14.28 | 3.09 | 2.20 − 03 | 5.61 − 03 |
| 1 kcal/h | 1.162 | 27.6 | 0.278 | 1 | 1.102 − 03 | 3.97 | 0.857 | 6.12 − 04 | 1.559 − 03 |
| 1 Btu/s | 1.055 + 03 | 2.50 + 04 | 252 | 908 | 1 | 3.60 + 03 | 778 | 0.556 | 1.415 |
| 1 Btu/h | 0.293 | 6.96 | 7.00 − 02 | 0.252 | 2.78 − 04 | 1 | 0.216 | 1.543 − 04 | 3.93 − 04 |
| 1 ft.lbf/s | 1.356 | 32.2 | 0.324 | 1.167 | 1.285 − 03 | 4.63 | 1 | 7.14 − 04 | 1.818 − 03 |
| 1 Chu/s | 1.899 + 03 | 4.51 + 04 | 454 | 1.634 + 03 | 1.8 | 6.48 + 03 | 1.401 + 03 | 1 | 2.55 |
| 1 Hp | 74 | 1.774 + 04 | 178.2 | 642 | 0.707 | 2.54 + 03 | 550 | 0.393 | 1 |

1 W = 1 J/s
1 ft.pdl/s = 1 lbm.ft$^2$/s$^3$
1 ft.lbf/s = 1 slug.ft$^2$/s$^3$
NOTE: 1 + 04 corresponds to 1 x $10^4$
1 − 04 corresponds to 1 x $10^{-4}$

## Energy Conversions

| | J | Cal | Btu | Chu | kW.h | Hp.h | ft.lbf | ft.pdl | erg. |
|---|---|---|---|---|---|---|---|---|---|
| 1 J | 1 | 0.239 | 9.48 − 04 | 5.27 − 04 | 2.78 − 07 | 3.72 − 07 | 0.738 | 23.7 | 1 + 07 |
| 1 Cal | 4.18 | 1 | 3.97 − 03 | 2.20 − 03 | 1.162 − 06 | 1.559 − 06 | 3.09 | 99.3 | 4.18 + 07 |
| 1 Btu | 1.055 + 03 | 252 | 1 | 0.556 | 2.93 − 04 | 3.93 − 04 | 778 | 2.50 + 04 | 1.055 + 10 |
| 1 Chu | 1.889 + 03 | 454 | 1.8 | 1 | 5.28 − 04 | 7.07 − 04 | 1.401 + 03 | 4.51 + 04 | 1.899 + 10 |
| 1 kW.h | 3.60 + 06 | 8.60 + 05 | 3.41 + 03 | 1.898 + 03 | 1 | 1.341 | 2.66 + 06 | 8.54 + 07 | 3.60 + 13 |
| 1 Hp.h | 2.68 + 06 | 6.42 + 05 | 2.54 + 03 | 1.41 + 03 | 0.746 | 1 | 1.980 + 06 | 6.37 + 07 | 2.68 + 13 |
| 1 ft.lbf | 1.356 | 0.324 | 1.285 − 03 | 7.14 − 04 | 3.77 − 07 | 5.05 − 07 | 1 | 32.2 | 1.356 + 07 |
| 1 ft.pdl | 4.21 − 02 | 1.007 − 02 | 3.99 − 05 | 2.22 − 05 | 1.171 − 08 | 1.570 − 08 | 3.11 − 02 | 1 | 4.21 + 05 |
| 1 erg | 1 − 07 | 2.39 − 08 | 9.48 − 11 | 5.27 − 11 | 2.78 − 14 | 3.72 − 14 | 7.38 − 08 | 2.37 − 06 | 1 |

1 J = 1 W.s = 1 N.m
1 ft. lbf = 1 slug.ft$^2$/s$^2$
1 ft. poundal = 1 lbm.ft$^2$/s$^2$
1 erg = 1 dyne.cm
NOTE: 1 + 04 corresponds to 1 x $10^4$
1 − 04 corresponds to 1 x $10^{-4}$

## Pressure Conversions

| | Pa | bar | atm | Torr | in $H_2O$ | in Hg | psi | pdl/ft$^2$ | lbf/ft$^2$ |
|---|---|---|---|---|---|---|---|---|---|
| 1 Pa | 1 | 1 + 05 | 9.87 − 06 | 7.50 − 03 | 4.20 − 03 | 2.95 − 04 | 1.450 − 04 | 0.672 | 2.09 − 02 |
| 1 bar | 1 + 05 | 1 | 0.987 | 750 | 402 | 29.5 | 14.5 | 6.72 + 04 | 2.09 + 03 |
| 1 atm | 1.013 + 05 | 1.013 | 1 | 760 | 407 | 29.9 | 14.70 | 6.81 + 04 | 2.12 + 03 |
| 1 Torr | 133.3 | 1.33 − 03 | 1.316 − 03 | 1 | 0.535 | 3.94 − 02 | 1.934 − 02 | 89.6 | 2.78 |
| 1 in $H_2O$ | 249 | 2.49 − 03 | 2.46 − 03 | 1.868 | 1 | 7.36 − 02 | 3.61 − 02 | 167.4 | 5.20 |
| 1 in Hg | 3.39 + 03 | 3.39 − 02 | 3.34 − 02 | 25.4 | 13.60 | 1 | 0.491 | 2.28 + 03 | 70.7 |
| 1 psi | 6.90 + 03 | 6.90 − 02 | 6.80 − 02 | 51.7 | 27.7 | 2.04 | 1 | 4.63 + 03 | 144 |
| 1 pdl/ft$^2$ | 1.488 | 1.488 − 05 | 1.469 − 05 | 1.116 − 02 | 5.97 − 03 | 4.40 − 04 | 2.16 − 04 | 1 | 3.11 − 02 |
| 1 lbf/ft$^2$ | 47.9 | 4.79 − 04 | 4.72 − 04 | 0.359 | 0.1922 | 1.414 − 02 | 6.94 − 03 | 32.2 | 1 |

1 Pa = 1 N/m$^2$
1 poundal/ft$^2$ = 1 lbm./(ft.s$^2$)
1 lbf/ft$^2$ = 1 slug/(ft.s$^2$)
1 Torr = 1 mm Hg
NOTE: 1 + 04 corresponds to 1 x $10^4$
1 − 04 corresponds to 1 x $10^{-4}$

## Conversion Factors

| | *SI to British* | *British to SI* |
|---|---|---|
| **Area** | 1 square meter = 10.76 $ft^2$ | 1 square foot = 92.9 x $10^{-3} m^2$ |
| **Calorific Value** | 1 J/kg = 4.300 x $10^{-4}$ Btu/lbm | 1 Btu/lbm = 2326 J/kg |
| | 1 $J/m^3$ = 26.8 x $10^{-6}$ $Btu/ft^3$ | 1 $Btu/ft^3$ = 37.25 x $10^3$ $J/m^3$ |
| **Density** | 1 $kg/m^3$ = 0.0624 $lbm/ft^3$ | 1 $lbm/ft^3$ = 16.02 $kg/m^3$ |
| **Energy** | 1 joule = 0.948 x $10^{-3}$ Btu | 1 Btu = 1055 J |
| | 1 J = 0.527 x $10^{-3}$ Chu | 1 Chu = 1899 J |
| **Force** | 1 newton = 0.225 lbf | 1 pound force = 4.448 N = 4.44k $kg.m.s^{-2}$ |
| | 1 N = 7.233 pdl (= lbm. $ft/s^2$) | 1 poundal = 0.138 N |
| **Length** | 1 meter = 3.28 ft | 1 foot = 0.3048m |
| | 1 meter = 39.4 in. | 1 inch = 25.4 x $10^{-3}$m |
| **Mass** | 1 kilogram = 2.205 lbm | 1 pound mass = 0.4536 kg |
| **Molar Mass** | (generally quoted as "Molecular weight," gram/mole) | |
| | 1 kg/mol = 1000 lbm/lbmole | 1 lbm/lbmole = $10^{-3}$ kg/mol |
| **Power** | 1 watt = 1.34 x $10^{-3}$ Hp | 1 Horse power = 745.7 W |
| **Specific Heat** | 1 J/(kg.K) = 0.239 x $10^{-3}$ Btu/(lbm.R) | 1 Btu/(lbm.R) = 4187 J/(kg.K) |

# LEGEND FOR SPECIFICATIONS

## Use Symbols/Codes

| | | | | | |
|---|---|---|---|---|---|
| AA | Air Supply | G-4 | Generator drive, closed cycle | P-W** | Pump, water |
| C-1 | Compressor drive, air | G-5 | Generator drive, emergency set | PI | Process industry |
| C-2 | Compressor drive, gas | G-6 | Generator drive, package plant | PR-1 | Propulsion, aircraft |
| D | De-Icer | G-7 | Generator drive, standby set | PR-2 | Propulsion, marine |
| E | Expander | G-8 | Generator drive, marine | PR-3 | Propulsion, railway |
| G-1 | Generator drive, electric utility | L-1 | Lift, aircraft | PR-4 | Propulsion, vehicle, land, nonrail |
| G-2 | Generator drive, industrial | L-2 | Lift, marine | R | Regasification of liquefied natural gas |
| G-3 | Generator drive, combined cycle | P-O* | Pump, oil | GG | Gas Generator Power Supply |

These are typical but not exclusive uses. For additional applications, and detailed engine data, consult with the manufacturer. You will find their addresses and telephone numbers in the heading of their specifications, and/or in Section 4 PRODUCT DIRECTORY.

*Previous Symbol/Code P-1
**Previous Symbol/Code P-2

## What The Designations Mean

**Model:** These are the manufacturer's designation: Numbers, letters or combination of both.

**Use:** See explanation above.

**Power Rating:** Indicated by hp, mw or lbt. The hp or mw ratings are at International Standards Organization (ISO) Conditions: sea level and 15°C (59°F), as indicated.

**Heat Rate:** BTU per hp-hr, BTU per kw-hr, BTU per lbt-hr, SFC, lb per lbt-hr, as indicated.

**Flow:** Mass flow, pounds per second or kg per sec, as indicated.

**Heat recovery:** If heat recovery unit is available with turbine, mfg. has indicated by symbol "Y".

**Temperature:** Temperature as indicated by °C or °F.

**Dry weight:** Kg or lbs. If heat exchanger for exhaust heat recovery is included, mfg. has indicated.

**Dimensions:** The dimensional length, width, height in MM or INCHES as indicated by mfg.

# INTERNATIONAL DIALING

## World Time Chart*

**To determine the time in the countries listed below, add the number of hours shown under your time zone (or subtract, if preceded by a minus sign) to your local time.**
**EST = Eastern Standard Time, CST = Central Standard Time, MST = Mountain Standard Time and PST = Pacific Standard Time. (Differences based on Standard Time in the U.S., which is observed from the last Sunday in October until the last Sunday in April. This may vary in other countries.)**

| Time Difference To — | U.S. Time Zones | | | |
|---|---|---|---|---|
| | EST | CST | MST | PST |
| Afghanistan | 9½ | 10½ | 11½ | 12½ |
| Alaska (Anchorage)† | −5 | −4 | −3 | −2 |
| Algeria | 5 | 6 | 7 | 8 |
| American Samoa | −6 | −5 | −4 | −3 |
| Andorra | 6 | 7 | 8 | 9 |
| Angola | 6 | 7 | 8 | 9 |
| Anguilla | 1 | 2 | 3 | 4 |
| Antarctica | 1 | 2 | 3 | 4 |
| Antiqua | 1 | 2 | 3 | 4 |
| Argentina | 2 | 3 | 4 | 5 |
| Ascension Island | 5 | 6 | 7 | 8 |
| Australia (Sydney)† | 16 | 17 | 18 | 19 |
| Austria | 6 | 7 | 8 | 9 |
| Azores | 4 | 5 | 6 | 7 |
| Bahamas | 0 | 1 | 2 | 3 |
| Bahrain | 8 | 9 | 10 | 11 |
| Balearic Islands | 6 | 7 | 8 | 9 |
| Bangladesh | 11 | 12 | 13 | 14 |
| Barbados | 1 | 2 | 3 | 4 |
| Belgium | 6 | 7 | 8 | 9 |
| Belize | −1 | 0 | 1 | 2 |
| Benin | 6 | 7 | 8 | 9 |
| Bequia | 1 | 2 | 3 | 4 |
| Bermuda | 1 | 2 | 3 | 4 |
| Bhutan | 10½ | 11½ | 12½ | 13½ |
| Bolivia | 1 | 2 | 3 | 4 |
| Botswana | 7 | 8 | 9 | 10 |
| Brazil (Rio de Janeiro)† | 2 | 3 | 4 | 5 |
| British Virgin Islands | 1 | 2 | 3 | 4 |
| Brunei | 13 | 14 | 15 | 16 |
| Bulgaria | 7 | 8 | 9 | 10 |
| Burma | 11½ | 12½ | 13½ | 14½ |
| Burundi | 7 | 8 | 9 | 10 |
| Cameroon | 6 | 7 | 8 | 9 |
| Canada (Ottawa)† | 0 | 1 | 2 | 3 |
| Canary Islands | 5 | 6 | 7 | 8 |
| Cape Verde Islands | 4 | 5 | 6 | 7 |
| Carriacou | 1 | 2 | 3 | 4 |
| Cayman Islands | 0 | 1 | 2 | 3 |
| Central African Republic | 6 | 7 | 8 | 9 |
| Chad | 6 | 7 | 8 | 9 |
| Channel Islands | 5 | 6 | 7 | 8 |
| Chatham Island | 17¾ | 18¾ | 19¾ | 20¾ |
| Chile | 2 | 3 | 4 | 5 |
| China | 13 | 14 | 15 | 16 |
| Christmas & Cocos Is. | 12 | 13 | 14 | 15 |
| Columbia | 0 | 1 | 2 | 3 |
| Comoros | 8 | 9 | 10 | 11 |
| Congo | 6 | 7 | 8 | 9 |
| Cook Islands | −5½ | −4½ | −3½ | −2½ |
| Costa Rica | −1 | 0 | 1 | 2 |
| Cuba | 0 | 1 | 2 | 3 |
| Cyprus | 7 | 8 | 9 | 10 |
| Czechoslovakia | 6 | 7 | 8 | 9 |
| Democratic Kampuchea | 12 | 13 | 14 | 15 |
| Denmark | 6 | 7 | 8 | 9 |
| Djibouti | 8 | 9 | 10 | 11 |
| Dominica | 1 | 2 | 3 | 4 |
| Dominican Republic | 1 | 2 | 3 | 4 |
| Easter Island | −1 | 0 | 1 | 2 |
| Ecuador | 0 | 1 | 2 | 3 |
| Egypt | 7 | 8 | 9 | 10 |
| El Salvador | −1 | 0 | 1 | 2 |
| England | 5 | 6 | 7 | 8 |
| Equatorial Guinea | 6 | 7 | 8 | 9 |
| Ethiopia | 8 | 9 | 10 | 11 |
| Faeroe Islands | 5 | 6 | 7 | 8 |
| Falkland Islands | 2 | 3 | 4 | 5 |
| Fiji | 17 | 18 | 19 | 20 |
| Finland | 7 | 8 | 9 | 10 |
| France | 6 | 7 | 8 | 9 |
| French Antilles | 1 | 2 | 3 | 4 |
| French Guiana | 2 | 3 | 4 | 5 |
| Gabon | 6 | 7 | 8 | 9 |

| Time Difference To — | U.S. Time Zones | | | |
|---|---|---|---|---|
| | EST | CST | MST | PST |
| Gambia | 5 | 6 | 7 | 8 |
| German Dem. Rep. | 6 | 7 | 8 | 9 |
| Germany, Fed. Rep. of | 6 | 7 | 8 | 9 |
| Ghana | 5 | 6 | 7 | 8 |
| Gibraltar | 6 | 7 | 8 | 9 |
| Greece | 7 | 8 | 9 | 10 |
| Greenland (Godthaab)† | 2 | 3 | 4 | 5 |
| Granada | 1 | 2 | 3 | 4 |
| Guam | 15 | 16 | 17 | 18 |
| Guatemala | −1 | 0 | 1 | 2 |
| Guinea | 5 | 6 | 7 | 8 |
| Guinea-Bissau | 5 | 6 | 7 | 8 |
| Guyana | 2 | 3 | 4 | 5 |
| Haiti | 0 | 1 | 2 | 3 |
| Hawaii | −5 | −4 | −3 | −2 |
| Honduras | −1 | 0 | 1 | 2 |
| Hong Kong | 13 | 14 | 15 | 16 |
| Hungary | 6 | 7 | 8 | 9 |
| Iceland | 5 | 6 | 7 | 8 |
| India | 10½ | 11½ | 12½ | 13½ |
| Indonesia (Jakarta)† | 12 | 13 | 14 | 15 |
| Iran | 8½ | 9½ | 10½ | 11½ |
| Iraq | 8 | 9 | 10 | 11 |
| Ireland | 5 | 6 | 7 | 8 |
| Isle of Man | 5 | 6 | 7 | 8 |
| Israel | 7 | 8 | 9 | 10 |
| Italy | 6 | 7 | 8 | 9 |
| Ivory Coast | 5 | 6 | 7 | 8 |
| Jamaica | 0 | 1 | 2 | 3 |
| Japan | 14 | 15 | 16 | 17 |
| Jordan | 7 | 8 | 9 | 10 |
| Kenya | 8 | 9 | 10 | 11 |
| Kiribati | 17 | 18 | 19 | 20 |
| Korea, Rep. of | 14 | 15 | 16 | 17 |
| Kuwait | 8 | 9 | 10 | 11 |
| Laos | 12 | 13 | 14 | 15 |
| Lebanon | 7 | 8 | 9 | 10 |
| Lesotho | 7 | 8 | 9 | 10 |
| Liberia | 5 | 6 | 7 | 8 |
| Libya | 7 | 8 | 9 | 10 |
| Liechtenstein | 6 | 7 | 8 | 9 |
| Luxembourg | 6 | 7 | 8 | 9 |
| Macao | 13 | 14 | 15 | 16 |
| Madagascar | 8 | 9 | 10 | 11 |
| Madeira Islands | 5 | 6 | 7 | 8 |
| Malawi | 7 | 8 | 9 | 10 |
| Malaysia (Kuala Lumpur)† | 12½ | 13½ | 14½ | 15½ |
| Maldives | 10 | 11 | 12 | 13 |
| Mali | 5 | 6 | 7 | 8 |
| Malta | 6 | 7 | 8 | 9 |
| Mauritania | 5 | 6 | 7 | 8 |
| Mauritius | 9 | 10 | 11 | 12 |
| Mexico (Mexico City)† | −1 | 0 | 1 | 2 |
| Midway | −6 | −5 | −4 | −3 |
| Monaco | 6 | 7 | 8 | 9 |
| Mongolia | 13 | 14 | 15 | 16 |
| Montserrat | 1 | 2 | 3 | 4 |
| Morocco | 5 | 6 | 7 | 8 |
| Mozambique | 7 | 8 | 9 | 10 |
| Mustique | 1 | 2 | 3 | 4 |
| Nauru | 17 | 18 | 19 | 20 |
| Nepal | 10½ | 11½ | 12½ | 13½ |
| Netherlands | 6 | 7 | 8 | 9 |
| Netherlands Antilles | 1 | 2 | 3 | 4 |
| Nevis | 1 | 2 | 3 | 4 |
| New Caledonia | 16 | 17 | 18 | 19 |
| New Zealand | 18 | 19 | 20 | 21 |
| Nicaragua | −1 | 0 | 1 | 2 |
| Niger | 6 | 7 | 8 | 9 |
| Nigeria | 6 | 7 | 8 | 9 |
| Niue | −6 | −5 | −4 | −3 |
| Norfolk Island | 16½ | 17½ | 18½ | 19½ |
| Northern Ireland | 5 | 6 | 7 | 8 |
| Norway | 6 | 7 | 8 | 9 |

| Time Difference To — | U.S. Time Zones | | | |
|---|---|---|---|---|
| | EST | CST | MST | PST |
| Oman | 9 | 10 | 11 | 12 |
| Pakistan | 10 | 11 | 12 | 13 |
| Palm Island | 1 | 2 | 3 | 4 |
| Panama | 0 | 1 | 2 | 3 |
| Papua New Guinea | 15 | 16 | 17 | 18 |
| Paraguay | 2 | 3 | 4 | 5 |
| Peru | 0 | 1 | 2 | 3 |
| Philippines | 13 | 14 | 15 | 16 |
| Poland | 6 | 7 | 8 | 9 |
| Portugal | 5 | 6 | 7 | 8 |
| Puerto Rico | 1 | 2 | 3 | 4 |
| Qatar | 8 | 9 | 10 | 11 |
| Reunion Island | 9 | 10 | 11 | 12 |
| Romania | 7 | 8 | 9 | 10 |
| Rwanda | 7 | 8 | 9 | 10 |
| St. Helena | 5 | 6 | 7 | 8 |
| St. Kitts | 1 | 2 | 3 | 4 |
| St. Lucia | 1 | 2 | 3 | 4 |
| St. Pierre & Miquelon | 2 | 3 | 4 | 5 |
| St. Vincent | 1 | 2 | 3 | 4 |
| Saipan | 15 | 16 | 17 | 18 |
| San Marino | 6 | 7 | 8 | 9 |
| Sao Tome | 5 | 6 | 7 | 8 |
| Saudi Arabia | 8 | 9 | 10 | 11 |
| Scotland | 5 | 6 | 7 | 8 |
| Senegal | 5 | 6 | 7 | 8 |
| Seychelles Islands | 9 | 10 | 11 | 12 |
| Sierra Leone | 5 | 6 | 7 | 8 |
| Singapore | 12½ | 13½ | 14½ | 15½ |
| Solomon Islands | 16 | 17 | 18 | 19 |
| Somali | 8 | 9 | 10 | 11 |
| South Africa | 7 | 8 | 9 | 10 |
| Spain | 6 | 7 | 8 | 9 |
| Spanish Sahara | 5 | 6 | 7 | 8 |
| Sri Lanka | 10½ | 11½ | 12½ | 13½ |
| Sudan | 7 | 8 | 9 | 10 |
| Suriname | 1½ | 2½ | 3½ | 4½ |
| Svalbard | 6 | 7 | 8 | 9 |
| Swaziland | 7 | 8 | 9 | 10 |
| Sweden | 6 | 7 | 8 | 9 |
| Switzerland | 6 | 7 | 8 | 9 |
| Syrian Arab Republic | 7 | 8 | 9 | 10 |
| Tahiti | −5 | −4 | −3 | −2 |
| Taiwan | 13 | 14 | 15 | 16 |
| Tanzania | 8 | 9 | 10 | 11 |
| Tasmania | 15 | 16 | 17 | 18 |
| Thailand | 12 | 13 | 14 | 15 |
| Togo | 5 | 6 | 7 | 8 |
| Tonga Islands | 18 | 19 | 20 | 21 |
| Trinidad & Tobago | 1 | 2 | 3 | 4 |
| Tunisia | 6 | 7 | 8 | 9 |
| Turkey | 7 | 8 | 9 | 10 |
| Turks and Caicos Isls. | 0 | 1 | 2 | 3 |
| Tuvalu | 17 | 18 | 19 | 20 |
| Uganda | 8 | 9 | 10 | 11 |
| Union Island | 1 | 2 | 3 | 4 |
| U.S.S.R. (Moscow)† | 8 | 9 | 10 | 11 |
| United Arab Emirates | 9 | 10 | 11 | 12 |
| United Kingdom | 5 | 6 | 7 | 8 |
| U.S. Trust Terr. Isls. (Palau)† | 15 | 16 | 17 | 18 |
| Upper Volta | 5 | 6 | 7 | 8 |
| Uruguay | 2 | 3 | 4 | 5 |
| Vanuatu | 16 | 17 | 18 | 19 |
| Vatican City | 6 | 7 | 8 | 9 |
| Venezuela | 1 | 2 | 3 | 4 |
| Vietnam | 12 | 13 | 14 | 15 |
| Virgin Islands (U.S.) | 1 | 2 | 3 | 4 |
| Wake | 17 | 18 | 19 | 20 |
| Wales | 5 | 6 | 7 | 8 |
| Western Samoa | −6 | −5 | −4 | −3 |
| Yemen Arab Rep. | 8 | 9 | 10 | 11 |
| Yemen, Peo. Dem. Rep. of | 8 | 9 | 10 | 11 |
| Yugoslavia | 6 | 7 | 8 | 9 |
| Zaire (Kinshasa)† | 6 | 7 | 8 | 9 |
| Zambia | 7 | 8 | 9 | 10 |
| Zimbabwe | 7 | 8 | 9 | 10 |

†Dagger indicates locations with more than one time zone.
*Source: Bell System's *International Phone Rates From the U.S. Mainland.*

## Country and City Codes*

**Dial 011 or 01 + Country Code + City Code + Local Number**

The following is a guide for placing international calls from the United States. Please note that international dialing varies in other countries. From areas equipped with international dialing: 1) Dial the International Access Code: 011. 2) Next, dial the Country Code. 3) Then, the City Code. 4) The Local Number.

To dial a person, collect, credit card, billed to a third number or any international call requiring the assistance of an operator, follow the previous instructions but use "01" for the International Access Code. After the call is dialed, the operator will come on the line to ask for your special instructions. If you cannot dial operator assisted calls from your area, or if your area doesn't have international dialing, just dial "0" (operator) and the operator will handle your call.

| Country / City | Code |
|---|---|
| **ARGENTINA** | **54** |
| Buenos Aires | 1 |
| Rosario | 41 |
| **AUSTRALIA** | **61** |
| Canberra | 62 |
| Melbourne | 3 |
| Sydney | 2 |
| **AUSTRIA** | **43** |
| Graz | 316 |
| Linz | 732 |
| Vienna | 222 |
| **BAHRAIN**** | **973** |
| **BELGIUM** | **32** |
| Antwerp | 31 |
| Brussels | 2 |
| Liege | 41 |
| **BELIZE** | **501** |
| Belize City** | |
| Belmopan | 08 |
| **BOLIVIA** | **591** |
| La Paz | 2 |
| **BRAZIL** | **55** |
| Belo Horizonte | 31 |
| Brasilia | 61 |
| Sao Paulo | 11 |
| **CHILE** | **56** |
| Penco | 42 |
| Santiago | 2 |
| Valparaiso | 31 |
| **COLOMBIA** | **57** |
| Bogota** | |
| Cali | 3 |
| Medellin | 4 |
| **COSTA RICA**** | **506** |
| **CYPRUS** | **357** |
| Limassol | 51 |
| Nicosia | 21 |
| Paphos | 61 |
| **DENMARK** | **45** |
| Aarhus | 6 |
| Copenhagen | 1 or 2 |
| Odense | 9 |
| **ECUADOR** | **593** |
| Cuenca | 4 |
| Guayaquil | 4 |
| Quito | 2 |
| **EL SALVADOR**** | **503** |
| **FINLAND** | **358** |
| Helsinki | 0 |
| Tampere | 31 |
| **FRANCE** | **33** |
| Bordeaux | 56 |
| Lille | 20 |
| Lyon | 7 |
| Marseille | 91 |
| Nice | 93 |
| Paris | 1 |
| Strasbourg | 88 |
| Toulouse | 61 |
| **GERMAN DEM. REP.** | **37** |
| Berlin | 2 |
| Dresden | 51 |
| Leipzig | 41 |
| **GERMANY, FED. REP. OF** | **49** |
| Berlin | 30 |
| Bonn | 228 |
| Essen | 201 |
| Frankfurt | 611 |
| Hamburg | 40 |
| Munich | 89 |
| **GREECE** | **30** |
| Athens | 1 |
| Iraklion | 81 |
| Larissa | 41 |
| Patrai | 61 |
| Piraeus | 1 |
| Thessaloniki | 31 |
| Volos | 421 |
| **GUAM**** | **671** |
| **GUATEMALA** | **502** |
| Guatemala City | 2 |
| Quezaltenango** | |
| **GUYANA** | **592** |
| Bartica | 05 |
| Georgetown | 02 |
| **HAITI** | **509** |
| Gonaive | 2 |
| Port Au Prince | 1 |
| **HONDURAS**** | **504** |
| **HONG KONG** | **852** |
| Hong Kong | 5 |
| **INDONESIA** | **62** |
| Jakarta | 21 |
| **IRAN** | **98** |
| Esfahan | 31 |
| Tabriz | 41 |
| Teheran | 21 |
| **IRAQ** | **964** |
| Baghdad | 1 |
| Mosul | 60 |
| **IRELAND** | **353** |
| Cork | 21 |
| Dublin | 1 |
| Limerick | 61 |
| **ISRAEL** | **972** |
| Haifa | 4 |
| Jerusalem | 2 |
| Tel Aviv | 3 |
| **ITALY** | **39** |
| Bologna | 51 |
| Florence | 55 |
| Genoa | 10 |
| Milan | 2 |
| Naples | 81 |
| Palermo | 91 |
| Rome | 6 |
| Turin | 11 |
| **IVORY COAST**** | **225** |
| **JAPAN** | **81** |
| Kitakyushu | 93 |
| Kobe | 78 |
| Kyoto | 75 |
| Nagoya | 52 |
| Osaka | 6 |
| Sapporo | 11 |
| Tokyo | 3 |
| Yokohama | 45 |
| **KENYA** | **254** |
| Mombasa | 11 |
| Nairobi | 2 |
| Nakuru | 37 |
| **KOREA, REP. OF***** | **82** |
| Pusan | 51 |
| Seoul | 2 |
| Taegu | 53 |
| **KUWAIT**** | **965** |
| **LIBERIA**** | **231** |
| **LIBYA** | **218** |
| Tripoli | 21 |
| **LIECHTENSTEIN** | **41** |
| All points | 75 |
| **LUXEMBOURG**** | **352** |
| **MALAYSIA** | **60** |
| Ipoh | 5 |
| Kelang | 3 |
| Kuala Lumpur | 3 |
| **MEXICO** | **52** |
| La Paz | 682 |
| Mexico City | 5 |
| Nogales | 631 |
| **MONACO** | **33** |
| All points | 93 |
| **NETHERLANDS** | **31** |
| Amsterdam | 20 |
| Rotterdam | 10 |
| **NETHERLANDS ANTILLES** | **599** |
| Aruba | 8 |
| Curacao | 9 |
| **NEW ZEALAND** | **64** |
| Auckland | 9 |
| Wellington | 4 |
| **NICARAGUA** | **505** |
| Chinandega | 341 |
| Leon | 31 |
| Managua | 2 |
| **NIGERIA** | **234** |
| Lagos | 1 |
| **NORWAY** | **47** |
| Bergen | 5 |
| Oslo | 2 |
| Trondheim | 75 |
| **PANAMA**** | **507** |
| **PARAGUAY** | **595** |
| Asuncion | 21 |
| **PERU** | **51** |
| Arequipa | 54 |
| Callao | 14 |
| Lima | 14 |
| **PHILIPPINES** | **63** |
| Cebu | 32 |
| Davao | 35 |
| Manila | 2 |
| **PORTUGAL** | **351** |
| Lisbon | 19 |
| Porto | 29 |
| **QATAR**** | **974** |
| **ROMANIA** | **40** |
| Bucharest | 0 |
| **SAN MARINO** | **39** |
| All points | 541 |
| **SAUDI ARABIA** | **966** |
| Jeddah | 2 |
| Mecca | 2 |
| Riyadh | 1 |
| **SINGAPORE**** | **65** |
| **SOUTH AFRICA** | **27** |
| Cape Town | 21 |
| Johannesburg | 11 |
| **SPAIN** | **34** |
| Barcelona | 3 |
| Madrid | 1 |
| Seville | 54 |
| Valencia | 6 |
| **SRI LANKA** | **94** |
| Colombo | 1 |
| Kandy | 8 |
| Moratuwa | 72 |
| **SURINAME**** | **597** |
| **SWEDEN** | **46** |
| Goteborg | 31 |
| Malmo | 40 |
| Stockholm | 8 |
| **SWITZERLAND** | **41** |
| Basel | 61 |
| Berne | 31 |
| Zurich | 1 |
| **TAIWAN** | **886** |
| Kaohsiung | 7 |
| Taipei | 2 |
| **THAILAND** | **66** |
| Bangkok | 2 |
| **TUNISIA** | **216** |
| Tunis | 1 |
| **TURKEY** | **90** |
| Adana | 711 |
| Ankara | 41 |
| Istanbul | 11 |
| Izmir | 51 |
| **U.S.S.R.** | **7** |
| Kiev | 044 |
| Leningrad | 812 |
| Minsk | 017 |
| Moscow | 095 |
| Tallinn | 0142 |
| **UNITED ARAB EMIRATES** | **971** |
| Abu Dhabi | 2 |
| Ajman | 6 |
| **UNITED KINGDOM** | **44** |
| Belfast | 232 |
| Birmingham | 21 |
| Cardiff | 222 |
| Edinburgh | 31 |
| Glasgow | 41 |
| Leeds | 532 |
| London | 1 |
| **VENEZUELA** | **58** |
| Caracas | 2 |
| Maracaibo | 61 |
| Valencia | 41 |
| **YUGOSLAVIA** | **38** |
| Belgrade | 11 |
| Sarajevo | 71 |
| Skoplje | 91 |
| Zagreb | 41 |

*Source: Bell System's *International Telephone Rates from the U.S. Mainland.*
**City Codes not required.
***Military bases cannot be dialed directly. For cities not listed dial "0" (operator).

# GAS TURBINE SPECIFICATIONS

| | | POWER RATING | | | | COMPRESSOR SHAFT | | POWER SHAFT | |
|---|---|---|---|---|---|---|---|---|---|
| | | NORMAL | | MAXIMUM | | | | | |
| MODEL | USE | hp, mw or lbt | Heat Rate | hp, mw or lbt | Heat Rate | Compressor Stages | Turbine Stages | Turbine Stages | RPM |
| **ACEC, Gas Turbine Systems Division, BP4, B6000 Charleroi (Belgium)** | | | | | | | | | |
| | | | (Btu/kW-hr) | | (BTU/kW-hr) | | | | |
| W251 | G1-2-3-5-6-7 | 41,280mw* | 11,120 | 44,540mw* | 11,100 | 18 | | 3 | 5,000 |
| *With gas fuel and silencers level A. | | | | | | | | | |
| **AEG-KANIS Turbinenfabrik GmbH, Sales Department, Altendorferstr. 39-85, 4300 Essen 1/West Germany** | | | | | | | | | |
| | | (mw) | (Btu/kW-hr) | (mw) | (Btu/kW-hr) | | | | |
| 5361P | G-2 | 25.42 | 12,230 | — | — | 17 | 2 | — | 5,105 |
| 5361P | G-6 | 24.8 | 12,450 | 26.62 | 12,340 | 17 | 2 | — | 5,105 |
| 6521B | G-2 | 36.73 | 11,120 | — | — | 17 | 3 | — | 5,105 |
| 6521B | G-6 | 35.86 | 11,280 | 38.95 | 11,180 | 17 | 3 | — | 5,105 |
| 7111E | G-2 | 76.9 | 10,590 | — | — | 17 | 3 | — | 3,600 |
| 7111E | G-6 | 75.0 | 10,790 | 81.15 | 10,750 | 17 | 3 | — | 3,600 |
| 9151E | G-2 | 109.3 | 10,700 | — | — | 17 | 3 | — | 3,000 |
| 9151E | G-6 | 106.7 | 10,850 | 115.9 | 10,810 | 17 | 3 | — | 3,000 |
| 9111B | G-2 | ,85.2 | 10,990 | — | — | 17 | 3 | — | 3,000 |
| 9111B | G-6 | 82.6 | 11,210 | 91.6 | 11,090 | 17 | 3 | — | 3,000 |
| 3142T | C-2 | 10.63 | 13,730 | — | — | 15 | 1 | 1 | 6,500 |
| 5262A | C-2 | 19.1 | 13,920 | — | — | 15 | 1 | 1 | 4,670 |
| 5352B | C-2 | 25.4 | 12,740 | — | — | 16 | 1 | 1 | 4,670 |
| LM2500 | C-2 | 20.5 | 9,534 | — | — | 16 | 2 | 6 | 3,600 |
| All ratings on Destillate Fuel. All types of gas turbines designed by General Electric Company, Gas Turbine Department, Schenectady, New York, U.S.A., under joint manufacturing agreement. | | | | | | | | | |
| **AiResearch Manufacturing Co., of Arizona. A division of the Garrett Corp., Box 5217, Phoenix, Arizona 85010** | | | | | | | | | |
| IE831-800 | G5[26] | 690[19] | 0.66 | 800 | 0.64 | — | — | 3A | 41,730 |
| IME831-800 | G5[26] | 690[19] | 0.66 | 800 | 0.64 | — | — | 3A | 41,730 |
| ME831-800 | G5[26] | 383[20] | — | 600 | — | — | — | 3A | 41,730 |
| ME990-3 | G5[26] | 5,700 | 0.487 | — | — | 2C | 2A | 3A | 7,200 |
| JE990-50A | G5 | 5,800 | 8,700[21] | — | — | 2C | 2A | 3A | 7,200 |
| JE990-50B | G5 | 5,600 | 0.475 | — | — | 2C | 2A | 3A | 7,200 |
| IGG990-50A | [27] | 6,450[22] | 7,835[23] | — | — | 2C | 2A | — | — |
| IGG990-50B | [27] | 6,250[22] | 0.478[24] | — | — | 2C | 2A | — | — |
| JFS100-13A | [26] | 90hp | 1.38 | — | — | 1C | 1A | 1A | Zero to 60,400 |
| BRU | G4 | 6KWE | — | 10.5KWE | — | — | — | 1A | 36,000 |
| GTCP30-67 | G5 | 32.3hp | 60lb/hr | 70hp | 92lb/hr | — | — | 1R | 52,870 |
| GTCP85-180 | A, G5 | 155lb/min[1] | 306hb/hr | 177hp + 98lb/min | 345lb/hr | — | — | 1R | 42,000 |
| GTCP85-184 | G5 | 155lb/min[1] | 306lb/hr | 125hp 118lb/min | 343lb/hr | — | — | 1R | 42,000 |
| JFS100-4 | [26] | 102hp | 1.27 | — | — | 1C | 1A | 1A | Zero to 61,000 |
| JFS100-22 | [26] | 85hp | 1.35 | — | — | 1C | 1A | 1A | Zero to 58,600 |
| JFS100-34 | [26] | 75hp | 1.49 | — | — | 1C | 1A | 1A | Zero to 62,200 |
| JFS190-1A | [26] | 166hp | 0.96 | — | — | 1C | 1A | 1A | Zero to 59,000 |

Ratings are at Sea Level 59 °F Day Except as Noted. [1] Full bleed with zero horsepower. [19] Continuous Duty. [20] Navy 100 °F, S.L. day. [21] BTU/HP-hr. [22] Isen. GHP. [23] BTU/GHP-hr. [24] LB/GHP-hr. [26] Shaft power. [27] Gas Generator.

# GAS TURBINE SPECIFICATIONS

| Pressure Ratio | Number of Combustors | EXHAUST Flow lb/sec. or kg/sec. | EXHAUST Temp. | Heat Exch. for Exhaust Heat Recovery | Dry Weight | DIMENSIONS L | DIMENSIONS W | DIMENSIONS H |
|---|---|---|---|---|---|---|---|---|
| | | | | | | Dimensions of Gas Turbine and Generator Sets | | |
| 10 | 8 | 350lb./sec. | 964°F | Y | 352,000lb | 750in | 200in | 160in |
| | | (kg/sec.) | (°F) | | | | (Inches) | |
| 9.8 | 10 | 123.0 | 915 | — | 398,000 | 924 | 132 | 144 |
| 9.8 | 10 | 120.3 | 919 | — | 570,000 | 1,380 | 228 | 420 |
| 11.3 | 10 | 139.4 | 1,016 | — | 528,000 | 1,068 | 132 | 144 |
| 11.3 | 10 | 138.1 | 1,020 | — | 700,000 | 1,476 | 288 | 408 |
| 11.5 | 10 | 281.3 | 1,000 | — | 587,000 | 840 | 348 | 156 |
| 11.5 | 10 | 279 | 1,000 | — | 1,070,000 | 1,584 | 852 | 372 |
| 9.5 | 14 | 405.1 | 968 | — | 650,000 | 996 | 180 | 228 |
| 9.5 | 14 | 401.6 | 971 | — | 1,425,000 | 1,680 | 576 | 456 |
| 9.5 | 14 | 347.8 | 944 | — | 1,850,000 | 996 | 180 | 228 |
| 9.5 | 14 | 344.3 | 950 | — | 1,850,000 | 1,680 | 576 | 456 |
| 7.1 | 6 | 52.7 | 979 | Y | 120,000 | 420 | 132 | 144 |
| 6.8 | 12 | 98.1 | 975 | Y | 253,000 | 600 | 132 | 144 |
| 8.6 | 12 | 122.6 | 916 | Y | 257,000 | 600 | 132 | 144 |
| 18 | 1 | 65.3 | 939 | — | 52,000 | 333.6 | 110.4 | 138 |
| | | | | | | | (MM) | |
| 11.0 | 1 | 7.8lb/sec | 928 | — | 1,750 | 65 | 39 | 34 |
| 11.0 | 1 | 7.8lb/sec | 928 | — | 1,750 | 65 | 39 | 34 |
| 11.0 | 1 | 7.8lb/sec | 825 | — | 1,680 | 65 | 39 | 34 |
| 12 | 1 | 45lb/sec | 890 | — | 5,500 | 123 | 62 | 48 |
| 12 | 1 | 44lb/sec | 875 | — | 7,000 | 132 | 44 | 44 |
| 12 | 1 | 44lb/sec | 880 | — | 7,000 | 132 | 44 | 44 |
| 12 | 1 | 44lb/sec | 1,215 | — | 5,000 | 85 | 43 | 43 |
| 12 | 1 | 44lb/sec | — | — | 5,000 | 85 | 43 | 43 |
| 3.3 | 1 | 1.7lb/sec | 1,200 | — | 82 | 28 | 12 | 15 |
| 1.9 | — | 0.8lb/sec | 326 | Rec | 175 | 30 | 19 | 21 |
| — | 1 | 1.2lb/sec | 1,300 | — | 87 | 25.8 | 19.9 | 17.1 |
| 3.7 | 1 | 4.1lb/sec | 1,200 | — | 290 | 37 | 32 | 23 |
| 3.7 | 1 | 4.1lb/sec | 1,200 | — | 315 | 37 | 28 | 24 |
| — | 1 | 1.75lb/sec | — | — | 88.1 | 22 | 12 | 12 |
| — | 1 | 1.7lb/sec | — | — | 82 | 21 | 12 | 14 |
| — | 1 | 1.7lb/sec | — | — | 100 | 24 | 15.6 | 10.6 |
| 3.95 | 1 | 2.35lb/sec | 1,165 | — | 95 | 20 | 13 | 14 |

Turbomachinery International Handbook

Turbomachinery Specifications

## GAS TURBINE SPECIFICATIONS

| MODEL | USE | POWER RATING — NORMAL: hp, mw or lbt | NORMAL: Heat Rate | MAXIMUM: hp, mw or lbt | MAXIMUM: Heat Rate | COMPRESSOR SHAFT: Compressor Stages | COMPRESSOR SHAFT: Turbine Stages | POWER SHAFT: Turbine Stages | POWER SHAFT: RPM |
|---|---|---|---|---|---|---|---|---|---|
| **Alsthom-Atlantique, 38, Avenue Kleber, 75795 Paris Cedex 16, France** | | | | | | | | | |
| M5332(B) | C1-C2, P1-P2, Pi | 33,550 hp | 8,910 hp | — | — | 16 | 1 | 1 | 4,670 |
| PG5361(P) | G,3,5,6,7 | 25,280 kW | 12,300 kWh | 27,130 kW | 12,200 kW | 17 | — | 2 | 5,105 |
| PG6521(B) | G1,3,5,6,7 | 36,630 kW | 11,160 kWh | 39,790 kW | 11,060 kWh | 17 | — | 3 | 5,105 |
| PG9111(B) | G1,3,5,6,7 | 84,700 kW | 11,050 kWh | 93,900 kW | 10,930 kWh | 17 | — | 3 | 3,000 |
| PG9151(E) | G1,3,5,6,7 | 108,800 kW | 10,770 kWh | 118,200 kW | 10,740 kWh | 17 | — | 3 | 3,000 |
| VEGA 205(P) | G1,3,5,6,7 | 73,100 kW | 8,512 kWh | — | — | 2 PG 5,361 + 1 Steam Turbine | | | |
| VEGA 206(P) | G1,3,5,6,7 | 106,540 kW | 7,687 kWh | | | 2 PG 6,521 + 1 Steam Turbine | | | |
| VEGA 209(E) | G1,3,5,6,7 | 305,300 kW | 7,698 kWh | | | 2 PG 9,151 + 1 Steam Turbine | | | |
| VEGA 109(E) | G1,3,5,6,7 | 153,980 kW | 7,714 kWh | | | 1 PG 9,151 + 1 Steam Turbine | | | |
| VEGA 109(E) | G1,3,5,6,7 | 149,870 kW | 7,913 kWh | | | 1 PG 9,151 + 1 Steam Turbine — Single Shaft | | | |
| All ratings on gas fuel — Other types of unit on request. | | | | | | | | | |
| **Avco Lycoming, 550 South Main Street, Stratford, Connecticut 06497** | | | | | | | | | |
| TF25 | C-1,-2,-L-2 | 2,500hp | 11,334hp-hr | 3,000hp | 10,862hp-hr | 7.1 | 1 | 2 | 14,500 |
| TF40 | P-1,-2 PR-2,-3,-4 | 4,000hp | 9,628hp-hr | 4,600hp | 9,464hp-hr | 7.1 | 2 | 2 | 15,400 |
| **Bet-Shemesh Engines Ltd., Mobile Post Haela, Bet Shemesh, Israel** | | | | | | | | | |
| | | (MW) | (Btu/kW-hr) | (MW) | (Btu/kW-hr) | | | | |
| M2TL | G5,6,7 | 0.76 | 25,200[1] | 0.8 | 24,200 | 1[2] | 1 | 1 | 18,000[3] |
| M5TL | G5,6,7 | 0.9 | 18,000 | 0.95 | 17,200 | 2[5] | 2 | 2 | 21,200 |
| [1]For fuel of L.H.V. of 18,360 Btu/lb. [2]Radial. [3]Build-in reduction gear for 3,000, 1,800, 1,500 rpm. [4]With reduction gears. [5]One axial and one radial. | | | | | | | | | |
| **BBC Brown, Boveri & Co., Ltd., Baden, Switzerland, Brown, Boveri & Cie. AG, Mannheim, Fed. Rep. of Germany** | | | | | | | | | |
| (Open-Cycle Turbines[1]) | | MW | (Btu/kW-h) | MW | (Btu/kW-h) | | | | |
| 9 (50/60Hz) | G1 | 35.4 | 11,810 | 38.5 | 11,645 | 15 | — | 4 | 4500 |
| 11 (only 60Hz) | G1 | 72.5 | 10,765 | 78.9 | 10,665 | 17 | — | 5 | 3,600 |
| 13 (only 50Hz) | G1 | 88.9 | 10,730 | 97.3 | 10,565 | 18 | — | 5 | 3,000 |
| 13E (only 50Hz) | G1 | 145.3 | 10,000 | 157.4 | 987 | 21 | — | 5 | 3,000 |
| (Air-Storage Turbines) | | | | | | | | | |
| L-GT 11/8 (60Hz) | G1 | 220 | 4,150[3] | — | — | 20 | — | 11 | 3,600 |
| L-GT 13/10 (50Hz)[7] | G1 | 290 | 5,560[3] | — | — | 25 | — | 11 | 3,000 |
| (Combined Gas/Steam Turbine Plants) (Combined Cycle Power Stations) | | | | | | | | | |
| KA9-1[11] | G1/3 | 51.3 | 8,061 | 55.9 | 7,971 | 15 | — | 4 | 4,473 or 4,485 |
| KA9-2[11] | G1/3 | 104.4 | 7,921 | 113.7 | 7,842 | 15 | — | 4 | 4,473 or 4,485 |
| KA9-3[11] | G1/3 | 157.5 | 7,874 | 171.6 | 7,796 | 15 | — | 4 | 4,473 or 4,485 |
| KA9-4[11] | G1/3 | 210.8 | 7,846 | 229.7 | 7,768 | 15 | — | 4 | 4,473 or 4,485 |
| KA11-1[12] | G1/3 | 105.2 | 7,399 | 114.6 | 7,325 | 17 | — | 5 | 3,600 |
| KA11-2[12] | G1/3 | 212.3 | 7,329 | 231.4 | 7,256 | 17 | — | 5 | 3,600 |
| KA11-3[12] | G1/3 | 318.5 | 7,329 | 347.1 | 7,256 | 17 | — | 5 | 3,600 |
| KA11-4[12] | G1/3 | 424.3 | 7,335 | 464.4 | 7,262 | 17 | — | 5 | 3,600 |
| KA13-1[13] | G1/3 | 128.0 | 7,413 | 139.3 | 7,677 | 17 | — | 5 | 3,000 |
| KA13-2[13] | G1/3 | 258.2 | 7,349 | 280.9 | 7,276 | 17 | — | 5 | 3,000 |
| KA13-3[13] | G1/3 | 385.3 | 7,384 | 419.5 | 7,310 | 17 | — | 5 | 3,000 |
| KA13-4[13] | G1/3 | 512.5 | 7,401 | 557.7 | 7,327 | 17 | — | 5 | 3,000 |

[1]ISO conditions at shaft for all units: pressure ratio and exhaust data are stated for base load operation. Ratings are stated for burning destillate fuel. Crude oil and heavy oil ratings may be slightly reduced, natural gas ratings slightly higher. [2]All dimensions and weights include combustion chamber, thermal block and diffusor. [3]Without considering of the storage process and air preheater. [4]Charging ratio (charging time: generating time). [5]Without air preheater (heat recovery). [6]Including generator and compressor group. [7]Data refer to Huntorf Plant, Fed. Rep. of Germany. [10]All data based on ISO conditions at terminal of generators and 0.04 bar at steam turbine outlet. Power ratings stated for burning destillate fuel, they may be increased with fired heat recovery boiler(s). [11]KA 9 type is available for 50 Hz and 60 Hz. [12]KA 11 only for 60 Hz available. [13]KA 13 only for 50 Hz available. [14]All plants with non-fired heat recovery steam generator(s).

## GAS TURBINE SPECIFICATIONS

| Pressure Ratio | Number of Combustors | EXHAUST Flow lb/sec. or kg/sec. | EXHAUST Temp. | Heat Exch. for Exhaust Heat Recovery | Dry Weight | DIMENSIONS L | DIMENSIONS W | DIMENSIONS H |
|---|---|---|---|---|---|---|---|---|
| | | | | | | | (Inches) | |
| 8.2 | 12 | 257 lb/sec | 930 °F | — | 257,000 lb | 600 | 130 | 150 |
| 10.3 | 10 | 267 lb/sec | 919 °F | — | 570,000 lb | 1,385 | 228 | 424 |
| 11.4 | 10 | 301 lb/sec | 1020 °F | — | 700,000 lb | 1,476 | 288 | 408 |
| 9.4 | 14 | 752 lb/sec | 950 °F | — | 1,425,000 lb | 1,880 | 576 | 405 |
| 11.4 | 14 | 877 lb/sec | 972 °F | — | 1,900,000 lb | 1,380 | 930 | 468 |
| — | — | — | — | Unfired Boiler | | Condenser Back Pressure — 250 mm bar | | |
| | | | | Unfired Boiler | | | | |
| | | | | Unfired Boiler | | Base Unit | | |
| | | | | Unfired Boiler | | | | |
| | | | | Unfired Boiler | Standby | Condenser Back Pressure: 70 mm/bar | | |
| | | | | | Peak Unit | Fast Start | | |
| | | | | | | | (Inches) | |
| 6.5 | 1 | 21.4lb/sec | 1,102F | Y | 1,183lb | 50.1 | 34.4 | 43.8 |
| 8.4 | 1 | 27.4lb/sec | 1,084F | Y | 1,325lb | 52.2 | 34.4 | 43.8 |
| | | (kg/sec) | (°C) | | (kg) | (mm) | (mm) | (mm) |
| 3.9 | 1 | 8.16 | 550 | — | 330[4] | 2,110 | 764 | 772 |
| 5.5 | 1 | 5.89 | 550 | — | 408[4] | 2,438 | 731.7 | 731.5 |
| | | (kg/sec.) | (°C) | | (kg) | (mm) | (mm) | (mm) |
| 9.0 | 1 | 163 | 510 | — | 79,000[2] | 10,000[2] | 3,200[2] | 7,600[6] |
| 11.0 | 1 | 290 | 520 | — | 165,000[2] | 11,200[2] | 7,200[5] | 5,100[2] |
| 11.8 | 1 | 373 | 495 | — | 270,000[2] | 14,500[2] | 4,600[5] | 9,800[2] |
| 14.3 | 1 | 508 | 525 | — | 426,000 | 18,800 | — | 14,000 |
| 1:1[4] | 2 | 313 | 140 | Y | — | 48,100[6] | 4,500 | 9,400 |
| 4:1[4] | 2 | 417 | 400[5] | — | — | 42,600[6] | 4,600 | 10,000 |
| 8.85 | 1 | 161 | 515 | [14] | | | | |
| 8.85 | 1 | 322 | 515 | [14] | | | | |
| 8.85 | 1 | 483 | 515 | [14] | | | | |
| 8.85 | 1 | 644 | 515 | [14] | | | | |
| 11.0 | 1 | 290 | 520 | [14] | | | | |
| 11.0 | 1 | 580 | 520 | [14] | | | | |
| 11.0 | 1 | 870 | 520 | [14] | | | | |
| 11.0 | 1 | 1,160 | 520 | [14] | | | | |
| 10.4 | 1 | 367 | 480 | [14] | | | | |
| 10.4 | 1 | 734 | 480 | [14] | | | | |
| 10.4 | 1 | 1,101 | 480 | [14] | | | | |
| 10.4 | 1 | 1.468 | 480 | [14] | | | | |

# GAS TURBINE SPECIFICATIONS

| | | POWER RATING | | | | COMPRESSOR SHAFT | | POWER SHAFT | |
|---|---|---|---|---|---|---|---|---|---|
| | | NORMAL | | MAXIMUM | | | | | |
| MODEL | USE | hp, mw or lbt | Heat Rate | hp, mw or lbt | Heat Rate | Compressor Stages | Turbine Stages | Turbine Stages | RPM |
| **Centrax Ltd., Gas Turbine Division, Shaldon Road, Newton Abbot, Devon, TQ12 4SQ England.** | | | | | | | | | |
| CS600-2 | G-4 | 785hp | 14,300hp-hr | 865hp | 13,900hp-hr | 5+1 | 2 | — | 22,000 |
| CS600-2Z | G-4 | 910hp | 14,400hp-hr | 1,000hp | 13,900hp-hr | 6+1 | 2 | — | 22,000 |
| **Scandiavian Engineering Limited A/S**: 8-20 Vojensvej, DK!2610, Copenhagen, Denmark. Contact: Mr. G. E. Frandsen. Phone: (01) 412333. Telex: 15646. | | | | | | | | | |
| **Machinery Oy:** P.O. Box 129, SF 00101, Helsinki 10, Finland. Contact: Mr. P. Husa. Phone: 716711. Telex: 12-1819 amom sf | | | | | | | | | |
| **E. Anduze & Cie S.A.**: 47 Rue Servan, 75011 Paris, France. Contact: Mr. G. Romoli. Phone: 805-90-76. Telex: 220259. | | | | | | | | | |
| **Ergen S.P.A.:** Via Roberto Cozzi 8, 20125 Milano, Italy. Contact: Mr. U. Laurella. Phone: (02) 642041/051. Telex: 31225 IMEL. | | | | | | | | | |
| **Commonwealth Aircraft Corporation Limited, 304 Lorimer Street, Port Melbourne, Victoria 3207, Australia.** | | | | | | | | | |
| AVON MK109[1] | PR-1 | 6,000lbt | .905lb/lbt-hr[6] | 7,350lbt | .930lb/lbt-hr[6] | 12 | 2 | — | — |
| ATAR 09C[2] | PR-1 | 7,277lbt[4] | .980lb/lbt-hr[6] | 13,232lbt[5] | 2.0lb/lbt-hr[6] [5] | 9 | 2 | — | — |
| VIPER MK 22/11[3] | PR-1 | 2,070lbt | 1.065lb/lbt-hr[6] | 2,500lbt | 1.11lb/lbt-hr[6] | 7 | 1 | — | — |
| (Above engines manufactured under license as follows:) [1] AVON — Rolls Royce Ltd., England [2] ATAR — S.N.E.C.M.A., France [3] VIPER — Rolls Royce Ltd., England | | | | | | | | | |
| (Ratings:) [4] Maximum continuous dry [5] Maximum with after burner [6] lb/lbt-hr | | | | | | | | | |
| **Cooper Rolls, North Sandusky Street, Mount Vernon, Ohio 43050 U.S.A.** | | | | | | | | | |
| Coberra 2348 | * | 16,150 | 9,215 | — | — | 17 | 3 | 2 | 5,200 |
| Coberra 2366 | * | 16,000 | 9,296 | — | — | 17 | 3 | 1 | 3,600 |
| Coberra 2448 | * | 18,000 | 9,222 | — | — | 17 | 3 | 2 | 5,200 |
| Coberra 2556 | * | 20,850 | 8,651 | — | — | 17 | 3 | 2 | 4,950 |
| Coberra 3045 | * | 16,350 | 7,618 | — | — | 5-11 | 2-2 | 2 | 5,900 |
| Coberra 6056 | * | 30,550 | 7,312 | — | — | 7-6 | 1-1 | 2 | 4,950 |
| Coberra 6462 | * | 34,000 | 7,000 | — | — | 7-6 | 1-1 | 2 | 4,800 |
| *Coberra Turbines are suitable for the following uses: C-1, C-2, G-2, G-6, P-0, P-W, PI | | | | | | | | | |
| **Cooper Rolls Corp.**, 6889 Rexwood Road, Mississauga, Ontario, Canada. Contact: R. W. Thompson. Phone: (416) 678-2030. | | | | | | | | | |
| **Cooper Rolls Inc.**, Mount Vernon, Ohio 43050. Contact: G. W. Stokes. Phone: (614) 397-0204. | | | | | | | | | |
| **Cooper Rolls Ltd.**, 46 Pall Mall, London, England. Contact: K. A. Thilman. Phone: 01-839-5161. | | | | | | | | | |
| **Creusot-Loire, Department Turbines et Compresseurs, 15, rue Pasquier 75383 Paris CEDEX 08** | | | | | | | | | |
| | | (kW) | (Btu/kW-hr) | (kW) | (Btu/kW-hr) | | | | |
| CA-3 | C1-C2, PO,PW,PRI | 3,230 | 12,360 | 4,045 | 12,180 | 14 | 2 | 2 | 13,820 |
| CA-5 | C1-C2, G1-G2 G5-G6, G7-G8 PO,PW,PR1 | 4,800 | 11,354 | 5,350 | 11,403 | 13 | 2 | 2 | 11,500 |
| SK30 | C1-C2, G1-G7 | 28,100 | 11,130 | 29,600 | 11,020 | 5+7 | 1+1 | 3 | 3,000/ 3,600 |
| **Curtiss-Wright Corp., Power Systems Group, One Passaic Street, Wood-Ridge, NJ 07075.** | | | | | | | | | |
| (Industrial Turbines) | | | | | | | | | |
| MOD POD 20HEI | C-2 | 27,650hp | 7,100hp-hr | 27,650hp | 7,100hp-hr | 16 | 2 | 6 | 3600 |
| MOD POD 25 | C-2 | 33,000hp | 8,630hp-hr | 38,100hp | 8,520hp-hr | 12 | 2 | 2 | 3600 |
| (Packaged Power Plants) | | | | | | | | | |
| MOD POD 20HEI | G-8 | 20.0MW | 9,800kW-hr | 20.1MW | 9,795kW-hr | 16 | 2 | 6 | 3600/3000 |
| MOD POD 25 | G-1 | 24.9MW | 11,850kW-hr | 26.7MW | 11,770kW-hr | 12 | 2 | 2 | 3600/3000 |
| MOD POD 50 | G-1 | 49.8MW | 11,850kW-hr | 53.4MW | 11,770kW-hr | 12 | 2 | 2 | 3600/3000 |
| (Combined Cycles) | | | | | | | | | |
| TEC 150 | G-3 | 150MW | 9,100kW-hr | 165MW | 9,000kW-hr | 12 | 2 | 2 | 3600/3000 |
| TEC 165 | G-3 | 160MW | 8,525kW-hr | — | — | 13 | 2 | 2 | 3600/3000 |
| TEC 200 | G-3 | 210MW | 9,300kW-hr | 230MW | 9,200kW-hr | 12 | 2 | 2 | 3600/3000 |
| TEC 330 | G-3 | 320MW | 8,525kW-hr | — | — | 13 | 2 | 2 | 3600/3000 |
| TEC 500 | G-3 | 480MW | 8,525kW-hr | — | — | 13 | 2 | 2 | 3600/3000 |
| TEC 750 | G-3 | 750MW | 7,210kW-hr | — | — | 18 | 2 | 3 | 3600/3000 |
| (Turbines That Fly) | | | | | | | | | |
| J65-W-5D | PR-1 | 6,750lbt | 16,930lbt-hr | 7,650lbt | 17,200lbt-hr | 13 | 2 | — | 8300 |
| J65-W-15 | PR-1 | 7,200lbt | 16,280lbt-hr | 9,000lbt | 17,100lbt-hr | 13 | 2 | — | 8630 |
| J65-W-20 | PR-1 | 7,400lbt | 17,110lbt-hr | 8,400lbt | 17,480lbt-hr | 13 | 2 | — | 8600 |

# GAS TURBINE SPECIFICATIONS

| Pressure Ratio | Number of Combustors | EXHAUST Flow lb/sec. or kg/sec. | EXHAUST Temp. | Heat Exch. for Exhaust Heat Recovery | Dry Weight | DIMENSIONS L | DIMENSIONS W | DIMENSIONS H |
|---|---|---|---|---|---|---|---|---|
| | | | | | | (mm) | (mm) | (mm) |
| 5.25 | 1 | 10.7lb/sec | 550C | — | 1,484lb | 1,320 | 914 | 1,422 |
| 6.30 | 1 | 12.0lb/sec | 535C | — | 1,500lb | 1,365 | 914 | 1,422 |
| **Gowrings Continental BV:** P.O. Box 5760, 12 Industriestraat, Strijen, Holland. Contact: Mr. J. A. Steenman. Phone: 01854-27 77. Telex: 21444 | | | | | | | | |
| **Gelpor:** Casal De Sta. Rita, Pavilhao 5, Ral-Sintra, Portugal. Contact: Mr. C. Couto. Phone: Lisbon 2480178. Telex: 12757 PLAGE P (for Gelpor). | | | | | | | | |
| **AB Erbings Maskinaffar:** Ulgardavagen 5, Box 261, 182 52 Djursholm, Stockholm, Sweden. Contact: Mr. S. Meiton. Phone: Stockholm 08-755-1590. Telex: 12513 Erbings S. | | | | | | | | |
| **Max Fischer Ingenieurbureau:** Bahnhofstrasse 100, 8021 Zurich, Switzerland. Contact: Mr. Max Fischer. Phone: Zurich 01-211-7781. Telex: 54338. | | | | | | | | |
| | | | | | | | (Inches) | |
| 6.4:1 | 8 | 119lb/sec | 1,200C | — | 2,405lb | 137 | 42.4 | 42.4 |
| 5.5:1 | 1 | 150lb/sec | 1,300C | — | 3,176lb | 259 | 37.2 | 36.3 |
| 4.16:1 | 1 | 44.2lb/sec | 1,350C | — | 604lb | 64 | 24.5 | 24.5 |
| **Commonwealth Aircraft Corporation Ltd.:** 304 Lorimer St., Port Melbourne, Victoria 3207, Australia. Contact: D. R. Rees, Marketing Manager. Phone: 64-0771. Telex: AA 30721 AUSTRALIA. | | | | | | | | |
| 8.8 | 8 | 168 | 766 | — | 50,000 | 288 | 120 | 120 |
| 8.8 | 8 | 168 | 768 | — | 50,000 | 288 | 120 | 120 |
| 9.0 | 8 | 167 | 840 | — | 50,000 | 288 | 120 | 120 |
| 9.0 | 8 | 169 | 875 | — | 52,000 | 252 | 120 | 120 |
| 18.5 | 10 | 126 | 808 | — | 48,000 | 228 | 120 | 120 |
| 18.4 | 1 | 207 | 835 | — | 50,000 | 252 | 120 | 120 |
| 19.2 | 1 | 198 | 887 | — | 58,000 | 252 | 120 | 120 |
| | | | | | (kg) | (mm) | (mm) | (mm) |
| 9.3 | 6 | 15.1kg/sec | 535°C | OPT | 1,200 | 3,150 | 2,050 | 1,600 |
| 12 | 1 (annular) | 19.4kg/sec | 580°C | OPT | 620 | 1,830 | 805 | 920 |
| 11 | 8 | 108kg/sec | 555°C | OPT | 230,000 | 9,000 | 1,630 | 5,880 |
| | | | | | | | (Inches) | |
| 18 | 1 | 144lb/sec | 888°F | Y | 103,000lb | 257 | 84 | 84 |
| 11 | 8 | 240lb/sec | 1,027°F | Y | 95,000lb | 360 | 126 | 134 |
| 18 | 1 | 144lb/sec | 888°F | Y | 315,000lb | 846 | 126 | 150 |
| 11 | 8 | 240lb/sec | 1,027 F | Y | 330,000lb | 846 | 126 | 150 |
| 11 | 8 | 480lb/sec | 1,027 F | Y | 390,000lb | 1,320 | 126 | 198 |
| 12 | 8 | 920lb/sec | 300°F | Y | — | — | — | — |
| 7 | 2 | 1,265lb/sec | 275°F | Y | — | — | — | — |
| 12 | 8 | 920lb/sec | 150°F | Y | — | — | — | — |
| 7 | 4 | 2,530lb/sec | 275°F | Y | — | — | — | — |
| 7 | 6 | 3,795lb/sec | 275°F | Y | — | — | — | — |
| 17 | 4 | 2,250lb/sec | 300°F | Y | — | — | — | — |
| 7.3 | 1 | 125lb/sec | 1,200°F | — | 2,800lb | 110 | 38 | 38 |
| 7.5 | 1 | 129lb/sec | 1,315°F | — | 2,825lb | 115 | 38 | 38 |
| 6.8 | 1 | 129lb/sec | 1,280°F | — | 2,800lb | 113 | 38 | 38 |

# GAS TURBINE SPECIFICATIONS

| MODEL | USE | POWER RATING — NORMAL: hp, mw or lbt | POWER RATING — NORMAL: Heat Rate | POWER RATING — MAXIMUM: hp, mw or lbt | POWER RATING — MAXIMUM: Heat Rate | COMPRESSOR SHAFT: Compressor Stages | COMPRESSOR SHAFT: Turbine Stages | POWER SHAFT: Turbine Stages | POWER SHAFT: RPM |
|---|---|---|---|---|---|---|---|---|---|
| **Detroit Diesel Allison, Box 894, Dept. U-5, Indianapolis, Indiana 46206 USA.** | | | | | | | | | |
| GT-404 | G-7 | 380shp | — | — | — | 1C | 1 | 1 | 2,880 |
| 250-KB | P-2 | 370shp | 11.960[1] shp-hr[1] | 400shp | 11.923[1] shp-hr[1] | 6-1C | 2 | 2 | 6,016 |
| 501-KB[a] | G-2,5,7,8 | 4,380shp | 8,990shp-hr[1] | 5,305shp | 8.540shp-hr[1] | 14 | 4 | — | 13,820[a] |
| 501-KB5[a] | G-2,5,7,8 | 5,276[2]<br>5,156 | 8,350[1]<br>8,428[1] | 5,983[2]<br>5,847 | 8,061[1]<br>8,137[1] | 14 | 4 | — | 14,200[a] |
| 501-KC | C-2,P1,P2,PI | 4,330shp | 9,255shp-hr[1] | 5,420shp | 13,095shp-hr[1] | 14 | 2 | — | —[3] |
| 501-KF | PR-2 | 4,330shp | 9,255shp-hr[1] | 5,420shp | 13,095shp-hr[1] | 14 | 2 | 2 | 13,820 |
| 570-KA | G-2, 5,7,8 | 6,445shp | 8.473hp-hr[1] | 7,170shp | 8.510hp-hr[1] | 13 | 2 | 2 | 11,500 |
| 570-KC | C-2,P1,P2,PI | 6,445shp | 8,473hp-hr[1] | 7,170shp | 8,510hp-hr[1] | 13 | 2 | 2 | 11,500 |
| 570-KF | PR-2 | 6,445shp | 8,473hp-hr[1] | 7,170shp | 8.510hp-hr[1] | 13 | 2 | 2 | 11,500 |

[1] BTU/SHP HR. [2] Natural Gas [3] Based on use of applicable power turbine [4] Gas generator [5] Twin rotating heat exchanger [6] Normal rating.
a) Single shaft engine.

**Detroit Diesel Allison** — **Europe**: Div. of General Motors Continental, S.A. Nederland, Parmentierplein 1. Mail: Postbox 5061, 3008 AB Rotterdam, Holland.
Phone: 010-290-0000. Telex: 28355 GMCNL.
**Detroit Diesel Allison** — **Europe**: Div. of General Motors Limited, P.O. Box No. 6, London Road, Wellingborough, Northamptonshire NN8 2DL, England.
Phone: 0933-71122. Telex: 31329 DDAIEUG.
**Detroit Diesel Allison** — **Europe**: Div. of General Motors Nordiska, AB, Motorvagen 1, Fack, S-104 60 Stockholm, Sweden.
Phone: 08-440180. Telex: 1569 GMNORD.
**Detroit Diesel Allison** — **Europe**: Div. of General Motors Norge A/A. Mail: P.O. Box 205, 2001 Lillestrom, Norway.
Phone: 0271-38-60 or 0271-58-60. Telex: 11682, Lillestrom.
**Detroit Diesel Allison** — **Europe**: General Motors Denmark, GMODC Branch, Borgmester Christiansens Gade 40, DK 2450 Copenhagen SV, Denmark.
Phone: 01-30-22-11. Telex: 6727 Copenhagen.

| MODEL | USE | NORMAL: hp, mw or lbt | NORMAL: Heat Rate | MAXIMUM: hp, mw or lbt | MAXIMUM: Heat Rate | Compressor Stages | Turbine Stages | Power Turbine Stages | RPM |
|---|---|---|---|---|---|---|---|---|---|
| **Dresser Industries, Inc., Dresser Clark Div., P.O. Box 560, Olean, New York 14760** | | | | | | | | | |
| | | (hp) | (Btu/hp-hr) | (hp) | (Btu/hp-hr) | | | | |
| DC-990 | * | 6,000 | 8,430 | 6,790 | 8,129 | 2 | 2 | 3 | 6,860 |
| DJ-125R | * | 15,900 | 9,500 | 19,000 | 8,570 | 17 | 3 | 1 | 5,330 |
| DJ-160R | * | 17,750 | 9,347 | 20,250 | 8,740 | 17 | 3 | 1 | 5,330 |
| DJ-270G-20 | * | 17,200 | 7,590 | 25,800 | 6,900 | 16 | 2 | 2 | 5,000 |
| DJ-270G | * | 29,200 | 7,030 | 32,500 | 6,960 | 16 | 2 | 2 | 5,400 |
| DJ-290R | * | 29,500 | 7,334 | 36,200 | 6,700 | 7/6 | 2 | 2 | 5,000 |
| DJ-290RC | * | 33,500 | 7,136 | 39,800 | 6,680 | 7/6 | 2 | 2 | 5,400 |
| DJ-290G | * | 36,900 | 7,068 | 39,500 | 6,850 | 5/14 | 3 | 2 | 5,000 |

*Application Symbols for ALL Clark Turbines: C-1, C-2, P-0, P-W, G-1, G-2, G-3, G-5, G-6, G-7

| MODEL | USE | NORMAL: hp, mw or lbt | NORMAL: Heat Rate | MAXIMUM: hp, mw or lbt | MAXIMUM: Heat Rate | Compressor Stages | Turbine Stages | Power Turbine Stages | RPM |
|---|---|---|---|---|---|---|---|---|---|
| **Fabrique Nationale Herstal, Societe Anonyme (en abrege FN), 33 rue Voie de Liege, B-4400 Herstal, Belgium** | | | | | | | | | |
| FN 531 (LI)[1] | G5[6] | 360[2] [3] | .78 | 400[3] | .75 | 2 | 1 | 1 | 32,000 |

[1] Caterpillar License. [2] At 80 °F and sea level. [3] Horsepower. [4] Inches. [5] Pounds. [6] Shaft power.

| MODEL | USE | NORMAL: hp, mw or lbt | NORMAL: Heat Rate | MAXIMUM: hp, mw or lbt | MAXIMUM: Heat Rate | Compressor Stages | Turbine Stages | Power Turbine Stages | RPM |
|---|---|---|---|---|---|---|---|---|---|
| **Fiat TTG S.p.A., Via Cuneo, 20, Casella Postale (P.O. Box) 500, 10100 Torino, Italy** | | | | | | | | | |
| | | | (Btu/kW-hr) | (MW) | (Btu/kW-hr) | | | | |
| TG7 LI | C2, G1,2,3,5,6 | 8.2MW | 14,850 | 8.95 | 14,520 | 16 | — | 5 | 6,000 |
| TG16 LI | C2,G1,2,3,5,6 | 18.2MW | 13,175 | 19.5 | 12,975 | 15 | — | 5 | 4,850 |
| TG16BLI | C2,P1 | 23,500hp | 10,380 | — | — | 15 | 3 | 2 | 4,200 |
| TG20 LI | G1,2,3,5,6,7 | 37.85MW | 11,590 | 41.18 | 11,300 | 18 | — | 3 | 4,920 |
| TG20 LI* | | 42.5MW | 10,600 | 45.9 | 10,585 | 19 | — | 3 | 5,400 |
| TG50 LI | G1,2,3,5,6,7 | 100MW | 10,990 | 108 | 10,890 | 20 | — | 4 | 3,000 |

*Not yet in production. [1] LI — Technical cooperation with Westinghouse Electric Corporation. [2] Base Load and Natural Gas. No inlet and exhaust losses. [3] Other gaseous or liquid fuels may be used after control of characteristics. [4] Peak load ratings. [5] Gas turbine only.

| MODEL | USE | NORMAL: hp, mw or lbt | NORMAL: Heat Rate | MAXIMUM: hp, mw or lbt | MAXIMUM: Heat Rate | Compressor Stages | Turbine Stages | Power Turbine Stages | RPM |
|---|---|---|---|---|---|---|---|---|---|
| **Garrett Turbine Engine Co., A Division of The Garrett Corp., Box 5217, Phoenix, Arizona 85010** | | | | | | | | | |
| | | | (Btu/kW-hr) | | (Btu/kW-hr) | | | | |
| IM831-800 | G-2,7 | 735 | 11,425 | 800 hp | 12,180 | 2 | 3 | — | — |
| ME 9901 | PR-1 | 5,600 | 9,016 | 6,250 | 8,832 | 2 | 2 | 3 | 7,200 |

# GAS TURBINE SPECIFICATIONS

<table>
<tr><th rowspan="2">Pressure Ratio</th><th rowspan="2">Number of Combustors</th><th colspan="2">EXHAUST</th><th rowspan="2">Heat Exch. for Exhaust Heat Recovery</th><th rowspan="2">Dry Weight</th><th colspan="3">DIMENSIONS</th></tr>
<tr><th>Flow lb/sec, or kg/sec.</th><th>Temp.</th><th>L</th><th>W</th><th>H</th></tr>
<tr><td></td><td></td><td></td><td></td><td></td><td>(lbs/kg)</td><td>(in/mm)</td><td>(in/mm)</td><td>(in/mm)</td></tr>
<tr><td>4.35</td><td>1</td><td>4.58lb/sec</td><td>550F</td><td>RE[6]</td><td>1,800/816</td><td>46.7/1,186</td><td>50.5/1,283</td><td>39.5/1,003</td></tr>
<tr><td>7</td><td>1</td><td>3.4lb/sec</td><td>1,120F[6]</td><td>—</td><td>150/68</td><td>41.0/1,041</td><td>19.0/483</td><td>23.0/584</td></tr>
<tr><td>9.3</td><td>6</td><td>33.2lb/sec</td><td>958F[6]</td><td>—</td><td>1,285/583</td><td>90.0/2,286</td><td>33.2/844</td><td>31.1/791</td></tr>
<tr><td>9.3</td><td>6</td><td>34.3lb/sec</td><td>991F[6]</td><td>—</td><td>1,285/583</td><td>90/2,286</td><td>33.2/844</td><td>31.1/791</td></tr>
<tr><td>9.3</td><td>6</td><td>33.2lb/sec</td><td>—</td><td>—</td><td>1,100/499</td><td>71.0/1,803</td><td>33.0/838</td><td>41.0/1,041</td></tr>
<tr><td>9.3</td><td>6</td><td>33.2lb/sec</td><td>995F[6]</td><td>—</td><td>2,500/1,134</td><td>105.0/2,667</td><td>55.1/1,402</td><td>54.2/1,378</td></tr>
<tr><td>12.0</td><td>1</td><td>42.8lb/sec</td><td>1,080F[6]</td><td>—</td><td>1,350/612</td><td>70.2/1,783</td><td>31.2/793</td><td>36.1/917</td></tr>
<tr><td>12.0</td><td>1</td><td>42.8lb/sec</td><td>1,080F[6]</td><td>—</td><td>1,350/612</td><td>70.2/1,783</td><td>36.0/914</td><td>37.6/955</td></tr>
<tr><td>12.0</td><td>1</td><td>42.8lb/sec</td><td>1,080F[6]</td><td>—</td><td>1,350/612</td><td>73.7/1,873</td><td>31.2/793</td><td>36.1/917</td></tr>
<tr><td colspan="9">Detroit Diesel Allison — Europe: Div. of General Motors Deutschland GmbH, Kupferstrasse 1. Mail: Postfach 1507, 6090 Ruesselsheim, Republic of Germany.<br>Phone: 06142/6021. Telex: 417541.<br>Detroit Diesel Allison — Europe: Div. of General Motors Suisse, S.A., Salzhausstrasse 21, 2501 Biel Bienne, Switzerland.<br>Phone: 032-21-51-11. Telex: 34217 Bienne.<br>Detroit Diesel Allison — Europe: Div. of General Motors France, S.A., 56/68 Ave Louis Roche, 92231, Gennevilliers, France.<br>Phone: 790-7000. Telex: 620050.<br>Detroit Diesel Allison — Australia: Div. of General Motors — Holden's Limited, Princes Highway. Mail: P.O. Box 163, Dandenong, Victoria, Australia 3175.<br>Phone: (03) 792-01111. Telex: GM Network MAD DANNENONG.<br>Detroit Diesel Allison — Southeast Asia: 15 Benoni Sector, Jurong Town, Singapore 22.<br>Phone: 654697, 2610801. Telex: RS21608 A/B GMSING.<br>Detroit Diesel Allison Operations: Athens Towers, Messoghion 2/4, Suite 705, Athens, 610 Greece.<br>Phone: 7785-344, 7706-669, 7787-281. Telex: 215759 DDAI.<br>Detroit Diesel Allison Operations: 2655 Villa Creek Drive, Suite 130, Dallas, Texas 75234 USA.<br>Phone: (214) 241-2754. Telex: 910-860-5549.</td></tr>
<tr><td></td><td></td><td>(lb/sec)</td><td>(°F)</td><td></td><td>(lb)</td><td></td><td>(Inches)</td><td></td></tr>
<tr><td>12:1</td><td>1</td><td>44.2</td><td>870</td><td></td><td>7,500</td><td>120</td><td>48</td><td>48</td></tr>
<tr><td>8.8</td><td>8</td><td>168</td><td>764</td><td></td><td>44,900</td><td>432</td><td>138</td><td>132</td></tr>
<tr><td>8.8</td><td>8</td><td>168</td><td>838</td><td></td><td>44,900</td><td>432</td><td>138</td><td>132</td></tr>
<tr><td>15</td><td>1</td><td>125</td><td>780</td><td></td><td>47,000</td><td>460</td><td>163</td><td>170</td></tr>
<tr><td>18</td><td>1</td><td>149</td><td>973</td><td></td><td>47,500</td><td>460</td><td>163</td><td>170</td></tr>
<tr><td>19</td><td>1</td><td>196</td><td>832</td><td></td><td>47,900</td><td>470</td><td>163</td><td>170</td></tr>
<tr><td>19.3</td><td>1</td><td>199</td><td>862</td><td></td><td>48,900</td><td>470</td><td>163</td><td>170</td></tr>
<tr><td>22</td><td>1</td><td>220</td><td>825</td><td></td><td>52,000</td><td>530</td><td>163</td><td>180</td></tr>
<tr><td></td><td></td><td></td><td></td><td></td><td></td><td></td><td>(INCHES)</td><td></td></tr>
<tr><td>5.6</td><td>2</td><td>4.4lb/sec</td><td>1,050F</td><td>—</td><td>400[5]</td><td>40[4]</td><td>28[4]</td><td>27[4]</td></tr>
<tr><td></td><td></td><td>(kg/sec.)</td><td>(°C)</td><td></td><td>(kg[5])</td><td></td><td>(mm[5])</td><td></td></tr>
<tr><td>6</td><td>6</td><td>61</td><td>422</td><td>Y</td><td>18,000</td><td>4,800</td><td>2,200</td><td>2,300</td></tr>
<tr><td>7</td><td>6</td><td>123</td><td>399</td><td>Y</td><td>40,000</td><td>7,700</td><td>3,100</td><td>2,900</td></tr>
<tr><td>7</td><td>6</td><td>123</td><td>413</td><td>Y</td><td>45,000</td><td>8,500</td><td>3,100</td><td>3,300</td></tr>
<tr><td>10</td><td>8</td><td>164</td><td>514</td><td>Y</td><td>60,000</td><td>9,000</td><td>3,100</td><td>3,300</td></tr>
<tr><td>14</td><td>8</td><td>160</td><td>515</td><td>Y</td><td>61,000</td><td>9,000</td><td>3,100</td><td>3,300</td></tr>
<tr><td>12</td><td>18</td><td>393</td><td>550</td><td>Y</td><td>160,000</td><td>14,200</td><td>4,300</td><td>4,500</td></tr>
<tr><td></td><td></td><td>(lb/sec)</td><td>(°F)</td><td></td><td>(lb)</td><td>(in)</td><td>(in)</td><td>(in)</td></tr>
<tr><td>11.0</td><td>1</td><td>7.9</td><td>930</td><td>No</td><td>1,750</td><td>48</td><td>36</td><td>36</td></tr>
<tr><td>12.0</td><td>1</td><td>44.0</td><td>1,900</td><td>No</td><td>6,260</td><td>120</td><td>60</td><td>48</td></tr>
</table>

# GAS TURBINE SPECIFICATIONS

**GEC Gas Turbines Limited, Cambridge Road, Whetstone, Leicester LE8 3LH, England.**

| MODEL | USE | POWER RATING: NORMAL: hp, mw or lbt | POWER RATING: NORMAL: Heat Rate | POWER RATING: MAXIMUM: hp, mw or lbt | POWER RATING: MAXIMUM: Heat Rate | COMPRESSOR SHAFT: Compressor Stages | COMPRESSOR SHAFT: Turbine Stages | POWER SHAFT: Turbine Stages | POWER SHAFT: RPM |
|---|---|---|---|---|---|---|---|---|---|
| | | (MW) | (Btu/kW-hr) | (MW) | (Btu/kW-hr) | | | | |
| EAS133[1] | | 12.2 | 12,360 | 16.2 | 11,560 | 17 | 3 | 1 | 6,500[3] |
| EAS134 | | 13.6 | 12,220 | 17.0 | 11,670 | 17 | 3 | 1 | 6,500[3] |
| EAS135 | | 14.9 | 12,140 | 18.1 | 11,710 | 17 | 3 | 1 | 6,500[3] |
| ESP.1[1] | | 12.0 | 10,190 | 14.4 | 9,375 | 5/11 | 2/2 | 2 | 6,700[3] |
| ERB124[1] | | 24.4 | 9,660 | 26.8 | 9,520 | 7/6 | 1/1 | 2 | 6,500[3] |
| E0-1C[1 2] | | 24.0 | 10,936 | 29.4 | 10,767 | 5/7 | 1/1 | 2 | 3,000/ 3,600 |
| ELM125[1] | | 22.0 | 9,220 | 24.3 | 9,290 | 16 | 2 | 6 | 3,600 |
| ELM150[1 2] | | 33.0 | 9,160 | 37.0 | 9,050 | 5/14 | 2/1 | 2 | 3,600 |
| EM85 | | 7.1 | 15,140 | — | — | 12 | 2 | 1 | 6,300[3] |
| EM610 | | 65.0 | 11,790 | 70.6 | 11,620 | 13 | — | 2 | 3,000 |
| Combined Cycle Electricity Generation | | | | | | | | | |
| EM610 | G3 | 96.8 | 7,900 | Higher Ratings and Multiple Gas Turbine Modules Available | | 13 | — | 2 | 3,000 |

[1]Increased ratings available for standby/emergency duty. [2]Double-ended machines available. [3]Maximum continuous speed.

**MAJOR OFFICES WORLDWIDE**: GEC Gas Turbines Limited, Cambridge Road, Whetstone, Leicester, LE8 3LH England, Contact: Mr. J. C. McMillan. Phone: Leicester (0533) 863434. Telex: 34331. • GEC Gas Turbines Limited, 132-135 Long Acre, London, WC2E 9AH, England. Contact: Mr. J. W. Mosenthal. Phone: 01-836-3444. Telex: 21403 •

**General Electric Company, Gas Turbine Division, Schenectady, New York 12345**

| MODEL | USE | POWER RATING: NORMAL: hp, mw or lbt | POWER RATING: NORMAL: Heat Rate | POWER RATING: MAXIMUM: hp, mw or lbt | POWER RATING: MAXIMUM: Heat Rate | COMPRESSOR SHAFT: Compressor Stages | COMPRESSOR SHAFT: Turbine Stages | POWER SHAFT: Turbine Stages | POWER SHAFT: RPM |
|---|---|---|---|---|---|---|---|---|---|
| (Mechanical Drive Gas Turbines) | | (hp) | (Btu/hp-hr) | (MW) | (Btu/kW-hr) | | | | |
| M3142[3] | C1,C2,P1 | 14,600 | 9,530[1] | — | — | 15 | 1 | 1 | 6,500 |
| M5251[3] | C1,C2,P1 | 25,000 | 9,640[1] | — | — | 16 | — | 2 | 4,860 |
| M5262(A)[3] | C1,C2,P1 | 26,250 | 9,780[1] | — | — | 15 | 1 | 1 | 4,670 |
| M5352(B)[3] | C1,C2,P1 | 35,000 | 8,830[1] | — | — | 16 | 1 | 1 | 4,670 |
| M3132R[3] | C1,C2,P1 | 14,000 | 7,410[1] | — | — | 15 | 1 | 1 | 6,500 |
| M5252R(A)[3] | C1,C2,P1 | 25,200 | 7,390[1] | — | — | 15 | 1 | 1 | 4,670 |
| M5322R(B)[3] | C1,C2,P1 | 32,000 | 7,070[1] | — | — | 16 | 1 | 1 | 4,670 |
| LM2500-20[3] | C1,C2,P1 | 17,600 | 7,565[1] | — | — | 16 | 2 | 6 | 3,000 |
| LM2500-30[3] | C1,C2,P1 | 29,500 | 7,080[1] | — | — | 16 | 2 | 6 | 3,600 |
| LM5000[3] | C1,C2,P1 | 44,700 | 7,020[1] | — | — | 19 | 3 | 3 | 3,600 |
| (Gas Turbine Generator Sets) | | (MW) | (Btu/kW-hr) | | | | | | |
| G3142[4] | G2 | 10.2 | 13,540[1] | — | — | 15 | 1 | 1 | 6,500 |
| G5261[4] | G2 | 18.9 | 13,400[1] | — | — | 16 | — | 2 | 5,100 |
| G3132R[4] | G2 | 9.75 | 10,520[1] | — | — | 15 | 1 | 1 | 6,500 |
| LM2500-20[4] | G2,3 | 12.86 | 10,420[1] | — | — | 16 | 2 | 6 | 3,000* |
| LM2500-30[4] | G2,3 | 21.56 | 9,760[1] | — | — | 16 | 2 | 6 | 3,600** |
| LM5000[4] | G2,3 | 32.53 | 9,650[1] | — | — | 19 | 3 | 3 | 3,600** |
| (Package Power Plants) | | | | | | | | | |
| PG5361[4] | G1,3,5,6,7 | 24.8 | 12,450[1] | 26.62 | 12,340 | 17 | — | 2 | 5,105 |
| PG6521(B)[4] | G1,3,5,6,7 | 35.86 | 11,280[1] | 38.95 | 11,180 | 17 | — | 3 | 5,105 |
| PG7111(E)[4] | G1,3,5,6,7 | 75.0 | 10,790[1] | 81.15 | 10,750 | 17 | — | 3 | 3,600 |
| PG9151(E)[4] | G1,3,5,6,7 | 106.7 | 10,850[1] | 115.9 | 10,810 | 17 | — | 3 | 3,000 |
| (Combined Steam Turbine and Gas Turbine Generator with Heat Recovery Steam Generator[7]) | | | | | | | | | |
| STAG 205P[5] | G1,3,6 | 67.4 | 8,700[6] | — | — | 17 | — | 2 | 5,105 |
| STAG 405P[5] | G1,3,6 | 138.4 | 8,700[6] | — | — | 17 | — | 2 | 5,105 |
| STAG 206B[5] | G1,3,6 | 96.2 | 8,070[6] | — | — | 17 | — | 3 | 5,105 |
| STAG 406B[5] | G1,3,6 | 192.4 | 8,070[6] | — | — | 17 | — | 3 | 5,105 |
| STAG 107E[5] | G1,3,6 | 99.3 | 7,700[6] | — | — | 17 | — | 3 | 3,600 |
| STAG 407E[5] | G1,3,6 | 396.7 | 7,650[6] | — | — | 17 | — | 3 | 3,600 |
| STAG 109E[5] | G1,3,6 | 140.4 | 7,940[6] | — | — | 17 | — | 3 | 3,000 |
| STAG 409E[5] | G1,3,6 | 564.7 | 7,890[6] | — | — | 17 | — | 3 | 3,000 |

[1]Heat Rates: Gas Fuel = 21,510 Btu/lb. LHV; Distillate Fuel = 18,550 Btu/lb. LHV. [2]Ratings are at International Standards Organization (I.S.O.) Conditions: Sea Level and 15°C, 59°F. Normal is base rating and Maximum is peak rating. [3]Ratings are for natural gas fuel. [4]Ratings are for distillate fuel. [5]STAG plants each have one steam generator with the following gas turbines: STAG 205P — Two PG5361's; STAG 405P — Four PG5361's; STAG 206B — Two PG6521(B)'s; STAG 406B — Four PG6521(B)'s; STAG 107E — One PG7111(E); STAG 407E — Four PG7111(E)'s; STAG 109E — One PG9151(E); STAG 409E — Four PG9151(E)'s. [6]Ratings are with HHV and include both steam and gas turbines. [7]Representative STAG configurations are listed; however, other combinations of gas turbines for 50 Hz or 60 Hz are available. Symbols: STAG — Combined steam and gas turbine generator.

# GAS TURBINE SPECIFICATIONS

| Pressure Ratio | Number of Combustors | EXHAUST Flow lb/sec. or kg/sec. | EXHAUST Temp. | Heat Exch. for Exhaust Heat Recovery | Dry Weight | DIMENSIONS L | DIMENSIONS W | DIMENSIONS H |
|---|---|---|---|---|---|---|---|---|
| | | (kg/sec.) | (°C) | | (kg) | (mm) | (mm) | (mm) |
| 9.8 | 8 | 76.0 | 412 | Y | 22,000 | 7,100 | 3,480 | 3,124 |
| 9.8 | 8 | 76.0 | 449 | Y | 22,000 | 7,100 | 3,480 | 3,124 |
| 9.5 | 8 | 77.0 | 466 | Y | 22,000 | 7,100 | 3,480 | 3,124 |
| 17 | 1 | 60 | 438 | Y | 22,000 | 7,100 | 3,480 | 3,124 |
| 20 | 1 | 90 | 477 | Y | 23,000 | 6,500 | 4,000 | 3,860 |
| 11 | 8 | 110 | 552 | Y | 25,460 | 7,310 | 3,450 | 3,200 |
| 18 | 1 | 67.6 | 513 | Y | 14,000 | 6,070 | 3,520 | 3,737 |
| 31 | 1 | 122 | 428 | Y | 55,000 | 9,200 | 3,200 | 3,685 |
| 6.0 | 1 | 50 | 460 | Y | 47,900 | 8,960 | 3,124 | 6,096 |
| 8.2 | 10 | 288 | 549 | Y | 174,000 | 8,550 | 4,800 | 4,470 |
| 8.2 | 10 | — | — | — | — | — | — | — |

English Electric Corporation, 140 North Loop, Suite 146, Houston, Texas 77009 U.S.A. Contact: Mr. P. L. Banks. Phone: (713) 861-2375. Telex: 775835 • GEC Gas Turbine Services Limited, Al Najim Saudi International Company, P.O. Box 1017, Dammam, Saudi Arabia. Contact: Mr. I. Duff. Phone 833-9476/7 832-5427. Telex: 601079 • GEC Singapore Limited, P.O. Box 4046 Magnet House, Bukit Timah Road, Singapore 2158. Contact: Mr. B. Thornborough. Phone: 663011. Telex: RS21508 • GEC De Mexico, Salamanca 34 P-H, Mexico 7 D.F. Contact: Mr. H. Codd, Phone: 5332848 or 5251394, Telex: 1777576.

| Pressure Ratio | Number of Combustors | EXHAUST Flow lb/sec. or kg/sec. | EXHAUST Temp. | Heat Exch. for Exhaust Heat Recovery | Dry Weight | DIMENSIONS L | DIMENSIONS W | DIMENSIONS H |
|---|---|---|---|---|---|---|---|---|
| | | (lb/sec.) | (°F) | | (lb) | | | |
| 6.0 | 6 | 115 | 979 | — | 120,000 | 420 | 130 | 144 |
| 8.0 | 10 | 202 | 975 | — | 165,000 | 458 | 130 | 150 |
| 6.9 | 12 | 215 | 975 | — | 253,000 | 600 | 130 | 150 |
| 8.2 | 12 | 268 | 915 | — | 257,000 | 600 | 130 | 150 |
| 7.2 | 6 | 115 | 668 | REGEN | 120,000 | 420 | 130 | 144 |
| 7.0 | 12 | 215 | 638 | REGEN | 253,000 | 600 | 130 | 150 |
| 8.3 | 12 | 250 | 667 | REGEN | 257,000 | 600 | 130 | 150 |
| 14.8 | 1 | 127 | 776 | — | 52,000 | 333 | 110 | 139 |
| 18.7 | 1 | 149 | 955 | — | 52,000 | 333 | 110 | 139 |
| 30 | 1 | 271 | 796 | — | 88,500 | 696 | 124 | 144 |
| 7.1 | 6 | 115 | 979 | — | 242,000 | 765 | 130 | 150 |
| 8.0 | 10 | 213 | 955 | — | 318,000 | 816 | 126 | 150 |
| 7.2 | 6 | 115 | 668 | REGEN | 242,000 | 765 | 130 | 150 |
| 14.8 | 1 | 127 | 788 | — | 220,000 | 833 | 135 | 150 |
| 18.7 | 1 | 149 | 973 | — | 220,000 | 633 | 135 | 150 |
| 30 | 1 | 271.2 | 788 | — | 314,200 | 1,428 | 240 | 372 |
| 10.1,10.3 | 10 | 267 | 919 | — | 570,000 | 1,385 | 228 | 424 |
| 11.3,11.5 | 10 | 301 | 1,020 | — | 700,000 | 1,476 | 288 | 408 |
| 11.5,11.7 | 10 | 609 | 1,004 | — | 1,070,000 | 1,590 | 850 | 370 |
| 11.4,11.6 | 14 | 877 | 972 | — | 1,900,000 | 1,380 | 930 | 468 |
| 10.1 | 20 10/TB | 534 | 916 | Boiler | | | | |
| 10.1 | 40 10/TB | 1,068 | 916 | Boiler | | | | |
| 11.5 | 20 10/TB | 608 | 1,020 | Boiler | | | | |
| 11.5 | 40 10/TB | 1,216 | 1,020 | Boiler | | | | |
| 11.5 | 10 10/TB | 614 | 1,004 | Boiler | | | | |
| 11.5 | 40 10/TB | 2,456 | 1,004 | Boiler | | | | |
| 11.4 | 14 14/TB | 877 | 972 | Boiler | | | | |
| 11.4 | 56 14/TB | 3,508 | 972 | Boiler | | | | |

*Also available at 3,600 RPM.
**Also available at 3,000 RPM.

## GAS TURBINE SPECIFICATIONS

| MODEL | USE | POWER RATING NORMAL hp, mw or lbt | POWER RATING NORMAL Heat Rate | POWER RATING MAXIMUM hp, mw or lbt | POWER RATING MAXIMUM Heat Rate | COMPRESSOR SHAFT Compressor Stages | COMPRESSOR SHAFT Turbine Stages | POWER SHAFT Turbine Stages | POWER SHAFT RPM |
|---|---|---|---|---|---|---|---|---|---|
| **General Electric Company, Marine and Industrial Engine Division, Neumann Way, Cincinnati, Ohio 45215** | | | | | | | | | |
| | | (hp[3]) | (Btu/hp-hr) | (hp[3]) | (Btu/hp-hr) | | | | |
| 7LM5000 PA101[1] | All | 44,700 | 6,732 | 51,000 | 6,575 | 5-14 | 2-1 | 3 | 3,600 |
| 7LM5000 PA101[1] | All | 44,700 | 6,804 | 51,000 | 6,706 | 5-14 | 2-1 | 3 | 3,000 |
| 7LM5000 PA102[2] | All | 44,700 | 6,782 | 51,000 | 6,622 | 5-14 | 2-1 | 3 | 3,600 |
| 7LM5000 PA102[2] | All | 44,700 | 6,855 | 51,000 | 6,755 | 5-14 | 2-1 | 3 | 3,000 |
| 7LM500[1] | All[5] | 5,450 | 8,062 | 6,000 | 8,010 | 14 | 2 | 4 | 7,000 |
| 7LM500[2] | All[5] | 5,450 | 8,096 | 5,750 | 8,063 | 14 | 2 | 4 | 7,000 |
| Gas Turbines | | | | | | | | | |
| 7LM2500[1] | All[5] | 17,600shp | 7,340 | 22,500shp | 6,990 | 16 | 2 | 6 | 3,000 |
| 7LM2500[2] | All[5] | 17,600shp | 7,385 | 22,500shp | 7,035 | 16 | 2 | 6 | 3,000 |
| 7LM2500[1] | All[5] | 29,500shp | 6,875 | 32,000shp | 6,910 | 16 | 2 | 6 | 3,600 |
| 7LM2500[2] | All[5] | 29,500shp | 6,920 | 32,000shp | 6,955 | 16 | 2 | 6 | 3,600 |
| Gas Generators | | | | | | | | | |
| 7LM2500[1] | All[5] | 20,000ighp | 6,530 | 25,000ighp | 6,225 | 16 | 2 | — | — |
| 7LM2500[2] | All[5] | 20,000ighp | 6,570 | 25,000ighp | 6,270 | 16 | 2 | — | — |
| 7LM2500[1] | All[5] | 33,700ighp | 6,060 | 36,600ighp | 6,055 | 16 | 2 | — | — |
| 7LM2500[2] | All[5] | 33,700ighp | 6,100 | 36,600ighp | 6,095 | 16 | 2 | — | — |
| **Hants and Sussex Aviation Ltd., The City Airport, Portsmouth, Hants, England PO3 5PJ** | | | | | | | | | |
| A250 | Air Start Supply | 120lb/min | — | — | — | 1 | | 1 | 25,000 |
| 1S250 | Shaft Drive | 250 | — | — | — | 1 | | 1 | 25,000 |
| **Hispano Suiza, Rue Du Capitaine Guynemer, B.P. 60, 92270 Bois Columbes (France)** | | | | | | | | | |
| | | (hp) | (Btu/kW-hr) | | | | | | |
| THM 1103 | P1,2,G2,C2,G5 | 5,900 | 11,310 | — | — | 9 | 2 | 2 | 7,600 |
| THM 1103R | d° | 5,630[1] | 8,344 | — | — | 9 | 2 | 2 | 7,600 |
| THM 1203 | d° | 7,510 | 10,604 | — | — | 10 | 2 | 2 | 7,800 |
| THM 1203R | d° | 7,240[1] | 7,953[1] | — | — | 10 | 2 | 2 | 7,800 |
| THM 1304 | d° | 10,860 | 9,430 | — | — | 11 | 2 | 2 | 8,000 |
| THM 1304R | d° | 10,500 | 7,953[1] | — | — | 11 | 2 | 2 | 8,000 |
| THM 1304-10 | P1,2,G2,C2,G5 | 13,400 | 8,776Btu/hp-hr | — | — | 11 | 2 | 2 | 8,000 |

General Electric footnotes: [1]Gas Fuel. [2]Liquid Fuel. [3]shp — Shaft Horsepower. ighp — Isentropic Gas Horsepower. All[5]: C-1, C-2, G-1, G-2, G-3, G-5, G-6, G-7, G-8, L-1, P-0, P-W, P1, PR-1, PR-2, PR-3 except D, E, G-4, R.

Hispano Suiza: **Major offices:** Hispano Suiza Inc., Shadow Wood Drive, Houston, Texas 77043 U.S.A. Phone: (713) 467-9812. Tlx: 6888 127 HS INC UW. Hispano Suiza Do Brazil Equipamentos Ltda., Rua Da Gloria 344, Gloria 20241 Rio De Janeiro. Phone: (021) 242-4004. Tlx: 21 989 HS BRBR.

[1]Power rating and heat rate with regenerator. [2]Weight and dimensions are with base plate and auxiliaries, excluding air filters, L.O. coolers, stack, silencers or regenerator.

## GAS TURBINE SPECIFICATIONS

| Pressure Ratio | Number of Combustors | EXHAUST Flow lb/sec, or kg/sec. | EXHAUST Temp. | Heat Exch. for Exhaust Heat Recovery | Dry Weight | DIMENSIONS L | DIMENSIONS W | DIMENSIONS H |
|---|---|---|---|---|---|---|---|---|
| | | (lb/sec.) | (°F) | | (lb) | (in) | (in) | (in) |
| 28.9 | 1 | 288 | 816 | — | 33,500 | 11,195 | 3,454 | 3,810 |
| 29.7 | 1 | 296 | 820 | — | 33,500 | 11,195 | 3,454 | 3,810 |
| 28.9 | 1 | 288 | 831 | — | 33,500 | 11,195 | 3,454 | 3,810 |
| 29.7 | 1 | 296 | 835 | — | 33,500 | 11,195 | 3,454 | 3,810 |
| | | | | | | (in) | (in) | (In) |
| 14 | 1 | 35.4 | 955 | — | 1,356 | 99.2 | 36 | 36 |
| 14 | 1 | 35.3 | 971 | — | 1,356 | 99.2 | 36 | 36 |
| 15 | 1 | 126 | 775 | — | 10,300 | 252 | 84 | 84 |
| 15 | 1 | 126 | 790 | — | 10,300 | 252 | 84 | 84 |
| 18 | 1 | 149 | 955 | — | 10,300 | 252 | 84 | 84 |
| 18 | 1 | 149 | 975 | — | 10,300 | 252 | 84 | 84 |
| 15 | 1 | 127 | 1,145 | — | 4,400 | 132 | 60 | 60 |
| 15 | 1 | 127 | 1,160 | — | 4,400 | 132 | 60 | 60 |
| 18 | 1 | 149 | 1,460 | — | 4,400 | 132 | 60 | 60 |
| 18 | 1 | 149 | 1,485 | — | 4,400 | 132 | 60 | 60 |
| | | | | | | | (Inches) | |
| 3.7 | 2 | 5.2lb/sec | 1,202F | — | | 51.5 | 40 | 36 |
| 3.7 | 2 | 5.2lb/sec | 1,202F | — | | 51.5 | 40 | 36 |
| | | (lb/sec) | (°F) | | (kg) | | (mm) | |
| 6.7 | 2 | 65 | 905 | Op + | 33,900 | 8,590 | 2,750 | 3,110 |
| 6.7 | 2 | 65 | — | — | — | — | — | — |
| 7.6 | 2 | 76 | 928 | Op + | 36,340 | 8,590 | 2,750 | 3,110 |
| 7.6 | 2 | 76 | — | — | — | — | — | — |
| 9.2 | 2 | 94 | 914 | Op + | 50,880 | 9,000 | 2,750 | 3,110 |
| 9.2 | 2 | 94 | — | — | — | — | — | — |
| 10.7 | 2 | 105.8 | — | — | 50,880 | 9,000 | 2,750 | 3,110 |

# GAS TURBINE SPECIFICATIONS

| | | POWER RATING | | | | COMPRESSOR SHAFT | | POWER SHAFT | |
|---|---|---|---|---|---|---|---|---|---|
| | | NORMAL | | MAXIMUM | | | | | |
| MODEL | USE | hp, mw or lbt | Heat Rate | hp, mw or lbt | Heat Rate | Compressor Stages | Turbine Stages | Turbine Stages | RPM |
| **Hitachi Ltd., Tokyo, Japan** | | | | | | | | | |
| M1502(B)[5] | C-2 | 5,050hp | 10,570hp-hr | — | — | 15 | 1 | 1 | 10,290 |
| M3142[5] | C-2 | 14,600hp | 9,530hp-hr | — | — | 15 | 1 | 1 | 6,500 |
| M5262(A)[5] | C-2 | 26,250hp | 9,780hp-hr | — | — | 15 | 1 | 1 | 4,670 |
| M5332(B)[5] | C-2 | 33,550hp | 8,910hp-hr | — | — | 16 | 1 | 1 | 4,670 |
| M7652[5] | C-2 | 65,400hp | 9,250hp-hr | — | — | 15 | 1 | 1 | 3,020 |
| M3132R[5] | C-2 | 14,000hp | 7,410hp-hr | — | — | 15 | 1 | 1 | 6,500 |
| M5252R(A)[5] | C-2 | 25,200hp | 7,390hp-hr | — | — | 15 | 1 | 1 | 4,670 |
| M5322R(B)[5] | C-2 | 32,000hp | 7,180hp-hr | — | — | 16 | 1 | 1 | 4,670 |
| G3142[6] | G-1 | 10.20MW | 13,540kW-hr | — | — | 15 | 1 | 1 | 6,500 |
| G-5341[6] | G-1 | 24.05MW | 12,310kW-hr | 25.95MW | 12,180kW-hr | 17 | — | 2 | 5,105 |
| G7821[6] | G-1 | 60.00MW | 10,960kW-hr | 66.30MW | 10,860kW-hr | 17 | — | 3 | 3,600 |
| G7981[6] | G-1 | 74.29MW | 10,700kW-hr | 80.30MW | 10,660kW-hr | 17 | — | 3 | 3,600 |
| G9111[6] | G-1 | 85.20MW | 10,990kW-hr | 94.40MW | 10,870kW-hr | 17 | — | 3 | 3,000 |
| G3132R[6] | G-1 | 9.75MW | 10,520kW-hr | — | — | 15 | 1 | 1 | 6,500 |
| PG5341[6] | G-6 | 23.45MW | 12,490kW-hr | 25.30MW | 12,370kW-hr | 17 | — | 2 | 5,105 |
| PG7821[6] | G-6 | 58.50MW | 11,130kW-hr | 64.60MW | 11,030kW-hr | 17 | — | 3 | 3,600 |
| PG7981[6] | G-6 | 72.90MW | 10,790kW-hr | 78.80MW | 10,750kW-hr | 17 | — | 3 | 3,600 |
| PG9111[6] | G-6 | 82.60MW | 11,210kW-hr | 91.60MW | 11,090kW-hr | 17 | — | 3 | 3,000 |
| PG7791R[6] | G-6 | 55.90MW | 9,630kW-hr | 61.20MW | 9,350kW-hr | 17 | — | 3 | 3,600 |

[1] Normal ratings are base load at sea level, 59°F ambient temperature. [2] Maximum ratings are peak load at sea level, 59°F ambient temperature.
[3] Weights and dimensions do not include stacks, silencers or regenerators. [4] Regenerator stack temperature. [5] Ratings are for natural gas fuel.
[6] Ratings are for distillate oil fuel. [7] Peak rating values. [8] Weight and dimensions are for complete package unit. [9] Includes regenerator and combustion air piping.

| MODEL | USE | hp, mw or lbt | Heat Rate | hp, mw or lbt | Heat Rate | Compressor Stages | Turbine Stages | Turbine Stages | RPM |
|---|---|---|---|---|---|---|---|---|---|
| **I.A.M. Rinaldo Piaggio, Via Cibrario 4, 16154, Genova (Italy)** | | | | | | | | | |
| | | | (SFC) | | (SFC) | | | | |
| Viper 11 | PR-1 | 2,165 lb. | 1.045 | 2,500 lb. | 1.07 | 7a | 1a | N/A | N/A |
| Viper Mk526 | PR-1 | 3,177 lb. | 1.002 | 3,360 lb. | 1.007 | 8a | 1a | N/A | N/A |
| Viper Mk540 | PR-1 | 3,195 lb. | 0.993 | 3,365 lb. | 1.004 | 8a | 1a | N/A | N/A |
| Viper Mk632-43 | PR-1 | 3,695 lb. | 0.944 | 3,970 lb. | 0.963 | 8a | 2a | N/A | N/A |
| T55L11A/D | PR-1 | 3,000hp | 0.562 | 3,750hp | 0.530 | 7a, 1c | 2a | 2a | 16,000 |
| T55L712 | PR-1 | 3,000hp | 0.548 | 3,400hp | 0.530 | 7a, 1c | 2a | 2a | 16,000 |
| T53L13A/B | PR-1 | 1,250hp | 0.600 | 1,400hp | 0.590 | 5a, 1c | 2a | 2a | 21,074[1] |
| GEM 2-3 | PR-1 | 808hp | 0.525 | 961hp | 0.512 | 4a, 1c | 2a | 2a | 27,000 |

[1] The engine has a reduction gear incorporated; output RPM = 6,630 (maximum).

| MODEL | USE | hp, mw or lbt | Heat Rate | hp, mw or lbt | Heat Rate | Compressor Stages | Turbine Stages | Turbine Stages | RPM |
|---|---|---|---|---|---|---|---|---|---|
| **Ingersoll-Rand Company, Turbo Products Division, Phillipsburg, N.J. 08865 U.S.A.** | | | | | | | | | |
| GT-22B | C1, C2, P0 PI, G-2 | 4,250hp | 9,430hp-hr | 5,900hp | 8,852hp-hr | 14 | 2 | 2 | 13,820 |
| GT-22C | C1, C2, P0 P1, G-2 | 5,000hp | 8,850hp-hr | 6,100hp | 8,750hp-hr | 14 | 2 | 2 | 13,820 |
| GT-52 | C1, C2, P0 PI, G-2 | 15,800hp | 9,525hp-hr | 18,170hp | 8,745hp-hr | 17 | 3 | 1 | 5,000 |
| GT-53 | C1, C2, P0 PI, G-2 | 17,550hp | 9,500hp-hr | 19,700hp | 8,612hp-hr | 17 | 3 | 1 | 5,000 |
| GT-54 | C1, C2, P0 PI, G-2 | 18,850hp | 9,580hp-hr | 21,500hp | 8,644hp-hr | 17 | 3 | 1 | 5,250 |
| GT-61 | C1, C2, P0 PI, G-2 | 28,150hp | 7,154hp-hr | 33,000hp | 6,780hp-hr | 16 | 2 | 2 | 5,250 |
| GT-61DR | C1, C2, P0 PI, G-2 | 16,775hp | 7,732hp-hr | 26,000hp | 6,850hp-hr | 16 | 2 | 2 | 5,250 |
| GT-62 | C1, C2, P0 PI, G-2 | 29,150hp | 7,400hp-hr | 35,300hp | 6,878hp-hr | 13 | 2 | 2 | 5,000 |
| GT-71 | C1, C2, P0 PI, G-2 | 45,485hp | 6,745hp-hr | 56,400hp | 6,287hp-hr | 19 | 3 | 3 | 4,000 |

# GAS TURBINE SPECIFICATIONS

| Pressure Ratio | Number of Combustors | EXHAUST Flow lb/sec. or kg/sec. | EXHAUST Temp. | Heat Exch. for Exhaust Heat Recovery | Dry Weight | DIMENSIONS L | DIMENSIONS W | DIMENSIONS H |
|---|---|---|---|---|---|---|---|---|
| | | | | | | | (Inches) | |
| 7.0 | 1 | 45lb | 967F | — | 70,000lb | 325 | 100 | 130 |
| 6.0 | 6 | 116lb | 979F | — | 120,000lb | 420 | 130 | 144 |
| 6.9 | 12 | 216lb | 975F | — | 253,000lb | 600 | 130 | 150 |
| 8.2 | 12 | 259lb | 930F | — | 257,000lb | 600 | 130 | 150 |
| 8.2 | 12 | 529lb | 932F | — | 510,000lb | 612 | 300 | 216 |
| 7.2 | 6 | 116lb | 668F[4] | — | 120,000lb | 420 | 130 | 144 |
| 7.0 | 12 | 215lb | 638F[4] | — | 253,000lb | 600 | 130 | 150 |
| 8.3 | 12 | 251lb | 679F[4] | — | 257,000lb | 600 | 130 | 150 |
| 7.1 | 6 | 116lb | 979F | — | 242,000lb | 765 | 130 | 150 |
| 10.0/10.2[7] | 10 | 263/263lb[7] | 904/954F[7] | — | 398,000lb | 930 | 130 | 150 |
| 9.5/9.7[7] | 10 | 535/536lb[7] | 947/1,016F[7] | — | 587,000lb | 838 | 348 | 158 |
| 11.3/11.5[7] | 10 | 612/613lb[7] | 976/1,038F[7] | — | 587,000lb | 838 | 348 | 158 |
| 9.4/9.6[7] | 14 | 768/770lb[7] | 945/1,015F[7] | — | 650,000lb | 996 | 180 | 228 |
| 7.2 | 6 | 116lb | 668F[4] | — | 242,000lb | 765 | 130 | 150 |
| 10.0/10.2[7] | 10 | 260/261lb[7] | 908/958F[7] | — | 570,000lb[8] | 1,385[8] | 228[8] | 424[8] |
| 9.5/9.7[7] | 10 | 530/531lb[7] | 954/1,023F[7] | — | 1,095,000lb[8] | 1,394[8] | 736[8] | 384[8] |
| 11.3/11.5[7] | 10 | 606/607lb[7] | 981/1,043F[7] | — | 1,070,000lb[8] | 1,590[8] | 850[8] | 370[8] |
| 9.4/9.6[7] | 14 | 761/762[7]lb[7] | 950/1,020F[7] | — | 1,425,000lb[8] | 1,680[8] | 576[8] | 405[8] |
| 9.5/9.7[7] | 10 | 528/529lb[7] | 745[4]/769F[4 7] | — | 1,560,000lb[8] | 1,390[8] | 1,180[8] | 384[8] |
| | | (lb./s) | (°C) | | (lb.) | (mm) | (mm) | (mm) |
| 4.3 | 1 | 42.5 | 670 | N/A | 585 | 1,626 | 772 | — |
| 5.2 | 1 | 52.7 | 700 | N/A | 830 | 2,690 | 705 | — |
| 5.5 | 1 | 52.2 | 690 | N/A | 790 | 2,006 | 694 | — |
| 5.95 | 1 | 59 | 714 | N/A | 810 | 2,460 | 703 | — |
| 7.9 | 1 | 26.5 | 582 | N/A | 710 | 1,119 | 616 | — |
| 7.89 | 1 | 26.45 | 560 | N/A | 750 | 1,119 | 616 | — |
| 7.1 | 1 | 12.35 | 627 | N/A | 540 | 1,209 | 584 | — |
| 11.6 | 1 | 7.1 | 557 | N/A | 405 | 1,095 | 575 | 590 |
| | | | | | (lb) | | (Inches) | |
| 9.7 | 6 | 33.8lb/sec | 994F | Op | 10,000 | 228 | 95 | 122 |
| 8.8 | 6 | 34.56lb/sec | 1078F | Op | 20,000 | 228 | 95 | 122 |
| 10.0 | 8 | 167.6lb/sec | 768F | Op | 22,000 | 432 | 138 | 154 |
| 10.0 | 8 | 168.0lb/sec | 829F | Op | 22,000 | 432 | 138 | 154 |
| 11.0 | 8 | 169.3lb/sec | 885F | Op | 22,000 | 432 | 138 | 154 |
| 18.0 | 1 | 146.0lb/sec | 938F | Op | 25,000 | 384 | 138 | 154 |
| 18.0 | 1 | 126.0lb/sec | 790F | Op | 25,000 | 384 | 138 | 154 |
| 20.0 | 1 | 195.8lb/sec | 850F | Op | 25,000 | 384 | 138 | 154 |
| 31.3 | 1 | 287.6lb/sec | 784F | Op | 30,000 | 456 | 138 | 154 |

# GAS TURBINE SPECIFICATIONS

| | | POWER RATING | | | | COMPRESSOR SHAFT | | POWER SHAFT | |
|---|---|---|---|---|---|---|---|---|---|
| | | NORMAL | | MAXIMUM | | | | | |
| MODEL | USE | hp, mw or lbt | Heat Rate | hp, mw or lbt | Heat Rate | Compressor Stages | Turbine Stages | Turbine Stages | RPM |
| **Ishikawajima-Harima Heavy Industries Co., Ltd., Ohtemachi 2-2-1, Chiyoda-ku, Tokyo, Japan** | | | | | | | | | |
| | | (MW)[1] | (Btu/kW-hr) | (MW) | (Btu/kW-hr) | | | | |
| IM100-2 | G1-G8,L-1 | 1.07 | 14,700 | 1.13 | 14,400 | 10 | 2 | 1 | 19,500 |
| IM100-4 | G1-G8,L1 | 1.30 | 13,000 | 1.35 | 12,600 | 10 | 2 | 2 | 19,500 |
| IM400 | G1-G8 | 3.21 | 12,060 | 3.96 | 11,450 | 14 | 4 | — | 13,820 |
| IM400-6 | G1-G8 | 3.84 | 11,300 | 4.36 | 10,900 | 14 | 4 | — | 14,200 |
| IM600 | G1-G8,C-2 | 4.80 | 11,360 | 5.35 | 11,400 | 13 | 2 | 2 | 11,500 |
| IM1500 | G1-G8 | 11.0 | 12,800 | 12.5 | 12,700 | 17 | 3 | 1 | 3,600/<br>3,000 |
| LM2500 | G1-G8,PR-1,C-2 | 22.0 | 9,280 | 23.9 | 9,330 | 16 | 2 | 6 | 3,600/<br>3,000 |
| IM2500 | C-2,PR-1,G1-G8 | 21.1 | 9,620 | 22.8 | 9,600 | 16 | 2 | 3 | 5,500 |
| IM5000 | G1-G8,C-2,PR-1 | 35.4 | 9,040 | 38.0 | 8,930 | 19 | 3 | 3 | 3,600/<br>3,000 |
| TWIN-IM5000 | G1-G8 | 70.8 | 9,040 | 76.0 | 8,930 | 19 | 3 | 3 | 3,600<br>3,000 |
| [1]Gas turbine output. [2]Including generator. [3]All rating are stated for distillate fuel. | | | | | | | | | |
| **John Brown Engineering Ltd., Clydebank, Dunbartonshire, Scotland G81 1YA, Contact: J. R. Barlow, Phone: 041-952-2030** | | | | | | | | | |
| Gas Fuel | | (MW) | (kW-hr) | | | | | | |
| Scotstag 103J | | 14.5 | 9,180 | — | — | 15 | 1 | 1 | 6,500 |
| Scotstag 105P | | 35.5 | 8,760 | — | — | 17 | — | 2 | 5,105 |
| Scotstag 205P | | 71.7 | 8,670 | — | — | 17 | — | 2 | 5,105 |
| Scotstag 106B | | 51.0 | 8,000 | — | — | 17 | — | 3 | 5,100 |
| Scotstag 206B | | 102.5 | 7,970 | — | — | 17 | — | 3 | 5,100 |
| Scotstag 306B | | 154.0 | 7,950 | — | — | 17 | — | 3 | 5,100 |
| Scotstag 107E | | 110.0 | 7,410 | — | — | 17 | — | 3 | 3,600 |
| Scotstag 207E | | 222.0 | 7,370 | — | — | 17 | — | 3 | 3,600 |
| Scotstag 307E | | 334.0 | 7,330 | — | — | 17 | — | 3 | 3,600 |
| Scotstag 109E | | 154.0 | 7,500 | — | — | 17 | — | 3 | 3,000 |
| Scotstag 209E | | 309.0 | 7,450 | — | — | 17 | — | 3 | 3,000 |
| MS 3002 | | 10.45 | 13,320 | — | — | 15 | 1 | 1 | 6,500 |
| MS 5001 | | 25.89 | 12,110 | 27.75 | 12,020 | 17 | — | 2 | 5,105 |
| MS 5002 | | 35,000hp | 8,830hp-hr | — | — | 16 | 1 | 1 | 4,670 |
| MS 6001 | | 37.5 | 11,000 | 40.68 | 10,930 | 17 | — | 3 | 5,100 |
| MS 7001 | | 78.4 | 10,510 | 84.80 | 10,480 | 17 | — | 3 | 3,600 |
| MS 9001 | | 111.4 | 10,620 | 121.42 | 10,610 | 17 | — | 3 | 3,000 |
| LM 2500 | | 20.1 | 9,800 | — | — | 16 | 2 | 6 | 3,600 |
| **Kawasaki Heavy Industries, Ltd., Jet Engine Division, 1-1, Kawasaki-Cho, Akashi-shi, Hyogo, Japan** | | | | | | | | | |
| KT25B | G-2,5,6,7<br>PR-1 | 2,250hp | 11,510hp-hr | 2,500hp | 11,380hp-hr | 7 + 1[1] | 1 | 2 | 14,500 |
| Super KTF25 | G-2,5,6,7<br>PR-1 | 2,500hp | 11,400hp-hr | 3,000hp | 10,920hp-hr | 7 + 1[1] | 1 | 2 | 14,500 |
| Super KTF35 | G-2,5,6,7<br>PR-1 | 3,500hp | 9,810hp-hr | 4,050hp | 9,490hp-hr | 7 + 1[1] | 2 | 2 | 15,400 |
| S5A-01 | G5 | 27hp | 20,730hp-hr | 27hp | 20,730hp-hr | 1 | 1 | — | 3,600 |
| S3A-01 | G5 | 118hp | 16,580hp-hr | 118hp | 16,580hp-hr | 1 | 2 | — | 6,000 |
| S1A-01 | G1-G8 | 256hp | 14,930hp-hr | 256hp | 14,930hp-hr | 2 | 2 | — | 1,800<br>1,500 |
| S1A-02 | G1-G8 | 306hp | 13,680hp-hr | 306hp | 13,680hp-hr | 2 | 2 | — | 1,800<br>1,500 |
| S1T-02 | G1-G8 | 592hp | 14,100hp-hr | 592hp | 14,100hp-hr | 2X2 | 2X2 | — | 1,800<br>1,500 |
| S2A-01 | G1-G8 | 937hp | 11,820hp-hr | 937hp | 11,820hp-hr | 2 | 3 | — | 1,800<br>1,500 |
| M1A-01 | G1-G8 | 1,578hp | 11,730hp-hr | 1,578hp | 11,730hp-hr | 2 | 3 | — | 1,800<br>1,500 |
| M1A-03 | G1-G8 | 1,972hp | 11,820hp-hr | 1,972hp | 11,820hp-hr | 2 | 3 | — | 1,800<br>1,500 |
| M1T-01 | G1-G8 | 3,156hp | 11,900hp-hr | 3,156hp | 11,900hp-hr | 2X2 | 3X2 | — | 1,800<br>1,500 |

# GAS TURBINE SPECIFICATIONS

| Pressure Ratio | Number of Combustors | EXHAUST Flow lb/sec. or kg/sec. | EXHAUST Temp. | Heat Exch. for Exhaust Heat Recovery | Dry Weight | DIMENSIONS L | DIMENSIONS W | DIMENSIONS H |
|---|---|---|---|---|---|---|---|---|
| | | (kg/sec) | (°C) | | (kg) | (mm) | (mm) | (mm) |
| 8.4 | 1 | 6.21 | 495 | — | 260 | 2,100 | 900 | 900 |
| 8.4 | 1 | 6.58 | 485 | — | 350 | 2,100 | 900 | 900 |
| 9.3 | 6 | 15.0 | 515 | — | 580 | 2,300 | 900 | 800 |
| 9.3 | 6 | 15.6 | 533 | — | 580 | 2,300 | 900 | 800 |
| 11.2 | 1 | 18.6 | 583 | — | 610 | 1,800 | 800 | 900 |
| 12.6 | 10 | 71.2 | 410 | — | 16,500 | 8,000 | 2,800 | 2,800 |
| 18.1 | 1 | 67.6 | 523 | — | 4,700 | 6,600 | 2,100 | 2,100 |
| 17.5 | 1 | 64.8 | 512 | — | 21,000 | 7,000 | 2,800 | 3,000 |
| 28.8 | 1 | 127 | 433 | — | 40,000 | 8,700 | 3,500 | 3,500 |
| 28.8 | 1 | 254 | 433 | — | 266,000[2] | 27,900[2] | 3,500[2] | 3,500[2] |
| | | (lb/sec) | (°C) | | | | | |
| 7.2 | 6 | 115 | 530 | Y | — | — | — | — |
| 10.2 | 10 | 269 | 525 | Y | — | — | — | — |
| 10.2 | 10 | 269 x 2 | 525 | Y | — | — | — | — |
| 11.5 | 10 | 304 | 552 | Y | — | — | — | — |
| 11.5 | 10 | 304x2 | 552 | Y | — | — | — | — |
| 11.5 | 10 | 304x3 | 552 | Y | — | — | — | — |
| 11.5 | 10 | 604 | 542 | Y | — | — | — | — |
| 11.5 | 10 | 604x2 | 542 | Y | — | — | — | — |
| 11.4 | 10 | 604x3 | 542 | Y | — | — | — | — |
| 11.4 | 14 | 884 | 524 | Y | — | — | — | — |
| 11.4 | 14 | 884x2 | 524 | Y | — | — | — | — |
| 7.2 | 6 | 115 | 526 | — | 280,000 | 700 | 160 | 150 |
| 10.2 | 10 | 269 | 515 | — | 400,000 | 830 | 125 | 150 |
| 8.2 | 12 | 268 | 490 | — | 260,000 | 600 | 130 | 144 |
| 11.5 | 10 | 304 | 547 | — | 420,000 | 915 | 125 | 155 |
| 11.5 | 10 | 604 | 538 | — | 580,000 | 1,090 | 145 | 235 |
| 11.4 | 14 | 884 | 520 | — | 650,000 | 1,260 | 175 | 235 |
| 18 | Ann. | 144 | 504 | — | 220,000 | 630 | 130 | 135 |
| 6.3 | 1 | — | — | N | 1,020lb[2] | 49.5 | 30.4 | 42.6 |
| 6.5 | 1 | — | — | N | 1,183lb[2] | 50.1 | 34.4 | 43.8 |
| 7.6 | 1 | — | — | N | 1,325lb[2] | 52.2 | 34.4 | 43.8 |
| | | (kg/sec) | (°C) | | (kg) | | (mm) | |
| 4 | 1 | 0.2 | 600 | — | 25 | 380 | 360 | 400 |
| 5.8 | 1 | 0.75 | 530 | — | 80 | 920 | 460 | 380 |
| 8.5 | 1 | 1.8 | 470 | — | 390 | 1,000 | 1,000 | 800 |
| 9 | 1X2 | 1.8 | 505 | — | 390 | 1,000 | 1,000 | 800 |
| 9 | 1 | 3.5 | 505 | — | 650 | 1,200 | 1,100 | 1,100 |
| 9 | 1 | 4.8 | 490 | — | 1,300 | 1,600 | 1,100 | 1,200 |
| 8 | 1 | 7.7 | — | — | 2,450 | 2,100 | 1,400 | 1,400 |
| 9.2 | 1X2 | 9.0 | 520 | — | 2,450 | 2,100 | 1,400 | 1,400 |
| 8 | 1X2 | 15.4 | 510 | — | 6,000 | 2,200 | 2,200 | 1,600 |

# GAS TURBINE SPECIFICATIONS

| MODEL | USE | POWER RATING | | | | COMPRESSOR SHAFT | | POWER SHAFT | |
|---|---|---|---|---|---|---|---|---|---|
| | | NORMAL | | MAXIMUM | | | | | |
| | | hp, mw or lbt | Heat Rate | hp, mw or lbt | Heat Rate | Compressor Stages | Turbine Stages | Turbine Stages | RPM |
| **Kawasaki Heavy Industries** (Continued) | | | | | | | | | |
| M1T-03 | G1-G8 | 3,944hp | 12,020hp-hr | 3,944hp | 12,020hp-hr | 2X2 | 3X2 | — | 1,800<br>1,500 |
| SK30[1] | G-2 | 25.3MW | 11,250kW-hr | 28.1MW | 11,180kW-hr | 5 + 7 | 1 + 1 | 3 | 3,600 |
| SK60[1] | G-2 | 50.6MW | 11,250kW-hr | 56.2MW | 11,180kW-hr | 5 + 7 | 1 + 1 | 3 | 3,600 |
| Olympus TM38[4] | PR-1 | 21,500hp[8] | 9,370hp-hr | 2,800hp[8] | 8,820hp-hr | 5 + 7 | 1 + 1 | 1 | 5,660[9] |
| TYNE RM1C[4] | PR-1 | 5,340hp[8] | 8,640hp-hr | — | — | 6 + 9 | 1 + 1 | 2 | 3,425[9] |
| Coberra 2348[4] | C-2, P-1 | 16,400hp[8] | 9,155hp-hr | — | — | 17 | 3 | 2 | 5,000 |
| Coberra 2448[4] | C-2, P-1 | 18,068hp[8] | 9,188hp-hr | — | — | 17 | 3 | 2 | 5,200 |

[1] 60 HZ, ISO condition, 100/100 mmAq intake and exhaust duct loss [2] At normal condition [3] Including packages for generator, auxiliaries, intake, exhaust and control equipment [4] No intake and exhaust loss, ISO condition [5] At maximum condition [6] Module total weight [7] Complete Module [8] British horse power [9] Gear box output shaft speed.

**New York Office:** 375 Park Avenue, New York, NY 10152 USA, (Room No. 3309, 33rd Floor, Seagram Building), Phone: (212) 759-4950. Telex: 23-420293.
**Houston Office:** 601 Jefferson Street, Suite 3670, Houston, Texas 77002 USA. Phone: (713) 654-8981. Telex: 744539.
**London Office:** 4th Floor, 3, St. Helen's Place, London, EC3A 6EB, England. Phone: (628) 9915-7. Telex: 51-886303

| MODEL | USE | Normal hp, mw or lbt | Normal Heat Rate | Maximum hp, mw or lbt | Maximum Heat Rate | Compressor Stages | Turbine Stages | Power Shaft Turbine Stages | RPM |
|---|---|---|---|---|---|---|---|---|---|
| **KHD-Luftfahrttechnik GmbH, Oberursel, Germany** | | | | | | | | | |
| | | | (Btu/kW-hr) | | | | | | |
| T 212 | C-2 | 0.64[1] | 635[2] | — | — | 1 | 2 | — | 64,000 |
| T 216 | G1, G5-8 | 80kW | 36,200 | — | — | 1 | 1 | — | 50,000 |

[1] 0.64 kg/s compressed air at 2.82 bar. [2] Btu/sec.

| MODEL | USE | Normal hp, mw or lbt | Normal Heat Rate | Maximum hp, mw or lbt | Maximum Heat Rate | Compressor Stages | Turbine Stages | Power Shaft Turbine Stages | RPM |
|---|---|---|---|---|---|---|---|---|---|
| **Klockner-Humboldt-Deutz A.G., 5 Koln 80, Germany** | | | | | | | | | |
| | | (MW) | (Btu/kW-hr) | (MW) | (Btu/kW-hr) | | | | |
| KA 108 | G-5 | — | — | 0.021 | 66,000 | C-1 | A-1 | — | 3,600 |
| TA 216 | P, G | — | — | 0.08 | 30,925 | C-1 | R-1 | — | 3,000<br>3,600 |
| KA 215 | G | 0.22 | 18,672 | 0.227 | 18,420 | C-2 | A-2 | — | 1,500<br>1,800 |
| UA 418 | G | 0.33 | 17,190 | 370 | 16,655 | A-1, C-1 | A-3 | — | 1,500<br>1,800 |
| KT 215 | G | 0.427 | 19,144 | 0.44 | 18,966 | 2xC-2 | 2xA-2 | — | 1,500<br>1,800 |
| KA 123 | G | 0.7 | 15,888 | 0.7 | 15,888 | C-2 | A-3 | | 1,500<br>1,800 |
| UA 224 | P-O,G | 1.0 | 14,710 | 1.3 | 11,490 | A-2, C-1 | A-2 | A-2 | 1,000-<br>10,000 |
| KA 134 | G | 1.177 | 15,697 | 1.177 | 15,697 | C-2 | A-3 | — | 1,500<br>1,800 |
| KA 334 | G | 1.470 | 16,961 | 1.470 | 16,961 | C-2 | A-3 | — | 1,500<br>1,800 |
| LA 136 | C,G,P,PR | 1.812 | 15,200 | 2.138 | 14,813 | A-7,C-1 | A-1 | A-2 | 1,500<br>1,800 |
| KT 134 | G | 2.26 | 16,187 | 2.35 | 16,032 | 2 x C-2 | 2x A-3 | — | 1,500<br>1,800 |
| KT 334 | G | 2.824 | 17,484 | 2.824 | 17,484 | 2xC-2 | 2xA-1 | — | 1,500<br>1,800 |
| LA 237 | C,G,P,PR | 2.900 | 12,911 | 3.117 | 12,815 | A-7,C-1 | A-2 | A-2 | 1,500<br>1,800 |
| DA 145 | C,G,P,PR | 4.100 | 11,720 | 4.543 | — | C-2 | A-2 | A-3 | 1,500<br>1,800 |

**Klockner-Humboldt-Deutz AG**, 5 Koln 80 Germany, P.O. Box 80059, Dept AG-V1, Contact: Mr. Stommel, Phone: 0221-822-4380, Telex: 8-812-255.

**Power rating at:** sea level, 15 °C, no air inlet and exhaust losses. All data including gear box.

## GAS TURBINE SPECIFICATIONS

| Pressure Ratio | Number of Combustors | EXHAUST Flow lb/sec. or kg/sec. | EXHAUST Temp. | Heat Exch. for Exhaust Heat Recovery | Dry Weight | DIMENSIONS L | DIMENSIONS W | DIMENSIONS H |
|---|---|---|---|---|---|---|---|---|
| 9.2 | 8 | 18.0 | 520 | — | 6.000 | 2,200 | 2,200 | 1,600 |
| 11 | 8 | 108.5[2] | 515[2] | — | 216,000[3] | 22,560[3] | 4,140[3] | 14,000[3] |
| 11 | 8 | 217[2] | 515[2] | — | 422.000[3] | 37,080[3] | 4,140[3] | 14,000[3] |
| 10 | 8 | 100[2] | 425[2] | — | 29.500[6] | 9,330[7] | 2,640[7] | 3,710[7] |
| 13 | 10 | 21[2] | 441[2] | — | 14,100[6] | 5,560[7] | 1,510[7] | 2,600[7] |
| 8.5 | 8 | 76 | 409 | — | 23,000[6] | 7,518[7] | 3,048[7] | 3,048[7] |
| 8.5 | 8 | 76 | 448 | — | 23,000[6] | 7,518[7] | 3,048[7] | 3,048[7] |

Dusseldorf **Office: Am Wehrhahn 45**, Dusseldorf, F.R., Germany. Phone: 357759, **353122-4. Telex: 41-8587421**.
**Oslo Office: Kirkegaten 26, P.O. Box** 746, Sentrum, Oslo 1, Norway. Phone: 20887**5. Telex: 19973.**
**Bangkok Office: Room No. 309,** Dusit Thani Building, 1-3 Rama IV Road, Lumpini, **Bangkok, Thailand.** Phone: 2331951. Telex: **TH2800**
**Manila Office: 11th Floor, Filcapital** Office, Condominium Building, AYALA Avenue, **Makati, Rizal, Philippines**. Phone: 891826, **891827. Telex: 2629.**
**Melbourne Office: 44 Market Street**, Melbourne 3000, Victoria AUSTRALIA. Phone: **(03)62-2752 Telex:** AA33320 KAWAJU.

| Pressure Ratio | Number of Combustors | Flow | Temp. | Heat Exch. | Dry Weight | L | W | H |
|---|---|---|---|---|---|---|---|---|
| | | (kg/sec.) | | | | (mm) | (mm) | (mm) |
| 4.88 | 1 | 0.89 | 605 | — | 45 | 930 | 322 | 345 |
| 2.8 | 1 | 0.9 | 665 | — | 83 | 824 | 617 | 815 |

| Pressure Ratio | Number of Combustors | Flow | Temp. | Heat Exch. | Dry Weight | L | W | H |
|---|---|---|---|---|---|---|---|---|
| | | (kg/sec.) | (°C) | | (kg) | (mm) | (mm) | (mm) |
| 4 | 1 | 0.2 | 600 | — | 25 | 380 | 360 | 400 |
| 2.8 | 1 | 0.9 | 650 | — | 83 | 820 | 620 | 820 |
| 9 | 1 | 1.83 | 518 | — | 390 | 1,000 | 1,000 | 800 |
| 5.3 | 1 | 2.55 | 470 | — | 340 | 1,320 | 710 | 730 |
| 9 | 2x1 | 3.56 | 518 | — | 650 | 1,200 | 1,100 | 1,100 |
| 9 | 1 | 4.88 | 493 | — | 1,300 | 2,000 | 1,200 | 1,200 |
| 8.2 | 1 | 6.94 | 450 | — | 490 | 2,140 | 720 | 920 |
| 8 | 1 | 7.83 | 505 | — | 2,500 | 2,100 | 1,400 | 1,400 |
| 9.2 | 1 | 9.05 | 545 | — | 2,500 | 2,100 | 1,400 | 1,400 |
| 6.5 | 1 | 9.6 | 600 | — | 2,520 | 2,300 | 1,400 | 1,630 |
| 8 | 2x1 | 15.7 | 505 | — | 5,700 | 2,200 | 2,200 | 1,600 |
| 9.2 | 2x1 | 17.96 | 545 | — | 5,700 | 2,200 | 2,200 | 1,600 |
| 8.4 | 1 | 12.49 | 610 | — | 2,585 | 2,350 | 1,400 | 1,630 |
| 12 | 1 | 19 | 465 | — | 3,970 | 3,800 | 1,100 | 1,100 |

## GAS TURBINE SPECIFICATIONS

| MODEL | USE | POWER RATING NORMAL hp, mw or lbt | NORMAL Heat Rate | MAXIMUM hp, mw or lbt | MAXIMUM Heat Rate | COMPRESSOR SHAFT Compressor Stages | COMPRESSOR SHAFT Turbine Stages | POWER SHAFT Turbine Stages | POWER SHAFT RPM |
|---|---|---|---|---|---|---|---|---|---|
| **A.S. Kongsberg Vapenfabrikk, Gas Turbine and Power Systems Division, P.O. Box 25, N-3601 Kongsberg, Norway** | | | | | | | | | |
| | | (MW) | | | | | | | |
| KG2-3C[2,4] | G-1,2,3,5,6, 7,8,P-2 | 1.47 | 19,800 | 1.67 | 19,200 | 1C | — | 1R | 18,000 |
| KG2-3D | G5,G-7 | — | — | 1.89 | 19,210 | 1C | — | 1R | 18,800 |
| KG2-3R*** | P1,C2 | 1.38 | 13,220 | — | — | 1C | — | 1R | 18,000 |
| KG5[2,5] | C-1,2,G-1,2,3, 5,6,7,8,P-1,2 PR-2 | 3.11 | 15,360 | 3.34 | 15,100 | 1C | 1R | 1A | 12,000 |
| KG831* | G-1,2,3,5,6, 7,8, P-2 | 0.53* | 15,620 | 0.596 | 15,050 | 2C | — | 3A | 41,730 |
| KG 831** | 11 | 0.548** | 15,330 | 0.596 | 14,960 | 2C | — | 3A | 41,730 |
| **Kraftwerk Union AG, Postfach 3220, D-8520 Erlangen, Federal Republic Germany** | | | | | | | | | |
| | | (MW) | (Btu/kW-hr) | (MW) | | | | | |
| V 84 | G-1,2,3 | 93.5[1] | 10,330[2] | 101.6[1] | 10,245[2] | 17 | — | 4 | 3,600 |
| V94 | G-1,2,3 | 133.2[1] | 10,440[3] | 144.6[1] | 10,365[2] | 16 | — | 4 | 3,000 |
| **Kvaerner Brug A/S, Steam and Gas Turbine Division, P.O. Box 3610 Gb Oslo 1, Norway, Phone: (472) 684050, Telex: 16866** | | | | | | | | | |
| | | (hp) | (Btu/hp-hr) | | | | | | |
| M3142(J) | MD | 14,600 | 9,530 | — | — | 15 | 1 | 1 | 6,500 |
| M5262(A) | MD | 26,250 | 9,780 | — | — | 15 | 1 | 1 | 4,670 |
| M5352(B) | MD | 35,000 | 8,830 | — | — | 16 | 1 | 1 | 4,670 |
| M5322R(B) | MD | 32,000 | 7,070 | — | — | 16 | 1 | 1 | 4,670 |
| LM2500-20 | MD | 17,600 | 7,340 | — | — | 16 | 2 | 6 | 3,000 |
| LM2500-30 | MD | 27,500 | 6,895 | — | — | 16 | 2 | 6 | 3,600 |
| LM2500 uprate | MD | 29,500 | 6,874 | — | — | 16 | 2 | 6 | 3,600 |
| | | (kW) | (Btu/kW-hr) | | | | | | |
| G5361(P) | G | 25,890 | 12,110 | — | — | 17 | — | 2 | 5,100 |
| G6521(B) | G | 37,500 | 11,000 | — | — | 17 | — | 3 | 5,100 |
| G9151(E) | G | 111,400 | 10,620 | — | — | 17 | — | 3 | 3,000 |
| LM2500-20 | G | 13,100 | 9,840 | — | — | 16 | 2 | 6 | 3,000 |
| LM2500-30 | G | 20,100 | 9,435 | — | — | 16 | 2 | 6 | 3,600 |
| LM2500 uprate | G | 21,650 | 9,410 | — | — | 16 | 2 | 6 | 3,600 |
| **Mitsubishi Heavy Industries, Ltd., 5-1 Marunouchi, 2-chome, Chiyoda-ku, Tokyo 100, Japan** | | | | | | | | | |
| | | | (kW-hr) | (MW) | (kW-hr) | | | | |
| MW-101L | G-1,2,3 | 8.7MW[2] | 14,960 | 9.3 | 14,640 | 17 | — | 5 | 6,000 |
| MW-191 | G-1,2,3 | 17.8MW[2] | 13,440 | 19.1 | 13,220 | 15 | — | 5 | 4,912 |
| MW-251 | G-1,2,3 | 36.8MW[2] | 11,800 | 40.0 | 11,520 | 18 | — | 3 | 4,894 |
| MW-501D | G-1,2,3 | 100MW[2] | 10,470 | 108 | 10,400 | 19 | — | 4 | 3,600 |
| MW-701B | G-1,2,3 | 90.8MW[2] | 11,160 | 97.8 | 11,070 | 17 | — | 4 | 3,000 |
| MW-701D | G-1,2,3 | 122MW[2] | 10,670 | 130 | 10,580 | 19 | — | 4 | 3,000 |
| MW-252B | C-1,2 | 26.5hp[3] | 12,440 | — | — | 15 | 1 | 1 | 4,850 |
| MW-252C | C-1,2 | 30.2hp[3] | 11,920 | — | — | 16 | 1 | 1 | 4,850 |
| JT8D-M9[5] | PR-1 | 12,600lbt | 10,488Btu/lb-hr | 14,500lbt | 10,948Btu/lb-hr | 13 | 4 | — | — |
| GCM1-B[6] | A | 183[7] | 26,680[8] | — | — | 1C | 2 | — | — |
| CT63-M-5A[9] | PR-1 | 270hp | — | 317hp | — | 6-1C | 2 | 2 | 6,000 |

Kongsberg notes: Packaged under OEM agreement with Garrett. Low calorific value gases are acceptable. Available with 1,200, and 1,500 and 1,800 RPM standard gearboxes. [2]Ratings are associated with natural gases and distillate liquid fuels crude and heavy oil ratings may be reduced. [3]Available with 1,500 and 1,800 RPM standard gearboxes. Special option available. [5]Exd oil tank. *Liq. fuel rating. **Gas fuel rating. ***Recuperated version: 80% recuperator — effectiveness

**Kongsberg Ltd.:** Priors Way, Maidenhead, Berks. SL6 2HP, England. Contact Mr. A. H. Bembridge. Phone (0628) 39292 Telex 847847.
**Kongsberg GmbH:** Hochhaus West Frankfurtherstrasse 176, 6078 Neuisenburg, West Germany. Contact Mr. Harald Pohlert. Phone 6102/37495 Telex 4185611.
**Kongsberg France:** 8D, Rue de la Ceinture, 78000 Versailles, France. Contact: Mr. Claude Lambert. Phone 03/950 8451. Telex 698023

Kraftwerk Union notes: [1]ISO-MW at turbine coupling with fuel-gas firing (LHV 50,056 kJ/kg). [2]Btu per kW-hr at turbine coupling. [3]Design ISO value.
**Main Contact:** Mr. John Joyce, Erlangen (FGR), Phone: (09131) 18-5612, Telex: 062929 kWu D.
**Sales:** Hammerbacher Str. 12 + 14, D-8520 Erlangen (FRG), Berliner Str. 295-299, D-6050 Offenbach (FRG).

Mitsubishi notes: [1] Ratings are for natural gas fuel. Normal is base rating and Maximum is peak rating. [2] Power at generator terminal with standard inlet and exhaust losses. [3] Power at turbine output coupling with no inlet and exhaust losses. [4] Exhaust gas temperature for base rating. [5] Pratt & Whitney Aircraft Div. of United Technology Corp., Licensee. [6] Own development. [7] Air hp. [8] BTU/air hp-hr. [9] Detroit Diesel Allison Div. of General Motors Corp. Licensee.

**New York Liaison Office:** 277 Park Ave., New York, NY 10017, U.S.A. Phone: 826-2188, 2189, 2966, 2967. Cable Address: HISHIJU NEWYORK-NY.
Telex: NEW YORK 223037, 012-7756.

# GAS TURBINE SPECIFICATIONS

| Pressure Ratio | Number of Combustors | EXHAUST Flow lb/sec. or kg/sec. | EXHAUST Temp. | Heat Exch. for Exhaust Heat Recovery | Dry Weight | DIMENSIONS L | DIMENSIONS W | DIMENSIONS H |
|---|---|---|---|---|---|---|---|---|
| | | | (°C) | | (kg) | (mm) | (mm) | (mm) |
| 4 | 1 | 13.1 | 603 | — | 2,750 | 2,000 | 1,600 | 2,200 |
| 4 | 1 | 13.7 | 638 | — | 2,750 | 2,000 | 1,600 | 2,200 |
| 4 | 1 | 12.9 | 557 | — | 3,100 | 2,530 | 1,600 | 3,860 |
| 6.3 | 1 | 21.7 | 510 | — | 3,760 | 3,700 | 1,900 | 2,200 |
| 11 | 1 | 3.5 | 525 | — | 770 | 1,600 | 1,000 | 900 |
| 11 | 1 | 3.5 | 522 | — | 770 | 1,600 | 1,000 | 900 |

**Kongsberg (Pte.)** 48, Tagore Lane Singapore 2678. Contact: Mr. Oivind Myrhaug. Phone 4565 955. Telex 36946.
**North American Turbine Corporation:** 11500 Charles Street, P.O. Box 40510, Houston, TX 77040. Contact: Mr. Thorbjorn Westrune. Phone: 713/466-6200. Telex 6869039.
**NATCO**: 2022 Powers Ferry Road, Suite 180, Atlanta, GA 30339. Phone 404/953-1438.
**NATCO**: 2300 East Higgins Road, Suite 2001 Elk Grove Village, IL 60007. Phone: 312/437-8430.
**NATCO**: 189 Viking Avenue, Brea, CA 92621. Phone: 714/529-0114. Telex 678-376.
**NATCO**: 135 Fort Lee Road, Leonia, NJ 07605. Phone: 201/461-8194. Telex 135115.

| Pressure Ratio | Number of Combustors | EXHAUST Flow lb/sec. or kg/sec. | EXHAUST Temp. | Heat Exch. for Exhaust Heat Recovery | Dry Weight | DIMENSIONS L | DIMENSIONS W | DIMENSIONS H |
|---|---|---|---|---|---|---|---|---|
| | | (kg/sec) | (°C) | | (kg) | (mm) | (mm) | (mm) |
| 10.4 | 2 | $345^3$ | $590^3$ | Y | 172,000 | 15,500 | 10,000 | 7,000 |
| 10.4 | 2 | $496^3$ | $590^3$ | Y | 275,000 | 18,000 | 12,500 | 7,500 |

**Service: Wiesenstr.** 35, D-4330 Muelheim/Ruhr (FRG). Huttenstr. 12-16, D-1000 Berlin 21 (FRG).
**Subsidiary: Utility Power** Corporation, 4303 First Street, Bradenton, FL 33508 (USA).

| Pressure Ratio | Number of Combustors | EXHAUST Flow lb/sec. or kg/sec. | EXHAUST Temp. | Heat Exch. for Exhaust Heat Recovery | Dry Weight | DIMENSIONS L | DIMENSIONS W | DIMENSIONS H |
|---|---|---|---|---|---|---|---|---|
| | | (lb/sec) | (°F) | | (lb) | | | |
| 6.0 | 6 | 115 | 979 | — | 120,000 | 420 | 130 | 144 |
| 6.9 | 12 | 215 | 975 | — | 253,000 | 600 | 130 | 150 |
| 8.2 | 12 | 257 | 915 | — | 257,000 | 600 | 130 | 150 |
| 8.3 | 12 | 250 | 667 | Regen | 257,000 | 600 | 130 | 150 |
| 18.0 | 1 | 124.9 | 775 | — | 52,000*** | 330 | 110 | 139 |
| 18.0 | 1 | 144.2 | 920 | — | 52,000*** | 330 | 110 | 139 |
| 18.0 | 1 | 147 | 955 | — | 52,000 | 330 | 110 | 139 |
| 10.1 | 10 | 269 | 916 | — | 398,000 | 930 | 130 | 150 |
| 11.7 | 10 | 304 | 1,017 | — | 528,000 | 1,068 | 132 | 144 |
| 11.5 | 14 | 884 | 968 | — | 1,015,000 | 1,150 | 135 | 264 |
| 18.0 | 1 | 124.9 | 775 | — | 220,000 | 620 | 125 | 139 |
| 18.0 | 1 | 144.2 | 920 | — | 220,000 | 620 | 125 | 139 |
| 18.0 | 1 | 147 | 955 | — | 220,000 | 620 | 125 | 139 |

| Pressure Ratio | Number of Combustors | EXHAUST Flow lb/sec. or kg/sec. | EXHAUST Temp. | Heat Exch. for Exhaust Heat Recovery | Dry Weight | DIMENSIONS L | DIMENSIONS W | DIMENSIONS H |
|---|---|---|---|---|---|---|---|---|
| | | (kg/sec) | (°C) | | (kg) | (mm) | (mm) | (mm) |
| 8 | 6 | 66.7 | $413^4$ | Y | 40,000 | 8,000 | 3,200 | 2,900 |
| 7 | 6 | 124 | $400^4$ | Y | 85,000 | 10,900 | 3,700 | 3,500 |
| 12 | 8 | 160 | $517^4$ | Y | 100,000 | 12,400 | 3,700 | 3,700 |
| 14 | 14 | 360 | $521^4$ | Y | 143,000 | 11,600 | 5,700 | 4,400 |
| 11 | 18 | 398 | $476^4$ | Y | 170,000 | 12,500 | 5,200 | 5,200 |
| 14 | 18 | 449 | $524^4$ | Y | 190,000 | 12,500 | 5,200 | 5,200 |
| 7 | 8 | 118 | 570 | Y | 92,000 | 8,800 | 3,600 | 4,300 |
| 8 | 8 | 133 | 557 | Y | 92,000 | 8,800 | 3,600 | 4,300 |
| 16.9 | 9 | 144 | — | — | 1,475 | 3,138 | 1,080 | 1,420 |
| 3.7 | 4 | 2.9 | 1,100 | — | 145 | 1,182 | 657 | 708 |
| 6.2 | 1 | 3.1lb/sec | — | — | 64 | 1,062 | 483 | 572 |

**London Liaison Office: c/o Mitsubishi** Corp., London Branch, Bow Bells House, Bread Street (Cheapside), London, EC4M 9BQ, England. Phone: 248-8821/8825.
Cable Address: HISHIJU LONDON-EC4. Telex: LONDON 888994.
**Duesseldorf Liaison Office:** 4 Duesseldorf Ratinger Strasse 45, West Germany. Phone: (0211) 80501-3. Cable Address: HISHIJU DUESSELDORF. Telex: DUESSELDORF 858-7666.
**Oslo Liaison Office:** Karl Johans Gate 27, Oslo 1, Norway. Phone: 413089, 411019. Cable Address: HISHIJU OSLO. Telex OSLO 11973.
**South East Asia Liaison Office:** Tower 3103, DBS Bldg. 6, Shenton Way Singapore 1, P.O. Box No. 3047, Republic of Singapore. Phone: 2216544, 2116545.
Cable Address: HISHIJU SINGAPORE. Telex: SINGAPORE 24544.

## GAS TURBINE SPECIFICATIONS

| MODEL | USE | POWER RATING — NORMAL hp, mw or lbt | POWER RATING — NORMAL Heat Rate | POWER RATING — MAXIMUM hp, mw or lbt | POWER RATING — MAXIMUM Heat Rate | COMPRESSOR SHAFT Compressor Stages | COMPRESSOR SHAFT Turbine Stages | POWER SHAFT Turbine Stages | POWER SHAFT RPM |
|---|---|---|---|---|---|---|---|---|---|
| **Mitsui Engineering & Shipbuilding Co., Ltd., Head Office: 6-4, Tsukiji 5-chome, Chuo-ku, Tokyo, Japan.** | | | | | | | | | |
| | | (MW) | (Btu/kW-hr) | (MW) | (Btu/kW-hr) | | | | |
| SB30C | G-1-2<br>G-5-8 | 4.87 | 14,848 | 5.11 | 13,935 | 11 | 4 | — | 9,410 |
| SB90C | G-1-2<br>G-5-8 | 16.17 | 13,364 | 16.81 | 13,299 | 10 | 4 | — | 5,475 |
| RB30C | G-1-2<br>G-5-8 | 4.69 | 10,405 | — | — | 11 | 4 | — | 9,410 |
| RB90C | G-1-2<br>G-5-8 | 15.61 | 9,852 | — | — | 10 | 4 | — | 5,475 |
| SB60C-M | G-1-2<br>G-5-8<br>C-1-2<br>P-1-2<br>PI | 13.07 | 11,123 | 14.14 | 10,986 | 16 | 2 | 2 | 5,680 |
| CB30C | G-3 | 7.5 | 9,850 | — | — | 11 | 4 | — | 9,410 |
| CB90C | G-3 | 24.3 | 9,450 | — | — | 10 | 4 | — | 5,475 |
| **MTU Motoren-u. Turbinen, Union Muchen GmbH, Dachauer Str. 665, 8000 Munchen 50, West Germany** | | | | | | | | | |
| MTU Tyne R.Ty. 20Mk22 | PR-1 | 3,874kW | — | 4,228kW | — | 15 | 1 | 3 | — |
| J79-MTU-J1K | PR-1 | 4,745kp | — | 7,235kp | — | 17 | 3 | — | 7,460 |
| J79-MTU-17A | PR-1 | 5,384kp | — | 8,119kp | — | 17 | 3 | — | 7,460 |
| Larzac 04 | PR-1 | — | — | 1,350kp | — | 6 | 1 | 1 | — |
| Turbo-Union RB 199 | PR-1 | — | — | 6,800kp | — | — | — | — | — |
| GE CF6 | PR-1 | 5,485kp | — | 23,815kp | — | 14 | 2 | 4 | — |
| **Noel Penny Turbines Limited, Siskin Drive, Toll Bar End, Coventry CV3 4FE, England.** | | | | | | | | | |
| NPT 401 | AJ[1] | 280lbt | 1.21SFC | 308lbt | 1.15SFC | A2, C1 | A2 | — | — |
| NPT 109 | C1 | 7,500 scfm at 6 psig | 0.91SFC | 6,000 scfm at 8 psig | 0.82SFC | C1 | A2 | A1 | — |
| NPT 409 | C1 | 7,500 scfm at 8 psig | 0.81SFC | 7,000 scfm at 11 psig | 0.77SFC | A2, C1 | A2 | A1 | — |
| NPT 100 | G5, C1 | — | — | — | — | C1 | A2 | — | — |
| **North American Turbine Corporation, P.O. Box 40510, Houston, Texas U.S.A. 77040** | | | | | | | | | |
| KG831[1] | G2,6,8, P1,2 | (690)<br>515kW | .64 | (800)<br>597kW | .60 | — | — | 3A | 41,700[3] |
| KG2 | G2,6,8, P1,2 | (2,110)<br>1,573kW | .77 | (2,430)<br>1,812kW | .74 | — | — | 1R | 18,000[4] |
| KG5 | G2,6,8, P1,2 | (4,400)<br>3,281kW | .57 | (4,670)<br>3,482kW | .55 | 1C | 1R | 1A | 12,000[5] |

Mitsui notes:

*1 = This figure includes the weight of heat exchanger.
** = This figure shows the weight of gas turbine only.

**Head Office:** 6-4, Tsukiji 5-chome, Chuo!ku, Tokyo. Phone: (03) 544-3512. Cable: "MITUIZOSEN TOKYO". Telex: J22821, J22924, 252-2661.
**Special Representative in New York,** Suite 2029, One World Trade Center, New York, NY 10048, USA. Phone: (212) 432-0240/4. Cable: "MITUIZOSEN NEW YORK". Telex: 232671, 420144 WU 129238.
**Special Representative in Duesseldorf,** Koenigsallee 92a, 4 Duesseldorf 1 F.R. West Germany. Phone: Duesseidorf 325701, 326125. Cable: "MITUIZOSEN DUESSELDORF". Telex: 8587297.
**Special Representative in London,** 7th Floor, Lee House, Wood Street, London Wall, London EC 2 Y 5 AH., England. Phone: 600-0241. Cable: "MITUIZOSEN LONDON EC4". Telex: 885971.

Noel Penny note:

[1]Pure jet.

North American Turbine notes:

All engines supplies by NATCO are available in their line of VIKING power modules. [1] Packaged under OEM-agreement with Garrett. [2] Ratings are associated with natural gases and distillate liquid fuels. Crude and heavy oil ratings may be reduced. Low caorific value gases are acceptable and selected crudes. [3] Available with 1500 and 1800 RPM standard gearboxes. Special options available. [4] Available with 1200, 1500 and 1800 RPM standard gearboxes. [5] Available in single and 2 shaft versions, output speeds 1500 and 1800 RPM standard gearboxes.

## GAS TURBINE SPECIFICATIONS

| Pressure Ratio | Number of Combustors | EXHAUST: Flow lb/sec. or kg/sec. | EXHAUST: Temp. | Heat Exch. for Exhaust Heat Recovery | Dry Weight | DIMENSIONS: L | DIMENSIONS: W | DIMENSIONS: H |
|---|---|---|---|---|---|---|---|---|
| | | (kg/sec) | (°C) | | (kg) | | (MM) | |
| 6.8 | 1 | 28.1 | 517 | — | 28,300 | 7,800 | 3,000 | 4,300 |
| 6.8 | 1 | 86.2 | 514 | — | 53,000 | 13,000 | 3,500 | 6,000 |
| 6.8 | 1 | 28.1 | — | Y | *[1]56,300 | — | — | — |
| 6.8 | 1 | 86.2 | — | Y | *[1]170,000 | — | — | — |
| 12.4 | 1 | 58.3 | 466 | — | 54,700 | 13,000 | 3,500 | 5,500 |
| 6.8 | 1 | 28.1 | — | Y | *[2]28,300 | — | — | — |
| 6.8 | 1 | 86.2 | — | Y | *[2]53,000 | — | — | — |

**Special Representative in Los Angeles**, c/o Mitsui & Co., (USA) Inc., Los Angeles Office, **Suite 2190, Crocker**-Bank Plaza, 611 **West 6th Street,** Los Angeles, **CA 90017, USA. Phone**: 213-680-1000, 213-628-5350. Telex: 67205.
**Special Representative in Hong Kong,** c/o Mitsui & Co., Ltd., Hong Kong Branch, Co**nnaught Centre, 33rd F**loor, Connaught Road, **Central, Hong Kong.**
Phone: 5-264291/3. **Telex: HX73916**.
**Special Representative in Singapore,** c/o Mitsui & Co., Ltd., Singapore Branch, 6th **Floor, Industrial & Com**mercial Bank Building, **Two Shenton Way, Singapore.**
Phone: 2204065. **Cable: MITUIZOSEN** C/O MITUICO SINGAPORE. Telex: RS24102 MESSPRE.
**Mitsui Zosen do Brasil Industriais Pesadas Ltda.**, 13th Floor, 63 Avenida Almirante **Barroso, Rio de Janeiro**, RJ, Brazil. Phone: 252-**3667, 242-4894. Telex: 021-22378.**

| Pressure Ratio | Number of Combustors | EXHAUST: Flow lb/sec. or kg/sec. | EXHAUST: Temp. | Heat Exch. for Exhaust Heat Recovery | Dry Weight | DIMENSIONS: L | DIMENSIONS: W | DIMENSIONS: H |
|---|---|---|---|---|---|---|---|---|
| | | | | | | (mm) | (mm) | (mm) |
| 13.5 | 10 | 21.1kg/sec | — | — | 995kg | 2,760 | 1,097 | 1,097 |
| 12.4 | 10 | 74.4kg/sec | — | — | 1,685kg | 5,301 | 992 | 992 |
| 13.5 | 10 | 77.0kg/sec | — | — | 1,724kg | 5,301 | 992 | 992 |
| 10.7 | 1 | 28kg/sec | — | — | 290kg | 1,180 | 760 | 760 |
| 23 | 1 | appr. 70kg/sec | — | — | appr. 1,000kg | appr. 3,200 | appr. 870 | appr. 870 |
| 29.4 | 1 | 668kg/sec | — | — | 3,775kg | 4,777 | 2,438 | 2,438 |

| Pressure Ratio | Number of Combustors | EXHAUST: Flow lb/sec. or kg/sec. | EXHAUST: Temp. | Heat Exch. for Exhaust Heat Recovery | Dry Weight | DIMENSIONS: L | DIMENSIONS: W | DIMENSIONS: H |
|---|---|---|---|---|---|---|---|---|
| | | | | | | | (Inches) | |
| 5.77 | — | 5.03 | 1,314F | — | 85lb | 940 | 381 | 381 |
| 3.54 | — | 3.50 | 1,080F | — | 476lb | 1,524 | 457 | 457 |
| 5.77 | — | 5.03 | 1,080F | — | 490lb | 1,524 | 457 | 457 |
| 2.5 | — | 2.2 | 842F | — | 310lb | 952 | 495 | 850 |

| Pressure Ratio | Number of Combustors | EXHAUST: Flow lb/sec. or kg/sec. | EXHAUST: Temp. | Heat Exch. for Exhaust Heat Recovery | Dry Weight | DIMENSIONS: L | DIMENSIONS: W | DIMENSIONS: H |
|---|---|---|---|---|---|---|---|---|
| | | | | | | | (Inches) | |
| 11 | 1 | 7.8lb/sec | 928 | OP | 1,700lb | 65 | 39 | 34 |
| 3.9 | 1 | 29lb/sec | 1,036 | OP | 3,020lb | 60 | 50.5 | 72 |
| 6.3 | 1 | 46lb/sec | 915 | OP | 6,515lb | 115 | 73 | 84 |

# GAS TURBINE SPECIFICATIONS

| MODEL | USE | POWER RATING NORMAL hp, mw or lbt | POWER RATING NORMAL Heat Rate | POWER RATING MAXIMUM hp, mw or lbt | POWER RATING MAXIMUM Heat Rate | COMPRESSOR SHAFT Compressor Stages | COMPRESSOR SHAFT Turbine Stages | POWER SHAFT Turbine Stages | POWER SHAFT RPM |
|---|---|---|---|---|---|---|---|---|---|
| **Norwalk-Turbo, Inc., 7 Northway Lane, Latham, New York 12110** | | | | | | | | | |
| | | (hp) | (Btu/hp-hr) | (hp) | (Btu/hp-hr) | | | | |
| T-9A | C-1,2,G-1,2,3, 5,7,8,P-0 | 860 | 10,875 | 970 | 10,525 | 4 | 1 | 1 | 33,000 |
| T-9B | C-1,2,G-1,2,3, 5,7,8,P-0 | 960 | 10,875 | 1,075 | 10,819 | 4 | 1 | 2 | 28,500 |
| T-9AT | C-1,2,G-1,2,3, 5,7,8,P-0 | 1,720 | 10,875 | 1,940 | 10,525 | 4 | 1 | 1 | 33,000 |
| T-9BT | C-1,2,G-1,2,3, 5,7,8,P-0 | 1,920 | 10,875 | 2,150 | 10,819 | 4 | 1 | 2 | 28,500 |
| T-40 | C-1,2,G-1,2,3, 5,7,8,P-0 | 4,000 | 9,480 | 4,600 | 9,464 | 8 | 2 | 2 | 15,400 |
| T-60 | C-1,2,G-1,2,3, 5,7,8,P-0 | 5,200 | 8,295 | 5,800 | 8,233 | 14 | 2 | 4 | 7,000 |
| *Each Engine | | | | | | | | | |
| Norwalk-Turbo, Inc., 7 Northway Lane, Latham, NY 12110, Contact: S. David Caplow, Tel: 518-783-1625, | | | | | | | | | |
| **Nuovo Pignone, Via Felice Matteucci, 2, 51027 Firenze, Italy** | | | | | | | | | |
| | | (kW) | (Btu/kW-hr) | (kW) | (Btu/kW-hr) | | | | |
| G 1642 | | 4,390 | 14,060 | — | — | | | | 10,290 |
| G 1522 | | 3,575 | 15,240 | — | — | | | | 10,290 |
| G 3142 | | 10,200 | 13,540 | — | — | | | | 6,500 |
| PG 5361 | | 24,800 | 12,450 | 26,620 | 12,340 | | | | 5,100 |
| PG 6521 | | 35,860 | 11,280 | 38,950 | 11,180 | | | | 5,100 |
| PG 9141 | | 106,700 | 10,850 | 115,900 | 10,810 | | | | 3,000 |
| LM 2500 | | 21,560 | 9,280 | — | — | | | | 3,600 |
| | | (HP) | (Btu/shp-hr) | | | | | | |
| M 1522 | | 5,200 | 10,570 | | | | | | 10,290 |
| M 1502R | | 5,050 | 7,970 | | | | | | 10,290 |
| M 1642 | | 6,400 | 9,800 | | | | | | 10,290 |
| M 1622R | | 6,000 | 7,500 | | | | | | 10,290 |
| M 3142 | | 14,600 | 9,530 | | | | | | 6,500 |
| M 3142R | | 14,000 | 7,410 | | | | | | 6,500 |
| M 5352 | | 35,000 | 8,830 | | | | | | 4,670 |
| M 5322R | | 32,000 | 7,070 | | | | | | 4,670 |
| LM 2500 | | 29,500 | 6,874 | | | | | | 3,600 |
| PGT 25 | | 29,000 | 6,960 | | | | | | 6,500 |
| **Onan Corporation, 1400 73rd Avenue N.E., Minneapolis, MN 55432** | | | | | | | | | |
| | | (MW) | (Btu/kW-hr)[3] | (MW)[1] | (Btu/kW-hr)[3] | | | | |
| 560 GTU | G-2,3,5,7 | 0.51 | $1.54 \times 10^4$ | 0.56 | $1.58 \times 10^4$ | 2 | — | 3 | 41,730 |
| 4000 GTU | G-2,3,5,7 | 4.10 | $1.19 \times 10^4$ | 4.38 | $1.17 \times 10^4$ | 2 | 2 | 3 | 7,200 |
| [1]At generator terminals. [2]Annular combustor. [3]Including generator and gearbox. | | | | | | | | | |
| **Prvni Brnenska Strojirna, (First Brno Engineering Works), Brno, Czechoslovakia** | | | | | | | | | |
| ST 1 CH | C-1[1] | 25,400[2] | — | 30,500[4] | — | — | — | 3 | 7,500 |
| ST 1.5 | G-7 | 1.5[3] | 0.922[11] | 1.8[5] | — | — | — | 2 | 14,250 |
| ST 15RD | G-6 | 15.0[3] | — | 18.0[5] | — | 8 | 2 | 1 | 3,000 |
| ST 30RD | G-6 | 30.0[3] | — | 36.0[5] | — | 8[6] | 2[6] | 1[6] | 3,000 |

[1] For chemical process. [2] Scfm (60°F, 30 inch Hg.). [3] MW on generator terminals. [4] Scfm 60°F, 30 inch Hg.) at lower ambient temperature.
[5] At lower ambient temperature. [6] Two power turbines, two jet engines. [7] Radial flow compressor. [8] Or other hydrocarbon gases.
[9] Without HR. [10]Including generator [11] Lb/hp-hr. [12] With gearbox 14250/1500 RPM and base frame.

## GAS TURBINE SPECIFICATIONS

| Pressure Ratio | Number of Combustors | EXHAUST Flow lb/sec. or kg/sec. | EXHAUST Temp. | Heat Exch. for Exhaust Heat Recovery | Dry Weight | DIMENSIONS L | DIMENSIONS W | DIMENSIONS H |
|---|---|---|---|---|---|---|---|---|
| | | (lb/sec) | (°F) | | (lb) | | (Inches) | |
| 7.1:1 | 1 | 6.65 | 1,100 | Y | 330 | 52 | 19 | 19 |
| 8.2:1 | 1 | 8.43 | 1,058 | Y | 380 | 59 | 19 | 19 |
| 7.1:1 | 1 | 13.3 | 1,100 | Y | 660 | 52* | 19* | 19* |
| 8.2:1 | 1 | 16.86 | 1,058 | Y | 760 | 59* | 19* | 19* |
| 8.4:1 | 1 | 26.6 | 1,120 | Y | 1,760 | 52.5 | 34.4 | 43.5 |
| 14.5:1 | 1 | 34.5 | 961 | Y | 1,276 | 86.1 | 34 | — |
| | | (lb/sec) | (°F) | | (lb) | | (Feet) | |
| 8.3 | | 52.6 | 940 | | 92,500 | 19 | 8 | 11 |
| 7.0 | | 45 | 1010 | | 92,500 | 19 | 8 | 11 |
| 7.1 | | 115 | 979 | | 242,000 | 63 | 11 | 12 |
| 10.2 | | 267 | 919 | | 398,000 | 77 | 11 | 12 |
| 11.5 | | 301 | 1020 | | 500,000 | 67 | 11 | 12 |
| 9.6 | | 877 | 972 | | 1,015,000 | 95 | 11 | 22 |
| 18.0 | | 147 | 963 | | 220,000 | 53 | 11 | 12 |
| 7.0 | | 45 | 1010 | | 38,000 | 19 | 8 | 11 |
| 7.4 | | 45 | 1015 | | 38,000 | 19 | 8 | 11 |
| 8.3 | | 52.6 | 940 | | 38,000 | 19 | 8 | 11 |
| 8.5 | | 52.6 | 950 | | 38,000 | 19 | 8 | 11 |
| 7.1 | | 115 | 980 | | 120,000 | 35 | 11 | 12 |
| 7.5 | | 115 | 994 | | 120,000 | 35 | 11 | 12 |
| 8.2 | | 268 | 915 | | 257,000 | 50 | 11 | 12 |
| 8.4 | | 250 | 940 | | 257,000 | 50 | 11 | 12 |
| 18.0 | | 147 | 955 | | 32,000 | 30 | 10 | 13 |
| 18.0 | | 147 | 968 | | 59,000 | 26 | 10 | 11 |
| | | (lb/sec) | (°F) | | (lb) | | (Inches) | |
| 11:1 | 1 | 7.75 | 923 | Y | 6,800 | 137.0 | 56.4 | 76.0 |
| 12:1 | 1[8] | 44.3 | 890 | Y | 50,000 | 347 | 100 | 108 |
| | | | | | | | (INCHES) | |
| 4 | 2 | 27.6lb/sec | 860F | HR | 60,000lb[9] | 216 | 80 | 63 |
| 3.85[7] | 1 | 33.0lb/sec | 968F | — | 17,650lb[12] | 98.5 | 79 | 83 |
| — | 14 | — | — | — | 73,500lb | 427 | 124 | 126 |
| — | 14[8] | — | — | — | 308,650lb[10] | 1,060[10] | 138 | 138 |

## GAS TURBINE SPECIFICATIONS

| MODEL | USE | POWER RATING — NORMAL: hp, mw or lbt | NORMAL: Heat Rate | MAXIMUM: hp, mw or lbt | MAXIMUM: Heat Rate | COMPRESSOR SHAFT: Compressor Stages | COMPRESSOR SHAFT: Turbine Stages | POWER SHAFT: Turbine Stages | POWER SHAFT: RPM |
|---|---|---|---|---|---|---|---|---|---|
| **Rolls-Royce Limited, Industrial and Marine Division, P.O. Box 72, Ansty, Coventry, CV7 9JR — U.K.** | | | | | | | | | |
| Mechanical Drive | | (eghp) | (Btu/eghp-hr) | | | | | | |
| Spey | C1,2,PO | 19,100 | 6,405 | | | 5-11 | 2-2 | | |
| Avon-1533 | C1,2,PO | 19,580 | 7,634 | | | 17 | 3 | | |
| Avon 1534 | C1,2,PO | 21,890 | 7,570 | | | 17 | 3 | | |
| Avon 1535 | C1,2,PO | 23,800 | 7,506 | | | 17 | 3 | | |
| RB211 | C1,2,PO | 39,000 | 6,090 | | | 7-6 | 1-1 | | |
| | | | (Btu/hp-hr) | | | | | | |
| SK30 | C1,2,PO | 38,600hp | 8,090 | | | 5-7 | 1-1 | | 3,000 or 3,600 |
| Electric Power Generation | | (MW) | (Btu/kW-hr) | (MW) | (Btu/kW-hr) | | | | |
| SK15 | G1,2,3,4, 5,6,7,8 | 16.3 | 11,790 | 18.2 | 11,540 | 17 | 3 | 2 | 3,000 or 3,600 |
| SK25 | G1,2,3,4, 5,6,7,8 | 23.5 | 10,190 | 24.5 | 10,070 | 7-1 | 1-1 | 3 | 3,000 or 3,600 |
| SK30 | G1,2,3,4, 5,6,7,8 | 28.1 | 11,130 | 29.6 | 11,020 | 5-7 | 1-1 | 3 | 3,000 or 3,600 |
| SK60 | G1,2,3,4, 5,6,7,8 | 56.2 | 11,130 | 59.2 | 11,020 | 5-7 | 1-1 | 3 | 3,000 or 3,600 |
| Combined Cycle | | (MW) | (Btu/kW-hp) | | | | | | |
| SK15-CS | G1,2,3,4,5,6,7 | 24.2 | 7,940 | | | 17 | 3 | 2 | 3,000 or 3,600 |
| SK25-CS | G1,2,3,4,5,6,7 | 32.4 | 7,390 | | | 7-6 | 1-1 | 3 | 3,000 or 3,600 |
| SK30-CS | G1,2,3,4,5,6,7 | 41.9 | 7,470 | | | 5-7 | 1-1 | 3 | 3,000 or 3,600 |
| SK60-CS | G1,2,3,4,5,6,7 | 83.8 | 7,470 | | | 5-7 | 1-1 | 3 | 3,000 or 3,600 |
| Marine Application | | (hp) | (Lb/hp-hr) | (hp) | (Lb/hp-hr) | | | | |
| Tyne RM1C | PR1 | 5,340 | 0.472 | | | 6-9 | 1-1 | 2 | 3,425 |
| Tyne RM2D | PR1 | 5,450 | 0.464 | 6,000 | 0.454 | 6-9 | 1-1 | 2 | 3,425 |
| SM1 | PR1 | 14,750 | 0.406 | 17,100 | 0.396 | 5-11 | 2-2 | 2 | 5,220 |
| SM2 | PR1 | 14,750 | 0.406 | 17,100 | 0.396 | 5-11 | 2-2 | 2 | 5,220 |
| SM3 | PR1 | 14,750 | 0.406 | 17,100 | 0.396 | 5-11 | 2-2 | 2 | 5,220 |
| Olympus TM3B | PR1 | 21,500 | 0.513 | 28,000 | 0.482 | 5-7 | 1-1 | 2 | 5,660 |
| Olympus TM3C | PR1 | 25,000 | 0.484 | 29,600 | 0.477 | 5-7 | 1-1 | 2 | 5,660 |
| RB211 | PR1 | 32,700 | 0.343 | 36,210 | 0.334 | 7-6 | 1-1 | 3 | 5,200 |

*Power unit only including air intake and exhaust system.

**Rolls-Royce Limited, Industrial and Marine Division:** P.O. Box 72, Ansty, Coventry, CV797R, U.K. Contact: D.W. Montford Phone (0203) 613211. Telex 3636
**Rolls-Royce Inc.**, 375 Park Avenue, New York, New York 10152 U.S.A. Contact John Sharo Phone (212) 935-9400 Telex 620839
**Rolls-Royce Inc., Industrial and Marine Division**, Suite 155 South, South 16800 Greenspoint Park Drive, Houston, TX. Contact: D. J. Stepson, Phone (713) 999-5900 Telex 795031 ROLLSROYCE HOU
**Rolls-Royce Inc.**, 10850 Riverside Drive: Suite 603, North Hollywood, CA 91602, Contact: J. Speller Phone (213) 985-5170 Telex 677275 ROYCALIF
**Rolls-Royce (Canada) Ltd.**, P.O. Box 1000, Montreal AMF, Dorval, Quebec, H4Y1BY Canada. Contact: B. K. Knight Phone (514) 631-3541 Telex 05821882
**Motores Rolls-Royce S.A.**: CA1Xa Postal 5593, Sao Paulo 0100SP, Brazil. Contact: J. Atack. Phone (011) 448-7722 Telex 2114115
**Rolls-Royce of Australia Pty. Ltd.** P.O. Box 38, Revesbv, N.S.W. Post Code 2212, Australia. Contact: J.P.S. Denholm. Phone Sydney 77-0641 Telex 21705
**Rolls-Royce (France) Ltd.** 122 Avenue Charles De Gaulle, 92522 Neuilly-Sur-Seine, Paris, France. Contact: N. Snell. Phone Paris 722-1440 Telex 620263

| MODEL | USE | POWER RATING — NORMAL: hp, mw or lbt | NORMAL: Heat Rate | MAXIMUM: hp, mw or lbt | MAXIMUM: Heat Rate | COMPRESSOR SHAFT: Compressor Stages | COMPRESSOR SHAFT: Turbine Stages | POWER SHAFT: Turbine Stages | POWER SHAFT: RPM |
|---|---|---|---|---|---|---|---|---|---|
| **Ruston Gas Turbines Ltd., P.O. Box 1, Lincoln, England, Tel: 0522 25212, Telex: 56231** | | | | | | | | | |
| | | (hp) | (Btu/hp-hr) | | | | | | |
| TA 1750 | C1,2,G1,2,3,5, 6,7,8,PO,W,P1 | 1,875 | 13,950 | — | — | 13 | 2 | 2 | 6,600 |
| TA 2500 | C1,2,G1,2,3,5, 6,7,8,PO,W,P1 | 2,500 | 12,000 | — | — | 14 | 2 | 2 | 7,950 |
| TB 5000 | C1,2,G1,2,3,5, 6,7,8,PO,W,P1 | 4,900 | 9,980 | — | — | 12 | 2 | 2 | 7,950 |
| Tornado (Twin Shaft) | C1,2,PO,W,P1 | 8,500 | 8,180 | — | — | 15 | 2 | 2 | 10,000 |
| Tornado (Single Shaft) | G1,2,3,5,6,7,8 | 8,350 | 8,144 | — | — | 15 | 2 | 2 | 11,085 |

## GAS TURBINE SPECIFICATIONS

| Pressure Ratio | Number of Combustors | EXHAUST Flow lb/sec. or kg/sec. | EXHAUST Temp. | Heat Exch. for Exhaust Heat Recovery | Dry Weight | DIMENSIONS L | DIMENSIONS W | DIMENSIONS H |
|---|---|---|---|---|---|---|---|---|
| | | (lb/sec) | (°F) | | (lb) | | (Inches) | |
| 18.5 | 10 | 129 | 1,090 | Y | 3,250 | 108 | 48 | 60 |
| 8.8 | 8 | 168 | 1,020 | Y | 3,560 | 138 | 36 | 36 |
| 8.8 | 8 | 168 | 1,120 | Y | 3,560 | 138 | 36 | 36 |
| 9.0 | 8 | 170 | 1,170 | Y | 3,560 | 138 | 36 | 36 |
| 19.2 | 1 | 199 | 1,320 | Y | 5,700 | 120 | 60 | 60 |
| | | | (°C) | | | | | |
| 11.0 | 8 | 239 | 780 | Y | 313,000* | 540 | 168 | 588 |
| 10.0 | 8 | 181 | 500 | Y | 376,000 | 720 | 168 | 288 |
| 19.2 | 1 | 207 | 490 | Y | 380,000 | 720 | 168 | 288 |
| 11.0 | 8 | 240 | 555 | Y | 515,000 | 912 | 168 | 588 |
| 11.0 | 8 | 480 | 555 | Y | 840,000 | 1,416 | 168 | 588 |
| 10.0 | 8 | 181 | 500 | Boiler | — | — | — | — |
| 19.2 | 1 | 207 | 490 | Boiler | — | — | — | — |
| 11.0 | 8 | 240 | 555 | Boiler | — | — | — | — |
| 11.0 | 8 | 480 | 555 | Boiler | — | — | — | — |
| | | | | | | | (Feet) | |
| 12.5 | 10 | 46 | 440 | Y | 31,000 | 18.2 | 7 | 8.6 |
| 13.0 | 10 | 46 | 455 | Y | 5,150 | 14 | 4.2 | 5.7 |
| 18.5 | 10 | 122 | 420 | Y | 42,150 | 24.5 | 7.5 | 10.1 |
| 18.5 | 10 | 122 | 420 | Y | 20,000 | 19.9 | 7.5 | 9.1 |
| 18.5 | 10 | 122 | 420 | Y | 18,500 | 21.5 | 6.7 | 7.7 |
| 10.5 | 8 | 235 | 430 | Y | 68,000 | 30 | 8.6 | 12 |
| 11.0 | 8 | 234 | 430 | Y | 68,000 | 30 | 8.6 | 12 |
| 20.1 | 1 | 202 | 492 | Y | 91,025 | 32 | 14 | 13 |
| | | (lb/sec) | (°F) | | (lb) | | (Inches) | |
| 4.3 | 1 | 25 | 979 | Y | 22,500 | 192 | 96 | 102 |
| 4.9 | 1 | 28 | 944 | Y | 20,000 | 228 | 96 | 102 |
| 7.0 | 4 | 46 | 944 | Y | 30,000 | 228 | 96 | 96 |
| 12.1 | 8 | 60 | 882 | Y | 42,000 | 288 | 96 | 114 |
| 12.1 | 8 | 60 | 887 | Y | 42,000 | 288 | 96 | 114 |

# GAS TURBINE SPECIFICATIONS

| MODEL | USE | POWER RATING: NORMAL hp, mw or lbt | NORMAL Heat Rate | MAXIMUM hp, mw or lbt | MAXIMUM Heat Rate | COMPRESSOR SHAFT: Compressor Stages | COMPRESSOR SHAFT: Turbine Stages | POWER SHAFT: Turbine Stages | POWER SHAFT: RPM |
|---|---|---|---|---|---|---|---|---|---|
| **Solar Turbines Incorporated, 2200 Pacific Highway, P.O. Box 80966, San Diego, CA 92138** | | | | | | | | | |
| | | (MW) | (kW-hr) | | | | | | |
| Saturn GSE-1000 | G5,7 | .9[1] | 15,300[1] | — | — | 8 | 3 | — | 22,300 |
| Saturn GSC-1200[4] | G2 | .8[3] | 15,796[3] | — | — | 8 | 3 | — | 22,300 |
| Centaur GSE-4000[4] | G5,7 | 3.22[2] | 13,343[2] | — | — | 11 | 3 | — | 14,950 |
| Centaur GSC-4000 | G2 | 2.88[3] | 13,492[3] | — | — | 11 | 3 | — | 14,950 |
| Mars GSE-10,000 | G5,7 | 9.23[2] | 10,600[2] | — | — | 15 | 2 | 2 | 8,600 |
| Mars GSP-10,000 | G2,7 | 8.63[3] | 10,690[3] | — | — | 15 | 2 | 2 | 8,600 |
| Mars GSC-10,000 | G2 | 7.93[3] | 10,890[3] | — | — | 15 | 2 | 2 | 8,600 |
| | | (hp[3]) | (hp-hr[3]) | | | | | | |
| MD 1200 | P1 | 1,200 | 10,920 | — | — | 8 | 2 | 1 | 22,300 |
| MD 10,000 | P1 | 11,150 | 7,760 | — | — | 15 | 2 | 2 | 9,500 |
| MD 4000 | P1 | 3,950 | 9,310 | — | — | 11 | 2 | 1 | 15,700 |
| CS 1200 | C2 | 1,200 | 10,920 | — | — | 8 | 2 | 1 | 22,300 |
| CS 4000 | C2 | 3,950 | 9,310 | — | — | 11 | 2 | 1 | 15,700 |
| CS 10,000 | C2 | 11,130 | 7,760 | — | — | 15 | 2 | 2 | 9,500 |

[1]At 74°F Sea level liquid fuel. [2]Liquid fuel, ISO. [3]Gas fuel, ISO. [4]Single shaft engine.

**Solar Turbines Incorporated**: 1615 West Crescent, Suite 260, Anaheim, CA 92801, Contact: E. E. Keffer. Phone: (714) 520-0511. TWX 910-591-2899.
**Solar Turbines Incorporated**: 33 W. Higgins Road, Suite 4000, South Barrington, IL 60010. Contact: R. M. Keller, Phone: 312-426-3200. TWX: 910-651-4726.
**Solar Turbines Canada**: Suite 3000, Bow Valley Square 2, 205 5th Ave., SW. P.O. Box 9087, Calgary, Alberta T2P OV7, Canada. Contact: C. B. Alexander Phone: (403) 261-2640 TWX: 03-825822.
**Solar Turbines Incorporated**: 7600 W. Tidwell Road, Suite 600, Houston, TX 7740, Contact: J. Fagg, Phone (713) 895-2300, TWX 910-881-2610.
**Solar Turbines Incorporated**: 499 South Capitol St. SW, Suite 422, Washington, D.C. 20003. Contact: D. L. Kearns. Phone: 202-484-5080. TWX: 710-822-9407.
**Solar Turbines Incorporated** Douglas Center, 2600 Douglas Center Road, Suite 1211, Coral Gables, FL 33134. Contact: L. A. Uriarte. Phone: (305) 442-2727. TWX: 810-848-6267.
**Solar Turbines Overseas**: 19 Knightsbridge Street, London SW1, England. Contact: J. T. Bueche. Phone: 235-5466. Telex: 917891.
**Solar Turbines International Company**: Suite 2005, Orchard Towers, 400 Orchard Road, Republic of Singapore 9. Contact: R. A. Rosner. Phone: 7372677. Telex: RS23009.
**Solar Turbines Overseas Ltd.**: Chaussee de la Hulpe, 185, 1170 Brussels, Belgium, Contact: I. D. Nasri. Phone: 6734076. Telex: 62644.
**Solar Turbines Australia**: 499 St. Kilda Road, 14th Floor, Melbourne, Victoria 3004 Australia. Contact: C. A. Kneass. Phone: 2671155. Telex: 79034183.

| MODEL | USE | NORMAL hp, mw or lbt | NORMAL Heat Rate | MAXIMUM hp, mw or lbt | MAXIMUM Heat Rate | Compressor Stages | Turbine Stages | Turbine Stages | RPM |
|---|---|---|---|---|---|---|---|---|---|
| **Stal-Laval Turbin AB, S-612 20 Finspong, Sweden** | | | | | | | | | |
| | | (MW) | (Btu/kW-hr) | (MW) | (Btu/kW-hr) | | | | |
| Gasteam 125[2] | G-3 | 113[3] | 7,500 | 122[4] | 7,420 | 7 + 9 | 1 + 1 | 2 | 3,000 |
| Gasteam 250[1] | G-3 | 228[3] | 7,420 | 246[4] | 7,340 | 7 + 9 | 1 + 1 | 2 | 3,000 |
| GT200 | G1,2,6 | 75[3] | 10,300 | 85[4] | 10,120 | 7 + 9 | 1 + 1 | 2 | 3,000 |
| GT120 | G1,2,6 | 70[3, 5] | 11,960 | 75[4, 5] | 11,780 | 10 + 11 | 1 + 2 | 3 | 3,000 |
| GT35 | G1,2,5,6,7 | 15.1[3] | 10,665 | 18.4[5] | 10,250[5] | 10 + 8 | 1 + 2 | 3 | 3,000/ 3,600 |
| GT35 | C1,2 | 14.9[3] | 10,830 | 16.3[5] | 10,780[5] | 10 + 8 | 1 + 2 | 2 | 6,000 |
| GT35 | P0,PW | 14.9 | 10,710 | 16.5 | 10,665 | 10 + 8 | 1 + 2 | 3 | 3,600 |

[1] Contains two GT 200 gas turbines, two unfired heat recovery boilers, one steam turbine. [2] Contains one GT 200 gas turbine, one unfired heat recovery boiler, and one steam turbine. [3] Base loading rating. [4] Max continuing rating. [5] Cooling water temperature + 15°C.

**Stal-Laval Turbin AB**, S-612 20 Finspong, Sweden, Tel: 0122-810-00, Contact: Borje Timerdal
**Stal-Laval, Inc.**: 525 Executive Boulevard, Elmsford, New York 10523 USA. Contact: Tom Stott. Phone: (914) 592-4710. Telex: 710-567-1227
**Stal-Laval Inc.**: 2323 Voss, Suite 360, Houston, Texas. Contact: Sune Jansson. Phone: (713) 780-4262. Telex: 910-881-2623.
**ASEA S.A.**: Apartado Postal M2434, Mexico 1 D.F. Contact: Stig Good. Phone: 565-0033. Telex: 017-2248.
**Stal-Laval Ltd.**: Crown House, London Road, Morden, Surrey SM4 5DX, Great Britain. Contact: Martin Stonard. Phone: 01-543 3476/9. Telex: 926 719.
**S. A. Stal-Laval (Benelus) N.V.**: De Keyserlei 5, Bus 22, Antwerp Tower, B-2000 Antwerp,, Belgium. Contact: Frank Verhaest. Phone: 031/31 94 10. Telex: 338 20.
**Stal-Laval, S.A.**: 62, rue Pergolese, F-75 116 Paris, France. Contact: Urban Dahlgren. Phone: 500 34 12. Telex: 610 282.

**ASEA Regional Office Arabian Gulf**: P.O. Box 11070, Al-Salimiah Bldg., Flat 32, 7th Floor, Hor-Al-Anz, Deira, Dubai, The United Arab Emirates. Contact: Lennart Rudblom. Phone: 66 61 51-53, Direct Line 66 60 83. Telex: 45759ASEAGFEM.
**Stal-Laval A/S**: Nedre Vollgage 8, Oslo 2, Norway. Contact: Richard Fosdahl. Phone: (02) 414 010. Telex: 16443.
**Oy Stal-Laval Ab**: Postfack 499, Kajsamiemigatan 1 B, SF-00100 Helsingfors 10, Finland. Contact: Bjarne Frilund. Phone: 65 94 66. Telex: 12 594 STAVAL SF.
**Elemac-ASEA Division**. P.O. Box 7937 JEDDAH Saudi Arabia. Contact: Johan Hanson. Phone: 60 22 27, 69 20 45. Telex: 400627 ASEAS.

# GAS TURBINE SPECIFICATIONS

| Pressure Ratio | Number of Combustors | EXHAUST Flow lb/sec. or kg/sec. | EXHAUST Temp. | Heat Exch. for Exhaust Heat Recovery | Dry Weight | DIMENSIONS L | DIMENSIONS W | DIMENSIONS H |
|---|---|---|---|---|---|---|---|---|
| | | (lb/sec) | (°F) | | (lb) | | (Inches) | |
| 6.2 | 1 | 13.8 | 860 | Y | 11,500 | 168 | 49 | 66 |
| 6.2 | 1 | 13.8 | 813 | Y | 17,000 | 252 | 70 | 75 |
| 9 | 1 | 39.36 | 880 | Y | 40,000 | 300 | 80 | 108 |
| 9 | 1 | 39.25 | 810 | Y | 160,000 | 300 | 80 | 108 |
| 16 | 1 | 82.98 | 884 | Y | 160,000 | 703 | 96 | 147 |
| 16 | 1 | 83.5 | 837 | Y | 160,000 | 703 | 96 | 147 |
| 16 | 1 | 81.28 | 815 | Y | 160,000 | 703 | 96 | 147 |
| 6.2 | 1 | 12.8 | 860 | Y | 8,230 | 158 | 70 | 76 |
| 16 | 1 | 80 | 785 | Y | 60,000 | 411 | 96 | 147 |
| 8.3 | 1 | 38.06 | 800 | Y | 27,000 | 249 | 95.5 | 99 |
| 6.2 | 1 | 12.8 | 860 | Y | 14,400 | 230 | 70 | 76 |
| 8.3 | 1 | 38.06 | 800 | Y | 37,000 | 282 | 94 | 99 |
| 16 | 1 | 80 | 785 | Y | 86,000 | 557 | 96 | 147 |
| | | (kg/sec) | (°C) | | (lbs) | | (mm) | |
| 16 | 8 | 350 | 454 | Y | — | — | — | — |
| 16 | 8 | 700 | 454 | Y | — | — | — | — |
| 16 | 8 | 350 | 454 | Y | — | 25,900 | 9,150 | 8,400 |
| 15 | 2 | 360[4] | 335[4] | Y | — | 25,850 | 12,700 | 12,450 |
| 12 | 7 | 93[3] | 385[3] | Y | 110,000 | 25,900 | 3,175 | 3,305 |
| 11 | 7 | 88[3] | 390[3] | Y | 85,000 | 11,050 | 3,175 | 3,305 |
| 11 | 7 | 85[3] | 385[3] | Y | 85,000 | 11,050 | 3,175 | 3,305 |

## GAS TURBINE SPECIFICATIONS

| MODEL | USE | POWER RATING NORMAL hp, mw or lbt | POWER RATING NORMAL Heat Rate | POWER RATING MAXIMUM hp, mw or lbt | POWER RATING MAXIMUM Heat Rate | COMPRESSOR SHAFT Compressor Stages | COMPRESSOR SHAFT Turbine Stages | POWER SHAFT Turbine Stages | POWER SHAFT RPM |
|---|---|---|---|---|---|---|---|---|---|
| **Stewart & Stevenson, Gas Turbine Division, P.O. Box 1637, Houston, TX 77001** | | | | | | | | | |
| | | (hp) | (Btu/hp-hr) | (hp) | | | | | |
| IM831 | GP | 735 | 11,470 | 800 | — | 2 | 3 | 0 | 1,800 or 1,500 |
| 501-KB | G | 4,450 | 8,970 | 5,420 | — | 14 | 4 | 0 | 14,200 |
| 501-KSS | C,P,MD | 4,330 | 9,253 | 5,420 | — | 14 | 2 | 2 | 14,200 |
| 501-KB5 | G | 5,256 | 8,356 | 5,960 | — | 14 | 4 | 0 | 14,200 |
| 501-K5SS | C,P,MP | 5,256 | 8,840 | 5,960 | — | 14 | 2 | 2 | 14,200 |
| 570-K | G,C,P,MP | 6,445 | 8,473 | 7,170 | — | 13 | 2 | 2 | 11,500 |
| 571-K | G,C,P,MP | 7,665 | 7,560 | 8,310 | — | 13 | 2 | 2 | 11,500 |
| LM2500-20 | G,C,P,MP | 17,600 | 7,385 | 23,920 | — | 16 | 2 | 6 | 3,600 or 3,000 |
| LM2500-33 | G,C,P,MP | 29,500 | 6,875 | 32,550 | — | 16 | 2 | 6 | 3,600 or 3,000 |
| **Sulzer Brothers LTD., Dept. Thermal Turbomachinery, CH-8401 Winterthur, Switzerland** | | | | | | | | | |
| | | (MW) | (Btu/kW-hr) | (MW) | (Btu/kW-hr) | | | | |
| 3 | G1,2,3,5,7,8 | 5.8 | 12,640 | 6.1 | 12,540 | 17 | — | 3 | 10,600 |
| R3 | G1,2,3,5,7,8 | 5.57 | 10,340 | 5.85 | 10,240 | 17 | — | 3 | 10,600 |
| S3 | C, P, PR1 | 5.8 | 12,640 | 6.1 | 12,540 | 17 | 2 | 1 | 10,600 |
| SR3 | C, P, PR1 | 5.57 | 10,340 | 5.85 | 10,240 | 17 | 2 | 1 | 10,600 |
| 7 | G1,2,3,5,7,8 | 10.7 | 13,430 | 11.2 | 13,280 | 13 | — | 6 | 6,400 |
| R7 | G1,2,3,5,7,8 | 10.3 | 10,630 | 10.8 | 10,500 | 13 | — | 6 | 6,400 |
| S7 | C,P,PR1 | 10.6 | 13,650 | 11.2 | 13,490 | 13 | 4 | 2 | 6,400 |
| SR7 | C, P, PR1 | 10.15 | 10,770 | 10.65 | 10,630 | 13 | 4 | 2 | 6,400 |
| 10 | G1,2,3,5,7,8 C, P | 20.7 | 10,310 | — | — | 10 | 2 | 2 | 7,700 |
| PRIMO 12 | G1,2,3,5,7,8 C,P | 12.2 | 12,060 | 15.8 | 11,600 | 17 | 3 | 2 | 6,500 |
| PRIMO 14 | G1,2,3,5,7,8 C,P | 15.5/ 13.7 | 11,680/ 11,930 | 17.4/ 17.5 | 11,450 | 17 | 3 | 2 | 6,500 |
| PRIMO 15 | G1,2,3,5,7,8 C,P | 16.2 14.9 | 11,680/ 11,850 | 18.0/ 18.4 | 11,450 | 17 | 3 | 2 | 6,500 |
| PRIMO 21 | C,P | 21.0 | 10,210 | 23.3 | 9,830 | 7 + 6 | 1 + 1 | 2 | 6,500 |
| **Teledyne CAE, 1330 Laskey Road, Toledo, Ohio 43612** | | | | | | | | | |
| J69-T-25 | PR-1 | 800lbt | 1.12SFC | 1,025lbt | 1.14SFC | 1c | 1 | — | — |
| J69-T-29 | PR-1 | 1,375lbt | 1.08SFC | 1,700lbt | 1.10SFC | 1A-1c | 1 | — | — |
| J69-T-41A | PR-1 | 1,650lbt | 1.09SFC | 1,920lbt | 1.10SFC | 1A-1c | 1 | — | — |
| J69-T-406 | PR-1 | 1,719lbt | 1.10SFC | 1,920lbt | 1.11SFC | 1A-1c | 1 | — | — |
| J100-CA0100 | PR-1 | 2,430lbt | 1.08SFC | 2,700lbt | 1.10SFC | 2A-1c | 2 | — | — |
| 490-4 | PR-1 | 2,750lbt | 0.67SFC | 29,600lbt | 0.70SFC | 2 Fan + 4A | 2 | — | — |
| J402-CA-400 | PR-1 | — | — | 660lbt | 1.20SFC | 1A-1c | 1 | — | — |
| J402-CA-700 | PR-1 | — | — | 640lbt | 1.19SFC | 1A-1c | 1 | — | — |
| 373-5 | PR-1 | — | — | 970lbt | 1.06SFC | 2A-1c | 1 | — | — |

**Sulzer Brothers Ltd.**, Thermal Turbomachinery: CH-8401 Winterthur. Contact: Hess/0561. Phone: (052) 812966. Telex: 76165.
**S.A. Sulzer Belgium N.V.**: 92, Square Plasky, B-1040 Bruxelles, Belgium. Phone: 0032/2-7349606, 0032/2-7343161. Telex: 21039 Sulzb.
**Escher Wyss GmbH,**: Postfach 1380, D-7980 Ravensburg, Germany. Phone: 0049/751-831. Telex: 0732901.

# GAS TURBINE SPECIFICATIONS

| Pressure Ratio | Number of Combustors | EXHAUST Flow lb/sec. or kg/sec. | EXHAUST Temp. | Heat Exch. for Exhaust Heat Recovery | Dry Weight | DIMENSIONS L | DIMENSIONS W | DIMENSIONS H |
|---|---|---|---|---|---|---|---|---|
| | | (lb/sec) | (°F) | | (lb) | | | |
| 11.0 | 1 | 7.9 | 930 | Y | 1,750 | | | |
| 9.3 | 6 | 34.5 | 939 | Y | 1,270 | | | |
| 9.3 | 6 | 34.5 | 939 | Y | 2,500 | | | |
| 9.3 | 6 | 34.5 | 991 | Y | 1,270 | | | |
| 9.3 | 6 | 34.5 | 991 | Y | 2,500 | | | |
| 12.0 | 1 | 41.3 | 1,050 | Y | 1,350 | | | |
| 12.2 | 1 | 43.7 | 1,000 | Y | 1,490 | | | |
| 13.5 | 1 | 124.8 | 780 | Y | 10,000 | | | |
| 18.0 | 1 | 149.0 | 955 | Y | 10,000 | | | |
| | | (kg/sec) | (°C) | | (lb) | (mm) | (mm) | (mm) |
| 8.85 | 4 | 32.1 | 505 | — | 60,000 | 7,350 | 3,100 | 2,690 |
| 9.02 | 4 | 32.0 | 391 | Y | 60,000 | 7,350 | 3,100 | 2,690 |
| 8.85 | 4 | 32.1 | 505 | — | 60,000 | 7,350 | 3,100 | 2,690 |
| 9.02 | 4 | 32.0 | 391 | Y | 60,000 | 7,350 | 3,100 | 2,690 |
| 7.55 | 9 | 64.6 | 493 | — | 138,000 | 11,700 | 3,580 | 3,810 |
| 7.70 | 9 | 64.6 | 342 | Y | 138,000 | 11,700 | 3,580 | 3,810 |
| 7.55 | 9 | 64.6 | 491 | — | 156,000 | 12,350 | 3,580 | 3,810 |
| 7.70 | 9 | 64.6 | 342 | Y | 156,000 | 12,350 | 3,580 | 3,810 |
| 13.5 | 1 | 73.1 | 507 | — | 61,700 | 8,000 | 3,150 | 3,400 |
| 8.8 | 8 | 76.3 | 403 | — | 62,000 | 12,275 | 4,200 | 4,000 |
| 8.8 | 8 | 79.3/ 76.3 | 464/440 | — | 62,000 | 12,275 | 4,200 | 4,000 |
| 9.0 | 8 | 79.5/ 77.3 | 478/459 | — | 62,000 | 12,275 | 4,200 | 4,000 |
| 18.4 | 1 | 88.9 | 453 | — | 64,000 | 10,710 | 4,200 | 4,000 |

**Compagne de Construction Mecanique Sulzer**: Cedex 59, F-75300 Paris-Brune, France. **Phone: 5392244.** Telex: ccmsf 200937 f.
**Sulzer Brothers (UK) Ltd.: Farnborough** (Hampshire), Great Britain. Phone: 0044/252-44311. **Telex: 858771**/2.
**Sulzer Brothers Inc.: 200 Park Avenue**, New York, N.Y. 10017. Phone: 001/212-949-0999. **Telex: WUI 620** 213 and domestic 127 297.

| | | | | | | (mm) | (mm) | (mm) |
|---|---|---|---|---|---|---|---|---|
| 3.9 | 1 | 20.5 | 1250 | — | 364 | 35.4 | 22.3 | 22.3 |
| 5.3 | 1 | 28.0 | 1340 | — | 341 | 44.8 | 22.3 | 24.8 |
| 5.5 | 1 | 29.8 | 1400 | — | 350 | 44.8 | 22.3 | 25.6 |
| 5.5 | 1 | 30.5 | 1400 | — | 360 | 44.8 | 22.5 | 23.7 |
| 6.3 | 1 | 44.9 | 1345 | — | 430 | 48.2 | 24.7 | 26.1 |
| 10.8 | 1 | 60.7 | — | — | 606 | 45.3 | 22.8 | 28.3 |
| 5.6 | 1 | 9.6 | 1545 | — | 100 | 29.4 | 12.5 | 12.5 |
| 5.6 | 1 | 9.5 | 1520 | — | 107 | 29.6 | 12.5 | 12.5 |
| 8.7 | 1 | 13.9 | 1425 | — | 128 | 38.0 | 12.5 | 12.5 |

## GAS TURBINE SPECIFICATIONS

| MODEL | USE | POWER RATING NORMAL hp, mw or lbt | POWER RATING NORMAL Heat Rate | POWER RATING MAXIMUM hp, mw or lbt | POWER RATING MAXIMUM Heat Rate | COMPRESSOR SHAFT Compressor Stages | COMPRESSOR SHAFT Turbine Stages | POWER SHAFT Turbine Stages | POWER SHAFT RPM |
|---|---|---|---|---|---|---|---|---|---|
| **Thomassen, Holland, De Steeg 6217** | | | | | | | | | |
| All types of gas turbines designed by General Electric Company, Gas Turbine Division, Schenectady, New York, U.S.A., under manufacturing associates agreement. | | | | | | | | | |
| **Toshiba Corporation, Heavy Apparatus Division, Toshiba Mita Building, 13-12 Mita 3 Chome, Minatoku, Japan** | | | | | | | | | |
| | | (MW) | | | | | | | |
| G9151 | G2,8 | 109.3[1] | 10,700[2] | — | — | 17 | — | 3 | 3,000 |
| PG9151 | G1,3,5,6,7 | 106.7[1] | 10,850[2] | 115.9MW | 10,810 | 17 | — | 3 | 3,000 |
| (Combined Steam Turbine and Gas Turbine Generator with Heat Recovery Steam Generator) | | | | | | | | | |
| STAG109E[4] | G1,3,6 | 140.4[3] | 7,940[3] | — | — | 17 | — | 3 | 3,000 |
| **Turbomeca, Bordes 64320 Bizanos, France** | | | | | | | | | |
| | | (hp) | (Btu/hp-hr) | | | | | | |
| OREDON IV | G1,2,8 APU | 135 | 15,600 | — | — | C1 | 2 | — | 8,000 |
| ASTAZOU IV | G1,2,5,6,7,8 RW,Z | 440 | 12,900 | — | — | A1,C1 | 3 | — | 1,500 |
| BASTAN VI | C1,G1,2,5,6, 7,8, P-W,Z | 820 | 12,210 | — | — | A1,C1 | 3 | — | 1,500 |
| BASTAN VII | C1,G1,2,5,6, 7,8,P-W,Z | 1,165 | 10,960 | — | — | A2,C1 | 3 | — | 1,500 |
| BI-BASTAN VI | C1,G1,2,5,6, 7,8,Z | 1,550 | 12,480 | — | — | A1,C1 | 3 | — | 1,500 |
| BI-BASTAN VII | C1,G1,2,5,6, 7,8,Z | 2,205 | 11,300 | — | — | A2,C1 | 3 | — | 1,500 |
| TURMO III | C1,2,G2,PO, P-W,PR1,2,Z | 950 | 13,290 | — | — | A1,C1 | 3 | — | 5,500 |
| TURMO XII | C1,2,G2,PO P-W,PR-1,2,Z | 1,340 | 10,980 | — | — | A2,C1 | 3 | — | 5,500 |
| **United Technologies Corp., Power Systems Division, 1690 New Britain Ave., Farmington, CT 06032.** | | | | | | | | | |
| FT4C-3F, GF, LF, DF | C-2, G-1,-2 I/M P-1, PR-2 | 39,000hp[2] [5] | 7,968[6] | 48,000hp[2] [5] | 7,747[6] | 16 | 3 | 3 | 3,600 |
| TP4-2 (C-3F) | G-1,-2,-3,-6 PG | 57MW[2] | 10,900[1] | 71.3MW[2] | 10,600[1] | 16 | 3 | 3 | 3,600 |
| Combined Cycle[4] | G-3 PG | 162MW[2] | 8,380[3] | 162MW[2] | 8,380[3] | 16 | 3 | 3 | 3,600 |
| FT4A-9 GF, LF, DF | C-1 I/M P-1 | 27,300 hp[7] [5] | 8,790[6] | — | — | 15 | 3 | 2 | 3,600 |

Toshiba: [1]Ratings are at International Standard Organization (ISO) Conditions: Sea level and 15°C, 59°F. Normal is base load. [2]Ratings are for distillate fuel using LHV. [3]Rating is for distillate fuel using HHV. [4]"STAG" is GE trademark.

Turbomeca: Z = TOTAL ENERGY (a) Other output speeds available. [1]Without gear box.

United Technologies: [1] Heat Rate BTU/Kw-hr (LHV). [2] Ratings are at I.S.O. Conditions: Sea level and 15°C, 59°F. [3] Based on liquid fuel BTU/Kw-hr (HHV). [4] Combined cycle, 4 FT4C-3F's, 2 boilers, 1 steam turbine, consult manufacturer for other variations. [5] Output in shaft horsepower. [6] Heat rate in BTU/Shp-hr (LHV). [7] Typical. Includes enclosure.

## GAS TURBINE SPECIFICATIONS

| Pressure Ratio | Number of Combustors | EXHAUST Flow lb/sec. or kg/sec. | EXHAUST Temp. | Heat Exch. for Exhaust Heat Recovery | Dry Weight | DIMENSIONS L | DIMENSIONS W | DIMENSIONS H |
|---|---|---|---|---|---|---|---|---|
| | | | | | | | | |
| | | | | | (lb) | | (inches) | |
| 11.4 | 14 | 884 | 968 | — | — | — | — | — |
| 11.4 | 14 | 877 | 972 | — | 1,900,000 | 1,380 | 930 | 468 |
| 11.4 | 14 | 877 | 972 | Boiler | — | — | — | — |
| | | (kg/sec) | (°C) | | (kg) | (mm) | (mm) | (mm) |
| 3.75 | 1 | 0.89 | 575 | | 30[1] | 580[1] | 410[1] | 450[1] |
| 5.6 | 1 | 2.55 | 490 | | 340 | 1,324 | 710 | 725 |
| 5.5 | 1 | 4.39 | 505 | | 320 | 1,630 | 710 | 840 |
| 7.2 | 1 | 6.04 | 460 | | 350 | 1,740 | 710 | 840 |
| 5.5 | 1 | 2 x 4.39 | 505 | | 1,480 | 2,290 | 1,370 | 1,200 |
| 7.2 | 1 | 2 x 6.04 | 460 | | 1,540 | 2,450 | 1,370 | 1,200 |
| 4.9 | 1 | 5.41 | 510 | | 355 | 2,130 | 760 | 730 |
| 8.2 | 1 | 6.94 | 450 | | 490 | 2,140 | 720 | 920 |
| | | | | | | (mm) | (mm) | (mm) |
| 14.5 | 8 | 314lb/sec | 918F | — | 29,000lb | 8,306 | 3,100 | 2,769 |
| 14.5 | 8 | 314lb/sec | 918F | — | — | 37,250[2] | 8,230[2] | 13,716[2] |
| 13.4 | 8 | 296lb/sec | 856F | boiler | — | — | — | — |
| — | 8 | — | — | | 15,000 | 8,026 | 1,956 | 2,184 |

## GAS TURBINE SPECIFICATIONS

| MODEL | USE | POWER RATING: NORMAL hp, mw or lbt | NORMAL Heat Rate | MAXIMUM hp, mw or lbt | MAXIMUM Heat Rate | COMPRESSOR SHAFT: Compressor Stages | COMPRESSOR SHAFT: Turbine Stages | POWER SHAFT: Turbine Stages | POWER SHAFT: RPM |
|---|---|---|---|---|---|---|---|---|---|
| **USSR, Energo Machino Export, Mossilmouskaya UL, Moscow, USSR, 117330 35** | | | | | | | | | |
| (Package Power Plants) | | (MW) | | (MW) | | | | | |
| GTN-6 UTMZ | G1,2 | 6.3 | 0.8 | 7.2 | — | 11 | — | 2 | 6,100 |
| GTN-16 UTMZ | G1,2 | 16 | 0.65 | 20 | — | 15 | — | 1 | 6,500 |
| GTN-25 NZL | G1,2 | 25 | 0.63 | 30 | — | 7 + 7 | — | 1 | 3,900 |
| (Process Gas Turbines) | | | | | | | | | |
| GUBT-6[1] UTMZ | PI | 6 | — | — | — | — | — | 2 | 3,000 |
| GUBT-8[1] UTMZ | PI | 8 | — | — | — | — | — | 2 | 3,000 |
| GUBT-12[1] UTMZ | PI | 12 | — | — | — | — | — | 2 | 3,000 |
| GTT-3[2] PMWH | PI | 3 | — | — | — | 16 | — | 7 | 5,000 |
| GTT-12[2] PMWH | PI | 12 | — | — | — | 10 | — | 1 | 7,400<br>5,200<br>5,000 |
| (Leningrad Metal Works — LMZ) | | | | | | | | | |
| GT-100-750 | G1,2 | 95 | 0.643 | 115 | — | 13 | 3 | 5 | 3,000 |
| (Neva Machine Works Leningrad — NZL) | | | | | | | | | |
| GT-750-6 | G1,2 | 6 | 0.68 | 7.5 | — | 12 | 2 | 1 | 5,600 |
| GTK-10 | G1,2 | 10 | 0.65 | 13 | — | 11 | 1 | 1 | 4,800 |
| (Ural Turbomachine Works — UTMZ) | | | | | | | | | |
| GT-6-750 | G1,2 | 6 | 0.8 | 7.2 | — | 11 | 3 | 2 | 6,100 |
| GTK-16 | G1,2 | 16 | 0.7 | 20 | — | 13 | 3 | 2 | 4,600 |
| (Turbomachine Works Harkov — HTGZ) | | | | | | | | | |
| GT-35 | G1,2 | 31 | 0.78 | 40 | — | — | — | 4 | 3,000 |
| **Westinghouse Canada Inc., P.O. Box 510, Hamilton, Ontario L8N 3K2.** | | | | | | | | | |
| Mechanical Drives | | | | | | | | | |
| W101M[1] | C1,C2,PL | 11,520hp | 10,820[2] Btu/hp-hr | 14,000hp[4] | — | 16 | 5 | — | 6,000 |
| W101RM[1] | C1,C2,PL | 10,530hp | 8,750[2] Btu/hp-hr | 14,000hp[4] | — | 16 | 5 | — | 6,000 |
| W171M[1] | C1,C2,PL | 22,270hp | 9,780[2] Btu/hp-hr | 25,200hp[4] | — | 14 | 5 | — | 4,912 |
| W171RM[1] | C1,C2,PL | 20,280hp | 8,040[2] Btu/hp-hr | 25,200hp[4] | — | 14 | 5 | — | 4,912 |
| W191M[1] | C1,C2,PL | 25,030hp | 9,530[2] Btu/hp-hr | 27,200hp[4] | — | 15 | 5 | — | 4,912 |
| W191RM[1] | C1,C2,PL | 22,450hp | 7,910[2] Btu/hp-hr | 25,800hp[4] | — | 15 | 5 | — | 4,690 |
| CW182MAA[1] | C1,C2, P1,P2,PL | 13,700hp | 9,710[2] Btu/hp-hr | 23,800hp[4] | — | 15 | 1 | 1 | 7,250 |
| CW182MA[1] | C1,C2, P1,P2,PL | 15,800hp | 9,150[2] Btu/hp-hr | 23,800hp[4] | — | 16 | 1 | 1 | 7,250 |
| CW182MB[1] | C1,C2, P1,P2,PL | 18,500hp | 8,750[2] Btu/hp-hr | 23,800hp[4] | — | 17 | 1 | 1 | 7,250 |
| CW182RMAA[1] | C1,C2, P1,P2,PL | 10,600hp[3] (12,800hp[4]) | 6,960[2] Btu/hp-hr | 23,800hp[4] | — | 15 | 1 | 1 | 7,250 |
| CW182RMA[1] | C1,C2, P1,P2,PL | 13,200hp[3] (14,800hp)[4] | 6,810[2] Btu/hp-hr | 23,800hp[4] | — | 16 | 1 | 1 | 7,250 |
| CW182RMB[1] | C1,C2, P1,P2,PL | 15,200hp[3] (17,300hp) | 6,910[2] Btu/hp-hr | 23,800hp[4] | — | 17 | 1 | 1 | 7,250 |
| CW352MAA[1] | C1,C2, P1,P2,PL | 30,100hp | 9,460[2] Btu/hp-hr | 50,000hp[4] | — | 15 | 1 | 1 | 5,000 |

[1] Blast Furnace Gas-Expansion Turbine. [2] Gas turbine for a process nitgrogen acid Power Machine Works Habarovsk — PMWH. Power Shaft. [3] Pressure ahead of turbine, atm.

## GAS TURBINE SPECIFICATIONS

| | Pressure Ratio | Number of Combustors | Exhaust Flow lb/sec. or kg/sec. | Exhaust Temp. | Heat Exch. for Exhaust Heat Recovery | Dry Weight | Dimensions L | Dimensions W | Dimensions H |
|---|---|---|---|---|---|---|---|---|---|
| | | | | | | | | (inches) | |
| | 6 | 1 | 100lb/sec | 770 | — | 108,000 | 468 | 126 | 165 |
| | 11.5 | 1 | 187lb/sec | 767 | — | 163,000 | 510 | 129 | 158 |
| | 12.5 | 1 | 385lb/sec | 727 | — | 147,000 | 547 | 136 | 146 |
| | 2.5³ | — | 147lb/sec | — | — | 101,000 | 251 | 82 | 111 |
| | 3.0³ | — | 159lb/sec | — | — | 101,000 | 251 | 82 | 111 |
| | 4.0³ | — | 221lb/sec | — | — | 101,000 | 251 | 82 | 111 |
| | 3.64 | 1 | 60lb/sec | — | — | 573,000 | 248 | 110 | 90 |
| | 2.01 | | | | | | | | |
| | 4.38 | — | 171lb/sec | — | — | 138,700 | 336 | 137 | 140 |
| | 3.06 | | | | | | | | |
| | 26.5 | 12 | 997lb/sec | 745 | — | 1,300,000 | 945 | 289 | 289 |
| | 4.6 | 1 | 118lb/sec | 540 | H | 114,500 | 323 | 134 | 127 |
| | 4.6 | 1 | 192lb/sec | 600 | H | 123,000 | 329 | 134 | 127 |
| | 6 | [illegible] | 100lb/sec | 770 | — | 112,000 | 333 | 133 | 151 |
| | 7.5 | 1 | 220lb/sec | 772 | — | 220,000 | 468 | 186 | 154 |
| | 6.5 | 16 | 468lb/sec | 815 | — | 533,000 | 941 | 161 | 168 |
| | 6.6:1 | 6 | 134lb/sec | 820 °F | — | 80,000lb | 324 in | 108 in | 128 in |
| | 6.9:1 | 6 | 133lb/sec | 590 °F | Y | 80,000lb | 324 in | 108 in | 128 in |
| | 7.0:1 | 6 | 242lb/sec | 802 °F | — | 171,000lb | 432 in | 121 in | 150 in |
| | 7.2:1 | 6 | 241lb/sec | 807 °F | Y | 171,000lb | 432 in | 121 in | 150 in |
| | 7.9:1 | 6 | 273lb/sec | 773 °F | — | 171,000lb | 432 in | 121 in | 150 in |
| | 7.7:1 | 8 | 258lb/sec | 791 °F | Y | 171,000lb | 432 in | 121 in | 150 in |
| | 7.1:1 | 8 | 103lb/sec | 1040 °F | — | 112,500lb | 450 in | 130 in | 159 in |
| | 7.9:1 | 8 | 115lb/sec | 1000 °F | — | 112,500lb | 450 in | 130 in | 159 in |
| | 9.0:1 | 8 | 131lb/sec | 965 °F | — | 112,500lb | 450 in | 130 in | 159 in |
| | 6.2:1 | 8 | 87.5lb/sec | 615 °F | Y | 115,700lb | 450 in | 130 in | 169 in |
| | 7.2:1 | 8 | 101.5lb/sec | 640 °F | Y | 115,700lb | 450 in | 130 in | 159 in |
| | 8.2:1 | 8 | 115lb/sec | 665 °F | Y | 115,700lb | 450 in | 130 in | 159 in |
| | 7.3:1 | 8 | 222lb/sec | 1050 °F | — | 210,000lb | 534 in | 130 in | 159 in |

# GAS TURBINE SPECIFICATIONS

| MODEL | USE | POWER RATING — NORMAL: hp, mw or lbt | POWER RATING — NORMAL: Heat Rate | POWER RATING — MAXIMUM: hp, mw or lbt | POWER RATING — MAXIMUM: Heat Rate | COMPRESSOR SHAFT: Compressor Stages | COMPRESSOR SHAFT: Turbine Stages | POWER SHAFT: Turbine Stages | POWER SHAFT: RPM |
|---|---|---|---|---|---|---|---|---|---|
| **Westinghouse Canada Inc.** (Continued) | | | | | | | | | |
| CW352MA[1] | C1,C2, P1,P2,PL | 35,000hp | 8,900[2] Btu/hp-hr | 50,000hp[4] | — | 16 | 1 | 1 | 5,000 |
| CW352MB[1] | C1,C2, P1,P2,PL | 39,800hp | 8,600[2] Btu/hp-hr | 50,000hp[4] | — | 17 | 1 | 1 | 5,000 |
| CW352RMAA[1] | C1,C2, P1,P2,PL | 20,800hp[5] (28,000hp)[6] | 6,750[2] Btu/hp-hr | 50,000hp[4] | — | 15 | 1 | 1 | 5,000 |
| CW352RMA[1] | C1,C2, P1,P2,PL | 27,900hp[5] (32,700hp)[6] | 6,700[2] Btu/hp-hr | 50,000hp[4] | — | 16 | 1 | 1 | 5,000 |
| CW352RMB[1] | C1,C2, P1,P2,PL | 33,100hp[5] (37,300hp)[6] | 6,860[2] Btu/hp-hr | 50,000hp[4] | — | 17 | 1 | 1 | 5,000 |
| Electric Power Generation[8] | | | | | | | | | |
| W101PG[7] | G1,G2,G3,G6 | 8,080kW | 15,320[3] Btu/kW-hr | 10,000kW[4] | — | 16 | 5 | — | 6,000 |
| W191PG[7] | G1,G2,G3,G6 | 17,700kW | 13,390[3] Btu/kW-hr | 19,500kW[4] | — | 15 | 5 | — | 4,912 |
| W191PDG[7] | G1,G2,G3,G6 | 19,900kW | 13,180 Btu/kW-hr | 23,500kW | — | 15 | 5 | — | 4,912 |
| W191G Super[1] | G1,G2,G3,G6 | 19,250kW | 12,930[3] Btu/kW-hr | 19,500kW[4] | — | 15 | 5 | — | 4,912 |
| W301G[1] | G1,G2,G3,G6 | 32,300kW | 13,060[3] Btu/kW-hr | 36,000kW[4] | — | 15 | 4 | — | 3,600 |
| W301G Super[1] | G1,G2,G3,G6 | 35,125kW | 12,600[3] Btu/kW-hr | 36,000kW[4] | — | 15 | 4 | — | 3,600 |

[1] Westinghouse Model Number: 1 = Single Shaft, 2 = Two shaft, M = Mechanical Drive, G = Generator Drive, R = Regenerative Cycle, Super = Supercharged. [2] Heat Rates are in Btu/hp-hr based upon natural gas fuel. [3] Heat rates are in Btu/kw-hr based upon natural gas fuel. [4] Maximum cold weather rating. [5] Power at optimum thermal efficiency. [6] Max. ISO power. [7] Packaged power plant ratings.
[8] Power output will be slightly decreased and heat rates slightly increased on packaged power plants due to inlet and exhaust duct losses. Please refer to factory for weights and dimensions of packaged power plants.

| MODEL | USE | NORMAL: hp, mw or lbt | NORMAL: Heat Rate | MAXIMUM: hp, mw or lbt | MAXIMUM: Heat Rate | Compressor Stages | Turbine Stages | Power Turbine Stages | RPM |
|---|---|---|---|---|---|---|---|---|---|
| **Westinghouse Electric Corp., Combustion Turbine Sys. Div., P.O. Box 251, M.C. C-180, Concordville, PA 19331** | | | | | | | | | |
| Mechanical Drive | | | | | | | | | |
| W251M[10] | C1,2,P0,P-W,P1 | 51,900hp[1] [3] | 8,450[4] [9] | — | — | 18 | — | 3 | 4,894 |
| Packaged Power Plants | | (kW) | | (kW) | | | | | |
| W251G[11] | G1,2,6 | 40,200[7] [3] | 11,250[9] [5] | 43,360[8] [3] | 11,230[9] [5] | 19 | — | 3 | 5,400 |
| W251G[12] | G1,2,6 | 40,200[7] [3] | 11,250[9] [5] | 43,360[8] [3] | 11,230[9] [5] | 19 | — | 3 | 5,400 |
| W501G[11] | G1,2,6 | 96,440[7] [3] | 10,660[9] [5] | 104,220[8] [3] | 10,580[9] [5] | 19 | — | 4 | 3,600 |
| W501G[12] | G1,2,6 | 95,170[7] [3] | 10,810[9] [5] | 102,810[8] [3] | 10,730[9] [5] | 19 | — | 4 | 3,600 |
| Combined Cycle Plants[14] | | | | | | | | | |
| PACE2511[13] | G3 | 55,900[7] [3] | 8,090[9] [5] | — | — | — | — | — | — |
| PACE2512[13] | G3 | 111,800[7] [3] | 8,090[9] [5] | — | — | — | — | — | — |
| PACE2513[13] | G3 | 167,700[7] [3] | 8,090[9] [5] | — | — | — | — | — | — |
| PACE2514[13] | G3 | 223,600[7] [3] | 8,090[9] [5] | — | — | — | — | — | — |
| PACE5011[13] | G3 | 134,700[7] [3] | 7,630[9] [5] | — | — | — | — | — | — |
| PACE5012[13] | G3 | 269,400[7] [3] | 7,630[9] [5] | — | — | — | — | — | — |
| PACE5013[13] | G3 | 404,100[7] [3] | 7,630[9] [5] | — | — | — | — | — | — |
| PACE5014[13] | G3 | 538,800[7] [3] | 7,630[9] [5] | — | — | — | — | — | — |
| PACE5015[13] | G3 | 673,500[7] [3] | 7,630[9] [5] | — | — | — | — | — | — |

[1] ISO gas base. [2] ISO oil base. [3] For other fuels refer to factory. [4] Btu/hp-hr. [5] Btu/kW-hr. [6] ISO oil peak. [7] Base. [8] Peak. [9] LHV of fuel. [10] Engine only at turbine coupling — no inlet or exhaust losses. [11] 60 HZ. [12] 50HZ. [13] 60 or 50 HZ. [14] Net nominal performance. Supplementary firing, optimized configurations available on all models listed.

**Westinghouse Electric Corp.**, Combustion Turbine Sys. Div., P.O. Box 251, M.C. C-180, Concordville, PA 19331, Contact: William G. Rogal, Tel: 215-358-4904.

| MODEL | USE | NORMAL: hp, mw or lbt | NORMAL: Heat Rate | MAXIMUM: hp, mw or lbt | MAXIMUM: Heat Rate | Compressor Stages | Turbine Stages | Power Turbine Stages | RPM |
|---|---|---|---|---|---|---|---|---|---|
| **Williams International, P.O. Box 200, Walled Lake, Michigan 48088.** | | | | | | | | | |
| WR 2-6 | PR-1 | — | — | 125lbt | 1.21SFC | 1 | 1 | — | — |
| WR 24-7 | PR-1 | — | — | 200lbt | 1.25SFC | 2 | 1 | — | — |
| F107-WR-101 | PR-1 | — | — | 600lbt | — | — | — | — | — |
| F107-WR-400 | PR-1 | — | — | 600lbt | — | — | — | — | — |

Ratings are at International Standards Organization (I.S.P.) Conditions: Sea level and 15°C, 59°F. [1] 87AHP + 15 SHP. [2] 125 AHP + 8 SHP. [3] 600 lbt class.

# GAS TURBINE SPECIFICATIONS

| Pressure Ratio | Number of Combustors | EXHAUST Flow lb/sec. or kg/sec. | EXHAUST Temp. | Heat Exch. for Exhaust Heat Recovery | Dry Weight | DIMENSIONS L | DIMENSIONS W | DIMENSIONS H |
|---|---|---|---|---|---|---|---|---|
| 8.2:1 | 8 | 248lb/sec | 1010 °F | — | 210,000lb | 534 in | 130 in | 159 in |
| 9.3:1 | 8 | 282lb/sec | 970 °F | — | 210,000lb | 534 in | 130 in | 159 in |
| 5.9:1 | 8 | 172lb/sec | 615 °F | Y | 214,000lb | 534 in | 130 in | 159 in |
| 7.3:1 | 8 | 213lb/sec | 650 °F | Y | 214,000lb | 534 in | 130 in | 159 in |
| 8.5:1 | 8 | 249lb/sec | 685 °F | Y | 214,000lb | 534 in | 130 in | 159 in |
| 6.6:1 | 6 | 134lb/sec | 820 °F | — | 95,000lb | 324 in | 109 in | 128 in |
| 7.5:1 | 6 | 273lb/sec | 780 °F | — | 171,000lb | 432 in | 121 in | 150 in |
| 7.6:1 | 6 | 270lb/sec | 844 °F | — | 171,000lb | 432 in | 121 in | 150 in |
| 8.6:1 | 8 | 297lb/sec | 758 °F | — | 171,000lb | 432 in | 121 in | 150 in |
| 6.8:1 | 12 | 438lb/sec | 800 °F | — | 234,000lb | 455 in | 181 in | 219 in |
| 7.5:1 | 12 | 481lb/sec | 785 °F | — | 234,000lb | 455 in | 181 in | 219 in |

**Westinghouse Canada Inc., Western** Canada Region: Home Oil Tower, Suite #807, 324-8th Avenue Southwest, Calgary, Alberta T2P 2Z2. Contact: A. M. Robertson, Turbine Sales Manager. Phone: 403 265-1204. Telex: 03-824634.
**Westinghouse Canada Inc., Southwestern** U.S.A. Region: P.O. Box 4250, Houston, Texas 77001. Contact: Mr. W. R. Hutson, Turbine Sales Manager. Phone: 713-771-1301; Telex: 775837.
**Westinghouse Canada, Inc., London**, England Office: Regal House, London Road, Twickenham, Middx. TW1 3QT. Contact: Mr. K. C. Sooth, Manager Turbine Sales Overseas. Phone: 01-891 1151 Telex: 27926.
For other locations, reference Westinghouse Electric Corp. Listings.

| Pressure Ratio | Number of Combustors | Flow | Temp. | Heat Exch. | Dry Weight | L | W | H |
|---|---|---|---|---|---|---|---|---|
| | | (lb/sec) | (°F) | | (lb) | (in) | (in) | (in) |
| 12.4 | 8 | 363 | 940[7] | — | 220,000 | 612 | 122 | 126 |
| 14.0 | 8 | 350 | 1,042[8] | — | 826,000 | 1,344 | 578 | 380 |
| 14.0 | 8 | 350 | 1,042[8] | — | 826,000 | 1,344 | 576 | 360 |
| 14.2 | 14 | 807 | 1,005[8] | — | 1,600,000 | 1,584 | 720 | 576 |
| 14.2 | 14 | 807 | 1,005[8] | — | 1,900,000 | 1,836 | 720 | 588 |
| — | — | — | — | — | — | — | — | — |
| — | — | — | — | — | — | — | — | — |
| — | — | — | — | — | — | — | — | — |
| — | — | — | — | — | — | — | — | — |
| — | — | — | — | — | — | — | — | — |
| — | — | — | — | — | — | — | — | — |
| — | — | — | — | — | — | — | — | — |
| — | — | — | — | — | — | — | — | — |
| — | — | — | — | — | — | — | — | — |

| Pressure Ratio | Number of Combustors | Flow | Temp. | Heat Exch. | Dry Weight | L | W | H |
|---|---|---|---|---|---|---|---|---|
| | | | | | | | (Inches) | |
| 4.2 | 1 | 2.3 | 1,400F | Open | 28.0lb | 22.1 | 10.8 dia | |
| 5.4 | 1 | 2.8 | — | Open | 42lb | 19.5 | 10.8 dia | |
| — | — | — | — | — | 145lb | 48.5 | 12.1 | |
| — | — | — | — | — | 144lb | 36.9 | 12.1 | |

## GAS TURBINE GENERATING SET SPECIFICATIONS

| Model | Normal Rating kW At ISO Condition 15°C, 59°F Sea Level | Service E-Emergency S-Standby | Heat Rate Normal Rating | Type Fuel L-Liquid G-Gas | Type Starter P-Pneumatic E-Electric H-Hydraulic | Location-Enclosure I-Indoors O-Outdoors | Noise Level dB (A) 10M Distance | Unit is S-Stationary T-Transportable |
|---|---|---|---|---|---|---|---|---|
| **ACEC, Gas Turbines Systems Division, BP4, B6000, Charleroi (Belgium)** | | | | | | | | |
| | (MW) | | (Btu/kW-hr) | | | | | |
| W251 | 41,280 | E,S | 11,120 | G&L | E & H or D & H | I,O | 76 | S |
| **Alturdyne, 8050 Armour, San Diego, California 92111 U.S.A.** | | | | | | | | |
| ALT 60-400 | 60 kW | S | | L | E | I,O | 80 | S,T |
| ALT 90-60 | 90 kW | S | | L | E | I,O | 80 | S,T |
| ALT 125-60 | 125 kW | S | | L | E | I,O | 80 | S,T |
| ALT 210-A | 210 HP | S | | L | E | O | 80 | T |
| ALT 210-H | 210 | S | | L | E | O | 80 | T |
| ALT 225-60 | 225 kW | S | | L,G | E | I,O | 80 | S,T |
| [1]In production. [2]125 units in service. [3]Air compressor. [4]Hydraulic starter. | | | | | | | | |
| MAJOR OFFICES WORLDWIDE: U.S.A.: 8050 Armour, San Diego, CA 92111, tel: 714-565-2131, telex.: 910-335-2000. Contact: Frank Verbeke, Pres. • U.S.A.: Box 412, Vienna, VA 22180, tel. 703-281-0791. Contact: Joe Campbell, Sales. | | | | | | | | |
| **Avco Industrial Engine Operation, 17220 Park Row, Houston, Texas 77084 U.S.A.** | | | | | | | | |
| | | | (kj/kW-hr) | | | | | |
| TF25 | 1,865 kW 2,500 hp | | 16,014 | G,D,K[1] | P,E,H | Op.[2] | Op.[2] | T |
| TF40 | 2,984 kW 4,000 hp | | 13,615 | G,D,K[1] | P,E,H | Op.[2] | Op.[2] | T |
| *Dimension in English (inches & pounds). [1]Gaseous, diesel, wide cut kerosene. [2]Optional. A[3]Relates to motor starting requirements. | | | | | | | | |
| **Centrax Ltd., Gas Turbine Division, Shaldon Road, Newton Abbot, Devon TQ12 4SQ, England** | | | | | | | | |
| | | | (btu/kW-hr) | | | | | |
| 74S | 600 | E,S | 19,718 | L,G | E | I,O | 1* | S,T |
| 74SZ | 700 | E,S | 19,698 | L,G | E | I,O | 1 | S,T |
| CX350 | 3,450 | E,S | 12,727 | L,G | P,E,H | I,O | 1 | S,T |
| CX570 | 4,700 | E,S | 12,497 | L,G | P,E,H | I,O | 1 | S,T |
| CX571 | 5,200 | E,S | 11,612 | L,G | P,E,H | I,O | 1 | S,T |
| *Notes 1. "As specified." O.A. = On Applications. | | | | | | | | |
| **Creusot-Loire, 15 rue Pasquier, 75383 Paris Cedex 08, France** | | | | | | | | |
| | | | (Kj/kW-hr) | | | | | |
| CA-5 | 4,800 | E,S | 11,960 | L, G | P, H | I, O | To suit | S,T |
| SK30 | 28,000 | E,S | 11,600 | L, G | P, H | I, O | To suit | S |

# GAS TURBINE GENERATING SET SPECIFICATIONS

| VOLTAGE REGULATION | | FREQUENCY CONTROL | | | DIMENSIONS | | | | |
|---|---|---|---|---|---|---|---|---|---|
| Transient Recovery with Load Change | Excursion at 100% Step Load Change | Transient Recovery with Load Change | Excursion at 100% Step Load Change | Time to get on line —Seconds | Length | Width | Height | Weight | Comments |
| | | | | | Dimensions of Gas Turbines and Generator Sets | | | | |
| | | | | | (Feet) | (Feet) | (Feet) | (Lb.) | |
| — | — | — | — | 480 sec. | 63 | 16 | 13 | 352,000 | |
| | | | | | (Feet) | (Feet) | (Feet) | (Pounds) | |
| ½% | 3% | 2 sec | 3% | 10 | 5 | 3 | 3 | 1,000 | |
| ½% | 3% | 2 sec | 3% | 10 | 6 | 3 | 4 | 2,000 | |
| ½% | 3% | 2 sec. | 3% | 10 | 6 | 3 | 4 | 2,500 | |
| — | — | — | — | — | 4 | 2 | 3 | 600 | |
| — | — | — | — | — | 4 | 2 | 3 | 800 | |
| ½% | 3% | 2 sec. | 3% | 10 | 8 | 3 | 4 | 3,000 | |
| <2 sec. | A³ | <5 sec. | A⁴ | <30 sec. | 50.1 | 34.4 | 43.8 | 1183 | |
| <2 sec. | A³ | <5 sec. | A⁴ | <30 sec. | 52.2 | 34.4 | 43.8 | 1325 | |
| A*Use IEEE 2?2 guide for contract specifications. | | | | | | | | | |
| | | | | | (mm) | (mm) | (mm) | (Kg) | |
| 2 secs | 17% | 2 secs | 2% | 60 secs | 4,013 | 1,257 | 1,920 | 5,000 | |
| 2 secs | 17% | 2 secs | 2% | 60 secs | 4,013 | 1,257 | 1,930 | 5,000 | |
| 2.5 secs | 3.8% | 0.8 secs | 3.5% | 45 secs | 8,500 | 2,500 | 2,600 | 22,000 | |
| O.A. | O.A. | 1.5 secs | 3% at 25% | 45 secs | 8,000 | 2,500 | 2,600 | 26,000 | |
| O.A. | O.A. | O.A. | O.A. | 45 secs | 8,000 | 2,500 | 2,600 | 32,000 | |
| | | | | | (mm) | (mm) | (mm) | (kg) | |
| <1 sec. | <25% | 1 sec. | 5% | 60 | 1.830 | 0.805 | 0.920 | 620 | Packaged plant available |
| [illegible] | <25% | 1 sec. | 7% | 150 | 9.00 | 1.63 | 5.88 | 230,000 | Packaged plant |

# GAS TURBINE GENERATING SET SPECIFICATIONS

**Detroit Diesel Allison, Box 894, Dept. U-5, Indianapolis, Indiana 46206 U.S.A.**

| Model | Normal Rating kW At ISO Condition 15°C, 59°F Sea Level | Service E-Emergency S-Standby | Heat Rate Normal Rating | Type Fuel L-Liquid G-Gas | Type Starter P-Pneumatic E-Electric H-Hydraulic | Location-Enclosure I-Indoors O-Outdoors | Noise Level dB (A) 1M Distance | Unit is S-Stationary T-Transportable |
|---|---|---|---|---|---|---|---|---|
| | (kW) | | | | | | | |
| GT404 | 251[1] | Either | 11,612 | — | — | Either | 85 | Either |
| 570-KA | 4,500[1] | Either | 12,150 | L or G | P,H | Either | 85 | Either |
| 501-KB | 3,066[1] | Either | 12,844 | L or G | P,H | Either | 85 | Either |
| 501KB5 | 3,934[1] | Either | 11,198[1] | L or G | P,H | Either | 85 | Either |
| | 3,845[2] | | 11,302[2] | | | | | |

[1]Natural Gas [2]DFII Fuel

**MAJOR WORLDWIDE OFFICES**: Holland: Parmentierplein 1, Postbox 5061, 3008 AB Rotterdam, tel. 010-290-000, telex. 28355 GMCNL • Great Britain: P. O. Box No. 6, London Road, Wellingborough, Northhamptonshire, NN8, 2DL, England, tel. 0933-71122, telex. 31320 DDAIEUG, • Sweden: Motorvagen 1, Fack, S-104 60 Stockholm, tel. 08-440180, telex. 1569 GMNORD • Norway: P.O. Box 205, 2001 Lillestrom, tel. 0271-38-60 or 0271-58-60, telex. 11882 Lillestrom • Denmark: General Motors Danmark, GMODC Branch, Borgmester Christiansens Gade 40, DK 2450 Copenhagen SV, tel. 01-30-22-11, telex. 6727 Copenhagen • Republic of Germany: Div. of General Motors Deutschland GmbH, Kupferstrasse 1, Postfach 1507, 6090 Ruesselsheim, Republic of Germany, tel. 06142/6021, telex. 417541 • Switzerland: Div. of General Motors Suisse, S.A., Salzhausstrasse 21, 2501 Biel Bienne, tel. 032-21-61-11, telex. 34217, Bienne • France: Div. of General Motors France, S.A., 56/68 Ave Louis Roche, 92231, Gennevilliers, tel. 790-7000, telex. 620050 • Australia: Div. of General Motors-Holden's Limited, Princes Highway, P. O. Box 163, Dandenong, Victoria, Australia 3175, tel. (03) 792-0111, telex. GM Network MAD DANDENONG • New Zealand: Div. of General Motors

**GEC Gas Turbines Limited, Cambridge Road, Whetstone, Leicester LE8 3LH England**

| Model | Normal Rating kW At ISO Condition 15°C, 59°F Sea Level | Service E-Emergency S-Standby | Heat Rate Normal Rating | Type Fuel L-Liquid G-Gas | Type Starter P-Pneumatic E-Electric H-Hydraulic | Location-Enclosure I-Indoors O-Outdoors | Noise Level dB (A) 1M Distance | Unit is S-Stationary T-Transportable |
|---|---|---|---|---|---|---|---|---|
| | | | (Btu/kW-hr) | | | | | |
| EAS-133 | 11,300[1] | 14,600[3] | 13,260 | L,G | P,E | I,O | To Suit | S,T |
| EAS-134 | 12,800[1] | 16,000[3] | 13,050 | L,G | P,E | I,O | To Suit | S,T |
| EAS-135 | 14,000[1] | 16,700[3] | 13,100 | L,G | P,E | I,O | To Suit | S,T |
| ERB-1 | 23,000[1] | 25,290[3] | 10,230 | L,G | P,H | I,O | To Suit | S,T |
| EO-1C | 22,940[1] | 27,330[3] | 11,460 | L,G | P,E | I,O | To Suit | S,T |
| ELM125 | 21,525[1] | 23,750[3] | 9,420 | L,G | P,H | I,O | To Suit | S,T |
| ELM150 | 32,200[1] | 36,100[3] | 9,410 | L,G | P | I,O | To Suit | S,T |

[1]Generator Output. [2]Normal. [3]Emergency.

**MAJOR OFFICES WORLDWIDE**: GEC Gas Turbines Limited, Cambridge Road, Whetstone, Leicester, LE8 3LH England. Contact: Mr. J. C. McMillan. Phone: Leicester (0533) 863434. Telex: 34381. • GEC Gas Turbines Limited, 132-135 Long Acre, London, WC2E 9AH, England, Contact: Mr. J. W. Mosenthal. Phone: 01-836-3444. Telex: 21409 • English Electric Corporation, 140 North Loop, Houston, Texas 77009 U.S.A. Contact: Mr. P. L. Banks, Phone: (713) 861-2375. Telex: 775835 • GEC Gas Turbine Services Limited, Al Hajim Saudi International Company, P.O. Box 1017, Dammam, Saudi Arabia, Contact: Mr. I. D. Duff, Phone: 833-9476/7 832-5427. Telex: 601079 • GEC Singapore Limited, P.O. Box 4846 Magnet House, Bukit Timah Road, Singapore 2156. Contact: Mr. B. Thornborough. Phone: 663011, Telex: RS21508 • GEC De Mexico, Salamanca 34 P-H, Mexico 7DF, Contact: Mr. H. Dodd, Phone: 5332848 or 5251394. Telex: 1777578.

**George Engine Company, 1401 Destrehan Avenue, P.O. Box 8, Harvey, LA 70058**

| Model | Normal Rating kW At ISO Condition 15°C, 59°F Sea Level | Service E-Emergency S-Standby | Heat Rate Normal Rating | Type Fuel L-Liquid G-Gas | Type Starter P-Pneumatic E-Electric H-Hydraulic | Location-Enclosure I-Indoors O-Outdoors | Noise Level dB (A) 1M Distance | Unit is S-Stationary T-Transportable |
|---|---|---|---|---|---|---|---|---|
| | | | (Btu/kW-hr) | | | | | |
| Stand-by | 3,650 | E-S | 12,400 | L-G | H-P | I-O | 90 | S-T |
| Base load | 3,000 | — | 13,100 | L-G | H-P | I-O | 90 | S-T |

[1]Includes integral control room. [2]Plus top-mounted equipment. [3]Waste heat boiler can be supplied as optional equipment. [4]All listed equipment driven by Allison 501-KB.

**Hispano Suiza, Rue Du Capitaine Guynemer B.P. 60 92270 Bois Colombes, France**

| Model | Normal Rating kW At ISO Condition 15°C, 59°F Sea Level | Service E-Emergency S-Standby | Heat Rate Normal Rating | Type Fuel L-Liquid G-Gas | Type Starter P-Pneumatic E-Electric H-Hydraulic | Location-Enclosure I-Indoors O-Outdoors | Noise Level dB (A) 1M Distance | Unit is S-Stationary T-Transportable |
|---|---|---|---|---|---|---|---|---|
| | | | (Kcal/kW-hr) | | | | | |
| THM 1103 | 4,400 | E,S | 3,826 | L,G | P,E,H | I,O | 85 | Either |
| THM 1203 | 5,600 | E,S | 3,583 | L,G | P,E,H | I,O | 85 | Either |
| THM 1304 | 8,100 | E,S | 3,188 | L,G | P,E,H | I,O | 85 | Either |
| THM 1304-10 | 10,000 | E,S | 2,965 | L,G | P,E,H | I,O | 85 (at 1m) | Either |

*100% load rejection. **Emergency Conditions

**MAJOR OFFICES WORLDWIDE**: Hispano Suiza Inc., 10633 Shadow Wood Drive, Houston, Texas 77043, Phone: (713) 467-9612, Telex: 6868 127 HSINC UW • Hispano Suiza Do Brazil Equipamentos Ltd., Rua Da Gloria, 344-Gloria 20241, Rio de Janeiro, Phone: (021) 242-4004, Telex: 21 989 HSBRBR.

# GAS TURBINE GENERATING SET SPECIFICATIONS

| VOLTAGE REGULATION | | FREQUENCY CONTROL | | | DIMENSIONS | | | | |
|---|---|---|---|---|---|---|---|---|---|
| Transient Recovery with Load Change | Excursion at 100% Step Load Change | Transient Recovery with Load Change | Excursion at 100% Step Load Change | Time to get on line —Seconds | Length | Width | Height | Weight | Comments |
| | | | | | (in/mm) | (in/mm) | (in/mm) | (lbs/kg) | |
| — | — | — | — | 10 | 46.7/1,186 | 50.5/1,283 | 39.5/1,003 | 1,800/816 | |
| — | — | — | — | 30-45 | 70.2/1,783 | 31.2/793 | 36.1/917 | 1,350/612 | |
| — | — | — | — | 45-60 | 90.0/2,286 | 33.2/843 | 31.1/790 | 1,285/583 | |
| — | — | — | — | 45-60 | 90.0/2,286 | 33.2/843 | 31.1/790 | 1,285/583 | |

New Zealand, Ltd., Princes Street, Private Bag, Upper Hutt, tel. 70555 Upper Hutt, telex. N.Z. 3771, Wellington • Singapore: 15 Benoni Sector, Jurong Town, Singapore 22, tel. 654697, 2610801, telex. RS21606 A/B GMSING • Greece: Athens Towers, Messoghion 2/4, Suite 705, Athens, 610, tel. 7785-334, 7706-669, 7787-281, telex. 215759. DDAI • U.S.A.: 2655 Villa Creek Drive, Suite 130, Dallas, Texas 75234, tel. (214) 241-2754, telex. 910-860-5549 • U.S.A.: Detroit Diesel Allison International Operations, 25200 Telegraph Road, Southfield, Michigan 48034, tel. (313) 424-4820, telex. 810-224-4898 • Columbia: c/o GMIC Zone Office, P. O. Box 53362, Bogota, tel. 232-68-04, telex. 396-414-99 • Mexico: Div. of General Motors de Mexico S.A. de C.V., Emerson 432, 3rd Floor, Mexico 5, D.F., Apartado 107 Bis., Mexico 1, D.F., tel. 250-4244, telex. 017-73835 • Brazil: Divisao da General Motors do Brazil, S.A., Avenido Paulista, 1106 6/7 Andar, Caixa Postal 6230, CEP - 01310 Sao Paulo - SP Brazil, tel. 289-3099, telex. (011) 23909.

| Transient Recovery with Load Change | Excursion at 100% Step Load Change | Transient Recovery with Load Change | Excursion at 100% Step Load Change | Time to get on line —Seconds | Length | Width | Height | Weight | Comments |
|---|---|---|---|---|---|---|---|---|---|
| | | | | | (meters) | (meters) | (meters) | (Kg) | |
| | | | | 180 | 7.100 | 3.480 | 3.124 | 22,000 | |
| | | | | 180 | 7.100 | 3.480 | 3.124 | 22,000 | |
| | | | | 180 | 7.100 | 3.480 | 3.124 | 22,000 | |
| | | | | 255 | 6.500 | 4.000 | 3.860 | 23,000 | |
| | ON APPLICATION | | | 460[3]180[3] | 7.310 | 3.450 | 3.200 | 25,460 | |
| | | | | 300[3]240[3] | 6.070 | 3.520 | 3.737 | 14,000 | |
| | | | | 540[3]330[3] | 9.200 | 3.200 | 3.685 | 55,000 | |

| Transient Recovery with Load Change | Excursion at 100% Step Load Change | Transient Recovery with Load Change | Excursion at 100% Step Load Change | Time to get on line —Seconds | Length | Width | Height | Weight | Comments |
|---|---|---|---|---|---|---|---|---|---|
| | | | | | (feet) | (feet) | (feet) | (lbs) | |
| | Varies with generator | | | 60 | 40[1] | 8 | 8[2] | 65,000 | |
| | Varies with generator | | | 60 | 40[1] | 8 | 8[2] | 65,000 | |

| Transient Recovery with Load Change | Excursion at 100% Step Load Change | Transient Recovery with Load Change | Excursion at 100% Step Load Change | Time to get on line —Seconds | Length | Width | Height | Weight | Comments |
|---|---|---|---|---|---|---|---|---|---|
| | | | | | (meters) | (meters) | (meters) | (Kg) | |
| 1 sec. | 10% | 2 to 5 sec. | 10%* | 60 sec.** | 14.5 | 2.75 | 4.1 | 44,000 | |
| 1 sec. | 10% | 2 to 5 sec. | 10%* | 60 sec.** | 14.5 | 2.75 | 4.1 | 44,000 | |
| 1 sec. | 10% | 2 to 5 sec. | 10%* | 60 sec.** | 15.5 | 2.75 | 4.1 | 50,000 | |
| 1 sec. | 10% | 2 to 5 sec. | 10%* | 60 sec.** | 14.5 | 2.75 | 4.1 | 50,000 | |

# GAS TURBINE GENERATING SET SPECIFICATIONS

| Model | Normal Rating kW At ISO Condition 15°C, 59°F Sea Level | Service E-Emergency S-Standby | Heat Rate Normal Rating | Type Fuel L-Liquid G-Gas | Type Starter P-Pneumatic E-Electric H-Hydraulic | Location-Enclosure I-Indoors O-Outdoors | Noise Level dB (A) 10M Distance | Unit is S-Stationary T-Transportable |
|---|---|---|---|---|---|---|---|---|
| **Ishikawajima-Harima Heavy Industries Co., Ltd., Shin Ohtemachi Bldg., Ohtemachi 2-2-1, Chiyoda-ku, Tokyo 100, Japan** | | | | | | | | |
| IM100-2 | 1.07 [1] | Either | 14,500 [2] | L,G | E | Either | 85 [3] | Either |
| IM100-4 | 1.3 [1] | Either | 12,900 [2] | L,G | E | Either | 85 [3] | Either |
| IM400 | 3.27 [1] | Either | 12,100 [2] | L,G | P | Either | 85 [3] | Either |
| IM400-6 | 3.84 [1] | Either | 11,300 | L,G | P | Either | 85 [3] | Either |
| IM600 | 4.80 [1] | Either | 11,360 | L,G | P | Either | 85 [3] | Either |
| IM1500 | 11.0 [1] | Either | 12,700 [2] | L,G | P | Either | 85 [3] | Either |
| LM2500 | 22.0 [1] | Either | 9,280 | L,G | P | Either | 85 [3] | Either |
| IM5000 | 35.4 [1] | Either | 9,620 [2] | L,G | P | Either | 85 [3] | S |

[1] MW (equivalent to gas turbine output) [2] Btu/kW-hr, at normal rating. [3] 100% load rejection. [4] Dimensions are for gas turbines alone. [5] Typical stationary type.

**MAJOR OFFICES WORLDWIDE**: U.S.A.: 277 Park Avenue, Suite 1321, New York, NY 10172, tel. 223-8200, telex: 232670 (23670 IHI UR) 420539 (420539 IHI UI), cable address IHICONO NEW YORK, contact: Mr. Takahashi • U.S.A.: Suite 3870, 1001 Fannin, Houston, Texas 77002, tel. 652-2072, telex: 762704 (IHICO HOU), cable address IHICO HOUSTON, contact: Mr. Hanawa • Great Britain: Europe House, World Trade Centre, London, E19AA, England, tel. 709-9493-9, telex: 886377 (IHICO LONDON), 883786 (IHICO LONDON), cable ad-

| Model | Normal Rating kW | Service | Heat Rate (Btu/kW-hr) | Type Fuel | Type Starter | Location-Enclosure | Noise Level | Unit is |
|---|---|---|---|---|---|---|---|---|
| **Kawasaki Heavy Industries, Ltd., Jet Engine Division, 1-1 Kawasaki-Cho, Akashi City, Hyogo — Pref., Japan** | | | | | | | | |
| GP200 | 150 | E | 22,890 | L | E | I,O | 80 | S,T |
| GP250 | 180 | E,S | 22,890 | L | E | I,O | 80 | S,T |
| GP500 | 350 | E,S | 23,700 | L | E | I,O | 80 | S |
| GP750 | 600 | E,S | 17,980 | L | E | I,O | 80 | S |
| GP1250 | 1,000 | E,S | 17,600 | L | E,H,P | I,O | 80 | S |
| GP2500 | 1,900 | E,S | 17,800 | L | E,H,P | I,O | 80 | S |
| **Klockner Humboldt-Deutz AG, 5 Koln 80, Germany** | | | | | | | | |
| KA108 | 20.6 | E | 66,000 | L | E | I,O | 75 | S,T |
| TA218 | 80 | E,S | 30,925 | L | E, hand | I,O | 73 | S,T |
| KA215 | 220 | S | 18,672 | L,G | E,P | I,O | 73 | S,T |
| UA418 | 330 | E,S | 17,190 | L,G | E,P | I,O | 75 | S,T |
| KT215 | 427 | S | 19,144 | L,G | E,P | I,O | 75 | S,T |
| KA123 | 700 | S | 15,888 | L,G | E,P | I,O | 78 | S,T |
| UA224 | 1,000 | E,S | 14,770 | L | E | I,O | 80 | S,T |
| KA134 | 1,177 | S | 15,697 | L,G | E,P | I,O | 80 | S,T |
| KA334 | 1,470 | S | 16,961 | L,G | E,P | I,O | 80 | S,T |
| LA136 | 1,812 | S | 15,200 | L,G | E,P,H | I,O | 80 | S,T |
| KT134 | 2,260 | S | 16,187 | L,G | E,P | I,O | 83 | S,T |
| KT334 | 2,824 | S | 17,484 | L,G | E,P | I,O | 83 | S,T |
| LA237 | 2,900 | S | 12,911 | L,G | E,P,H | I,O | 83 | S,T |
| DA145 | 4,100 | E,S | 11,720 | L,G | P | I,O | 85 | S,T |

**Klockner Humboldt-Deutz AG**, 5 Koln 80, Germany, P.O. Box 800509, Dept. AG-V1, Contact: Mr. Stommel, Phone: 0221-822-4380, Telex: 8-812-255

| Model | Normal Rating kW | Service | Heat Rate (Btu/kW-hr) | Type Fuel | Type Starter | Location-Enclosure | Noise Level | Unit is |
|---|---|---|---|---|---|---|---|---|
| **Mitsui Engineering & Shipbuilding Co., Ltd., 6-4, Tsukiji 5-Chome, Chuo-Ku, Tokyo, Japan** | | | | | | | | |
| SB30C | 4,870 | S | 14,848 | L,G | P,E | O,I | 80 | T |
| SB90C | 16,170 | S | 13,364 | L,G | P,E | O,I | 80 | S |
| SB60C-M | 13,070 | E,S | 11,123 | L,G | P | O,I | 80 | S |

**MAJOR OFFICES WORLDWIDE**: Japan: 6-4, Tsukiji 5-chome, Chuo-ku, Tokyo, tel. (03) 544-3131, telex. J22821,0J22924, 252-2861, cable, "MITUIZOSEN TOKYO" • U.S.A.: Special Representative in New York, Suite 2029, One World Trade Center, New York, N.Y. 10048, tel. (212) 432-0240/4, telex. 232671, 420144 WU 129238, cable, "MITUIZOSEN NEW YORK" • U.S.A.: Special Representative in Los Angeles, c/o Mitsui & Co., Inc., Crocker-Bank Plaza, 611 West 6th Street, Suite 2190, Los Angeles, California 90017, tel. 213-680-1000, 213-628-5350, telex. 6-7205 • West Germany: Special Representative in Duesseldorf, Koenigsalle 92a, 4 Duesseldorf 1 F.R. West Germany, tel. Duesseldorf 325701, 326125, telex. 8587297, cable, "MITUIZOSEN DUESSELDORF" • Great Britain: Special Representative in London 7th Floor, Lee House, Wood Street, London Wall, London EC2Y 5AH, England, tel.

| Model | Normal Rating kW | Service | Heat Rate | Type Fuel | Type Starter | Location-Enclosure | Noise Level | Unit is |
|---|---|---|---|---|---|---|---|---|
| **Noel Penny Turbines Ltd., Siskin Drive, Toll Bar End, Coventry, CV3 4FE, England** | | | | | | | | |
| NPT100 | 100 | E | 0.82 | L | E | I or O | 75@10 ft. | T |
| NPT400 | 250 | E | 0.77 | L | E | I or O | 75@10 ft. | T |

# GAS TURBINE GENERATING SET SPECIFICATIONS

| Voltage Regulation: Transient Recovery with Load Change | Voltage Regulation: Excursion at 100% Step Load Change | Frequency Control: Transient Recovery with Load Change | Frequency Control: Excursion at 100% Step Load Change | Time to get on line —Seconds | Dimensions: Length | Dimensions: Width | Dimensions: Height | Dimensions: Weight | Comments |
|---|---|---|---|---|---|---|---|---|---|
| | | | | | (in.) | (in.) | (in.) | (lb.) | |
| 3 sec. | 15%[3] | 10 sec.[3] | 10%[3] | 60 | 56 | 21 | 24 | 350 | [4] |
| 3 sec. | 15%[3] | 10 sec.[3] | 10%[3] | 60 | 67 | 24 | 26 | 440 | [4] |
| 3 sec. | 15%[3] | 10 sec.[3] | 3%[3] | 60 | 90 | 33.5 | 31.5 | 1,270 | [4] |
| 3 sec. | 15%[3] | 10 sec.[3] | 3%[3] | 60 | 90 | 33.5 | 31.5 | 1,270 | [4] |
| 1.5 sec. | 15%[3] | 10 sec.[3] | 10%[3] | 60 | 71 | 31 | 35 | 1,340 | [4] |
| 3 sec. | 15%[3] | 10 sec.[3] | 10%[3] | 120 | 315 | 98 | 95 | 36,400 | [4] |
| 1.5 sec. | 15%[3] | 12 sec.[3] | 10%[3] | 120 | 260 | 83 | 83 | 10,360 | [4] |
| 1.5 sec. | 15%[3] | 10 sec.[3] | 10%[3] | 180 | 352 | 138 | 132 | 63,000 | [4] |

dress, IHICO LONDON E1, contact: Mr. Niwa • Kuwait: 7th Floor, Office No. 3, Souk Al-Kauwait Building, Kuwait, P. O. Box No. 2464 SAFAT, Kuwait Arabian Gulf, tel. 449121-3, telex 22635 (IHICO KT 22635KT), contact: Mr. [illegible] • Australia: 5th Floor, 213-219 Miller Street, North Sydney, N.S.W. 2060, P.O. Box No. 688-North Sydney, tel. 438-4777, telex 20480 (IHICO AA20480), cable address, IHIHEAVY SYDNEY, contact: Mr. Uchiyoma.

| Voltage Regulation: Transient Recovery with Load Change | Voltage Regulation: Excursion at 100% Step Load Change | Frequency Control: Transient Recovery with Load Change | Frequency Control: Excursion at 100% Step Load Change | Time to get on line —Seconds | Dimensions: Length | Dimensions: Width | Dimensions: Height | Dimensions: Weight | Comments |
|---|---|---|---|---|---|---|---|---|---|
| | | | | | (meters) | (meters) | (meters) | (Kg) | |
| 2 sec. | 20% | 2 sec. | 4% max. | 40 sec. | 2.5 | 1.3 | 1.7 | 2,800 | Package |
| 2 sec. | 20% | 2 sec. | 4% max. | 40 sec. | 2.5 | 1.3 | 1.7 | 2,800 | Without |
| 2 sec. | 20% | 2 sec. | 4% max. | 40 sec. | 3.1 | 1.5 | 2.0 | 5,800 | Silencer |
| 2 sec. | 20% | 2 sec. | 4% max. | 40 sec. | 4.1 | 1.8 | 2.3 | 8,000 | Indoor T. |
| 2 sec. | 20% | 2 sec. | 4% max. | 40 sec. | 5.1 | 1.9 | 2.5 | 11,500 | " |
| 2 sec. | 20% | 2 sec. | 4% max. | 40 sec. | 6.1 | 2.8 | 3.1 | 21,500 | " |

| Voltage Regulation: Transient Recovery with Load Change | Voltage Regulation: Excursion at 100% Step Load Change | Frequency Control: Transient Recovery with Load Change | Frequency Control: Excursion at 100% Step Load Change | Time to get on line —Seconds | Dimensions: Length | Dimensions: Width | Dimensions: Height | Dimensions: Weight | Comments |
|---|---|---|---|---|---|---|---|---|---|
| (sec.) | (%) | (sec.) | (%) | | (meters) | (meters) | (meters) | (Kg) | |
| — | — | — | — | — | 0.38 | 0.36 | 0.4 | 25 | |
| 1 | 6 | 1.5 | 1.5 | 30 | 0.82 | 0.62 | 0.82 | 83 | |
| 0.5 | 10 | 1 | 1 | 40 | 1.0 | 1.0 | 0.8 | 390 | |
| <1 | <20 | 1 | 3 | 30 | 1.32 | 0.71 | 0.73 | 340 | |
| 0.5 | 10 | 1 | 1 | 40 | 1.2 | 1.1 | 1.1 | 650 | All data |
| 0.6 | 10 | 1 | 1 | 40 | 2.0 | 1.2 | 1.2 | 1,300 | including |
| — | — | — | — | 60 | 2.14 | 0.72 | 0.92 | 490 | gearbox |
| 0.5 | 10 | 1 | 1 | 40 | 2.1 | 1.4 | 1.4 | 2,500 | |
| 0.5 | 10 | 1 | 1 | 40 | 2.1 | 1.4 | 1.4 | 2,500 | |
| — | — | — | — | 60 | 2.3 | 1.4 | 1.63 | 2,520 | |
| 0.5 | 10 | 1 | 1 | 40 | 2.2 | 2.2 | 1.6 | 5,700 | |
| 0.5 | 10 | 1 | 1 | 40 | 2.2 | 2.2 | 1.6 | 5,700 | |
| — | — | — | — | 60 | 2.35 | 1.4 | 1.63 | 2,585 | |
| — | — | — | — | 180 | 3.8 | 1.1 | 1.1 | 3,970 | |

| Voltage Regulation: Transient Recovery with Load Change | Voltage Regulation: Excursion at 100% Step Load Change | Frequency Control: Transient Recovery with Load Change | Frequency Control: Excursion at 100% Step Load Change | Time to get on line —Seconds | Dimensions: Length | Dimensions: Width | Dimensions: Height | Dimensions: Weight | Comments |
|---|---|---|---|---|---|---|---|---|---|
| | | | | | (meters) | (meters) | (meters) | (Kg) | |
| 1 | 5% | 4 sec. | 3% | 135 | 7.8 | 3.0 | 4.3 | 28,300 | |
| 1 | 5% | 5 sec. | 4% | 150 | 13.0 | 3.5 | 6.0 | 53,000 | |
| 2 | 8% | 8 sec. | 5% | 110 | 13.0 | 3.5 | 5.5 | 54,700 | |

600-0241, telex 885071, cable, "MITUIZOSEN LONDON EC4" • Hong Kong: Special Representative in Hong Kong, c/o Mitsui & Co., Ltd., Connaught centre, 33rd Floor, Connaught Road, Central, Hong Kong, tel. 5-264201/3, telex HX73916 • Singapore: Special Representative in Singapore, c/o Mitsui & Co., Ltd., 6th Floor, Industrial & Commercial Bank building, Two Shenton Way, Singapore, tel. 2204065, telex RS24102 MESSPRE cable, MITUIZOSEN c/o MITUICO SINGAPORE • Brazil: Mitsui Zosen do Brasil Industrials Pesadas Ltda, 13th Floor, 89 Avenida Almirante Barroso, Rio de Janeiro, RJ, Brazil, tel. 252-3667, 242-4898, telex 021-22378.

| Voltage Regulation: Transient Recovery with Load Change | Voltage Regulation: Excursion at 100% Step Load Change | Frequency Control: Transient Recovery with Load Change | Frequency Control: Excursion at 100% Step Load Change | Time to get on line —Seconds | Dimensions: Length | Dimensions: Width | Dimensions: Height | Dimensions: Weight | Comments |
|---|---|---|---|---|---|---|---|---|---|
| | | | | | (Inches) | (Inches) | (Inches) | (Pounds) | |
| 5 sec. | 1% | 5 sec. | 15 | 30 | 84 | 24 | 36 | 2500 | |
| 5 sec. | 1% | 5 sec. | 15 | 30 | 84 | 24 | 36 | 4200 | |

# GAS TURBINE GENERATING SET SPECIFICATIONS

| Model | Normal Rating kW At ISO Condition 15°C, 59°F Sea Level | Service E-Emergency S-Standby | Heat Rate Normal Rating | Type Fuel L-Liquid G-Gas | Type Starter P-Pneumatic E-Electric H-Hydraulic | Location-Enclosure I-Indoors O-Outdoors | Noise Level dB (A) 10M Distance | Unit is S-Stationary T-Transportable |
|---|---|---|---|---|---|---|---|---|
| **North American Turbine Corp., Box 40501, Houston, Texas 77040 U.S.A.** | | | | | | | | |
| | | | (Btu/kW-hr) | | | | | |
| NATCO 5 | 550 | E,S | 17,100 | L,G | P,E | Either | 65 | Either |
| NATCO 2 | 1,800 | E,S | 21,000 | L,G | P,E | Either | 70 | Either |
| NATCO 831 | 3,250 | E,S | 15,200 | L,G | P,E | Either | 70 | Either |
| **Norwalk-Turbo, Inc., 7 Northway Lane, Latham, NY 12110** | | | | | | | | |
| | | | (Btu/kW-hr) | | | | | |
| TG-9A | 641 | E,S | 15,530 | L,G | P,E,H | I,O | 85 | S,T |
| TG-9B | 716 | E,S | 15,510 | L,G | P,E,H | I,O | 85 | S,T |
| TG-9AT | 1,282 | E,S | 15,500 | L,G | P,E,H | I,O | 85 | S,T |
| TG-9BT | 1,432 | E,S | 15,480 | L,G | P,E,H | I,O | 85 | S,T |
| TG-40 | 2,983 | E,S | 12,713 | L,G | P,E,H | I,O | 85 | S,T |
| TG-60 | 3,878 | E,S | 11,124 | L,G | P,E,H | I,O | 85 | S,T* |

Norwalk-Turbo, Inc., Latham, NY 12110, Contact: S. David Caplow, Tel (518) 783-1625.

| Model | Normal Rating kW | Service | Heat Rate | Type Fuel | Type Starter | Location-Enclosure | Noise Level | Unit is |
|---|---|---|---|---|---|---|---|---|
| **Onan Corp., 1400 73rd Avenue, N.E., Minneapolis, Minnesota 55432 U.S.A.** | | | | | | | | |
| | | | (Btu/kW-hr) | | | | | |
| 560GTU | 560[1] | S,E | $1.58 \times 10^4$ | L,G,Dual | P,E | Either | 70 | S |
| 4,000GTU | 4,380[1] | S,E | $1.17 \times 10^4$ | L,G,Dual | P | Either | 70 | S |

[1]kW at Generator Terminals

Onan Corporation, 1400 73rd Avenue, N.E., Minneapolis, MN 55432. Phone 612-574-5000. Telex: 29 0476. Contact U.S.: Jerry Barton. International: Joseph Carroll.

| Model | Normal Rating kW | Service | Heat Rate | Type Fuel | Type Starter | Location-Enclosure | Noise Level | Unit is |
|---|---|---|---|---|---|---|---|---|
| **Rolls Royce Ltd., Industrial and Marine Division, P.O. Box 72, Ansty, Coventry, CV7 9JR, England** | | | | | | | | |
| | | | (Btu/kW-hr) | | | | dB(A)[1] | |
| SK15 | 16,300 | E,S | 11,790 | L,G | P,E,H | I,O | 54 | S,T |
| SK25 | 23.500 | E,S | 10,190 | L,G | P,E,H | I,O | 54 | S |
| SK30 | 28,100 | E,S | 11,130 | L,G | P,E,H | I,O | 57 | S |
| SK60 | 56,200 | E,S | 11,130 | L,G | P,E,H | I,O | 57 | S |

[1]These are standard levels, but any requirement can be met. Noise level = dB(A) 100M distance. [2]From dead start. Does not include 2 minutes necessary for purging while on gas fuel only.

Rolls-Royce Inc.: 375 Park Avenue, New York, New York 10152 U.S.A. Contact: John Sharo. Phone (212) 935-9400. Telex 620839.

Rolls-Royce Inc.: Industrial and Marine Division, Suite 155 South 16800 Greenspoint Park Drive Houston, Texas. Contact: D. J. Stepson. Phone (713) 999-5900. Telex 795031 ROLLSROYCE HOU.

Rolls-Royce Inc.: 10850 Riverside Drive, Suite 603, North Hollywood, California 91602. Contact: J. Speller. Phone: (213) 985-5170. Telex 677275 ROYCALIF.

| Model | Normal Rating kW | Service | Heat Rate | Type Fuel | Type Starter | Location-Enclosure | Noise Level | Unit is |
|---|---|---|---|---|---|---|---|---|
| **Solar Turbines Incorporated, 2200 Pacific Highway, San Diego, California 92138 U.S.A.** | | | | | | | | |
| | | | (Btu/kW-hr) | | | | | |
| GSE-10,000 | 9,230 | E,S | 10,600 | L,G | P,D/H*, E/H# | Either | 85 | S |
| GSE-4000 | 3,220 | E,S | 13,343 | L,G | P,E/4#, T/H% | Either | 85 | Either |
| GSE-1000 | 900 | E,S | 15,300 | L | P,E | Either | 85 | Either |

[1]Faster start available. *Diesel-engine-driven hydraulic. #Electric-motor-driven hydraulic. %Gas turbine-driven hydraulic. **25% load increments only.

# GAS TURBINE GENERATING SET SPECIFICATIONS

| VOLTAGE REGULATION | | FREQUENCY CONTROL | | | DIMENSIONS | | | | |
|---|---|---|---|---|---|---|---|---|---|
| Transient Recovery with Load Change | Excursion at 100% Step Load Change | Transient Recovery with Load Change | Excursion at 100% Step Load Change | Time to get on line —Seconds | Length | Width | Height | Weight | Comments |
| | | | | | (Feet) | (Feet) | (Feet) | (Pounds) | |
| <1 sec. | <20% | 1 sec. | <3 Hz | 10 | 10'4'' | 4'3'' | 5'6'' | 6,500 | |
| <1 sec. | <20% | 1 sec. | <3 Hz | 60 | 16'6'' | 5'6'' | 6'8'' | 22,800 | |
| <1 sec. | <20% | 1 sec. | <3 Hz | 90 | 12'11'' | 7'0'' | 9'11'' | 45,500 | |
| | | | | | (Feet) | (Feet) | (Feet) | (lb.) | |
| CONSULT WITH FACTORY | | | | | 14'6'' | 5'0'' | 6'0'' | 8,800 | |
| | | | | | 14'6'' | 5'0'' | 6'0'' | 9,200 | |
| | | | | | 23'0'' | 9'0'' | 8'6'' | 25,000 | |
| | | | | | 23'0'' | 9'0'' | 8'6'' | 26,000 | |
| | | | | | 22 | 6 | 7 | 15,000 | |
| | | | | | 23 | 6 | 7 | 21,000 | |
| | | | | | (Inches) | (Inches) | (Inches) | (Pounds) | |
| .5 sec | ± 20% | 1.5 sec. | ± 5% | 10 | 137 | 56.4 | 76.0 | 6,800 | |
| | | On Application | | 75 | 347 | 100 | 108 | 50,000 | |
| | | | | | (Feet) | (Feet) | (Feet) | (Lb) | |
| | | On request | | 150[2] | 60 | 14 | 24 | 376,000 | All rating on Natural Gas Fuel |
| | | On request | | 150[2] | 60 | 14 | 24 | 380,000 | |
| | | On request | | 150[2] | 76 | 14 | 49 | 515,000 | |
| | | On request | | 150[2] | 118 | 14 | 49 | 840,000 | |
| | | | | | (Feet) | (Feet) | (Feet) | (Pounds) | |
| <1 sec. | <25**% | 4 sec. | 4% | 300[1] | 54'5'' | 7'11'' | 12'3'' | 150,000 | |
| <1 sec. | <25% | 3 sec. | 3% | 135 | 25' | 6'6'' | 9' | 40,000 | |
| <1 sec. | <25% | 3 sec. | 3% | 60-80 | 14' | 4'6'' | 5'7'' | 11,500 | |

# GAS TURBINE GENERATING SET SPECIFICATIONS

| Model | Normal Rating kW At ISO Condition 15°C, 59°F Sea Level | Service E-Emergency S-Standby | Heat Rate Normal Rating | Type Fuel L-Liquid G-Gas | Type Starter P-Pneumatic E-Electric H-Hydraulic | Location-Enclosure I-Indoors O-Outdoors | Noise Level dB (A) 10M Distance | Unit is S-Stationary T-Transportable |
|---|---|---|---|---|---|---|---|---|
| **Stewart & Stevenson, Gas Turbine Division, P.O. Box 1637, Houston, Texas 77001 U.S.A.** | | | | | | | | |
| | | | (Btu/kW-hr) | | | | | |
| GT500-831 | 560 | E,S | 15,800 | L,G | P,E | Either | 85 | S,T |
| GT4000-501K | 3,820 | E,S | 12,040 | L,G | P,H | Either | 85 | S,T |
| GT5000-501K5 | 4,200 | E,S | 10,790 | L,G | P,H | Either | 85 | S,T |
| GT7000-570K | 5,050 | E,S | 12,070 | L,G | P,H | Either | 85 | S,T |
| GT8000-571K | 5,850 | E,S | 10,620 | L,G | P,H | Either | 85 | S,T |
| GT30000-LM2500 | 23,660 | E,S | 9,530 | L,G | P,H | Either | 85 | S |
| **Sulzer Brothers Ltd., Thermal Turbomachinery, CH-8401 Winterthur, Switzerland** | | | | | | | | |
| | | | (Btu/kW-hr) | | | | | |
| 3 | 5,800 | E,S | 12,640 | L,G | P,E,H | I,O | 80* | S,T |
| 7 | 10,700 | E,S | 13,430 | L,G, | P,E,H | I,O | 80* | S |
| 10 | 20,700 | E,S | 10,310 | L,G | P,E | I,O | 80* | S |
| PRIMO12 | 12,200 | E,S | 12,060 | L,G | P,E | I,O | 80* | S,T |
| PRIMO14 | 15,500 | E,S | 11,680 | L,G | P,E | I,O | 80* | S,T |
| PRIMO15 | 16,200 | E,S | 11,680 | L,G | P,E | I,O | 80* | S,T |
| **Turbomeca, Bordes 64320, Bizanos, France** | | | | | | | | |
| | (kW) | | (Btu/kW-hr) | | | | | |
| Astazou IV | 340 | E,S | 18,100 | L,G | E | I,O | [1] | S,T |
| Bastan VI | 640 | E,S | 17,150 | L,G | E | I,O | [1] | S,T |
| Bastan VII | 990 | E,S | 14,550 | L,G | E | I,O | [1] | S,T |
| Bi-Bastan VI | 1225 | E,S | 17,500 | L,G | E | I,O | [1] | S,T |
| Bi-Bastan VII | 1885 | E,S | 14,900 | L,G | E | I,O | [1] | S,T |

*With enclosure; can be adapted to customer special request.

**MAJOR OFFICES WORLDWIDE**: Belgium: B-1040 Bruxelles, S.A. Sulzer Belgium N.V., 92 Square Plasky, tel. 0032/2-7349606 0032/2-7343161, telex. 21039 Sulzb • Germany: D-7980 Ravensburg, Escher Wyss GmbH, Postfach 1380, tel. 0049/751-831, telex. 0732901 • France: F-75300 Paris-Brune, Compagnie de Construction Mecanique Sulzer, Cedex 59, tel. 5392244, telex ccmsf2009371 • Great Britain: Farnborough (Hampshire) Sulzer Bros (UK) Ltd., tel. 0044/252-44311, telex, 858771/2 • Italy: 1-20122 Milano, Co Preflo-Escher Wyss S.p.A., Via San Senatore, 6/3, tel. 898644 800544 806337 806338, telex. 334 050 dpewmil • Norway: Oslo 2, Sulzer Brothers Norge A/S, Riddervoldsgate 7, tel. 447970, telex. 11619 Sulzon • South Africa: 2000 Johannesburg, Sulzer Bros. (South Africa) Ltd., P. O. Box 930, tel. 618-4125, telex. 8-0441 Sa • Libya: Tripoli (Libya), Universal Company for Libyan Development, P. O. Box 2395, tel. 32017, telex. 38227 • Canada: Pointe-Claire Dorval 700 (Quebec) Sulzer Bros. (Canada) Ltd., P. O. Box 210, tel. 001/514-695-8320, telex. 05-821577 • U.S.A.: 200 Park Avenue, Sulzer Bros. Inc., New York, New York 10017, tel. 001/212-949-0999, telex wui 620 219 and domestic 127297 • Brasil: 20000-Rio de Janeiro/RJ (Brasil) Sulzer do Brasil S.A., Caixa postal 2435-ZC-00, tel. 263-5577, telex. 2121540 Subr br • Mexico: Mexico 1, D.F., Sulzer Hermanos S.A., Apartado postal M-7103, tel. 546 8873, telex. 017-73928 • Iran: Tehran, 15, (Iran) Sulzer Iran, P. O. Box 2842, tel. 83 8010-19, telex 215 556 Sulzir • Japan: Tokyo 100, Sulzer Brothers (Japan) Ltd., C.P.O. Box 147, tel. 03-242-1551, telex. 22483 Sulzer j • Australia: North Sydney, NSW 2060 Sulzer Australia Pty. Limited, P. O. Box 362, tel. 9296900, telex. 22307.

[1]As requested

# GAS TURBINE GENERATING SET SPECIFICATIONS

| VOLTAGE REGULATION | | FREQUENCY CONTROL | | | DIMENSIONS | | | | |
|---|---|---|---|---|---|---|---|---|---|
| Transient Recovery with Load Change | Excursion at 100% Step Load Change | Transient Recovery with Load Change | Excursion at 100% Step Load Change | Time to get on line —Seconds | Length | Width | Height | Weight | Comments |
| | | | | | (Feet) | (Feet) | (Feet) | (Pounds) | |
| 1 sec. | 19% | 1.5 sec. | 3%@25% | 10 | 12.5 | 5 | 5.1 | 9,000 | |
| 1 sec. | 19% | 1.5 sec. | 2%@25% | 45 | 26.2 | 8 | 8.2 | 55,000 | |
| 1 sec. | 19% | 1.5 sec. | 2%@25% | 45 | 26.2 | 8 | 8.2 | 60,000 | |
| 1 sec. | 19% | 1.5 sec. | 3%@25% | 60 | 28.0 | 8 | 10.0 | 650,000 | |
| 1 sec. | 19% | 1.5 sec. | 3%@25% | 60 | 28.0 | 8 | 10.0 | 70,000 | |
| 1 sec. | 19% | 1.5 sec. | 3%@25% | 60 | 71 | 14 | 12.0 | 200,000 | |
| | | | | | (meters) | (meters) | (meters) | (Pounds) | |
| 0.3 sec. | ≤20% | ≤12'' | ≤10% | 180 | 7.35 | 3.1 | 2.69 | 60,000 | |
| 0.3 sec. | ≤20% | ≤10'' | ≤8% | 180 | 11.7 | 3.58 | 3.81 | 138.000 | |
| — | — | — | — | — | 8.0 | 3.15 | 3.4 | 61,700 | |
| 0.3 sec. | ≤20% | ≤10'' | ≤10% | 120 | 12.28 | 4.2 | 4.0 | 62,000 | |
| 0.3 sec. | ≤20% | ≤10'' | ≤10% | 120 | 12.28 | 4.2 | 4.0 | 62,000 | |
| 0.3 sec. | ≤20% | ≤10'' | ≤10% | 120 | 12.28 | 4.2 | 4.0 | 62,000 | |
| | | | | | (meters) | (meters) | (meters) | (Kg) | |
| <1 sec. | <20% | 1 sec. | 3% | <60 sec. | 3.0 | 1.2 | 1.7 | 3,700 | All ratings at gen terminals. All weights & dimensions for packaged set without duct & silencer. |
| <1 sec. | <20% | 1 sec. | 1% | <60 sec. | 3.8 | 1.5 | 1.7 | 4,000 | |
| <1 sec. | <20% | 1 sec. | 1% | <60 sec. | 4.4 | 1.5 | 1.7 | 4,500 | |
| <1 sec. | <20% | 1 sec. | 1% | <60 sec. | 5.8 | 1.8 | 2.1 | 8,900 | |
| <1 sec. | <20% | 1 sec. | 1% | <60 sec. | 6.0 | 1.8 | 2.1 | 10,500 | |

## TURBO COMPRESSOR SPECIFICATIONS

| MODEL | INLET VOLUME M³ OR FT³ | DISCHARGE PRESSURE $P_2$(BAR) | MAXIMUM NUMBER OF STAGES | RPM | H-HORIZONTAL V-VERTICAL SPLIT CASING | TYPE OF SEALS L=LABYRINTH O=OIL |
|---|---|---|---|---|---|---|
| **Allis-Chalmers Corporation, Compressor & Custom Pump Division, P. O. Box 512, Milwaukee, Wisconsin 53201 U.S.A.** | | | | | | |
| V, VS | 170,000 | 48 | — | — | H | L & O |
| VA | 1,000,000 | 10 | — | — | H | L |
| VC | 155,000 | 10 | — | — | H | L |
| VG | 72,000 | 25 | — | — | V | L |
| DH | 50,000 | 90 | — | — | V | O |
| VH | 75,000 | 350 | — | — | V | O |
| VT,VTS,VTZ | 14,000 | 35 | — | — | V | L & O |
| D,E | 200,000 | 2 | — | — | V | L |
| **Alsthom-Atlantique, Etablissement de La Courneuve, 141 Rue Rateau, 93123 La Courneuve, France** | | | | | | |
| | (M³) | | | | | |
| CMS 32 to | 1,500 | 35 | 9 | 12000 | H | L & O |
| CMS 100 | 7,500 | 25 | 8 | 4800 | H | L & O |
| CMR 42 to | 8,000 | 11 | 6 | 16000 | H | L |
| CMR 66 | 30,000 | 10 | 6 | 9000 | H | L |
| CBS 32 to | 1,500 | 200 | 9 | 12000 | V | O |
| CBS 80 | 45,000 | 150 | 8 | 7000 | V | [illegible] |
| SM | 3,000 to 100,000 | 10 | 214 | 40000 | V | L&O |
| **Alturdyne, 8050 Armour, San Diego, California 92111 U.S.A.** | | | | | | |
| GTC | 2,000-4,000 | 2-4 | 1 | 61,000 | H | L |
| **Atlas-Copco Turbonetics, Inc., 20 School Road, Voorheesville, NY 12186** | | | | | | |
| | | (psia) | | (Impeller/Driver) | | D = Dry Face |
| SC-8 | 1,000 ft³/min<br>1,700 m³/hr | 45 | 1 | 52,000/3,600 | V | L,D |
| SC-10 | 2,000 ft³<br>3,400 m³ | 45 | 1 | 33,000/3,600 | V | L,D |
| SC-14 | 5,000 ft³<br>8,500 m³ | 45 | 1 | 22,000/3,600 | V | L,D |
| SC-20 | 10,000 ft³<br>17,000 m³ | 45 | 1 | 14,000/3,600 | V | L,D |
| SC-28 | 20,000 ft³<br>34,000 m³ | 45 | 1 | 10,000/3,600 | V | L,D |
| SC-40 | 70,000 ft³<br>119,000 m³ | 45 | 1 | 8,000/1,800 | V | L,D |
| TP-6/6 | 1,000 ft³/min<br>1,700 m³/hr | 110 | 2 | 52,000/3,600 | V | L,D |
| TP-10/10 | 3,000 ft³<br>5,100 m³ | 110 | 2 | 33,000/3,600 | V | L,D |
| TP-14/14 | 10,000 ft³<br>17,000 m³ | 110 | 2 | 22,000/3,600 | V | L,D |

Units are capable of handling working pressures to 500 PSI (34 Bar).

# TURBO COMPRESSOR SPECIFICATIONS

| TYPE OF GAS COMPRESSORS CAN HANDLE | TYPE OF DESIGN C=CENTRIFUGAL A=AXIAL | DIMENSIONS | | | WEIGHT KG OR LB. | COMMENTS |
|---|---|---|---|---|---|---|
| | | L | W | H | | |
| Various | C | — | — | — | — | — |
| Air | A | — | — | — | — | — |
| Air | C | — | — | — | — | — |
| Air | C | — | — | — | — | — |
| Various | C | — | — | — | — | — |
| Various | C | — | — | — | — | — |
| Various | C | — | — | — | — | — |
| Air & Others | C | — | — | — | — | — |
| | | (mm) | (mm) | (mm) | (kg) | |
| Air or gas or | Centrifugal | 1725 | 1030 | 945 | 4900 | Dimensions in meters |
| hydrocarbons | Centrifugal | 3400 | 2300 | 2250 | 22500 | |
| Air | Centrifugal | 2050 | 4900 | 2600 | 22600 | |
| Air | Centrifugal | 3600 | 600 | 3450 | 50000 | |
| Gas | Centrifugal | 1585 | 1080 | 1050 | 4900 | |
| Gas | Centrifugal | 3450 | 2300 | 2250 | 22500 | |
| Air-Steam-Gas | Centrifugal | 2500 | 1500 | 2000 | 15000 | |
| Various | C | 6 | 10 | 12 | 25 | — |
| | | | (Inches) | | | |
| Air, Toxic, Corrosive | C | 6 | 3 | 4 | | |
| Air, Toxic, Corrosive | C | 6 | 3 | 4 | | |
| Air, Toxic, Corrosive | C | 6 | 3 | 4 | | |
| Air, Toxic, Corrosive | C | 10 | 4 | 5 | | |
| Air, Toxic, Corrosive | C | 10 | 4 | 5 | | |
| Air, Toxic, Corrosive | C | 10 | 4 | 5 | | |
| Air, Toxic, Corrosive | C | 10 | 4 | 5 | | |
| Air, Toxic, Corrosive | C | 10 | 4 | 5 | | |
| Air, Toxic, Corrosive | C | 10 | 4 | 5 | | |

## TURBO COMPRESSOR SPECIFICATIONS

| MODEL | INLET VOLUME M³ OR FT³ | DISCHARGE PRESSURE $P_2$(BAR) | MAXIMUM NUMBER OF STAGES | RPM | H-HORIZONTAL V-VERTICAL SPLIT CASING | TYPE OF SEALS L=LABYRINTH O=OIL |
|---|---|---|---|---|---|---|
| **Borsig GmbH, Gruppe Deutsche Babcock, P. O. Box 1000 Berlin 27, West Germany** | | | | | | |
| | (m³/hr) | (max) | | (max) | | [1] |
| GA 355 | 1,500-13,800 | 60 | 8 | 24,000 | H | L + O |
| GA 425 | max 19,700 | 60 | 8 | 16,800 | H | L + O |
| GA 500 | max 27,800 | 55 | 9 | 14,000 | H | L + O |
| GA 600 | max 38,600 | 50 | 10 | 12,000 | H | L + O |
| GA 710 | max 55,300 | 45 | 10 | 10,000 | H | L + O |
| GA 850 | max 78,800 | 37 | 10 | 8,400 | H | L + O |
| GA 1000 | max 111,000 | 30 | 10 | 7,000 | H | L + O |
| GA 1180 | max 154,500 | 14 | 10 | 6,000 | H | L + O |
| GA 1400 | max 221,000 | 8 | 10 | 5,000 | H | L + O |
| GC 300 | 1,000-6,600 | 360 | 8 | 24,000 | V | L + O |
| GC 355 | max 9,500 | 300 | 8 | 20,000 | V | L + O |
| GC 425 | max 13,600 | 240 | 9 | 16,800 | V | L + O |
| GC 500 | max 19,100 | 170 | 10 | 14,000 | V | L + O |
| GC 600 | max 26,600 | 70 | 10 | 12,000 | V | L + O |
| GC 710 | max 38,100 | 70 | 10 | 10,000 | V | L + O |
| GC 850 | max 54,300 | 50 | 10 | 8,400 | V | L + O |
| GC 1000 | max 76,600 | 50 | 10 | 7,000 | V | L + O |
| GK 223 | 2,700-5,200 | 10 | 3 | 3,000/3,600[2] | V | L |
| GK 313 | max 9,900 | 10 | 3 | 3,000/3,600 | V | L |
| GK 453 | max 19,600 | 10 | 3 | 3,000/3,600 | V | L |
| GK 633 | max 38,700 | 10 | 3 | 1,500/1,800 | H | L |
| GK 903 | max 85,000 | 10 | 3 | 1,500/1,800 | H | L |
| GK 454 | max 19,600 | 21 | 4 | 3,000/3,600 | V | L |
| GK 634 | max 38,700 | 21 | 4 | 1,500/1,800 | H | L |
| GK 904 | max 85,000 | 21 | 4 | 1,500/1,800 | H | L |
| GKS 315 | max 13,500 | 5 | 1 | 24,000 | V | L & C |
| GKS 450 | max 27,500 | 5 | 1 | 17,000 | V | L & C |
| GKS 630 | max 55,500 | 5 | 1 | 12,000 | V | L & C |
| GKS 900 | max 110,000 | 5 | 1 | 8,500 | V | L & C |
| GKS 1250 | max 225,000 | 5 | 1 | 6,000 | V | L & C |
| GKS 1800 | max 460,000 | 5 | 1 | 4,300 | V | L & C |
| VA | 85,000-540,000 | 5.5 | 14 | | H | L |
| VC | 25,000-210,000 | max 11 | 5 | | H | L |
| VS | 7,000-250,000 | max 5 | 1 | max 10,000 | V | L |
| VG | 15,000-63,000 | 6-26 | 4 | 1,500/1,800(8) | H | L |

[1]Oil seals depending on gas. [2]Synchronous motor speed. [3]Without motor. [4]Based on 6-stage compressors
[5]Also available: Type GD with external seals and pressurized lube oil system.

| MODEL | INLET VOLUME M³ OR FT³ | DISCHARGE PRESSURE $P_2$(BAR) | MAXIMUM NUMBER OF STAGES | RPM | H-HORIZONTAL V-VERTICAL SPLIT CASING | TYPE OF SEALS L=LABYRINTH O=OIL |
|---|---|---|---|---|---|---|
| **Cooper Energy Services, North Sandusky Street, Mount Vernon, Ohio 43050 U.S.A.** | | | | | | |
| | (acfm) | (psia) | | (nominal) | | |
| RFA-36 | 50,000 | 1,200/1,800 | 1 | 7,000 | V | O |
| RCBB-14 | 6,000 | 1,500 | 3 | 11,500 | V | O |
| RFBB-20 | 12,500 | 1,500 | 4 | 11,000 | V | O |
| RFBB-24 | 19,000 | 1,200 | 4 | 7,000 | V | O |
| RFBB-30 | 26,000 | 1,200/1,440 | 4 | 7,000 | V | O |
| RFB-36 | 40,000 | 1,200 | 1 | 7,000 | V | O |
| RFBB-36 | 40,000 | 1,200/1,440 | 5 | 7,000 | V | O |
| RFB-42 | 54,000 | 1,200 | 1 | 7,000 | V | O |
| RFBB-42 | 54,000 | 1,200/1,440 | 5 | 7,000 | V | O |

# TURBO COMPRESSOR SPECIFICATIONS

| TYPE OF GAS COMPRESSORS CAN HANDLE | TYPE OF DESIGN C=CENTRIFUGAL A=AXIAL | DIMENSIONS L | W | H | WEIGHT KG OR LB. | COMMENTS |
|---|---|---|---|---|---|---|
| | | (*) | (mm*) | (*) | (Kg*) | |
| | C | 1,600 | 1,100 | 950 | 4,300 | |
| | C | 1,800 | 1,280 | 1,120 | 5,500 | |
| | C | 2,100 | 1,560 | 1,340 | 7,600 | |
| Air, $N_2$, $CO_2$, | C | 2,400 | 1,860 | 1,600 | 14,500 | Multi stage single flow design (Type GA). |
| Hydrocarbons | C | 2,800 | 2,160 | 1,900 | 21,000 | |
| $NH_3$, $Cl_2$, Freon | C | 3,200 | 2,610 | 2,200 | 35,000 | Models 710 and up also available as |
| | C | 3,700 | 3,050 | 2,700 | 52,000 | double-flow design (Type GB)* |
| | C | | 3,600 | 3,200 | | |
| | C | | 4,400 | 3,800 | | |
| | C | | 1,000 | 850 | | |
| | C | 1,600 | 1,100 | 1,000 | 5,100 | |
| | C | 1,800 | 1,220 | 1,190 | 6,600 | Multi stage barrel compressors |
| $H_2$, Hydrocarbons, | C | 2,100 | 1,400 | 1,400 | 9,100 | (Type GC). Also available: |
| Air, $CO_2$ | C | 2,400 | 1,580 | 1,650 | 15,200 | Light compressors |
| | C | 2,800 | 1,760 | 1,900 | 23,200 | series (Type KC) and single |
| | C | 3,200 | | 2,300 | 33,000 | stage barrels (Type GCS) |
| | C | 3,700 | | 2,800 | 60,000 | |
| | C | 4,300 | 1,520 | 2,050 | 7,600³ | |
| Plant air | C | 5,200 | 1,680 | 2,500 | 9,300³ | |
| Instrument air | C | 6,400 | 2,020 | 3,050 | 17,400³ | Integral gear compressors "Bornado" |
| Process air | C | 9,000 | 3,220 | 3,350 | 27,100³ | |
| Nitrogen | C | 10,000 | 4,600 | 4,000 | 67,500 | |
| | C | 6,400 | 2,020 | 3,050 | 22,400 | |
| | C | 8,000 | 3,220 | 3,350 | 34,100 | |
| | C | 10,000 | 4,600 | 4,000 | 75,000 | |
| | C | 1,990 | | 1,580 | 3,200 | |
| | C | 2,030 | | 1,750 | 3,700 | |
| Steam, Air, | C | 2,070 | | 2,000 | 4,600 | Single stage |
| $CO_2$, others | C | 2,190 | | 2,710 | 6,100 | |
| | C | | | | | |
| | C | | | | | |
| Air | A | 2,840-4,320(7) | 2,250-3,810(7) | 2,800 4,320(7) | 17,000 58,000(7) | Axial |
| Air | C | 2,900-3,910 | 3,780-5,820 | 4,190-6,070 | 38,000-120,000 | Internally cooled |
| Air, Steam | C | 1,100-2,000 | 1,800-4,800 | 1,800-4,100 | 3,000-16,000 | Single stage |
| Air | C | 2,920-5,200(9) | 3,400-5,400(9) | 2,770-5,350(9) | 23,000-57,000(9) | Integral gear |
| Models VA, VC, VS, VG are manufactured by Borsig under License of Allis-Chalmers/USA. | | | | | | |
| | | (in) | (in) | (in)' | (lb) | |
| natural | C/A mixed flow | 80 | 122 | 140 | 60,000 | Pipeline type |
| natural | C | 54 | 80 | 100 | 9,500 | Pipeline type |
| natural | C | 60 | 96 | 106 | 22,000 | Pipeline type |
| natural | C | 103 | 129 | 140 | 60,000 | Pipeline type |
| natural | C | 103 | 129 | 140 | 61,000 | Pipeline type |
| natural | C | 110 | 152 | 144 | 95,000 | Pipeline type |
| natural | C | 110 | 152 | 144 | 105,000 | Pipeline type |
| natural | C | 122 | 152 | 144 | 11,000 | Pipeline type |
| natural | C | 122 | 152 | 144 | 12,500 | Pipeline type |

# TURBO COMPRESSOR SPECIFICATIONS

| MODEL | INLET VOLUME M³ OR FT³ | DISCHARGE PRESSURE $P_2$(BAR) | MAXIMUM NUMBER OF STAGES | RPM | H-HORIZONTAL V-VERTICAL SPLIT CASING | TYPE OF SEALS L=LABYRINTH O=OIL |
|---|---|---|---|---|---|---|
| **Cooper Energy Services** (Continued) | | | | | | |
| RBB | 6,000 | 6,500 | 9 | 12,000 | V | O |
| RCB | 13,500 | 2,500 | 9 | 11,000 | V | O |
| RDB | 17,500 | 2,000 | 8 | 7,500 | V | O |
| REB | 25,000 | 1,200 | 8 | 6,500 | V | O |

¹Top of seal oil tank-pipeline type.
**Cooper Energy Servies**, Mount Vernon, Ohio 43050. **Contact:** R. T. Harnsberger, Phone: (614) 397-0121.

**Creusot-Loire, 15 Rue Pasquier, 75383 Paris Cedex 08, France**

| MODEL | INLET VOLUME | DISCHARGE PRESSURE | STAGES | RPM | CASING | SEALS |
|---|---|---|---|---|---|---|
| RB-S | 11,800 | 50 | 9 | 13,000 | H | L,O |
| RC-S | 24,800 | 42 | 9 | 11,800 | H | L,O |
| RD-S | 35,800 | 35 | 9 | 8,500 | H | L,O |
| RE-S | 84,000 | 30 | 9 | 6,500 | H | L,O |
| RF-S | 115,000 | 23 | 9 | 5,500 | H | L,O |
| RG-S | 242,000 | 20 | 8 | 4,000 | H | L,O |
| RA-B | 4,000 | 400 | 9 | 16,500 | V | L,O |
| RB-B | 13,300 | 550 | 10 | 13,000 | V | L,O |
| RC-B | 21,700 | 200 | 10 | 11,800 | V | L,O |
| RD-B | 39,800 | 140 | 9 | 8,500 | V | L,O |
| | (m³/hour) | | | (Nominal) | | |
| RB-B | 10,000 | 90 | 2 | 13,500 | V | O |
| RFA 36* | 54,400 | 125 | 1 | 5,000 | V | O |
| RFB B24* | 26,400 | 100 | 4 | 5,000 | V | O |
| RFB B30* | 42,500 | 100 | 4 | 5,000 | V | O |
| RFB 36* | 54,400 | 80 | 1 | 5,000 | V | O |
| RFB B36* | 54,400 | 100 | 4 | 5,000 | V | O |
| RFB 42* | 76,500 | 80 | 1 | 5,000 | V | O |
| RFB B42* | 76,500 | 100 | 4 | 5,000 | V | O |
| RCB B14* | 8,800 | 105 | 4 | 9,000 | V | O |

*Under American license.

**Dresser Clark, Division Dresser Industries, Box 560, Olean, New York 14760 U.S.A.**

| MODEL | INLET VOLUME | DISCHARGE PRESSURE | STAGES | RPM | CASING | SEALS |
|---|---|---|---|---|---|---|
| | FT³/min | psi | | | | |
| B Line | to over 350000 | to over 10000 | 10 | — | V | O |
| M Line | to over 350000 | to over 800 | 10 | — | H | L or O |
| H Line | to over 40000 | to over 500 | 4 | — | H | L |
| P Line | to 55000 | to 1800 | 6 | — | V | O |
| Isopac Line | to 140000 | to 500 | 6 | — | H | L |
| Axial Line | to over 500000 | press. ratios roll | 16 axial | — | H | L |

**Elliott Company, North Fourth Street, Jeanette, Pennsylvania 15644 U.S.A.**

| MODEL | INLET VOLUME | DISCHARGE PRESSURE | STAGES | RPM | CASING | SEALS |
|---|---|---|---|---|---|---|
| | (ft³/min) | | | | | |
| 110M | 190M | 3 | 8 | 2,600 | H | L, O |
| 103M | 160M | 3 | 8 | 2,800 | H | L, O |
| 88M | 135M | 22.4 | 8 | 3,160 | H | L, O |
| 70M | 84M | 22.4 | 8 | 4,000 | H | L, O |
| 60M | 58M | 22.4 | 8 | 4,860 | H | L, O |
| 46M | 34M | 43 | 9 | 6,300 | H | L, O |
| 38M | 22M | 43 | 9 | 7,725 | H | L, O |
| 29M | 7.5M | 52 | 10 | 11,500 | H | L, O |
| 88MB | 135M | 55.2 | 8 | 3,160 | V | O |
| 70MB | 84M | 55.2 | 8 | 4,000 | V | O |
| 60MB | 58M | 55.2 | 8 | 4,860 | V | O |
| 46MB | 34M | 82.7 | 9 | 6,300 | V | O |
| 38MB | 22M | 103 | 9 | 7,725 | V | O |
| 32MB | 8M | 690 | 10 | 8,300 | V | O |
| 25MB | 5.5M | 690 | 12 | 10,000 | V | O |
| 20MB | 3.6M | 690 | 12 | 12,400 | V | O |
| 15MB | 2.4M | 690 | 12 | 15,300 | V | O |

## TURBO COMPRESSOR SPECIFICATIONS

| TYPE OF GAS COMPRESSORS CAN HANDLE | TYPE OF DESIGN C=CENTRIFUGAL A=AXIAL | DIMENSIONS | | | WEIGHT KG OR LB. | COMMENTS |
|---|---|---|---|---|---|---|
| | | L | W | H | | |
| natural | C | 66 | 54 | 54 | 25,000 | barrel compressors |
| natural | C | 90 | 62 | 62 | 45,000 | " |
| natural | C | 89 | 75 | 76 | 77,000 | " |
| natural | C | 138 | 90 | 92 | 135,000 | " |
| Cooper Energy Services Ltd. 6889 Rexwood Road, Mississauga, Ontario, Canada. Contact: R. W. Thompson. Phone (416) 678-2030. Cooper Energy Services International Inc. 46 Pall Mall Road, London, England. Contact: K. A. Thilman. Phone: 01-839-5161. | | | | | | |
| | | (mm) | (mm) | (mm) | (kg) | |
| All Types | C | 1,840 | 1,120 | 1,120 | 3,800 | |
| All Types | C | 2,180 | 1,750 | 1,520 | 6,000 | |
| All Types | C | 3,220 | 2,000 | 1,580 | 13,200 | |
| All Types | C | 3,640 | 2,550 | 2,120 | 35,200 | |
| All Types | C | 4,660 | 2,900 | 2,655 | 56,000 | |
| All Types | C | 5,300 | 3,750 | 3,500 | 87,500 | |
| All Types | C | 1,310 | 1,600 | 1,280 | 9,550 | min. inlet vol. 150m³/h |
| All Types | C | 1,740 | 1,530 | 1,700 | 7,000 | |
| All Types | C | 2,470 | 2,150 | 1,965 | 17,500 | |
| All Types | C | 3,020 | 2,500 | 2,620 | 22,100 | |
| Natural Gas | C | 1,640 | 1,420 | 1,700 | 5,500 | Pipeline Type |
| Natural Gas | C | 3,050 | 2,135 | 2,745 | 27,300 | Pipeline Type Cooper Bessemer License |
| Natural Gas | C | 3,050 | 1,525 | 1,830 | 27,300 | Pipeline Type Cooper Bessemer License |
| Natural Gas | C | 3,355 | 1,830 | 2,135 | 27,700 | Pipeline Type Cooper Bessemer License |
| Natural Gas | C | 3,965 | 2,135 | 2,135 | 43,200 | Pipeline Type Cooper Bessemer License |
| Natural Gas | C | 3,965 | 2,135 | 2,135 | 47,700 | Pipeline Type Cooper Bessemer License |
| Natural Gas | C | 3,965 | 2,440 | 2,135 | 50,000 | Pipeline Type Cooper Bessemer License |
| Natural Gas | C | 3,965 | 2,440 | 2,135 | 56,800 | Pipeline Type Cooper Bessemer License |
| Natural Gas | C | 2,135 | 1,525 | 1,525 | 4,300 | Pipeline Type Cooper Bessemer License |
| any | C | — | — | — | — | |
| any | C | — | — | — | — | |
| air, oxygen | C | — | — | — | — | |
| natural gas | C | — | — | — | — | |
| air | C | — | — | — | — | |
| air & other gases | A and A + C | — | — | — | — | |
| | | (in.) | (in.) | (in.) | | |
| All gases. Available | C | 258 | 176 | 177 | 315,000 | All M & P models are available for gas |
| materials of | C | 236 | 148 | 156 | 263,000 | turbine steam turbine, or M & G drives. |
| construction may | C | 211 | 131 | 142 | 251,000 | " |
| limit applications | C | 178 | 128 | 120 | 131,000 | " |
| " | C | 165 | 103 | 92 | 75,000 | " |
| " | C | 141 | 79 | 71 | 55,000 | " |
| " | C | 107 | 83 | 68 | 33,000 | " |
| " | C | 92 | 58 | 61 | 16,000 | " |
| " | C | 292 | 160 | 146 | 400,000 | " |
| " | C | 247 | 142 | 134 | 246,000 | " |
| " | C | 195 | 134 | 125 | 147,000 | " |
| " | C | 160 | 118 | 92 | 80,000 | " |
| " | C | 152 | 90 | 86 | 83,000 | " |
| " | C | 119 | 88 | 86 | 107,000 | " |
| " | C | 134 | 70 | 83 | 91,000 | " |
| " | C | 70 | 47 | 49 | 21,000 | " |
| " | C | 81 | 41 | 45 | 16,000 | " |

# TURBO COMPRESSOR SPECIFICATIONS

| MODEL | INLET VOLUME M³ OR FT³ | DISCHARGE PRESSURE $P_2$(BAR) | MAXIMUM NUMBER OF STAGES | RPM | H-HORIZONTAL V-VERTICAL SPLIT CASING | TYPE OF SEALS L=LABYRINTH O=OIL |
|---|---|---|---|---|---|---|
| **Elliott** (Continued) | | | | | | |
| | range | | | *range | | |
| P | 2M-210M | 5.2 | 1 | 2M-13M | V | L, 0 |
| | range | | | range | | |
| PH | 1M-50M | 48.3 | 1 | 1.5M-15M | V | L, 0 |
| 110SA2 | 1,100 | 7.6 | 2 | 3,600 | H | Carbon |
| 130SA2 | 1,300 | 8.0 | 2 | 3,600 | H | Carbon |
| 150SA2 | 1,500 | 2.1 | 2 | 3,600 | H | Carbon |
| 70SA3 | 635 | 6.9 | 3 | 3,600 | H | Carbon |
| 90SA3 | 925 | 8.7 | 3 | 3,600 | H | Carbon |
| 110SA3 | 1,100 | 8.7 | 3 | 3,600 | H | Carbon |
| 130SA3 | 1,300 | 8.7 | 3 | 3,600 | H | Carbon |
| 150SA3 | 1,500 | 8.7 | 3 | 3,600 | H | Carbon |
| 150DA3 | 1,500 | 8.7 | 3 | 3,600 | H | Carbon |
| 180DA3 | 1,750 | 8.7 | 3 | 3,600 | H | Carbon |
| 220DA3 | 2,120 | 8.7 | 3 | 3,600 | H | Carbon |
| 270DA3 | 2,620 | 8.7 | 3 | 3,600 | H | Carbon |
| 310DA3 | 3,050 | 8.7 | 3 | 3,600 | H | Carbon |
| 360DA3 | 3,500 | 8.7 | 3 | 3,600 | H | Carbon, L |
| 410DA3 | 4,050 | 8.7 | 3 | 3,600 | H | Carbon, L |
| 460DA3 | 4,550 | 8.7 | 3 | 3,600 | H | Carbon, L |
| 510DA3 | 5,000 | 8.7 | 3 | 3,600 | H | Carbon, L |
| 560DA3 | 5,500 | 8.7 | 3 | 3,600 | H | Carbon, L |
| 700DA3 | 7,000 | 8.7 | 3 | 3,600 | H | Carbon, L |
| 810DA3 | 8,100 | 8.7 | 3 | 3,600 | H | Carbon, L |
| 950DA3 | 9,350 | 8.7 | 3 | 3,600 | H | Carbon, L |
| 1000DA3 | 9,900 | 8.7 | 3 | 1,800 | H | Carbon, L |
| 1120DA3 | 11,200 | 8.7 | 3 | 1,800 | H | Carbon, L |
| 1360DA3 | 13,600 | 8.7 | 3 | 1,800 | H | Carbon, L |
| 1600DA3 | 16,000 | 8.7 | 3 | 1,800 | H | Carbon, L |
| 21A | 40-100M | 5.5 | 14 | 6,100 | H | L,0 |
| 26A | 70-141M | 5.5 | 14 | 5,100 | H | L,0 |
| 32A | 125-202M | 5.5 | 14 | 4,270 | H | L,0 |
| 37A | 190-285M | 5.5 | 14 | 3,600 | H | L,0 |
| 44A | 250-396M | 5.5 | 14 | 3,050 | H | L,0 |

*Length based on max no. of stages. **Length based on stage unit.

Major Offices: Elliott Company, North Fourth Street, Jeannette, PA 15644, Raymond King, Jim Ferrero, Tel: (412) 527-2811, Telex: 86-6643 • Elliott Company, P. O. Box 1008, 1071 Bristol Rd., Mountainside, NJ 07092, Clinton M. Seeman, Tel: (201) 233-9600, Telex: 138-365 • Elliott Company, 6100 N. Pulaski Rd., Chicago 60646, William J. Lyons, Tel: (312) 267-7800, Telex: 910-221-5607 • Elliott Company, P. O. Box 4281, Houston, Texas 77210, Ronald L. Wheeler, Tel: (713) 933-8700, Telex: 910-881-1172 • Elliott Company, 16900 E.

**General Electric Company, Mechanical Drive Turbine Dept., 166 Boulder Drive, Fitchburg, MA 01420**

| MODEL | INLET VOLUME | DISCHARGE PRESSURE | MAXIMUM NUMBER OF STAGES | RPM | SPLIT CASING | TYPE OF SEALS |
|---|---|---|---|---|---|---|
| | (cfm) | (psig) | | | | |
| MC | To over 200,000 | 650 | 9 | 18,000 | H | L,0 |
| 3MC | To over 200,000 | 650 | 9 | 15,000 | H | L,0 |
| 2MC | To over 200,000 | 650 | 9 | 15,000 | H | L,0 |
| DMC | To over 200,000 | 150 | 8 | 6,000 | H | L,0 |
| BC | To over 60,000 | Over 10,000 | 8 | 18,000 | V | L,0 |
| 2BC | To over 60,000 | Over 10,000 | 8 | 18,000 | V | L,0 |
| DBC | To over 100,000 | 1,450 | 8 | 18,000 | V | L,0 |
| PC | To over 60,000 | 1,700 | 4 | 12,000 | V | 0 |
| SR | To over 140,000 | 435 | 5 | 1,500-3,600 | H | L,0 |

General Electric Co., Mechanical Drive Turbine Dept., 166 Boulder Drive, Fitchburg, MA 01420, Contact: G. E. Walker, Tel: 617-343-1000, Telex: 928442

**Hispano Suiza, Rue Du Capitaine Guynemer, B.P. 60, 92270 Bois Colombes, France**

| MODEL | INLET VOLUME | DISCHARGE PRESSURE | MAXIMUM NUMBER OF STAGES | RPM | SPLIT CASING | TYPE OF SEALS |
|---|---|---|---|---|---|---|
| | (m³/hr) | (max.) | | (max.) | | |
| CGN | 300,000 | 100 | 1 | 8,000 | Barrel | Carbon |
| HS 70A | 1,200,000 | | | | | |
| CGN | 170,000 | 100 | 2 | 8,000 | Barrel | Carbon |
| HS 71 | 900,000 | | | | | |

Hispano Suiza Inc. 10633 Shadow Wood Drive, Houston, Texas 77043 U.S.A. Tel: (713) 467 9612 — Tlx: 6868 127 HSINC UW.
Hispano Suiza De Brazil Equipamentos Ltda., Rua Da Gloria 344, Gloria 20241 Rio De Janeiro — Tel: (021) 242 40 04 — Tlx: 21 989 HS BR BR.

## TURBO COMPRESSOR SPECIFICATIONS

| TYPE OF GAS COMPRESSORS CAN HANDLE | TYPE OF DESIGN C=CENTRIFUGAL A=AXIAL | DIMENSIONS L | W | H | WEIGHT KG OR LB. | COMMENTS |
|---|---|---|---|---|---|---|
| " | C | 90 | 128 | 123 | 34,000 | " |
| " | C | 104 | 102 | 93 | 21,250 | " |
| air and all inert, | C | 90 | 55 | 100 | 7,400 | All SA and DA models are available |
| non-toxic gases. | C | 90 | 55 | 100 | 7,400 | for 50Hz motor and steam turbine drive |
| " | C | 90 | 55 | 100 | 7,400 | " |
| " | C | 90 | 55 | 100 | 7,400 | " |
| " | C | 90 | 55 | 100 | 7,400 | " |
| " | C | 90 | 55 | 100 | 7,400 | " |
| " | C | 90 | 55 | 100 | 7,400 | " |
| " | C | 90 | 55 | 100 | 7,400 | " |
| " | C | 117 | 63 | 100 | 9,200 | " |
| " | C | 117 | 63 | 100 | 9,200 | " |
| " | C | 117 | 63 | 100 | 9,200 | " |
| " | C | 117 | 63 | 100 | 9,200 | " |
| " | C | 117 | 63 | 100 | 9,200 | " |
| " | C | 130 | 95 | 104 | 8,800 | " |
| " | C | 130 | 95 | 104 | 9,000 | " |
| " | C | 130 | 95 | 104 | 9,000 | " |
| " | C | 130 | 95 | 104 | 9,100 | " |
| " | C | 182 | 118 | 127 | 12,600 | " |
| " | C | 182 | 118 | 127 | 12,800 | " |
| " | C | 182 | 118 | 127 | 13,200 | " |
| " | C | 182 | 118 | 127 | 13,200 | " |
| " | C | 230 | 130 | 128 | 28,300 | " |
| " | C | 230 | 130 | 128 | 28,400 | " |
| " | C | 230 | 130 | 154 | 30,900 | " |
| " | C | 230 | 130 | 154 | 31,600 | " |
| air, $N_2$ | A | 129** | 104 | 87 | 29,100 | |
| HC Mixture | A | 152 | 119 | 101 | 50,000 | |
| " | A | 182 | 130 | 119 | 85,000 | |
| " | A | 218 | 148 | 142 | 140,000 | |
| " | A | 260 | 180 | 169 | 230,000 | |

Chestnut St., P. O. Box 1112, City of Industry, CA 91749, Thomas H. Harris, Tel: (213) 965-1688, Telex: 910-584-4899 • Elliott Company-Japan Operations, P. O. Box 1112, 16900 E. Chestnut St., City of Industry, CA 91749, Robert D. Barbour, Tel: (213) 965-1688, Telex: 670-405 • Elliott Turbomachinery Ltd., Knightsbridge House, 197 Knightsbridge, London, SW7 1RB, Eugene Dickert, Tel: 01-489-8111, Telex: 914688 or 914689 • Elliott Turbomachinery B.V., Bucaillestraat 6, Voorburg (ZH) The Netherlands, Adriaan Jansen, Tel: (070) 873434, Telex: The Hague 31688

| TYPE OF GAS COMPRESSORS CAN HANDLE | TYPE OF DESIGN | L | W | H | WEIGHT | COMMENTS |
|---|---|---|---|---|---|---|
| Process | Centrifugal | All models produced | | | | In-line Multistage |
| and | Centrifugal | in a range of | | | | In-line w/side streams |
| Industrial | Centrifugal | sizes to suit | | | | Back-to-back (Inter-cooled) |
| gases | Centrifugal | the application | | | | Double-flow |
| Process/Industrial | Centrifugal | General Electric also manufactures | | | | In-line multistage |
| | Centrifugal | a broad range of gas turbines, | | | | Back-to-back (Inter-cooled) |
| Gases | Centrifugal | steam turbines, motors, and gears; | | | | Double-flow |
| Nat. gas/Process gas | Centrifugal | as well as auxiliary equipment to | | | | |
| Air/non-Toxic gas | Centrifugal | completely support compression | | | | Integrally geared, inter-cooled |
| | | services. | | | | motor driven Pkg. |

| TYPE OF GAS COMPRESSORS CAN HANDLE | TYPE OF DESIGN | L | W | H | WEIGHT | COMMENTS |
|---|---|---|---|---|---|---|
| | | (mm) | (mm) | (mm) | (Kg) | |
| Natural | C | 1,590 | 2,200 | 1,750 | 9,870 | Seal ensures tightness during shutdown |
| Natural | C | 1,590 | 2,200 | 1,750 | 9,920 | d* |

# TURBO COMPRESSOR SPECIFICATIONS

| MODEL | INLET VOLUME M³ OR FT³ | DISCHARGE PRESSURE $P_2$(BAR) | MAXIMUM NUMBER OF STAGES | RPM | H-HORIZONTAL V-VERTICAL SPLIT CASING | TYPE OF SEALS L=LABYRINTH O=OIL |
|---|---|---|---|---|---|---|
| **Hitachi, Ltd.**, Ohtemachi 2-6-2, Chyoda-Ku, Tokyo, Japan. | | | | | | |
| | (m³/hr) | | | | | |
| DH-45 | 26,000 | 5~10 | 4 | 14,000/17,200 | H | L |
| DH-50 | 32,000 | 5~10 | 4 | 12,600/15,300 | H | L |
| DH-56 | 39,000 | 5~10 | 4 | 11,300/13,600 | H | L |
| DH-63 | 51,000 | 5~10 | 4 | 10,000/12,200 | H | L |
| DH-71 | 64,000 | 5~10 | 4 | 8,900/10,900 | H | L |
| DH-80 | 82,000 | 5~10 | 4 | 7,900/9,700 | H | L |
| DH-90 | 107,000 | 5~10 | 4 | 7,000/8,600 | H | L |
| DH-100 | 132,000 | 5~10 | 4 | 6,300/7,600 | H | L |
| DH-112 | 170,000 | 5~10 | 4 | 5,600/6,800 | H | L |
| DH-125 | 210,000 | 5~7 | 4 | 4,800/5,800 | H | L |
| DH-140 | 265,000 | 5~7 | 4 | 4,300/5,200 | H | L |
| DH-160 | 345,000 | 5~7 | 4 | 3,700/4,700 | H | L |
| IMB | 10,000~200,000 | 5~100 | 6 for 1 casing | | H/V | L |
| MCL | 500~350,000 | 45 | 9 | 2,500~18,000 | H | L,O |
| 2MCL | 500~350,000 | 45 | 9 | 3,500~15,000 | H | L,O |
| 3MCL | 500~350,000 | 45 | 9 | 3,500~15,000 | H | L,O |
| BCL | 200~100,000 | 750 | 8 | 3,000~18,000 | V | O |
| 2BCL | 200~100,000 | 750 | 8 | 3,000~18,000 | V | O |
| PCL | 500~100,000 | 100 | 4 | 3,000~12,000 | V | O |
| **Ingersoll-Rand Co.**, Turbo Products Division, Phillipsburg, New Jersey 08865 U.S.A. | | | | | | |
| "Multistage" | (ft³/min.) | (psig) | | | | |
| 24 | 8,000 | 600 | 9 | 12,380 | H | L/O |
| 24B | 8,000 | 1,350 | 9 | 12,380 | V | L/O |
| 24HYD | 2,600 | 3,375 | 7 | 12,380 | V | O |
| 33 | 10,300 | 600 | 9 | 11,200 | H | L/O |
| 33B | 10,300 | 1,350 | 9 | 11,200 | V | L/O |
| 33HYD | 4,200 | 3,375 | 7 | 11,200 | V | O |
| 42 | 20,000 | 600 | 9 | 8,450 | H | L/O |
| 42B | 20,000 | 1,350 | 9 | 8,450 | V | L/O |
| 52 | 30,000 | 400 | 9 | 7,520 | H | L/O |
| 52B | 30,000 | 1,120 | 9 | 7,520 | V | L/O |
| 65 | 70,000 | 400 | 7 | 5,100 | H | L/O |
| 87 | 110,000 | 250 | 7 | 4,200 | H | L/O |
| 112 | 180,000 | 125 | 6 | 3,300 | H | L |
| 140 | 250,000 | 125 | 6 | 2,700 | H | L |
| "CVM"[1] | | | | | | |
| EX | 40,000 | 175 | 4 | 10,400 | H | L |
| F | 60,000 | 175 | 4 | 8,430 | H | L |
| G | 110,000 | 175 | 4 | 6,750 | H | L |
| "Axial"[2] | | | | | | |
| 1000 | 75,000 | 85 | 15 | 8,000 | H | L |
| 1500 | 110,000 | 85 | 15 | 6,870 | H | L |
| 3000 | 160,000 | 85 | 15 | 5,590 | H | L |
| 4000 | 230,000 | 85 | 15 | 4,580 | H | L |
| 5000 | 340,000 | 85 | 15 | 3,750 | H | L |
| 6000 | 500,000 | 85 | 15 | 3,120 | H | L |
| "Centregal"[3] | | | | | | |
| MG-1 | 4,000 | 1,500 | 5 | 13,820 | V | Gas seals |
| MG-2 | 4,000 | 1,500 | 9 | 13,820 | V | Gas seals |
| MG-3 | 4,000 | 1,500 | 14 | 13,820 | V | Gas seals |
| "Pipeline"[4] | | | | | | |
| CDP-416 | 7,200 | 1,400 | 4 | 13,820 | V | O/Gas seals |
| CDP-20 | 10,000 | 1,400 | 1 | 11,000 | V | O/Gas seals |
| CDP-24 | 15,000 | 1,440 | 4 | 7,000 | V | O/Gas seals |
| CDP-30 | 25,000 | 1,440 | 4 | 7,000 | V | O/Gas seals |
| CDP-36 | 35,000 | 1,440 | 4 | 7,000 | V | O/Gas seals |
| CDP-42 | 46,000 | 1,440 | 4 | 7,000 | V | O/Gas seals |
| CVP-30 | 25,000 | 1,440 | 1 | 7,000 | V | O/Gas seals |
| CVP-36 | 35,000 | 1,800 | 1 | 7,000 | V | O/Gas seals |
| CVP-42 | 48,000 | 1,800 | 1 | 7,000 | V | O/Gas seals |

# TURBO COMPRESSOR SPECIFICATIONS

| | TYPE OF GAS COMPRESSORS CAN HANDLE | TYPE OF DESIGN C=CENTRIFUGAL A=AXIAL | DIMENSIONS | | | WEIGHT KG OR LB. | COMMENTS |
|---|---|---|---|---|---|---|---|
| | | | L | W | H | | |
| | Air, $N_2$ | C | 6,000 | 4,300 | | | |
| | Air, $N_2$ | C | 9,000 | 6,800 | | | |
| | Air, $N_2$ | C | 9,500 | 7,000 | | | |
| | Air, $N_2$ | C | 10,500 | 7,500 | | | |
| | Air, $N_2$ | C | 11,500 | 8,000 | | | |
| | Air, $N_2$ | C | 12,500 | 8,800 | | | |
| | Air, $N_2$ | C | 13,000 | 9,300 | | | Each model is manufactured in a range of sizes with different dimensions and weights. |
| | Air, $N_2$ | C | 14,000 | 10,100 | | | |
| | Air, $N_2$ | C | 15,000 | 10,900 | | | |
| | Air, $N_2$ | C | 16,500 | 11,800 | | | |
| | Air, $N_2$ | C | 17,500 | 12,600 | | | |
| | Air, $N_2$ | C | 20,000 | 13,900 | | | |
| | $O_2$ | C | | | | | |
| | Industrial Gas | C | | | | | Licensed by Nuovo Pignone |
| | Industrial Gas | C | | | | | Licensed by Nuovo Pignone |
| | Refrigerating Gas | C | | | | | Licensed by Nuovo Pignone |
| | | | Main drivers such as steam turbine, gas turbine, electric motor and gear are also manufactured by Hitachi Ltd. | | | | |
| | Industrial Gas | C | | | | | Licensed by Nuovo Pignone |
| | Industrial Gas | C | | | | | Licensed by Nuovo Pignone |
| | Natural Gas | C | | | | | Licensed by Nuovo Pignone |
| | | | (in.) | (in.) | (in.) | (lb.) | |
| | most | C | 82 | 42 | 40 | 10,500 | [1]Two casing/two intercooler standard design |
| | most | C | 82 | 53 | 47 | 14,400 | F and G Models are available with axial low pressure casing. |
| | most | C | 86 | 55 | 50 | 18,000 | |
| | most | C | 96 | 56 | 43 | 19,000 | |
| | most | C | 96 | 64 | 58 | 22,000 | |
| | most | C | 91 | 64 | 64 | 41,000 | |
| | most | C | 123 | 65 | 47 | 32,000 | [2]Fixed or variable stators |
| | most | C | 123 | 72 | 74 | 41,000 | |
| | most | C | 139 | 80 | 74 | 40,000 | |
| | most | C | 139 | 84 | 83 | 77,000 | |
| | most | C | 132 | 84 | 94 | 53,000 | |
| | most | C | 176 | 112 | 108 | 87,000 | |
| | most | C | 216 | 144 | 177 | 112,000 | |
| | most | C | 216 | 168 | 180 | 217,000 | |
| | Air | C | 125 | 80 | 160 | 81,000 | |
| | Air | C | 144 | 120 | 190 | 100,000 | |
| | Air | C | 215 | 140 | 240 | 135,000 | |
| | Air | A | 130 | 62 | 53 | 17,800 | [3]Multi-casing, multistage centrifugal compressors combined by integral gearing into a unitized structure. |
| | Air | A | 150 | 87 | 84 | 29,640 | |
| | Air | A | 165 | 108 | 69 | 44,660 | |
| | Air | A | 200 | 123 | 75 | 59,000 | |
| | Air | A | 247 | 148 | 108 | 102,660 | |
| | Air | A | 305 | 183 | 133 | 212,800 | |
| | Air or hydrocarbon | C | 58 | 32 | 35 | 8,500 | |
| | Air or hydrocarbon | C | 58 | 51 | 35 | 9,000 | [4]Natural gas transmission |
| | Air or hydrocarbon | C | 58 | 51 | 48 | 11,000 | |
| | Hydrocarbon | C | 49 | 54 | 76 | 15,500 | |
| | Hydrocarbon | C | 74 | 98 | 108 | 30,000 | |
| | Hydrocarbon | C | 109 | 172 | 138 | 192,000 | |
| | Hydrocarbon | C | 109 | 172 | 138 | 206,000 | |
| | Hydrocarbon | C | 109 | 176 | 138 | 215,000 | |
| | Hydrocarbon | C | 109 | 180 | 138 | 225,000 | |
| | Hydrocarbon | C | 136 | 134 | 134 | 152,000 | |
| | Hydrocarbon | C | 136 | 134 | 134 | 160,000 | |
| | Hydrocarbon | C | 136 | 134 | 134 | 175,000 | |

# TURBO COMPRESSOR SPECIFICATIONS

| MODEL | INLET VOLUME M³ OR FT³ | DISCHARGE PRESSURE $P_2$(BAR) | MAXIMUM NUMBER OF STAGES | RPM | H-HORIZONTAL V-VERTICAL SPLIT CASING | TYPE OF SEALS L=LABYRINTH O=OIL |
|---|---|---|---|---|---|---|
| **IHI Co., Ltd., Ohtemachi 2-2-1, Chiyoda-Ku, Tokyo, Japan** | | | | | | |
| R Series | M³/H | | | | | |
| 28 | 8,000 | 69 | 8 | 21,800 | H | L,O |
| 35 | 15,000 | 69 | 8 | 16,600 | H | L,O |
| 45 | 4,000 | 35 | 8 | 13,550 | H | L,O |
| 56 | 38,000 | 35 | 8 | 10,850 | H | L,O |
| 71 | 68,000 | 21 | 8 | 8,600 | H | L,O |
| 90 | 110,000 | 21 | 8 | 6,800 | H | L,O |
| 112 | 150,000 | 21 | 8 | 5,450 | H | L,O |
| RB Series | | | | | | |
| 28 | 8,000 | 560 | 8 | 21,800 | V | O |
| 35 | 15,000 | 560 | 8 | 16,600 | V | O |
| 45 | 24,000 | 560 | 8 | 13,550 | V | O |
| 56 | 38,000 | 560 | 8 | 13,550 | V | O |
| 71 | 68,000 | 100 | 8 | 8,600 | V | O |
| 90 | 110,000 | 100 | 8 | 6,800 | V | O |
| RI Series | | | | | | |
| 6 case sizes | 30,000-140,000 | 10 | 5 | 5,300-12,100 | H | L |
| ARI Series | | | | | | |
| 4 case sizes | 1,400,000-450,000 | 11 | 6 + 3 axial centr. | 3,800-6,200 | H | L |
| A Series | | | | | | |
| 12 case sizes | 50,000-1,170,000 | 8 | 17 | 1,700-9,550 | H | L |
| AV Series | | | | | | |
| 12 case sizes | 50,000-1,170,000 | 8 | 18 | 1,600-8,850 | H | |
| **Joy Machinery Co., Turbocompressor Div., 3101 Broadway, Buffalo, NY 14225 U.S.A.** | | | | | | |
| | (ft.³) | | | (Input) | | |
| TA18 | 1,300 to 2,300 | 10 | 3 | 3,000 & 3,600 | H | L |
| TA22 | 2,200 | 10 | 1 to 3 | 3,000 & 3,600 | H | L |
| TA26 | 2,600 | 15 | 1 to 4 | 3,000 & 3,600 | H | L |
| TA30 | 3,000 | 15 | 1 to 4 | 3,000 & 3,600 | H | L |
| TA35 | 3,500 | 15 | 1 to 4 | 3,000 & 3,600 | H | L |
| TA40 | 4,000 | 15 | 1 to 4 | 3,000 & 3,600 | H | L |
| TA50 | 5,000 | 15 | 1 to 4 | 3,000 & 3,600 | H | L |
| TA60 | 6,000 | 15 | 1 to 4 | 3,000 & 3,600 | H | L |
| TA70 | 7,000 | 15 | 1 to 4 | 3,000 & 3,600 | H | L |
| TA85 | 8,500 | 15 | 1 to 4 | 3,000 & 3,600 | H | L |
| TA110 | 10,000 | 15 | 1 to 4 | 1,000, 1,200, 1,800 | H | L |
| TA160 | 15,000 | 15 | 1 to 4 | 1,000, 1,200, 1,800 | H | L |
| MSG | up to 80,000 | 45 | 1 to 6 | 50&60HZ 1,000 to 3,600 | H | L |
| **Kawasaki Heavy Industries, Ltd., Prime Mover Division, 1-1, Higashikawasaki-cho, 3-chome, Chuo-ku, Kobe 650-91, Japan.** | | | | | | |
| | (acfm) | | | | | |
| RBS | to 6,000 | 41 | 10 | 14,600 | H | either |
| RCS | to 13,000 | 41 | 10 | 11,750 | H | either |
| RDS | to 22,500 | 34 | 10 | 9,100 | H | either |
| RES | to 40,000 | 28 | 8 | 7,100 | H | either |
| RFS | to 70,000 | 24 | 8 | 5,650 | H | either |
| RBB | to 6,000 | 724 | 10 | 14,600 | V | either |
| RCB | to 13,000 | 172 | 10 | 11,750 | V | either |
| RDB | to 17,500 | 79 | 9 | 9,100 | V | either |
| REB | to 25,000 | 47 | 9 | 7,100 | V | either |

# TURBO COMPRESSOR SPECIFICATIONS

| | TYPE OF GAS COMPRESSORS CAN HANDLE | TYPE OF DESIGN C=CENTRIFUGAL A=AXIAL | DIMENSIONS L | W | H | WEIGHT KG OR LB. | COMMENTS |
|---|---|---|---|---|---|---|---|
| | | | (mm) | (mm) | (mm) | (kg) | |
| | any | centrifugal | 1,810 | 1,540 | 1,110 | 1,370 | Additionally RZ-models with intercoolers and RS-models with side stream are available. |
| | any | centrifugal | 1,070 | 1,850 | 1,310 | 2,160 | |
| | any | centrifugal | 2,400 | 2,100 | 1,600 | 3,180 | |
| | any | centrifugal | 2,950 | 2,600 | 1,990 | 6,010 | |
| | any | centrifugal | 3,500 | 2,720 | 2,430 | 9,000 | |
| | any | centrifugal | 4,300 | 3,400 | 3,000 | 15,600 | |
| | any | centrifugal | 5,000 | 4,200 | 3,700 | 31,000 | |
| | any | centrifugal | 1,850 | 1,400 | 1,150 | 3,210 | |
| | any | centrifugal | 2,350 | 1,550 | 1,400 | 6,010 | |
| | any | centrifugal | 3,000 | 1,650 | 1,550 | 12,820 | |
| | any | centrifugal | 3,750 | 2,750 | 2,300 | 26,110 | |
| | any | centrifugal | 4,700 | 3,000 | 2,750 | 43,640 | |
| | any | centrifugal | 6,000 | 3,250 | 3,100 | 84,610 | |
| | air, $N_2$ | centrifugal | — | — | — | — | |
| | air, $N_2$ | axial & centrifugal | — | — | — | — | |
| | air, $N_2$ | axial | — | — | — | — | Fixed stator blade type |
| | air, $N_2$ | axial | — | — | — | — | Variable stator blade type |
| | | | (in.) | (in.) | (in.) | (lb.) | |
| | Air & Nitrogen | C | 106 | 55 | 75 | 12,000 | |
| | Air & Nitrogen | C | 131 | 80 | 57 | 13,700 | |
| | Air & Nitrogen | C | 131 | 80 | 57 | 15,900 | |
| | Air & Nitrogen | C | 131 | 80 | 57 | 16,000 | |
| | Air & Nitrogen | C | 131 | 80 | 57 | 16,200 | |
| | Air & Nitrogen | C | 155 | 93 | 70 | 22,700 | |
| | Air & Nitrogen | C | 155 | 93 | 70 | 23,600 | |
| | Air & Nitrogen | C | 180 | 98 | 74 | 31,700 | |
| | Air & Nitrogen | C | 180 | 98 | 74 | 32,300 | |
| | Air & Nitrogen | C | 180 | 98 | 74 | 33,400 | |
| | Air & Nitrogen | C | 200 | 118 | 111 | 60,000 to 70,000 | |
| | Air & Nitrogen | C | 223 | 127 | 115 | 78,000 to 92,000 | |
| | Air & Nitrogen | C | Frame sizes too numerous to list | | | | |
| | | | (in.) | (in.) | (in.) | (Max.) | |
| | All types | C | 82 | 44 | 38½ | 10,000 | Cooper-Bessemer license |
| | All types | C | 97 | 50 | 42 | 22,000 | " |
| | All types | C | 116 | 68 | 56½ | 33,000 | " |
| | All types | C | 127 | 96 | 76 | 70,000 | " |
| | All types | C | 157 | 111½ | 95 | 130,000 | " |
| | All types | C | 79 | 52 | 49 | 15,000 | " |
| | All types | C | 94 | 60 | 63 | 40,000 | " |
| | All types | C | 126 | 65 | 84 | 50,000 | " |
| | All types | C | 141 | 76 | 95 | 100,000 | " |

# TURBO COMPRESSOR SPECIFICATIONS

| MODEL | INLET VOLUME M³ OR FT³ | DISCHARGE PRESSURE $P_2$(BAR) | MAXIMUM NUMBER OF STAGES | RPM | H-HORIZONTAL V-VERTICAL SPLIT CASING | TYPE OF SEALS L=LABYRINTH O=OIL |
|---|---|---|---|---|---|---|
| **Lotema Corporation, P.O. Box 72, 5089 W. Western Reserve Road, Canfield, OH 44406** | | | | | | |
| | (ft.³/min.) | | | | | |
| GT016 | 880 to 3,700 | 5.7 | 2 | 48,000 | V | L |
| GT021 | 5,000 | 18 | 4 | 36,400 | V | L |
| GT026 | 9,400 | 64 | 6 | 29,400 | V | L |
| GT032 | 11,800 | 64 | 6 | 23,900 | V | L |
| GT040 | 23,500 | 64 | 6 | 19,100 | V | L |
| GT050 | 31,800 | 64 | 6 | 15,300 | V | L |
| GT063 | 58,900 | 64 | 6 | 12,100 | V | L |
| GT078 | 90,000 | 64 | 6 | 9,800 | V | L |
| GT098 | 142,000 | 64 | 6 | 7,800 | V | L |
| GT123 | 224,000 | 64 | 6 | 6,200 | V | L |
| GT153 | 347,000 | 64 | 6 | 5,000 | V | L |
| **M.A.N. Unternehmensbereich GHH STERKRADE, P.O. Box 110240, D-4200 Oberhausen 11, West Germany** | | | | | | |
| | (m³/h) | | | | | |
| RG | 600-140,000 | 80 | 10 | depending on application | H | L/O |
| RGD | 1,500-200,000 | 80 | 2 x 4 | " | H | L/O |
| RGVF | 5,000-100,000 | 80 | 1 | " | V | L/O |
| RGV | 500-100,000 | 350 | 8 | " | V | L/O |
| RGVP | max. 1,000 | 1,000 | 14 | " | V | L/O |
| RKZ | 15,000-150,000 | 80 | 9 | " | H | L/O |
| RKZD | 20,000-200,000 | 80 | 6 | " | H | L/O |
| AG | 50,000-1,000,000 | 4 | 11 | depending on application | H | L/O |
| AGR | 50,000-1,000,000 | 13 | 16 | | H | L/O |
| AK | 70,000-1,000,000 | 8 | 17 | " | H | L/O |
| AKF | 70,000-1,000,000 | 12 | 18 | " | H | L/O |
| AR | 70,000-500,000 | 10 | 14 | " | H | L/O |
| **Mannesmann Demag, Verdichter und Drucklufttechnik, Postfach 10 01 41, D-4100 Duisburg 1, W. Germany** | | | | | | |
| | (m³/hr) | | | | | |
| MH-Line | 2,000-300,000 | 70 | 7-9 | 20,000-4,000 | H | L/O |
| MV-Line | 1,000-100,000 | 800 | 6-8 | 18,000-5,200 | V | L/O |
| VK-Line | 30,000-200,000 | 13 | 3-4 | 1,800-900 | H | L |
| HVK-Line | 7,000-30,000 | 40 | 4 | 1,800-1,200 | H | L |
| PVK-Line | 4,500-30,000 | 15 | 3-4 | 3,600-1,500 | H | L |
| SEZ-Line | 1,000-200,000 | 1:2.8 | 1 | 3,600-1,200 | H | L/O |
| KG-Line | 1,000-25,000 | 1:2.8 | 1 | 3,600-1,800 | H | L |
| AX-Line | 70,000-1,000,000 | 6 | 6-18 | 8,700-3,000 | H | L |
| AR-Line | 70,000-540,000 | 10 | 6-9/2-3 | 8,700-3,000 | H | L |
| Mannesmann Demag, Verdichter und Drucklufttechnik, Postfach 10 01 41, D-4100 Duisburg 1, West Germany, Contact: Dept. 9160, Mr. Hornberger, Tel. (0203) 805-1, Telex: 855 855. | | | | | | |
| **Mitsubishi Heavy Industries, Ltd., 5-1 Marunouchi, 2 Chome, Chiyoda-ku, Tokyo 100 Japan** | | | | | | |
| | (m³/min.) | (Kg/cm²G) | | | | |
| MT-27 | 450 | 0-10 | 16 | 3,800-13,000 | L,O | L,O |
| MT-35 | 830 | 0-10 | 16 | 3,800-13,000 | L,O | L,O |
| MT-44 | 1,300 | 0-10 | 16 | 3,800-13,000 | L,O | L,O |
| MT-53 | 1,900 | 0-10 | 16 | 3,800-13,000 | L,O | L,O |
| MT-60 | 2,400 | 0-10 | 16 | 3,800-13,000 | L,O | L,O |
| MT-69 | 3,200 | 0-10 | 16 | 3,800-13,000 | L,O | L,O |
| MT-76 | 3,900 | 0-10 | 16 | 3,800-13,000 | L,O | L,O |
| MT-84 | 4,900 | 0-10 | 16 | 3,800-13,000 | L,O | L,O |
| MT-92 | 5,700 | 0-10 | 16 | 3,000-3,800 | L,O | L,O |
| MT-100 | 6,800 | 0-10 | 16 | 3,000-3,800 | L,O | L,O |
| MT-107 | 7,800 | 0-10 | 16 | 3,000-3,800 | L,O | L,O |
| MT-109 | 11,000 | 0-10 | 16 | 3,000-3,800 | L,O | L,O |
| MT-124 | 11,000 | 0-10 | 16 | 3,000-3,800 | L,O | L,O |

# TURBO COMPRESSOR SPECIFICATIONS

| | TYPE OF GAS COMPRESSORS CAN HANDLE | TYPE OF DESIGN C=CENTRIFUGAL A=AXIAL | DIMENSIONS | | | WEIGHT KG OR LB. | COMMENTS |
|---|---|---|---|---|---|---|---|
| | | | L | W | H | | |
| | Air, $N_2$, CO, $CO_2$ | Integral Gear | Depending on number of stages | | | | Discharge pressure is based on |
| | Hydrocarbon | Centrifugal | " | " | " | " | one bar suction. |
| | " | " | " | " | " | " | |
| | " | " | " | " | " | " | All models available in single and all |
| | " | " | " | " | " | " | multi stage configurations up to the |
| | " | " | " | " | " | " | stated max. no. of stages. |
| | " | " | " | " | " | " | |
| | " | " | " | " | " | " | Skid mounted design including |
| | " | " | " | " | " | " | driver and coolers. |
| | " | " | " | " | " | " | |
| | " | " | " | " | " | " | Carbon Ring and mechanical seals |
| | " | " | " | " | " | " | available for the larger models. |
| | Air, CN-gas, $NH_3$, wet-gas, natural gas, etc. | C | | depending on size | | | |
| | " | C | | " | | | double flow |
| | Air, steam and other gases | C | | " | | | |
| | Natural gas, $H_2$, Syngas and other gases | C | | " | | | |
| | Syngas | C | | " | | | |
| | Air, $O_2$, $N_2$, natural gas, all other gases | C | | " | | | with one or more intercoolings |
| | Air, $N_2$, Rawgas, other gases | C | | " | | | " |
| | | | | | | | (pressure ratio, max.) |
| | Air, NO gas and | A | | depending on size | | | 3 |
| | other gases | A | | " | | | 6 |
| | " | A | | " | | | 7.5 |
| | " | A | | " | | | 12 |
| | " | A | | " | | | 7.5 |
| | All gases | C | | | | | external intercooling and |
| | All gases | C | | | | | side streams possible |
| | Air, $N_2$ | C | | | | | external intercooled |
| | $N_2$ | C | | | | | external intercooled |
| | Air, $N_2$ | C | | | | | intercooled, packaged |
| | All gases | C | | | | | |
| | Air, $N_2$' | C | | | | | |
| | Air, industrial gases | A | | | | | |
| | Air | A/C | | | | | external intercooled |
| **Mannesmann Demag Corp.**, Compression Equipment Div., c/o Mannesmann Pipe & Steel Corp., 1990 Post Oak Boulevard, 18th Floor, Houston, TX 77056, Tel: 713-960-1900 | | | | | | | |
| **Mannesmann Demag Corp.**, Compression Equipment Div., 100 Park Avenue, 34th Floor, New York, NY 10017, Tel: 212-599-0660. Telex: Western Union 12-5911 | | | | | | | |
| | | | (mm) | (mm) | (mm) | (kg) | |
| | Air & Gas | A | 2,500-7,500 | 2,300-5,100 | 1,500-4,100 | 7,000-120,000 | |
| | Air & Gas | A | 2,500-7,500 | 2,300-5,100 | 1,500-4,100 | 7,000-120,000 | |
| | Air & Gas | A | 2,500-7,500 | 2,300-5,100 | 1,500-4,100 | 7,000-120,000 | |
| | Air & Gas | A | 2,500-7,500 | 2,300-5,100 | 1,500-4,100 | 7,000-120,000 | |
| | Air & Gas | A | 2,500-7,500 | 2,300-5,100 | 1,500-4,100 | 7,000-120,000 | |
| | Air & Gas | A | 2,500-7,500 | 2,300-5,100 | 1,500-4,100 | 7,000-120,000 | |
| | Air & Gas | A | 2,500-7,500 | 2,300-5,100 | 1,500-4,100 | 7,000-120,000 | |
| | Air & Gas | A | 2,500-7,500 | 2,300-5,100 | 1,500-4,100 | 7,000-120,000 | |
| | Air & Gas | A | 2,500-7,500 | 2,300-5,100 | 1,500-4,100 | 7,000-120,000 | |
| | Air & Gas | A | 2,500-7,500 | 2,300-5,100 | 1,500-4,100 | 7,000-120,000 | |
| | Air & Gas | A | 2,500-7,500 | 2,300-5,100 | 1,500-4,100 | 7,000-120,000 | |
| | Air & Gas | A | 2,500-7,500 | 2,300-5,100 | 1,500-4,100 | 7,000-120,000 | |
| | Air & Gas | A | 2,500-7,500 | 2,300-5,100 | 1,500-4,100 | 7,000-120,000 | |

# TURBO COMPRESSOR SPECIFICATIONS

| MODEL | INLET VOLUME $M^3$ OR $FT^3$ | DISCHARGE PRESSURE $P_2$(BAR) | MAXIMUM NUMBER OF STAGES | RPM | H-HORIZONTAL V-VERTICAL SPLIT CASING | TYPE OF SEALS L=LABYRINTH O=OIL |
|---|---|---|---|---|---|---|
| **Mitsubishi Heavy Industries** (Continued) | | | | | | |
| MACE | to 11,000 | 5-12 | 12 | 3,000-8,000 | H | L(0) |
| MTV | to 400 | to 200 | 9 | | V | L,0 |
| MTH | 4,000 | 50 | 9 | | H | L,0 |
| MTHC | 2,000 | 8 | 9 | | H | L,0 |
| MTBL | 600 | 2 | 9 | | H | L,0 |
| MTMF | 4,000 | 1 | 3 | | H | L,0 |
| **Mitsui Engineering & Shipbuilding Co., Ltd., 6-4, Tsukiji 5-chome, Chuo-Ku, Tokyo, Japan** | | | | | | |
| | ($M^3$/hr) | (kg/$cm^2$G) | | | | |
| H22,H28,H38 | 10,000 | 20-80 | 9 | | H | L,0 |
| H48 | 16,000 | 10-80 | 9 | | H | L,0 |
| H58 | 23,000 | 5-60 | 9 | | H | L,0 |
| H64 | 28,000 | 5-40 | 9 | | H | L,0 |
| H76 | 40,000 | 5-20 | 9 | | H | L,0 |
| H90 | 56,000 | 5-20 | 9 | | H | L,0 |
| H102,H120,H140 | 140,000 | 5-20 | 9 | | H | L,0 |
| V22,V28,V38 | 10,000 | 80-800 | 9 | | V | L,0 |
| V48 | 16,000 | 80-240 | 9 | | V | L,0 |
| V58 | 23,000 | 40-240 | 9 | | V | L,0 |
| V64 | 28,000 | 40-160 | 9 | | V | L,0 |
| V76 | 40,000 | 30-120 | 9 | | V | L,0 |
| V90 | 56,000 | 20-120 | 9 | | V | L,0 |
| V102,V120 | 100,000 | 20-120 | 9 | | V | L,0 |
| V140,V160 | 180,000 | 20-30 | 9 | | V | L,0 |
| Axial | | | | | | |
| MA40,MA60 | | | | | | |
| MA80,MA100 | 24,000-205,000 | 5-15 | 16 | 12,000-3,000 | H*[2] | L*[2] |
| MA120,MA140 | | | | | | |
| MA160,MA180 | | | | | | |
| MA200,MA220 | 205,000-1,000,000 | 5-10 | 16 | 4,300-1,800 | H*[2] | L*[2] |

[1]Dimensions and weight will vary in accordance with the number of stages and design pressure. Figures show maximum dimensions and weights.
[2]Vertical split casing with oil film or mechanical seal is available on request for other gases.

**Major Offices:** Head Office, 6-4, Tsukiji 5-Chome, Chuo-ku, Tokyo, Tel: (03) 544-3131, Cable: ''MITUIZOSEN TOKYO,'' Telex: J22821, J22924, 252-2661 • Special Representative in New York, Suite 2029, One World Trade Center, New York, NY 10048, U.S.A., Tel: (212) 432-0240/4, Cable: ''MITUIZOSEN NEW YORK,'' Telex: 232671, 420144 WU 129238 • Special Representative in Duesseldorf, Koenigsallee 92a, 4 Duesseldorf 1 F.R. West Germany, Tel: Duesseldorf 325701, 326125, Cable: ''MITUIZOSEN DUESSELDORF,'' Telex: 8587297 • Special Representative in London, 7th Floor, Lee House, Wood Street, London Wall, London EC 2 Y 5 AH., England, Tel: 600-0241, Cable: ''MITUIZOSEN LONDON EC4,''

| MODEL | INLET VOLUME | DISCHARGE PRESSURE | MAXIMUM NUMBER OF STAGES | RPM | SPLIT CASING | TYPE OF SEALS |
|---|---|---|---|---|---|---|
| **NEU — Turbomachines, P.O. Box 110, 47, rue Fourier 59003 Lille Cedex, France** | | | | | | |
| | ($m^3$) | | | | | |
| M | 3,000 to 60,000 | 25 | 3 | — | V | L |
| MS | to 60,000 | 10 | 4 | — | H | L |
| MD | to 120,000 | 8 | 2 | — | V | L |
| MDL | to 120,000 | 40 | 6 | — | V,H | L |
| L | to 100,000 | 40 | 4 | — | H | L or O |
| ML | to 60,000 | 70 | 6 | — | V,H | L |
| DL | to 200,000 | 5 | 2 | — | H | L or O |
| LD | to 100,000 | 70 | 8 | | H | L or O |
| DLD | to 200,000 | 20 | 4 | — | H | L or O |
| S | — | 100 | — | — | V | O |
| **Norwalk-Turbo, Inc., 7 Northway Lane, Latham, New York 12110 U.S.A.** | | | | | | |
| | (cfm) | (psig) | | | | |
| V-20 | 800 | 2,500 | 8 | 30,000 | V | Oil Shaft End Seals Labyrinth Inner Seals |
| V-26 | 2,400 | 2,500 | 8 | 23,000 | V | " |
| V-30 | 4,200 | 2,500 | 8 | 19,000 | V | " |
| V-35 | 6,000 | 2,500 | 8 | 16,400 | V | " |
| V-40 | 10,000 | 2,500 | 8 | 14,300 | V | " |
| V-48 | 14,000 | 1,500 | 8 | 12,000 | V | " |

**Norwalk-Turbo, Inc.** 7 Northway Lane, Latham, NY 12110, Contact: S. David Caplow, Tel: (518) 783-1625.

# TURBO COMPRESSOR SPECIFICATIONS

| TYPE OF GAS COMPRESSORS CAN HANDLE | TYPE OF DESIGN C=CENTRIFUGAL A=AXIAL | DIMENSIONS L | W | H | WEIGHT KG OR LB. | COMMENTS |
|---|---|---|---|---|---|---|
| Air & Gas | A&C | | | | | |
| Any types | C | | | | | |
| Any types | C | | | | | |
| Any types | C | | | | | |
| Any types | C | | | | | |
| Any types | C | | | | | |
| | | (mm) | (mm) | (mm) | (kg) | |
| any | C | 2,140 | 1,350 | 1,200 | 5,000 | |
| any | C | 2,440 | 1,600 | 1,500 | 7,400 | |
| any | C | 2,740 | 1,900 | 1,800 | 11,700 | |
| any | C | 2,920 | 2,050 | 1,970 | 14,200 | |
| any | C | 3,280 | 2,400 | 2,350 | 20,500 | |
| any | C | 3,690 | 2,800 | 2,790 | 30,000 | |
| any | C | 5,160 | 4,150 | 4,340 | 93,000 | |
| any | C | 2,250 | 1,460 | 1,310 | 12,200 | |
| any | C | 2,580 | 1,900 | 1,580 | 18,800 | |
| any | C | 2,910 | 1,900 | 1,850 | 28,800 | |
| any | C | 3,130 | 2,060 | 2,010 | 36,600 | |
| any | C | 3,600 | 2,380 | 2,330 | 47,000 | |
| any | C | 3,970 | 2,740 | 2,710 | 54,000 | |
| any | C | 4,960 | 3,520 | 3,520 | 110,000 | |
| any | C | 6,280 | 4,560 | 4,600 | | |
| air[1] | A | 2,100/4,500 | 800/2,100 | 900/2,300 | | |
| air[2] | A | 4,500/7,500 | 2,100/4,000 | 2,300/4,200 | | |

Telex: 885971 • Special Representative in Los Angeles c/o Mitsui & Co., (U.S.A.) Inc. Los Angeles Office, Suite 2190, Crocker-Bank Plaza, 611 West 6th Street, Los Angeles, California 90017, U.S.A., Tel: 213-680-1000, 213-628-5350, Telex: 67205 • Special Representative in Hong Kong, c/o Mitsui & Co., Ltd., Hong Kong Branch, Connaught centre, 33rd Floor, Connaught Road, Central, Hong Kong, Tel: 5-264291/3, Telex: HX73916 • Special Representative in Singapore, c/o Mitsui & Co., Ltd., Singapore Branch, 6th Floor, Industrial & Commercial Bank Building, Two Shenton Way, Singapore, Tel: 2204065, Cable: MITUIZOSEN c/o MITUICO SINGAPORE, Telex: RS24102 MESSPRE • Mitsui Zosen Do Brazil Industrials Pesadas Ltda 13th Floor, 63 Avenida Almirante Barroso, Rio de Janeiro, RJ., Brazil, Tel: 252-3667, 242-4898, Telex: 021-22378.

| TYPE OF GAS COMPRESSORS CAN HANDLE | TYPE OF DESIGN C=CENTRIFUGAL A=AXIAL | DIMENSIONS L | W | H | WEIGHT KG OR LB. | COMMENTS |
|---|---|---|---|---|---|---|
| Air, $N_2$, Steam, $CO_2$ | C | — | — | — | — | — |
| $CO_2$, CO.$N_2$' | C | — | — | — | — | — |
| Air, $N_2$, Steam, $CO_2$ | C | — | — | — | — | — |
| Air, $N_2$, Steam, $CO_2$ | C | — | — | — | — | — |
| Most gases | C | — | — | — | — | — |
| Air, $N_2$, Steam, $CO_2$ | C | — | — | — | — | — |
| Most gases | C | — | — | — | — | — |
| Most gases | C | — | — | — | — | — |
| Most gases | C | — | — | — | — | — |
| Most gases | C | — | — | — | — | — |
| | | (in.) | (in.) | (in.) | (lb.) | |
| Natural Gas, | C | 26 | 28 | 23 | 2,200 | |
| Propane, Air | | | | | | |
| Ammonia, Oxygen | C | 28 | 34 | 26 | 4,000 | |
| Hydrogen | C | 32 | 38 | 31 | 6,000 | |
| Hydrocarbon mixtures | | | | | | |
| | C | 38 | 44 | 37 | 12,000 | |
| | C | 44 | 50 | 42 | 17,000 | |
| | C | 50 | 59 | 50 | 29,000 | |

# TURBO COMPRESSOR SPECIFICATIONS

| MODEL | INLET VOLUME M³ OR FT³ | DISCHARGE PRESSURE $P_2$(BAR) | MAXIMUM NUMBER OF STAGES | RPM | H-HORIZONTAL V-VERTICAL SPLIT CASING | TYPE OF SEALS L=LABYRINTH O=OIL |
|---|---|---|---|---|---|---|
| **Nuovo Pignone, Via F. Matteucci, 2, Box 414, Florence, Italy** | | | | | | |
| | (m³/hr) | | | | | |
| MC | 500-350,000 | 45 | 9 | 2,500-18,000 | H | L,O |
| 3MC | 500-350,000 | 45 | 9 | 3,500-15,000 | H | L,O |
| 2MC | 500-350,000 | 45 | 9 | 3,500-15,000 | H | L,O |
| DMC | 80,000-400,000 | 10 | 8 | 3,500-8,000 | H | L,O |
| SR | 10,000-240,000 | 30 | 5 | 1,500-3,000* | H | L,O |
| BC | 150-100,000 | 750 | 8 | 3,000-18,000 | V | L,O |
| DBC | 80,000-200,000 | 100 | 8 | 3,000-18,000 | V | L,O |
| 2BC | 150-100,000 | 750 | 8 | 3,000-18,000 | V | L,O |
| PC | 500-100,000 | 100 | 4 | 3,000-12,000 | V | O |
| AN | 50,000-1,000,000 | 8 | 20 | 2,100-10,000 | H | L |
| AC | 100,000-600,000 | 25 | — | 2,100-14,000 | H | L |
| Pignone compressors are usually used in refineries, petrochemical plants, natural gas compression stations and nuclear power plants. Main drivers are steam and gas turbines (both manufactured by Nuovo Pignone) and electric motors. | | | | | | |
| **Solar Turbines Incorporated, P.O. Box 80966, 2200 Pacific Coast Highway, San Diego, California 92138 U.S.A** | | | | | | |
| | | (psia) | | | | Steel-backed carbon ring oil seals |
| C16 | 150-3,200 | 4,000 | 10 | 22,300 | V | |
| C28 | 1,000-7,500 | 750 | 4 | 22,300 | V | " |
| C30 | 1,000-15,000 | 2,000 | 8 | 18,000 | V | " |
| C50 | 3,000-16,000 | 1,500 | 5 | 14,000 | V | " |
| C60 | 4,000-25,000 | 1,750 | 1 | 9,500 | V | " |
| **Sulzer Brothers Ltd., Thermal Turbomachinery, P.O. Box CH-8023, Zurich Switzerland** | | | | | | |
| | (M³/hr) | | | | | |
| R 8 sizes | 650-200,000 | min. 3 max. 80 | 8 | 20,000-4,300 | H | L,O |
| RZ 8 sizes | 800-200,000 | min 4 max. 80 | 8 | 20,000-4,300 | H | L,O |
| RB 4 sizes | 650-45,000 | min. 64 max. 400 | 8 | 20,000-9,700 | V | L+O |
| RBZ 4 sizes | 800-45,000 | min 64 max. 400 | 8 | 20,000-9,700 | V | L+O |
| RI 5 sizes | 35,000-140,000 | 10 | 5 | 9,500-5,400 | H | L |
| ARI 5 sizes | 140,000-600,000 | 10 | 6A + 3C | 6,000-3,600 | H | L |
| A 12 sizes | 60,000-1,150,000 | 8 | 20 | 9,500-2,700 | H | L,O |
| AV 12 sizes | 100,000-1,150,000 | 8 | 20 | 8,800-2,500 | H | L,O |
| Major Offices: B-1040 Bruxelles, S.A. Sulzer Belgium N.V., 92, Square Plasky • F-75300 Paris-Brune, Compagnie de Construction Mechanique Sulzer, Cedex 59, Paris, France • Pointe-Claire Dorval 700 (Quebec), Sulzer Bros. (Canada) Ltd., Sulzer Bros. Inc., 200 Park Ave., New York, NY 10012. | | | | | | |
| **Tech Development, Inc., P. O. Box 1400-0, 6800 Poe Avenue, Dayton, Ohio 45414 U.S.A.** | | | | | | |
| | (cfm) | | | | | |
| 897 | 981 | 2.8[a] | 1 | 60,000 | — | — |
| 457 | 4,354 | 3.7[a] | 1 | 35,800 | — | — |
| 292 | 8,350 | 2.9[a] | 1 | 22,900 | — | — |
| 1109 | 20,400 | 5.9[a] | 1 | 18,000 | — | — |

[1]Driven by compressed air, max drive flow is 251 cfm at 200 psig. [2]Driven by compressed air, max drive flow is 863 cfm at 300 psig. [3]Driven by compressed air, max drive flow is 1804 cfm at 140 psig. [4]Driven by compressed air, max drive flow is 4629 cfm at 300 psig. [a]psig.

# TURBO COMPRESSOR SPECIFICATIONS

| | TYPE OF GAS COMPRESSORS CAN HANDLE | TYPE OF DESIGN C = CENTRIFUGAL A = AXIAL | DIMENSIONS L | W | H | WEIGHT KG OR LB. | COMMENTS |
|---|---|---|---|---|---|---|---|
| | Main industrial gases | C | Each model is manufactured in a range of sizes with different dimensions and weights. | | | | In line multistage |
| | Refrigerating gases | C | | | | | In line with side streams |
| | Main industrial gases | C | | | | | Back to back |
| | Main industrial gases | C | | | | | Double flow |
| | Air** | C | Intercooled motor driven package *Available also for 60 Hz motor drive. | | | | **Air standard design on request for other gases. |
| | Main industrial gases | C | | | | | In line barrel |
| | Main industrial gases | C | | | | | Double flow |
| | Main industrial gases | C | | | | | Back to back barrel |
| | Natural gas and other gases | C | | | | | Mainly for pipeline booster service |
| | Air* | A | | | | | *Air standard design — on request for other gases. |
| | Air* | A-C | | | | | *Air standard design — on request for other gases. |
| | | | (in.) | (in.) | (in.) | (Heaviest Configuration) | |
| | Natural Gas SG = .6 to 1.6 | Centrifugal | 46 | 64 | 56 | 7,750 lb | |
| | " | Centrifugal | 52 | 49 | 41 | 6,800 lb. | |
| | " | Centrifugal | 67 | 67 | 77 | 20,400 lb. | |
| | " | Centrifugal | 75 | 76 | 69 | 32,000 lb. | |
| | " | Centrifugal | 56 | 71 | 71 | 24,000 lb. | |
| | air, industrial gases | C | | | | | Side stream upon request with 1 or 2 external intercoolers |
| | air, industrial gases | C | | | | | |
| | air, industrial gases | C | | | | | |
| | air, industrial gases | C | | | | | with one external intercooler |
| | air, nitrogen | C | | | | | with integral intercoolers |
| | air, nitrogen | A + C | | | | | with integral intercoolers |
| | air, industrial gases | A | | | | | with fixed stator blades |
| | air, industrial gases | A | | | | | with adjustable stator blades |
| | | | (dia. in.) | (length in.) | (dia. in.) | (lb.) | |
| | air, $N_2$ | A | 3.00 | 2.00 | 3.73 | 0.7 | |
| | air, $N_2$ | A | 5.50 | 3.25 | 6.50 | 8.6 | |
| | air, $N_2$ | A | 8.00 | 4.65 | 9.80 | 27.8 | |
| | air, $N_2$ | A | 12.00 | 6.00 | 14.00 | 72.0 | |

Turbomachinery International Handbook

Turbomachinery Specifications

# TURBO COMPRESSOR SPECIFICATIONS

| MODEL | INLET VOLUME M³ OR FT³ | DISCHARGE PRESSURE $P_2$(BAR) | MAXIMUM NUMBER OF STAGES | RPM | H-HORIZONTAL V-VERTICAL SPLIT CASING | TYPE OF SEALS L=LABYRINTH O=OIL |
|---|---|---|---|---|---|---|
| **Transamerica Delaval, Inc., P. O. Box 8788, Trenton, NJ 08650 U.S.A.** | | | | | | |
| | (FT³) | $(P_s)$ | | | | |
| 22 | 7,000 | 5,000 | 10 | 15,100 | both | either |
| 26 | 10,000 | 5,000 | 10 | 12,700 | both | either |
| 31 | 14,000 | 5,000 | 10 | 10,700 | both | either |
| 37 | 20,000 | 3,000 | 10 | 9,000 | both | either |
| 44 | 28,000 | 2,000 | 10 | 7,550 | both | either |
| 52 | 40,000 | 2,000 | 10 | 6,390 | both | either |
| 62 | 56,000 | 1,000 | 10 | 5,350 | both | either |
| 74 | 79,000 | 750 | 10 | 4,500 | both | either |
| 88 | 110,000 | 750 | 10 | 3,800 | both | either |
| B12/12 | 3,600 | 1,200 | 5 | 16,000 | vertical | oil |
| B16/16 | 6,000 | 1,200 | 5 | 16,000 | vertical | oil |
| B18/18 | 8,000 | 1,200 | 5 | 8,200 | vertical | oil |
| B20/20 | 10,000 | 1,200 | 5 | 8,200 | vertical | oil |
| B24/24 | 15,000 | 1,200 | 5 | 6,500 | vertical | oil |
| B30/30 | 25,000 | 1,200 | 3 | 6,500 | vertical | oil |
| B36/36 | 36,000 | 1,200 | 3 | 6,000 | vertical | oil |
| PV24/24 | 17,000 | 1,800 | 1 | 7,500 | vertical | oil |
| PV30/30 | 30,000 | 1,800 | 1 | 6,500 | vertical | oil |
| PV36/36 | 45,000 | 1,800 | 1 | 6,500 | vertical | oil |

*Case size and weight will vary in accordance with the configuration and pressure rating.
**Single stage pipeline compressors.

**Major Offices:** Transamerica Delaval, P. O. Box 8788, Trenton, NJ 08650, Tel: (609) 890-5000 • Delavel-Stork V.O.F., 199/201 Old Marylebone Road, London NW1, England, B.M. Scully, Tel: (44) 01 262 0081, Telex: 851-24435 • Delaval-Stork V.O.F., 55 Jan Van Nassaustraat, 2596 BP The Hague, The Netherlands, W. Van Lenthe, Tel: (0031) 070 244484, Telex: 844-33283 • Delaval-Stork V.O.F., 28 Avenue Hoche, 75008 Paris, France, A. De La Bassetiere, Tel: (33) 01 561 0351 Telex: 842-280252 • Delaval-Stork V.O.F., P.O. Box 329.

| MODEL | INLET VOLUME | DISCHARGE PRESSURE | MAXIMUM NUMBER OF STAGES | RPM | SPLIT CASING | TYPE OF SEALS |
|---|---|---|---|---|---|---|
| **Worthington Compressors, Inc., 65 Springfield Ave., Springfield, New Jersey 07081 U.S.A.** | | | | | | |
| | (ft³) | (psig) | | | | |
| CAP-15 | 1,450 | 100-125 | 3 | 3,600 | H | L |
| CAP-17 | 1,700 | 100-125 | 3 | 3,600 | H | L |
| CAP-21 | 2,000 | 100-125 | 3 | 3,600 | H | L |
| CAP-23 | 2,300 | 100-125 | 3 | 3,600 | H | L |
| CAP-25 | 2,600 | 100-125 | 3 | 3,600 | H | L |
| CAP-31 | 3,050 | 100-125 | 3 | 3,600 | H | L |
| HB | 37-287 | 20-150 | 1 | 300-550 | H | O |
| VB | 44-168 | 20-150 | 1 | 450-625 | V | O |
| BDC | varies | max. 10,000 | 10 | 300-720 | H | O |
| Y | 406-1,724 | 100-500 | 3 | 450-600 | V | O |
| **Worthington Division, McGraw-Edison Company, 5 Washington Ave., Fairfield, New Jersey 07006** | | | | | | |
| | (Nominal acfm) | (Pressure ratio) | | | | |
| 400 | 700 | 3.2 | 1 | 48,000 | V | L |
| 600 | 2,000 | 3.2 | 1 | 40,000 | V | L |
| 800 | 3,500 | 3.2 | 1 | 34,000 | V | L |
| 1100 | 7,000 | 3.2 | 1 | 29,000 | V | L |
| 1500 | 12,000 | 3.2 | 1 | 21,000 | V | L |
| 1800 | 15,000 | 3.2 | 1 | 18,000 | V | L |

## TURBO COMPRESSOR SPECIFICATIONS

| TYPE OF GAS COMPRESSORS CAN HANDLE | TYPE OF DESIGN C=CENTRIFUGAL A=AXIAL | DIMENSIONS L | W | H | WEIGHT KG OR LB. | COMMENTS |
|---|---|---|---|---|---|---|
| | | (in.) | (in.) | (in.) | (lb.) | |
| all types | centrifugal | | | | | • |
| all types | centrifugal | | | | | • |
| all types | centrifugal | | | | | • |
| all types | centrifugal | | | | | • |
| all types | centrifugal | | | | | • |
| all types | centrifugal | | | | | • |
| all types | centrifugal | | | | | • |
| all types | centrifugal | | | | | • |
| all types | centrifugal | | | | | • |
| natural | centrifugal | 53 | 48 | 38 | 13,000 | |
| natural | centrifugal | 57 | 48 | 53 | 16,200 | |
| natural | centrifugal | 103 | 96 | 69 | 34,300 | |
| natural | centrifugal | 105 | 96 | 72 | 45,000 | |
| natural | centrifugal | 103 | 102 | 84 | 48,000 | |
| natural | centrifugal | 125 | 144 | 90 | 85,000 | |
| natural | centrifugal | 126 | 160 | 117 | 120,000 | |
| natural | centrifugal | 96 | 120 | 77 | 40,500 | •• |
| natural | centrifugal | 102 | 134 | 81 | 51,000 | •• |
| natural | centrifugal | 104 | 144 | 104 | 66,000 | •• |

7550 AH Hengelo ov The Netherlands, C.W. Pearson, Tel: (0031) 74-423542, Telex: 844-44564 • EPN-Division Maquinaria de Proceso Hegel 153-6 Piso Mexico City 5, Mexico, Tel: 531-8828, Telex: 001-774-415 • Transamerica Delaval, P. O. Box 610, Riyadh, Saudi Arabia, A.L. Foltz, Tel: 160 966 60394, Telex: 928-201206 • Domus 2000, 12th floor, 84 Ethnikis Aristaseos Str., P. O. Box 41, Psychico, Athens, Greece, S.F. Troulakis, Tel: 30 1 8028761, Telex: 863-219289.

| TYPE OF GAS COMPRESSORS CAN HANDLE | TYPE OF DESIGN C=CENTRIFUGAL A=AXIAL | L | W | H | WEIGHT KG OR LB. | COMMENTS |
|---|---|---|---|---|---|---|
| | | | | | (lb.) | |
| air | C | 11'-7" | 5'-6" | 7'-6" | 11,000 | induction mtr. or turbine drive non-lube |
| air | C | " | " | " | 11,500 | " |
| air | C | " | " | " | 12,000 | " |
| air | C | " | " | " | 13,000 | " |
| air | C | " | " | " | 13,500 | " |
| air | C | " | " | " | 23,000 | " |
| | | | largest | | heaviest | |
| | | 16'-5" | 3'-3" | 5'-7" | 9,000 | |
| | reciprocative | | smallest | | lightest | |
| air or process gas | (positive displacement) | 7'-6" | 1'-10" | 2'-10" | 1,100 | non-lube or lube |
| | | | largest | heaviest | | |
| | | 7'-3" | 3'-6" | 9'-4" | 6,000 | |
| | | | smallest | | lightest | |
| air or process gas | " | 4'-5" | 2'-6" | 5'-2" | 1,240 | " |
| | | | largest | | heaviest | |
| | | 45'-1" | 28'-6" | 10'-0" | 300,000 | |
| | | | smallest | | lightest | |
| air or process gas | " | 9'-1" | 9'-0" | 5,000 | " | |
| | | | largest | | heaviest | |
| | | 10'-8" | 13'-2" | 8'-5" | 26,500 | |
| | | | smallest | | lightest | |
| air or process gas | " | 6'-9" | 10'-0" | 6'-7" | 12,500 | " |

| TYPE OF GAS COMPRESSORS CAN HANDLE | TYPE OF DESIGN C=CENTRIFUGAL A=AXIAL | L | W | H | WEIGHT KG OR LB. | COMMENTS |
|---|---|---|---|---|---|---|
| | | | | | | (driver) |
| Air, Nitrogen, Oxygen, Hydro-carbon | C | | Depending on Model concult plant | | | Turbine or induction motor |
| " | C | | " | | | " |
| " | C | | " | | | " |
| " | C | | " | | | " |
| " | C | | " | | | " |
| " | C | | " | | | " |

# TURBO EXPANDER SPECIFICATIONS

| MODEL | INLET PRESSURE BAR-A | INLET TEMPERATURE °C | INLET FLOW KG PER HR | NUMBER OF STAGES | RPM | H-HORIZONTAL OR V-VERTICAL SPLIT CASING |
|---|---|---|---|---|---|---|
| **Alsthom-Atlantique, Etablissement de La Courneuve, 141 Rue Rateau, 93123 La Courneuve, Rateau, France** | | | | | | |
| | (ata) | (°C) | (Kp/h) | | | |
| TRA 32 | 1/100 | 450 | 15,000 | 1 or 2/8 | 15,000 | H or V |
| to | | | | | | |
| TRA 80 | 1/50 | 450 | 280,000 | 1 or 2/8 | 5,000 | H |
| **Atlas Copco Turbonetics, Inc., 968 Albany-Shaker Rd., Latham, New York 12110 U.S.A.** | | | | | | |
| | (Expander) | | | | | |
| Expander/Brake | To 1,200 psia | To 900°F | Dependent upon process conditions | 1 | Variable | V |
| Expander/ Compressor | To 1,200 psia | To 900°F | | 1 | Variable | V |
| Expander/ Generator | To 1,200 psia | To 900°F | | 1 | Variable | V |
| **Borsig GmbH, Gruppe Deutsche Babcock, P. O. Box 1000, Berlin 27, West Germany** | | | | | | |
| GX 355 | 100 | 500 | 145,000* | 12 | max 16,000 | H |
| GX 425 | 100 | 500 | 175,000* | 20 | max 16,000 | H |
| GX 500 | 100 | 500 | 225,000* | 35 | max 12,500 | H |
| GX 710 | 100 | 500 | 355,000* | 50 | max 10,000 | H |
| GX850 | 100 | 500 | 555,000* | 60 | max 8,000 | H |
| *Related to Air 85 Bar, 450°C. | | | | | | |
| **Dresser Clark, Division Dresser Industries, Inc., P. O. Box 560, Olean, New York 14760 U.S.A.** | | | | | | |
| | (psi) | (°F) | | | | |
| Type GT | To over 150 | To 660 | — | 1 | To over 12,000 | H |
| GTA | To over 1,000 | To 850 | — | Multi | To over 11,000 | H |
| GTHR | To 1,500 | 1,400 | — | Multi | To 15,000 | H |
| **Elliott Company, A subsidiary of United Technologies Corp., North Fourth St., Jeannette, Pennsylvania 15644 U.S.A.** | | | | | | |
| TH | 4.46 | To 760 | 45,700-840,000 | 1,2 | 3,600-8,000 | V |
| • Elliott Company, North Fourth Street, Jeannette, Pennsylvania 15644, Contact: Edward J. Bryson, Tel: 412-527-8429, Telex: 866643. • Elliott Company, P. [illegible]ox 1008, 1071 Bristol Rd., Mountainside, N.J. 07092, Contact: Clinton M. Seeman, Tel: (201) 233-9600, Telex: 710-997-9548 • Elliott Company, 6100 N. Pulaski Rd., Chica[illegible]0646, Contact: William J. Lyons, Tel: (312) 267-7800, TWX: 910-221-5607 • Elliott Company, P. O. Box 4281, Houston, Texas 77210, Contact: Ronald L. Wheeler, Tel: (713) 933-870[illegible], TWX: 910-881-1172 • Elliott Company, 22632 E. Golden Springs Drive, Diamond Bar, California 91765, Contact: Thomas H. Harris, Tel: (213) 723-0261, TWX: 910-584-4899 • Elliott Company | | | | | | |
| **First Brno Engineering Works, Velka Bites Plant — 595 12 Velka Bites Czechoslovakia** | | | | | | |
| | (bar-a) | (°C) | | | (1/min.) | |
| EXT-61-N 8,5/1,3 | 8.33 | −143 | 7,700 | 1 | 18,000 | V |
| EXT-69-N 10,2/1,3 | 10 | −90 | 8,640 | 1 | 22,500 | V |
| EXT-80-N 10/1,4 | 10 | −123 | 11,430 | 1 | 18,000 | V |
| EXT-108-N 5,4/1,4 | 5.37 | −128.5 | 13,600 | 1 | 17,000 | V |
| EXT-250-N 5,4/1,4 | 5.29 | −151 | 34,900 | 1 | 7,500 | V |
| EXT-320-N 5/1,3 | 4.9 | −151 | 41,400 | 1 | 7,500 | V |
| EXT-107-S 31/6 | 30.4 | −128 | 13,400 | 1 | 22,500 | V |
| EXT-130-S 34/6 | 33.3 | −112 | 16,800 | 1 | 22,500 | V |
| EXT-236-S 37/6,3 | 36.3 | −105 | 30,600 | 1 | 15,000 | V |
| EXT-366-S 37/6,3 | 36.3 | −105 | 47,300 | 1 | 15,000 | V |
| EXT-408-S 30/6 | 29.4 | −113 | 51,000 | 1 | 15,000 | V |
| EXT-20NK1 9,5/1,3 | 9.5 | −112 | 2,600 | 1 | 44,000 | V |

# TURBO EXPANDER SPECIFICATIONS

| | TYPE OF SEALS L = LABYRINTH O = OIL | TYPE OF GAS EXPANDER CAN HANDLE | TYPE OF DESIGN C = CENTRIFUGAL A = AXIAL | DIMENSIONS | | | COMMENTS |
|---|---|---|---|---|---|---|---|
| | | | | L | W | H | |
| | L or oil | gas | A | 1,345 | 1,030 | 945 | |
| | L or oil | gas | A | 3,000 | 2,300 | 2,250 | |
| | (M—Mechanical) | | | | | | Horsepower limitation on all models to 3000 |
| | L, O, & M | Most | C | | | | |
| | L, O, & M | Most | C | | Consult Factory | | Engineered process equipment Cryogenic capabilities. Geared two-stage units available. |
| | L, O, & M | Most | C | | | | |
| | | | | (mm) | (mm) | (mm) | |
| | L or O | Air, | A | 2,000 | 1,720 | 1,600 | |
| | L or O | $N_2 + CO_2$ + other, | A | 2,200 | 1,720 | 1,730 | |
| | L or O | $NH_3, C_3H_8, C_4H_{10}$ | A | 2,400 | 1,920 | 1,910 | |
| | L or O | and others | A | 3,000 | 2,320 | 2,130 | |
| | L or O | | A | 4,000 | 2,800 | 3,200 | |
| | L | Gas | A | | | | |
| | L | Gas | A | | | | |
| | L | Gas | A | | | | |
| | L | $N_2 + CO_2 + CO$ | A | — | — | — | |
| — Latin American Operations, P.O. Box 4281, Houston Texas 77210, Contact: John W. Gribble, Tel: (713) 933-8700, TWX: 910-881-1172 • Elliott Company — Japan Operations, 22632 E. Golden Springs Drive, Diamond, California 91765, Contact: Charles A. Cofas, Tel: (213) 723-0261, 910-584-4899 • Elliott Turbomachinery Ltd., Knightsbridge House, 197 Knightsbridge, London, SW7 1RB, Contact: Eugene J. Dickert, Tel: 01-489-8111, 914688 or 914689 • Elliott Turbomachinery B. V., Bucaillestraat 6, Voorburg (ZH) The Netherlands, Contact: Adriaan Jansen, Tel: (070) 873434, Telex: The Hague 31688. | | | | | | | |
| | | | | (mm) | (mm) | (mm) | |
| | L | nitrogen | centripetal | 3,600 | 1,900 | 1,500 | |
| | L | CO | centripetal | 4,450 | 1,900 | 1,500 | |
| | L | Oxygen | centripetal | 3,700 | 1,200 | 1,500 | |
| | L | nitrogen | centripetal | 3,750 | 1,900 | 1,500 | |
| | L | air | centripetal | 5,150 | 2,300 | 2,000 | |
| | L | air | centripetal | 5,200 | 2,300 | 2,000 | |
| | L | nitrogen | centripetal | 4,200 | 1,500 | 1,400 | |
| | L | air | centripetal | 4,200 | 1,500 | 1,400 | |
| | L | air | centripetal | 4,600 | 2,000 | 2,800 | |
| | L | air | centripetal | 4,800 | 2,000 | 2,800 | |
| | L | nitrogen | centripetal | 4,850 | 200 | 2,800 | |
| | L | air | centripetal | 3,200 | 1,500 | 2,500 | |

# TURBO EXPANDER SPECIFICATIONS

| MODEL | INLET PRESSURE BAR-A | INLET TEMPERATURE °C | INLET FLOW KG PER HR | NUMBER OF STAGES | RPM | H-HORIZONTAL OR V-VERTICAL SPLIT CASING |
|---|---|---|---|---|---|---|
| **IHI Co. Ltd.**, Ohtemachi 2-2-1, **Chiyoda-Ku, Tokyo,** Japan | | | | | | |
| E series | | | | | | |
| Expander/Blower | | | | | | |
| 8 case sizes | 11 | -100/-170 | 1,000/12,000 | 1 | Variable | V |
| Expander/Blower | | | | | | |
| 6 case sizes | 11 | -100/-170 | 12,000/45,000 | 1 | Variable | H |
| Expander/Generator | | | | | | |
| 6 case sizes | 11 | -100/-170 | 12,000/45,000 | 1 | To 15,000 | H |
| **Ingersoll Rand** Company, Turbo **Products Division,** Phillipsburg, New **Jersey 08865 U.S.A.** | | | | | | |
| | (psia) | (°F) | (lb./min.) | | | |
| E-516 | 140/250 | 1,400 | 1,800 | 5 | 10,400 | H |
| E-520 | 140/250 | 1,400 | 3,200 | 5 | 8,430 | H |
| E-522 | 140/250 | 1,400 | 3,800 | 5 | 8,000 | H |
| E-526 | 140/250 | 1,400 | 5,000 | 5 | 6,750 | H |
| E-530 | 140/250 | 1,400 | 7,400 | 5 | 5,500 | H |
| E-132 | 50 | 1,400 | 4,200 | 1 | 8,000 | V |
| E-138 | 50 | 1,400 | 7,100 | 1 | 5,700 | V |
| E-238 | 70 | 1,400 | 7,800 | 2 | 5,700 | V |
| E-148 | 50 | 1,400 | 12,200 | 1 | 4,400 | V |
| E-248 | 70 | 1,400 | 13,800 | 2 | 4,400 | V |
| E-156 | 50 | 1,400 | 17,200 | 1 | 3,600 | V |
| E-256 | 70 | 1,400 | 18,200 | 2 | 3,600 | V |
| **Lotema Corp**oration, P.O. Box 7**2, 5089 W. Western** Reserve Rd., Can**field, Ohio 44406** | | | | | | |
| | (BAR A) | (°C) | (Nm³/hr.) | | | |
| ETF 6 | 16 | −265 to 450 | 500 | 1 | 95,000 | V |
| ETF 8 | 16 | " | 750 | 1 | 72,000 | V |
| ETF/B/G 100 | 16 | " | 2,600 | 1 | 57,000 | V |
| ETF/B/G 150 | 80 | " | 26,000 | 1 | 38,000 | V |
| ETB/G 190 | 80 | " | 44,000 | 1 | 30,000 | V |
| ETB/G 240 | 80 | " | 74,600 | 1 | 24,000 | V |
| ETB/G 300 | 80 | " | 49,700 | 1 | 19,000 | V |
| ETB/G 360 | 80 | " | 78,500 | 1 | 16,000 | V |
| ETG 450 | 16 | " | 57,600 | 1 | 12,700 | V |
| ETB/G 560 | 16 | " | 89,200 | 1 | 10,200 | V |
| **M.A.N. Unter**nehmensbereich **GHH STERKRADE, P.** O. Box 110240, **D-4200 Oberhausen** 11, **West Germany** | | | | | | |
| | (bar-a) | (°C) | (kg/h) | | | |
| GTV or GTF | To 30 | To 350 | To 200,000 | 7 | To 14000 | H |
| GYA | To 75 | To 480 | To 300,000 | 2-5 | To 12,000 | H |
| GTR | To 20 | To 600 | To 300,000 | 3-10 | To 14,000 | H |
| GTRH | To 20 | To 780 | To 500,000 | 3-10 | To 14,000 | H |

# TURBO EXPANDER SPECIFICATIONS

| TYPE OF SEALS L = LABYRINTH O = OIL | TYPE OF GAS EXPANDER CAN HANDLE | TYPE OF DESIGN C = CENTRIFUGAL A = AXIAL | DIMENSIONS | | | COMMENTS |
|---|---|---|---|---|---|---|
| | | | L | W | H | |
| Carbon | Air & other gases | Centrifugal | — | — | — | |
| Carbon | Air & other gases | Centrifugal | — | — | — | |
| Carbon | Air & other gases | Centrifugal | — | — | — | |
| | | | (dimensions in inches) | | | |
| L | $N_2$ | A | 64 | 56 | 72 | Typical applications: Nitric acid, fluid cat cracker, coal gasification, pressurized boiler |
| L | $N_2$ | A | 74 | 66 | 88 | |
| L | $N_2$ | A | 78 | 74 | 94 | |
| L | $N_2$ | A | 86 | 84 | 106 | |
| L | $N_2$ | A | 96 | 94 | 118 | |
| L | $N_2$ | A | 112 | 70 | 120 | |
| L | $N_2$ | A | 128 | 81 | 135 | |
| L | $N_2$ | A | 138 | 81 | 135 | |
| L | $N_2$ | A | 212 | 130 | 165 | |
| L | $N_2$ | A | 224 | 130 | 165 | |
| L | $N_2$ | A | 192 | 160 | 210 | |
| L | $N_2$ | A | 207 | 160 | 210 | |
| L/O | Air, $N_2$, $H_2$, CO, HE, NG | Radial Inflow | Depending on the type of loading. | | | Loading |
| L/O | Hydrocarbons Refrigerants | " | | | | F—Fluid Brake, B—Compressor, G—Generator Integral |
| L/O | " | " | | | | |
| L/O | " | " | | | | |
| L/O | " | " | | | | |
| L/O | " | " | | | | |
| L/O | " | " | | | | |
| L/O | " | " | | | | Gear Multi Stage Designs Are Offered For Special Applications |
| L/O | " | " | | | | |
| L/O | " | " | | | | |
| L | Tail gas* | A | Depending on size | | | *usually high $N_2$ content with acidic impurities |
| L | Tail gas | A | Depending on size | | | |
| L | Tail gas | A | Depending on size | | | |
| L | Flue gas, helium and others | A | Depending on size | | | |

# TURBO EXPANDER SPECIFICATIONS

| MODEL | INLET PRESSURE BAR-A | INLET TEMPERATURE °C | INLET FLOW KG PER HR | NUMBER OF STAGES | RPM | H-HORIZONTAL OR V-VERTICAL SPLIT CASING |
|---|---|---|---|---|---|---|
| **Mafi-Trench Corporation, 3037 Industrial Parkway, Santa Maria, California 93455** | | | | | | |
| | (psia) | | | | | |
| 1 | 1,000 | −250°F to +250°F | 480 | 1 | 95,000 | V |
| 2 | 1,000 | −250°F to +250°F | 1,750 | 1 | 55,000 | V |
| 2½ | 1,000 | −250°F to +250°F to | 2,300 | 1 | 40,000 | V |
| 3 | 1,000 | −250°F +250°F | 4,700 | 1 | 34,000 | V |
| 3½ | 1,000 | −250°F −250°F to | 8,300 | 1 | 29,000 | V |
| 4 | 1,000 | +250°F +250°F | 11,800 | 1 | 24,000 | V |
| 5 | 1,000 | −250°F to +250°F | 21,000 | 1 | 17,000 | V |
| 10 | 700 | +400°F | 53,000 | 1 | 6,400 | V |
| Note: 3, 4, and 5 can also be motor driven compressors only. | | | | | | |
| **Mitsubishi Heavy Industries, Ltd., 5-1 Marunouchi, 2-Chome, Chiyoda-ku, Tokyo 100 Japan** | | | | | | |
| | (BAR-A) | (°C) | (Kg/min.) | | | |
| 10 Cases | to 200 | Amb. to 750 | 5,000 | up to 12 | | H |
| **Mitsui Engineering & Shipbuilding Co., Ltd., 6-4 Tsukiji 5-Chome, Chuo-ku, Tokyo, Japan** | | | | | | |
| | (BAR-A) | (°C) | (kg/hr) | | | |
| GE300-GE1200 | Max. 11.0 | Max. 540 | 36,000-103,000 | Max. 6 | 12,000-7,200 | H |
| GE800D-GE1200D | Max. 11.0 | Max. 700 | 75,700-103,000 | Max. 6 | 7,700-7,000 | H |
| | (Kg/cm²G*) | (°C) | (NM³/hr.dry) | | | |
| MAT140W | Max. 4.5 | 55 | Max. 520,000 | 3 | Max. 3,600 | H |
| MAT160W | Max. 4.5 | 55 | Max. 670,000 | 3 | Max. 3,600 | H |
| MAT180W | Max. 4.5 | 55 | Max. 850,000 | 3 | Max. 3,600 | H |
| MAT1600 | Max. 4.5 | 150 | Max. 410,000 | 3 | Max. 2,800 | H |
| MAT1800 | Max. 4.5 | 150 | Max. 520,000 | 3 | Max. 2,400 | H |
| MAT2100 | Max. 4.5 | 150 | Max. 670,000 | 3 | Max. 2,100 | H |
| MAT2300 | Max. 4.5 | 150 | Max. 850,000 | 3 | Max. 1,900 | H |
| Mitsui Eng. & Shipbuilding Co., 6-4 Tsukiji 5-Chome, Chuo-ku, Tokyo, Japan. Contact: M. Yamanoto. Phone: 03-544-3512, Telex: J 22821, J22924. | | | | | | |
| **NEU — Turbomachines, P.O. Box 110, 47, rue Fourier, 59003 Lille Cedex, France** | | | | | | |
| | (BAR-A) | (°C) | ($M_3$/hr) | | | |
| T F | 5 | 100 | 600,000 | 1-2 | — | H |
| T H | 100 | 50 | 10,000 | 1-2 | — | H |
| T V | 100 | 50 | 10,000 | 1-2 | — | V |

# TURBO EXPANDER SPECIFICATIONS

| TYPE OF SEALS L = LABYRINTH O = OIL | TYPE OF GAS EXPANDER CAN HANDLE | TYPE OF DESIGN C = CENTRIFUGAL A = AXIAL | DIMENSIONS L | W | H | COMMENTS |
|---|---|---|---|---|---|---|
| | | | (in) | (in) | (in) | LOAD: |
| L | Nat gas, $N_2$, $H_2$, Air | C | 38 | 21 | 50 | Compressor, Blower, Hydrobrake |
| L | Nat. gas, $N_2$, $H_2$, Air | C | 55 | 30 | 60 | Compressor, Blower, Hydrobrake |
| L | Nat. gas $N_2$, $H_2$, Air | C | 70 | 35 | 70 | Compressor |
| L | Nat. gas $N_2$, $H_2$, Air | C | 85 | 40 | 70 | Compressor, Generator |
| L | Nat. gas $N_2$,$H_2$,Air | C | 110 | 60 | 90 | Compressor, Generator |
| L | Nat. gas $N_2$, $H_2$, Air | C | 110 | 60 | 90 | Compressor, Generator |
| L | Nat. gas $N_2$, $H_2$, Air | C | 140 | 65 | 90 | Compressor, Generator |
| L | $C_4$ | C | 150 | 112 | 100 | Compressor, Generator |
| L | Any types | A | | | | |
| L | $N_2$ | A | | | | |
| L | $N_2$ | A | | | | |
| L | B.F.G WET | A | | | | |
| L | B.F.G. WET | A | | | | |
| L | B.F.G. WET | A | | | | |
| L | B.F.G. DRY | A | | | | |
| L | B.F.G. DRY | A | | | | Furnace Exhaust Gas Recovery Turbine |
| L | B.F.G. DRY | A | | | | |
| L | B.F.G. DRY | A | | | | |
| L + Carbon Rings | Blast Furnace Gas | C | — | — | — | |
| L | $N_2$—Air, Nat. Gas | C | — | — | — | |
| L | $N_2$—Air, Nat. Gas | C | — | — | — | |

# TURBO EXPANDER SPECIFICATIONS

**Rotoflow Corporation, 2235 Carmelina Ave., Los Angeles, California 90064 U.S.A.**

| MODEL | INLET PRESSURE BAR-A | INLET TEMPERATURE °C | INLET FLOW KG PER HR | NUMBER OF STAGES | RPM | H-HORIZONTAL OR V-VERTICAL SPLIT CASING |
|---|---|---|---|---|---|---|
| | (psia) | (°F) | (cubic ft./min.*) | | | |
| 10R | 1,500 (Std.) 3,000 | +450 to −450 | 235 | ½ | up to 75,000 | V |
| 20R | 1,500 (Std.) 3,000 | +450 to −450 | 420 | ½ | up to 55,000 | V |
| 30R | 1,500 (Std.) 3,000 | +450 to −450 | 1,000 | ½ | up to 39,000 | V |
| 40R | 1,500 (Std.) 3,000 | +450 to −450 | 1,700 | ½ | up to 26,000 | V |
| 50R | 1,500 (Std.) 3,000 | +450 to −450 | 3,800 | ½ | up to 20,000 | V |
| 60R | 1,500 (Std.) 3,000 | +450 to −450 | 10,500 | ½ | up to 15,000 | V |
| 100R | 720, 1,500 | 1,100°F Special | 100,000 | ½ | Up to 4,000 | V and H |

**Major Offices Worldwide**: Rotoflow Corporation — Sales & Service Contacts — Los Angeles, CA, Telex: 67-3687 Rotoflow LSA • ing. Mario E. Pavon, Tiber 68-404, Mexico, D.F. Mexico, Tel: 528-77-69, Telex: 001777583 INESME • C. Itoh & Company (Japan), 555 S. Flower St., Los Angeles, CA 90071, Tel: 213-687-0610, Telex: 6-7281 ITOAME LSA • Peter Burkardt (RAG), 8547 Gachnang, TG. CH (Switzerland), Tel: 054-947-84, Telex: 76557 ROINT CH • R. G. Jolliffe & Co., Ltd., 6 Annerly Station Road, London SE20 8PT England, Tel: 01-778-1000, Telex: 897200 TELEJOG • Technisch Bureau Van Aken, Jan Van Nassaustraat 117, Den Haag, Netherlands, Tel: 070-247566, Telex: 32029 Aken NL • Sherman Corporation, Himalaya House/23 Kasturba Ghandi Marg, New Delhi 110001, India, Tel: 40334, Telex: 314092 SRMN IN • DuFour Pere, Fils Et Cie, 11 Rue De L'Aspirant-Dargent, 92300, Paris, France, Tel: 758-54-50, Telex: 620293 Riolfor.

**Sulzer Brothers Ltd., Thermal Turbomachinery, P. O. Box CH 8023, Zurich, Switzerland**

| MODEL | INLET PRESSURE BAR-A | INLET TEMPERATURE °C | INLET FLOW KG PER HR | NUMBER OF STAGES | RPM | H-HORIZONTAL OR V-VERTICAL SPLIT CASING |
|---|---|---|---|---|---|---|
| | (bar-a) | (°C) | (KG/hr) | | | |
| AE | | | 25,000 | | 15,500 | |
| 7 sizes | 17/11 | 450/500 | 400,000 | 2-4 | 5,500 | H |
| AEV | | | 25,000 | | 15,500 | |
| 7 sizes | 17/11 | 450/500 | 400,000 | 2-4 | 5,500 | H |
| AEH | | | 75,000 | | 9,500 | |
| 2 sizes | 16 | 710 | 185,000 | 6 | 6,800 | H |

**Major Offices Worldwide**: B-1040 Bruxelles, S.A. Sulzer Belgium N.V., 92, Square Plasky • Dk-1554 Kopenhagen V., AAGE Christensen A/S, Studiestraede 63 • D-7980 Ravensburg, Escher Wyss GmbH, Postfach 1380 • F-75300 Paris-Brune, Compagnie de Construction Mecanique Sulzer, Cedex 59 • Pointe-Claire Dorval 700 (Quebec), Bros. (Canada) Ltd. P. O. Box 210 • New York, N.Y. 10017, Sulzer Bros. Inc., 200 Park Avenue • Tokyo 100 Sulzer Brothers (Japan) Ltd., C.P.O. Box 147

**Tech Development, Inc., 6800 Poe Ave., P. O. Box 1400-0, Dayton, Ohio 45414 U.S.A.**

| MODEL | INLET PRESSURE BAR-A | INLET TEMPERATURE °C | INLET FLOW KG PER HR | NUMBER OF STAGES | RPM | H-HORIZONTAL OR V-VERTICAL SPLIT CASING |
|---|---|---|---|---|---|---|
| | (psig) | (°F) | (lb./min.) | | | |
| 402 | 186 | 59 | 58 | 4 | 15,000 | V |
| 845A | 500 | 59 | 58 | 4 | 24,000 | V |
| 866A | 150 | 59 | 31 | 1 | 100,000 | V |
| 1134B | 60 | 59 | 10 | 1 | 20,000 | V |
| Series 10A | 60 | 70 | 6.8 | 1 | 36,000 | V |
| Series 51A | 90 | 70 | 112.0 | 1 | 2,700 | V |

**Worthington Division, McGraw-Edison Company, 5 Washington Ave., Fairfield, New Jersey 07006**

| MODEL | INLET PRESSURE BAR-A | INLET TEMPERATURE °C | INLET FLOW KG PER HR | NUMBER OF STAGES | RPM | H-HORIZONTAL OR V-VERTICAL SPLIT CASING |
|---|---|---|---|---|---|---|
| | (psia) | (°F) | (lb./min.) | | | |
| 200 | 150 max. | +500 to −200 | 25 max. | 1 | 90,000 | V |
| 300 | 150 max. | +500 to −200 | 75 max. | 1 | 60,000 | V |
| 400 | 900 max. | +500 to −200 | 1,000 max. | 1 | 48,000 | V |
| 600 | 900 max. | +500 to −200 | 2,500 max. | 1 | 32,000 | V |
| 900 | 150 max. | +500 to −200 | 1,000 max. | 1 | 21,000 | V |
| 1200 | 150 max. | +500 to −200 | 1,500 max. | 1 | 16,000 | V |
| 1500 | 150 max. | +500 to −200 | 2,000 max. | 1 | 12,000 | V |

NOTE: All models are available with or without Control/Lube system console.

## TURBO EXPANDER SPECIFICATIONS

| TYPE OF SEALS L = LABYRINTH O = OIL | TYPE OF GAS EXPANDER CAN HANDLE | TYPE OF DESIGN C = CENTRIFUGAL A = AXIAL | DIMENSIONS L | W | H | COMMENTS |
|---|---|---|---|---|---|---|
| | | | | (Inches) | | |
| L | Clean gas can be corrosive & can have liquid at inlet | C | 36 | 18 | 24 | Air Sep. Plants |
| L | '' | C | 43 | 22 | 27 | Natural Gas Drying |
| L | '' | C | 65 | 26 | 36 | Power Recovery |
| L | '' | C | 83 | 26 | 38 | Ethylene Recovery |
| L | '' | C | 136 | 56 | 65 | Hydrocarbon Liquids Recovery |
| L | '' | C | 167 | 64 | 105 | Hydrocarbon Liquids Recovery |
| L | '' | C | 194 | 128 | 136 | Large Power Recovery Binary Syst. |
| L | $N_2$, Tail gases | A | | | | |
| L | $N_2$, Tail gases | A | | | | with control valves |
| L | $N_2$, Tail gases | A | | | | cooled inlet |
| | | | | (Inches) | | pneumatic motor mode |
| L | Air, $N_2$ | A | 6.5 | 3.25 dia. | — | 50 HP output |
| L | Air, $N_2$ | A | 4.8 | 2.0 | 2.0 | 59 HP output |
| L | Air, $N_2$ | A | 4.3 | 2.50 dia. | — | 25 HP output |
| L | Air, $N_2$ | A | 6.0 | 5.0 dia. | — | 3 HP output |
| L | Air, $N_2$ | A | 5.7 | 3.2 dia. | — | 2.0 HP output |
| O | Air, $N_2$, NG, Freon | A | 11.8 | 6.1 dia. | — | 75.0 HP output |
| L | Nitrogen, Air | C | Depending on Model consult plant | | | Dynamometer or Blower |
| L | Nitrogen, Air | C | | | | Dynamometer or Blower |
| L | Nitrogen, Air, Hydro-carbon | C | | | | Compressor or Generator |
| L | Nitrogen, Air, Hydro-carbon | C | | | | Compressor or Generator |
| L | Nitrogen, Air | C | | | | Compressor or Generator |
| L | Nitrogen, Air | C | | | | Compressor or Generator |
| L | Nitrogen, Air | C | | | | Compressor or Generator |

# HYDRO TURBINE SPECIFICATIONS

| TYPE | EFFECTIVE HEAD | REQUIRED DISCHARGE | MAXIMUM POWER OUTPUT | TURBINE SPEED RANGE | INLET DIAMETER | RUNNER DIAMETER |
|---|---|---|---|---|---|---|
| **Allis-Chalmers,** Hydro-Turbine **Division, Box 712,** York, **Pennsylvania 17405, Contact:** Howard A. **Mayo, Jr. Phone: 717-792-3511** | | | | | | |
| Standard Propeller Units:[1] | | | | | | |
| Mini TUBE (fixed blade) | To 40 ft. (12m) | 20-1,000 ft.³/s (1-30m³/s) | 50-2,000kW$_G$ | Varies | 1.5-11 ft. | 1-7.5 ft. |
| Standard TUBE (adj. or fixed) | To 100 ft. (30m) | 50-1,800 ft.³/s (2-50m³/s) | 100-6,000kWG | Varies | 3.5-20 ft. (1.1-6m) | 2.5-13 ft. (0.75-4m) |
| Gated TUBE[2] | To 80 ft. (25m) | 1,000-3,000 ft³/s (30m³/s-90m³/s) | 100-15,000kW$_G$ | Varies | To 22 ft. (6.8m) | To 13ft. (4m) |
| Custom Designed Turbines: | | | | | | |
| Bulb Unit[2] | To 65 ft. (20m) | | To 50 MW$_G$ | | | To 33 ft. (8m) |
| Vertical Propeller[2] | To 200 ft. (60m) | | To 250 MW$_G$ | Varies | Varies | To 31 ft. (9.5m) |
| Francis | To 2,000 ft. (200m) | | To 800 MW$_G$ | | | To 33 ft. (8m) |
| Pump/Turbine (Single Stage) | To 3,300 ft. (1000m) | | To 500 MW$_G$ | | | To 21 ft. (6.5m) |
| Impulse (Pelton) | To 3,300 ft. (1000m) | | To 100 MW$_G$ | | | To 10 ft. (3m) |
| **Ateliers Bouvier**, 53 Rue Pierre **Semard, 38000 Grenoble**, France | | | | | | |
| | (Feet) | (cfs) | (kW) | (rpm) | | (Feet) |
| Kaplan | 15 to 60 | 90 to 1,600 | 5,000 | Up to 1,000 | | Max. 10 |
| Francis | 60 to 750 | 25 to 700 | 20,000 | Up to 1,500 | | Max. 5 |
| Pelton | 300 to 3,000 | 5 to 140 | 20,000 | Up to 1,500 | | Max. 5 |
| **Bouvier Hydropower**, Inc. General **Sales Office, 12** Bayard Lane, **Suffern, NY 10901** U.S.A., Contact: **Reese W. May, Phone: 914-357-2189**. | | | | | | |
| | (Feet) | (cfs) | (kW) | (rpm) | | (Feet) |
| Kaplan | 15 to 60 | 90 to 1,600 | 5,000 | Up to 900 | | Max. 8 |
| Francis | 100 to 750 | 25 to 400 | 5,000 | Up to 1,200 | | Max. 4 |
| Pelton | 300 to 3,000 | 5 to 95 | 5,000 | Up to 1,200 | | Max. 4 |
| **CE/N Hydro Power** Inc., 277 Park **Avenue, New York**, NY 10022 | | | | | | |
| | (Feet) | (cfs) | (kW) | (rpm) | | (Feet) |
| Kaplan | 40 to 220 | 250 to 24,000 | 1,000 to 120,000 | 70 to 375 | 3D (D runner diameter) | 3 to 25 |
| Francis | 160 to 1,150 | 80 to 25,000 | 1,000 to 750,000 | 80 to 600 | 3D | 3 to 28 |
| Peltron | 1,000 to 4,500 | 5 to 1,800 | 1,000 to 200,000 | 250 to 1,000 | N/A | 2 to 10 |
| Bulbs | 9 to 70 | 1000 to 15,000 | 3,000 to 80,000 | 50 to 180 | 1.8 D | 9 to 28 |
| Pump turbine | 30 to 4,500 | 300 to 9,500 | 600 to 250,000 kW | 88 to 750 | depends upon type and head | 3 to 20 |

[1]Complete unit from intake gate or valve to power transformer. [2]Kaplan, adjustable blade, fixed blade.

## HYDRO TURBINE SPECIFICATIONS

| TYPE | EFFECTIVE HEAD | REQUIRED DISCHARGE | MAXIMUM POWER OUTPUT | TURBINE SPEED RANGE | INLET DIAMETER | RUNNER DIAMETER |
|---|---|---|---|---|---|---|
| **Escher Wyss Limited Hydraulics, P.O. Box CH-8023, Zurich, Switzerland** | | | | | | |
| Bulb | 2.3-18.3m | | 40,706 kW | 50-289 | | |
| Francis | up to 700m + | | 340,000 kW | 107.1-1,500 | | |
| Kaplan | up to 70m | | 110,000 kW | 60-500 | | |
| Pelton | up to 1,233m | | 260,000 kW | 250-1,000 | | |
| Propellor | up to 70m | | | | | |
| Pump Turbines | up to 1,256m | | 250,000 kW | 83-1,500 | | |
| Storage Pumps' | up to 1,438.6m | | 234,000 kW | 176-1,500 | | |
| Straight Flow | 7.5-9m | | 1,900 kW | | | 1.85-2.05m |
| Straflo | up to 10.3m | | 8,350 kW | | | 3.0-3.7m |
| Large Straflo | 7m | | 20,000 kW | | | 7.6m |
| **Ganz-Mavag, H-1967 Budapest, P.O. Box 136, Budapest VIII, Konyves, Kalman Krt. 76, Contact: Pal Niederlander, Phone: 335-950.** | | | | | | |
| Kaplan | 2-34m | | 10-50 MW | | | max. 6.5m |
| Francis | 10-400m | | to 100 MW | | | max. 5.0m |
| Pelton | to 1000m | | to 100 MW | | | max. 3.0m |
| Bulb | 2-15m | | to 20 MW | | | max. 5.0m |
| **Hayward Tyler Ltd., P.O. Box 2, Luton, Beds, England, LUI 3LW** | | | | | | |
| Pelton* | 2,500 ft. | Dependant on duty | 3,000 HP | 1500/6300 rpm | To suit application | To suit application up to 27". |
| *Ideal application R.O. Plant energy recovery, or other high pressure duties. | | | | | | |
| **Kvaerner Brug A/S, Energy Division, P.O. Box 3610 Gb, Oslo 1, Norway, Tel. 47 2 67 69 70, Telex 71650 kb n** | | | | | | |
| | (meters) | ($m^3/s$) | (kW) | (rpm) | (meters) | (meters) |
| Pelton | 885 | 40 | 315,000 | 300 | 2.05 | 4.06 |
| Pelton | 1,126 | 27.4 | 260,000 | 428 | 1.65 | 3.12 |
| Pelton | 840 | 32.7 | 242,000 | 375 | 1.875 | 3.19 |
| Reversible Pump Turbine | 425 | 43 | 160,000 | 428 | 1.8 | 4.0 |
| Reversible Pump Turbine | 400 | 43 | 150,000 | 500 | 1.8 | 3.27 |
| Francis | 105 | 514.5 | 469,000 | 107.1 | 8.506 | 6.6 |
| Francis | 520 | 66 | 315,000 | 333 | 2.2 | 4.2 |
| Francis | 33 | 55 | 16,200 | 187 | 3.2 | 2.8 |
| Kaplan | 39 | 556 | 181,700 | 107.1 | 10.35 | 7.5 |
| Kaplan | 10.6 | 472 | 40,500 | 65.2 | Area 351.9 $m^2$ | 7.8 |
| Bulb | 23 | 446.6 | 88,000 | 100 | Area 159.8 $m^2$ | 7.00 |
| Bulb | 5 | 240 | 10,800 | 71.4 | 11.0 | 6.05 |
| Delivered by Nohab Turbinteknik, member of Kvaerner Group. | | | | | | |
| **Kvaerner Moss Inc., 800 Third Avenue, New York, NY 10022, Contact: James J. Victory V. P., Phone: 212-752-7310.** | | | | | | |
| | (Meters) | | | | | |
| Pelton | 550 & up | | Manufacture complete line of turbines from 1MW up to largest built. Cover all ranges of speeds heads and flows. Manufacturing companies are Kvaerner Brug A/S, Sorumsand A/S and Kvaerner Nohab A/S. All are located in Oslo, Norway. | | | |
| Francis | 10 to 550 | | | | | |
| Kaplan | Up to 10 | | | | | |
| Bulb | Up to 10 | | | | | |
| Tube | Up to 10 | | | | | |

# HYDRO TURBINE SPECIFICATIONS

| TYPE | EFFECTIVE HEAD | REQUIRED DISCHARGE | MAXIMUM POWER OUTPUT | TURBINE SPEED RANGE | INLET DIAMETER | RUNNER DIAMETER |
|---|---|---|---|---|---|---|
| **Mitsubishi Heavy Industries, Ltd., 5-1 Marunouchi, 2-chome, Chiyoda-ku, Tokyo 100, Japan** | | | | | | |
| | (M) | (M³/S) | | | (M) | (M) |
| Francis | 438 | 11 | 42,000 | 600 | 1.0 | 2.1 |
| Francis | 185 | 184.1 | 318,000 | 163.6 | 4.58 | 5.07 |
| Kaplan | 55 | 33.3 | 16,000 | 333 | — | 2.25 |
| Kaplan | 37.3 | 50 | 50,000 | 150 | 5.58 | 4.9 |
| Deriaz | 57 | 99.1 | 50,000 | 180 | 4.2 | 3.9 |
| Pelton | 559.7 | 5.18 | 27,000 | 600 | — | 1.57 |
| Tubler | 13.7 | 40 | 4,800 | 240/720 | — | 2.3 |
| Francis Pump Turbine | 557.4/576 | 64/45.8 | 307,000/294,000 | 400 | 2.31 | 4.9 |
| Francis Pump Turbine | 241.1/264.4 | 155.5/100.6 | 336,000/297,000 | 214 | 3.62 | 6.33 |
| Deriaz Pump Turbine | 117/122 | 97.1/74.2 | 102,000/103,000 | 231 | 3.68 | 4.15 |
| **Obermeyer Hydraulic Turbines Ltd., 10 Front Street, Collinsville, Connecticut 06022, Phone: 203-693-4292, Contact: Mathew W. Roth** | | | | | | |
| | (Feet) | (cfs) | | (rpm) | (Inches) | (Inches) |
| Slant Type | To 50 | 1 to 200 | 700kW | To 1,800 | 8 to 60 | 5 to 40 |
| Axial Flow Bulb Type | To 100 | 120-1,200 | 5MW | 900 output w/integral speed increaser | 60 to 150 | 40 to 100 |
| Francis | To 1,000 | 1 to 1,000 | 5MW | To 1,800 | 14 to 200 | 8 to 100 |
| Pelton | To 5,000 | .1 to 100 | 5MW | To 3,600 | 1 to 20 | 6 to 80 |
| **J. M. Voith GmbH, Postfach 1940, D-7920 Heidenheim, West Germany** | | | | | | |
| | (Meters) | (m³/s) | (kW) | (rpm) | (Meters) | (Meters) |
| Francis | 118.4 | 680 | 740,000 | 92.3 | 9.64 | 8.6 |
| Francis | 380 | 12.45 | 41,200 | 600 | 2.15 | 1.92 |
| Kaplan | 77.2 | 16.5 | 11,050 | 610 | 1.85 | 1.49 |
| Kaplan | 26.7 | 511.03 | 103,000 | 85.71 | 22.65 | 7.70 |
| Pelton | 1,095 | 20.62 | 197,510 | 500 | 1.45 | 2.855 |
| Pelton | 413 | 46.12 | 166,910 | 180 | 2.4 | 4.350 |
| Pumpenturbine | Tu: 348 | 87 | 271,000 | 375 | 2.55 | 4.350 |
| | Pu: 346 | 67 | 246,000 | | | |
| Pumpenturbine | Tu: 542 | 32 | 155,000 | 600 | 1.4 | 3.137 |
| | Pu: 552 | 20.5 | 121,000 | | | |
| Bulb | 13.6 | 334.8 | 41,220 | 103.4 | 11.7 | 6.0 |
| Bulb | 11.97 | 352 | 22,280 | 85.71 | 12.6 | 6.3 |
| **Worthington Division, McGraw Edison Co., 5310 Taneytown Pike, Taneytown, Maryland 21787, Contact: Larry Shafer/John Scheivert. Phone: 301-756-2602** | | | | | | |
| | (Feet) | (cfs) | (kW) | (rpm) | (Feet) | (Feet) |
| Axial Flow Models KLT/KMT | Up to 50 | Up to 200 | Up to 700 | 400 to 1,800 | 6 to 60 | 5 to 50 |
| Francis Models MFT/MNT | 50 to 200 | Up to 100 | Up to 1,000 | 400 to 1,800 | 6 to 42 | 8 to 48 |
| Double Discharge Models LNT | 150 to 700 | Up to 100 | Up to 1,000 | 600 to 1,800 | 6 to 36 | 10 to 35 |

# STEAM TURBINE SPECIFICATIONS

| ELECTRIC POWER GENERATION | | | | MECHANICAL DRIVE | |
|---|---|---|---|---|---|
| ELECTRIC UTILITY MW RANGE | | INDUSTRIAL | PROCESS | MARINE | OTHER |
| Fossil Fueled | Nuclear | HP or KW Range | HP or KW Range | HP or KW Range | HP or KW Range |
| **ACEC (Ateliers de Constructions Electriques de Charleroi)** | | | | | |
| 1-20 MW | — | 1-20 MW[a]<br>1-15 MW[b] | 1-20MWa<br>1-15MW[b] | — | |
| [a]Back-pressure. [b]Condensing. | | | | | |
| **AEG-KANIS TURBINENFABRIK GmbH** | | | | | |
| 100kW-160 MW | — | 100kW-160 MW | 30 kW-90 MW | Auxiliary Turbines<br>30kW-5,000kW | 30kW-90 MW |
| **ALSTHOM-ATLANTIQUE Large Turbine Generator Division** | | | | | |
| all ratings | all ratings to 1350 MW and above | | | | |
| **ALSTHOM-ATLANTIQUE Mechanical Division** | | | | | |
| | | Approx. 3MW and up | Approx. 3MW and up | | |
| **BLOHM + VOSS AG** | | | | | |
| Condensing 0.5 to 30/40 MW<br>Backpressure 0.5 to 60 MW | | Condensing 0.5 to 30/40 MW<br>Backpressure 0.5 to 60 MW | Condensing 0.5 to 30/40 MW<br>Backpressure 0.5 to 60 MW | Propulsion Turbines 7.5 to 40MW<br>Auxiliary Turbines above 0.5MW | |
| **BORSIG GmbH** | | | | | |
| 3-60 MW[*] | — | 3-60 MW[*] | 0.5-30 MW[*] | — | — |
| *back pressure, condensing, extraction, double extraction, bled steam, speed from 3,000-16,000r/m. | | | | | |
| **BBC Brown, Boveri & Company, Ltd., Baden, Switzerland**<br>**Brown, Boveri & Cie. AG, Mannheim, Fed. Rep. Germany**<br>**Brown, Boveri Corporation, North Brunswick, NJ** | | | | | |
| up to 1350MW | up to 1350MW | 2-150MW | 2-100MW | Auxiliary Turbines above 2 MW | Auxiliary turbines over 2MW for feed pump drives |
| **BROWN BOVERI TURBOMACHINERY, INC., ST. CLOUD, MN** | | | | | |
| 15-150 MW | — | 15-150 MW | 15-100 MW | — | — |

# STEAM TURBINE SPECIFICATIONS

| ELECTRIC POWER GENERATION | | | | MECHANICAL DRIVE | |
|---|---|---|---|---|---|
| ELECTRIC UTILITY MW RANGE | | INDUSTRIAL | PROCESS | MARINE | OTHER |
| Fossil Fueled | Nuclear | HP or KW Range | HP or KW Range | HP or KW Range | HP or KW Range |
| **CARLING TURBINE BLOWER CO.** | | | | | |
| — | — | Up to 750 kW | Fractional to 1000 hp | Fractional to 1000 hp | |
| **COPPUS ENGINEERING CORPORATION** | | | | | |
| — | — | ½ to 4000 hp up to 2500 kW | ½ to 4000 hp | ½ to 4000 hp up to 2500 kW | — |
| **CREUSOT-LOIRE** | | | | | |
| up to 60 MW Main and auxiliary | up to 60 MW | up to 60 MW | up to 60 MW | up to 45 MW Main and auxiliary | up to 60 MW for compressors, pumps, blower |
| **ELLIOTT CO.** | | | | | |
| Up to 55 MW | — | Up to 55 MW | Up to 60,000 HP | — | — |
| **GEC TURBINE GENERATORS LTD., Industrial & Marine Steam Turbine Div.** | | | | | |
| 5-140 MW[1] | Up to 80 MW[2] | 5-80 MW[3] | 5,000-60,000 HP[4] | 10,000-70,000SHP | |
| [1]900-1800 lb/in² (62-124 bar g), 900-1000 °F (482-538 °C). [2]Single Cylinder Turbines. [3]900-1800 lb/in² (62-124 bar g), 900-1000 °F (482-538 °C) Extr'n. 20-600 lb/in², BP 15-600 lb/in², Exhaust up to 28.5 ins. Hg. [4]4000-11000 rev/min., Steam Conditions up to 1800 lb/in²/1000 °F (124 bar g/538 °C) | | | | | |
| **GENERAL ELECTRIC CO.** | | | | | |
| BFP Drivers up to 50,000hp | — | Up to 10,000kW | Up to 100,000hp | Up to 10,000kW | Up to 100,000hp |
| **HAYWARD TYLER LTD.** | | | | | |
| | 1500 kW 2 stage | 3 kW-2.5 MW | ½-2500 hp | ½-2500 hp | ½-2500 hp |
| Inlet up to 900 PSIG Temp. 900 °F. Back Pressure 175 PSIG. Single stage, single, double row, and bucket wheel. | | | | | |
| **IHI, CO. LTD.** | | | | | |
| — | — | 1-5 MW | 1-20 MW | 8,000-70,000hp* for Marine Propulsion 500-4,000 kW for Generator 200-5,000 kW for cargo oil & Ballast Pump | 1-20 MW |
| **KAWASAKI HEAVY INDUSTRIES LTD.** | | | | | |
| | | 2-80 MW | 2-80 MW | Various HP | |

# STEAM TURBINE SPECIFICATIONS

| ELECTRIC POWER GENERATION | | | MECHANICAL DRIVE | | | |
|---|---|---|---|---|---|---|
| ELECTRIC UTILITY MW RANGE | | INDUSTRIAL | PROCESS | MARINE | OTHER | |
| Fossil Fueled | Nuclear | HP or KW Range | HP or KW Range | HP or KW Range | HP or KW Range | |
| **KKK AG, KUEHNLE, KOPP & KAUSCH** | | | | | | |
| Up to 6 MW | — | Up to 6,000 kW | 1 kW-6,000 kW | 1 kW-6,000 kW | 1 kw-6,000 kW | |
| **KRAFTWERK UNION** | | | | | | |
| 30-750 MW(a) | 300-1,350 (2,000) MW(a) | — | Heating Turbines for all requirements | 18,000-32,000 (90,000) Hp(a) | | |
| **M.A.N. Unternehmensereich GHH STERKRADE** | | | | | | |
| | | 2-80 MW | 2-80 MW | | 2-80 MW | |
| **MITSUBISHI HEAVY INDUSTRIES LTD.** | | | | | | |
| Up to 1,000 MW | 340 MW-1,175 MW | 1 MW-125 MW | Up to 100 MW | Up to 60,000 MW | | |
| **MITSUI** | | | | | | |
| 1-120 MW | — | 1-120 MW | Nominal 1-120 MW | 12-70 MW | | |
| **NEI-APE LTD. — W. H. ALLEN** | | | | | | |
| Multi-stage 1-30 MW Single stage 50-5000kW | 1-30MW 50-5000kW | 1-30MW 50-5000kW | 1-30MW 50-5000kW | 1-10MW 400-3000kW Auxiliary | | |
| **NUOVO PIGNONE** | | 2,000-80,000 kW | | | | |
| **PETER BROTHERHOOD LTD.** | | 400-15,000 kW | 400-15,000 kW | 400-3,000 kW | Single Stage 400-4,000 kW | |
| **PRVNI BRNENSKA STROJIRNA** | | | | | | |
| Up to 50 MW | 12 MW | Up to 70 MW 50 & 60 HZ | 800-32,000 kW | — | Auxiliary 30-2600 kW | |
| **SIEMENS A.G.** | | | | | | |
| Boiler feed-pump drives up to 60,000 BHP or 45 MW | Boiler feed pump drives up to 60,000 BHP or 45 MW | 1-150 MW or 1,600-200,000 BHP | 1-150 MW or 1,600-200,000 BHP | | | |
| **STAL-LAVAL TURBIN AB** | | | | | | |
| 200-1100 MW | 200-1100 MW | 4-60 MW[1 3] 50-200 MW[2 3] | 4-60 MW[1 3] 0.7-4 MW[4] | 10-33 MW | 50-200 MW[5] | |

[1]Contra-rotating, radial flow. [2]Axial Flow. [3]Combined heat and power generation. [4]Also electric power generation. [5]District heating.

# STEAM TURBINE SPECIFICATIONS

| ELECTRIC POWER GENERATION | | | | MECHANICAL DRIVE | |
|---|---|---|---|---|---|
| ELECTRIC UTILITY MW RANGE | | INDUSTRIAL | PROCESS | MARINE | OTHER |
| Fossil Fueled | Nuclear | HP or KW Range | HP or KW Range | HP or KW Range | HP or KW Range |
| **TERRY STEAM TURBINE CO.** | | | | | |
| To 10 MW | To 10 MW | To 20,000hp<br>To 10,000kW | To 20,000 hp<br>To 10,000kW | To 5,000kW<br>To 10,000 hp | To 20,000 hp<br>To 10,000kW |
| **THERMO ELECTRON CORP.** | | | | | |
| 20 MW | | 500 kW-20,000 kW | | | 500 kW-20,000 kW |
| **TOSHIBA CORPORATION** | | | | | |
| Up to 1,000 MW | Up to 1,100 MW | Up to 175 MW | Up to 38 MW | | Auxiliary up to 45 MW |
| **THE TRANE COMPANY, Process Division Steam Turbine Group** | | | | | |
| 200-10,000 kW<br>50-15,000 hp | 200-10,000 kW<br>50-15,000 hp | 200-10,000kW*<br>50-15,000 hp | 200-10,000 kW*<br>50-15,000 hp | 200-10,000 kW<br>50-15,000 hp | 200-10,000 kW*<br>50-15,000 hp |
| *Condensing, Back Pressure, Extraction Available with Custom Capabilities. | | | | | |
| **TRANSAMERICA DELAVAL** | | | | | |
| 10-80 MW<br>5,000-45,000 hp[a] | 5,000-20,000 hp[a] | 10-80 MW | 1,000-80,000 hp | 10,000-70,000shp<br>1,000-3,000 kW[b] | |
| [a]Auxiliary drives. [b]Shipboard lighting sets (SSTG) | | | | | |
| **TURBODYNE DIVISION — McGRAW-EDISON CO.** | | | | | |
| 50-60,000 kW<br>10-80,000 hp | — | 50-60,000 kW<br>10-80,000 hp | 50-60,000 kW<br>10-80,000 hp | 50-3,000 kW<br>10-5,000 hp | |
| **TURBONETICS ENERGY, INC., (A subsidiary of Mechanical Technology Inc.)** | | | | | |
| 1-4 | N/A | 500-5,000 hp | 500-5,000 hp | N/A | |
| **UTILITY POWER CORP.** | | | | | |
| 400-2,500 MW<br>FOR U.S. UTILITIES ONLY. | 400-2,500 MW | — | — | — | — |
| **WESTINGHOUSE CANADA INC., Turbine & Generator Division** | | | | | |
| 100 kW-60,000 kW | — | 100 kW-60,000 kW[a] | 50 hp-80,000 hp[a] | — | Boiler Feed Pump Turbines 7,000-30,000 hp |
| [a]Condensing, back pressure, also extraction available. | | | | | |
| **WESTINGHOUSE ELECTRIC CORP.** | | | | | |
| 60-1100 MW<br>50 & 60 Hz | 600-1,350 MW<br>50 & 60 Hz | — | — | — | Feed Pump Turbines 7,000-30,000 hp |

# APPENDIX M

## PERFORMANCE SPECIFICATIONS FOR SIMPLE CYCLE GAS TURBINE GENERATORS AND COMBINED CYCLE PLANTS

## ELECTRIC POWER GENERATION

# GAS TURBINE PERFORMANCE SPECIFICATIONS

| Model (series designation) | Year (1st year unit was (will be) available) | ISO Base Rating (kW) (6000 hrs a year or more) | Heat Rate (Btu/kw-hr) (lower heating value (LHV)) | ISO Peak Rating (kW) (good for up to 2000 hrs a year) | Heat Rate (Btu/kw-hr) (lower heating value (LHV)) | Pressure Ratio (overall compressor) |
|---|---|---|---|---|---|---|
| **ACEC** | | | | | | |
| W-251 | 1979 | 38 160 kW | 11 564 Btu/kw-hr | 41 510 kW | 11 267 Btu/kw-hr | 10.0 |
| W-1101 | 1980 | 95 770 kW | 11 174 Btu/kw-hr | 103 440 kW | 11 073 Btu/kw-hr | 12.0 |
| **AEG-Kanis** | | | | | | |
| G3142 | 1949 | 10 200 kW | 13 540 Btu/kw-hr | ------- | ------- | 7.2 |
| G3132R | 1951 | 9 750 kW | 10 520 Btu/kw-hr | ------- | ------- | 7.2 |
| G5261 | 1958 | 18 900 kW | 13 400 Btu/kw-hr | ------- | ------- | 8.1 |
| G5341 | 1958 | 24 700 kW | 12 230 Btu/kw-hr | ------- | ------- | 10.2 |
| PG5341 | 1961 | 24 110 kW | 12 450 Btu/kw-hr | 26 050 kW | 12 340 Btu/kw-hr | 10.2 |
| G7851 | 1971 | 61 750 kW | 10 920 Btu/kw-hr | -------- | ------- | 9.7 |
| PG7851 | 1970 | 60 000 kW | 11 130 Btu/kw-hr | 66 300 kW | 11 030 Btu/kw-hr | 9.7 |
| G7101 | 1977 | 75 000 kW | 10 590 Btu/kw-hr | ------- | ------- | 11.5 |
| PG7101 | 1976 | 72 900 kW | 10 790 Btu/kw-hr | 78 800 kW | 10 750 Btu/kw-hr | 11.5 |
| G9141 | 1978 | 109 000 kW | 10 650 Btu/kw-hr | ------- | ------- | 9.6 |
| PG9141 | 1978 | 107 000 kW | 10 850 Btu/kw-hr | 116 200 kW | 10 810 Btu/kw-hr | 9.6 |
| LM2500 | 1979 | 20 100 kW | 9 800 Btu/kw-hr | ------- | ------- | 18.0 |
| PG6501 | 1978 | 34 850 kW | 11 310 Btu/kw-hr | 38 000 kW | 11 190 Btu/kw-hr | 11.5 |
| G6501 | 1978 | 35 650 kW | 11 170 Btu/kw-hr | ------- | ------- | 11.5 |
| **Alsthom-Atlantique** | | | | | | |
| G5341 | 1958 | 25 350 kW | 12 020 Btu/kw-hr | ------- | ------- | 10.0 |
| PG5341 | 1961 | 24 500 kW | 12 310 Btu/kw-hr | 25 550 kW | 12 310 Btu/kw-hr | 10.0 |
| PG6501 | 1978 | 35 600 kW | 11 178 Btu/kw-hr | 38 800 kW | 11 170 Btu/kw-hr | 11.5 |
| G9111 | 1976 | 87 400 kW | 10 820 Btu/kw-hr | ------- | ------- | --- |
| PG9111 | 1976 | 84 700 kW | 11 050 Btu/kw-hr | 93 900 kW | 10 930 Btu/kw-hr | ---- |
| G9141 | 1978 | 107 500 kW | 10 620 Btu/kw-hr | ------- | ------- | ---- |
| PG9141 | 1978 | 104 500 kW | 10 820 Btu/kw-hr | 113 300 kW | 10 790 Btu/kw-hr | ---- |
| **Avco Industrial Engine Operation** | | | | | | |
| TF25 | 1968 | 1 835 kW | 15 041 Btu/kw-hr | 2 163 kW | 14 757 Btu/kw-hr | 6.5 |
| TF40 | 1975 | 2 950 kW | 12 880 Btu/kw-hr | 3 312 kW | 12 759 Btu/kw-hr | 8.4 |
| **BBC Brown, Boveri & Co Ltd<br>Brown, Boveri & Cie AG<br>Brown Boveri Turbomachinery Inc** | | | | | | |
| Type 9 | 1970 | 34 500 kW | 12 016 Btu/kw-hr | 37 500 kW | 11 890 Btu/kw-hr | 8.8 |
| Type 11 | 1971 | 72 500 kW | 10 766 Btu/kw-hr | 78 900 kW | 10 664 Btu/kw-hr | 11.0 |
| Type 13 | 1970 | 88 900 kW | 10 731 Btu/kw-hr | 97 300 kW | 10 578 Btu/kw-hr | 11.8 |
| LGT 13/10 | 1974 | 290 000 kW | 5 560 Btu/kw-hr | ------- | ------- | 65.0 |
| LGT 11/8 | 1981 | 220 000 kW | 4 150 Btu/kw-hr | ------- | ------- | 55.0 |
| HHT 350 | 198? | 350 000 kW | ------- | ------- | ------- | --- |
| HHT 1200 | 198? | 1 200 000 kW | ------- | ------- | ------- | --- |
| **Bet-Shemesh Engines** | | | | | | |
| M2TL | 1971 | 750 kW | 23 600 Btu/kw-hr | 790 kW | 23 200 Btu/kw-hr | 3.8 |
| M5TL | 1981 | 960 kW | 15 500 Btu/kw-hr | 1040 kW | 15 000 Btu/kw-hr | 5.5 |

# ELECTRIC POWER

| Air Flow (lb/sec) at peak load | Turbine Speed output shaft speed | Turbine Inlet Temp at peak rating | Exhaust Temp at peak rating | Dry Weight (Approx) | Length x Width x Height (Approx) | Comments and Footnotes |
|---|---|---|---|---|---|---|
| **ACEC** | | | | | | |
| 360 lb/sec | 5 000 rpm | 1940°F | 966°F | 340 000 lb | 130 x 40 x 25 ft | 50-Hz plants; gas fuel ratings |
| 868 lb/sec | 3 000 rpm | 2070°F | 1015°F | 730 000 lb | 130 x 50 x 40 ft | at gen terminals w. duct losses |
| **AEG-Kanis** | | | | | | |
| 113 lb/sec | 6 500 rpm | 1730°F | 979°F | 242 000 lb | 63 x 11 x 12 ft | all ratings on |
| 113 lb/sec | 6 500 rpm | 1750°F | 668°F | 242 000 lb | 63 x 11 x 12 ft | distillate fuel |
| 213 lb/sec | 5 100 rpm | 1720°F | 955°F | 318 000 lb | 68 x 10 x 12 ft | |
| 261 lb/sec | 5 105 rpm | 1730°F | 901°F | 398 000 lb | 77 x 11 x 12 ft | |
| 261 lb/sec | 5 105 rpm | 1805°F | 955°F | 570 000 lb | 115 x 19 x 35 ft | packaged plant |
| 529 lb/sec | 3 600 rpm | 1850°F | 944°F | 587 000 lb | 70 x 29 x 13 ft | |
| 527 lb/sec | 3 600 rpm | 1960°F | 1018°F | 1 095 000 lb | 96 x 29 x 16 ft | packaged plant |
| 590 lb/sec | 3 600 rpm | 1985°F | 977°F | 587 000 lb | 70 x 29 x 13 ft | |
| 587 lb/sec | 3 600 rpm | 2085°F | 1045°F | 1 070 000 lb | 132 x 71 x 31 ft | packaged plant |
| 885 lb/sec | 3 000 rpm | 2085°F | 1029°F | 650 000 lb | 83 x 15 x 19 ft | 50-Hz machine |
| 877 lb/sec | 3 000 rpm | 2085°F | 1034°F | 1 425 000 lb | 140 x 48 x 38 ft | packaged plant |
| 144 lb/sec | 3 600 rpm | 2167°F | 940°F | 220 000 lb | 53 x 11 x 13 ft | |
| 302 lb/sec | 5 105 rpm | 2085°F | 1055°F | 700 000 lb | 123 x 24 x 34 ft | packaged plant |
| 302 lb/sec | 5 105 rpm | 1985°F | 989°F | 528 000 lb | 89 x 11 x 12 ft | |
| **Alsthom-Atlantique** | | | | | | |
| 269 lb/sec | 5 100 rpm | 1730°F | 901°F | 398 000 lb | 77 x 11 x 12 ft | |
| 266 lb/sec | 5 100 rpm | 1730°F | 905°F | 428 000 lb | 87 x 11 x 12 ft | |
| 300 lb/sec | 5 705 rpm | 1985°F | 991°F | 700 000 lb | 123 x 24 x 34 ft | |
| 760 lb/sec | 3 000 rpm | 1840°F | 945°F | ------- | ------- | |
| 752 lb/sec | 3 000 rpm | 1840°F | 950°F | ------- | ------- | |
| 876 lb/sec | ------- | 1955°F | 953°F | ------- | ------- | |
| 868 lb/sec | 3 000 rpm | 1955°F | 958°F | ------- | ------- | |
| **Avco Industrial Engine Operation** | | | | | | |
| 22 lb/sec | 13 750 rpm | 1775°F | 1110°F | 1 185 lb | 5 x 3 x 4 ft | at output shaft |
| 28 lb/sec | 14 630 rpm | 1940°F | 1130°F | 1 325 lb | 5 x 3 x 4 ft | of GT, not elec gen |
| **BBC Brown, Boveri & Co Ltd / Brown, Boveri & Cie AG / Brown Boveri Turbomachinery Inc** | | | | | | |
| 355 lb/sec | 4473/4485 | ------- | 545°C | 79 000 kg | 10 x 3.2 x 7.6 m | ratings at output |
| 639 lb/sec | 3 600 rpm | ------- | 550°C | 165 000 kg | 11.2 x 7.2 x 5.1 m | shaft of GT, |
| 825 lb/sec | 3 000 rpm | ------- | 526°C | 270 000 kg | 14.5 x 4.6 x 9.8 m | distillate fuel |
| 919 lb/sec | 3000/7620 | ------- | 400°C | 2 566 000 kg | 43 x 4.6 x 10 m | air storage plant 50-Hz |
| 672 lb/sec | 3 600 rpm | ------- | 150°C | ------- | 48 x 4.5 x 9.4 m | air storage plant 60-Hz |
| ------- | ------- | ------- | ------- | ------- | ------- | closed-cycle helium plant |
| ------- | ------- | ------- | ------- | ------- | ------- | closed-cycle helium plant |
| **Bet-Shemesh Engines** | | | | | | |
| 17 lb/sec | 17 900 rpm | 1395°F | 977°F | 615 lb | 7 x 2 x 3 ft | built-in reduction gear |
| 13 lb/sec | 21 200 rpm | 1680°F | 1020°F | 900 lb | 8 x 2.4 x 2.4 ft | built-in reduction gear |

| series designation | 1st year unit was (will be) available | 6000 hrs a year or more | lower heating value (LHV) | good for up to 2000 hrs a year | lower heating value (LHV) | overall compressor |
|---|---|---|---|---|---|---|
| **Model** | **Year** | **ISO Base Rating (kW)** | **Heat Rate (Btu/kw-hr)** | **ISO Peak Rating (kW)** | **Heat Rate (Btu/kw-hr)** | **Pressure Ratio** |
| **Centrax Gas Turbine** | | | | | | |
| Type 33 | 1961 | 500 kW | 21 000 Btu/kw-hr | 550 kW | 19 800 Btu/kw-hr | 5.2 |
| Type 19H | 1974 | 500 kW | 21 000 Btu/kw-hr | 550 kW | 19 800 Btu/kw-hr | 5.2 |
| Type 74S | 1977 | 500 kW | 21 000 Btu/kw-hr | 550 kW | 20 000 Btu/kw-hr | 5.3 |
| Type 74SZ | 1977 | 560 kW | 21 460 Btu/kw-hr | 630 kW | 20 540 Btu/kw-hr | 6.3 |
| Type 350 | 1979 | 3 000 kW | 13 165 Btu/kw-hr | 3 370 kW | 12 626 Btu/kw-hr | 9.3 |
| Type 470 | 1980 | 4 400 kW | 12 424 Btu/kw-hr | 4 700 kW | 12 497 Btu/kw-hr | 11.2 |
| **Curtiss-Wright Power Systems** | | | | | | |
| Mod-Pod 20HE1 | 1973 | 20 000 kW | 9 800 Btu/kw-hr | 20 100 kW | 9 795 Btu/kw-hr | 18.0 |
| Mod-Pod 25 | 1974 | 24 900 kW | 11 850 Btu/kw-hr | 26 700 kW | 11 770 Btu/kw-hr | 11.0 |
| Mod-Pod 50 | 1974 | 49 800 kW | 11 850 Btu/kw-hr | 53 400 kW | 11 770 Btu/kw-hr | 11.0 |
| **Detroit Diesel Allison** | | | | | | |
| GT-404 | Mid-80's | 252 kW | 9 566 Btu/kw-hr | 252 kW | 9 566 Btu/kw-hr | 4.0 |
| 501-KB | 1962 | 3 066 kW | 12 844 Btu/kw-hr | 3 713 kW | 12 200 Btu/kw-hr | 9.3 |
| 570-KA | 1979 | 4 512 kW | 12 104 Btu/kw-hr | 5 019 kW | 12 157 Btu/kw-hr | 11.2 |
| 571-KB | 1982 | 5 110 kW | 11 363 Btu/kw-hr | 5 530 kW | 11 253 Btu/kw-hr | 12.0 |
| **Fiat Aviazione** | | | | | | |
| LM500 | 1980 | 3 600 kW | 11 537 Btu/kw-hr | ------- | ------- | 14.0 |
| LM2500-20 | 1980 | 12 850 kW | 10 041 Btu/kw-hr | ------- | ------- | 14.2 |
| LM2500 | 1979 | 20 100 kW | 9 435 Btu/kw-hr | ------- | ------- | 17.2 |
| **Fiat TTG** | | | | | | |
| TG7 | 1964 | 8 200 kW | 14 840 Btu/kw-hr | 8 950 kW | 14 520 Btu/kw-hr | 6.0 |
| TG16 | 1968 | 18 200 kW | 13 175 Btu/kw-hr | 19 500 kW | 12 975 Btu/kw-hr | 7.0 |
| TG20 | 1971 | 37 850 kW | 11 590 Btu/kw-hr | 41 180 kW | 11 310 Btu/kw-hr | 11.0 |
| TG50 | 1975 | 92 600 kW | 11 090 Btu/kw-hr | 100 000 kW | 10 990 Btu/kw-hr | 12.0 |
| **Garrett Corp** | | | | | | |
| IM831-800 | 1980 | 505 kW | 16 400 Btu/kw-hr | 565 kW | 15 920 Btu/kw-hr | 11.0 |
| IM831-800 | 1980 | 520 kW | 16 150 Btu/kw-hr | 565 kW | 15 700 Btu/kw-hr | 11.0 |
| IE990-50 | 1978 | 4 327 kW | 11 305 Btu/kw-hr | 4 621 kW | ------- | 12.0 |
| **GEC Gas Turbines** | | | | | | |
| EM85 | 1960 | 6 750 kW | 15 950 Btu/kw-hr | ------- | ------- | 6.0 |
| EM610 | 1975 | 74 000 kW | 12 030 Btu/kw-hr | 80 600 kW | 11 860 Btu/kw-hr | 9.6 |
| EAS-133 | 1965 | 11 300 kW | 13 260 Btu/kw-hr | 14 580 kW | 12 390 Btu/kw-hr | 9.5 |
| EAS-134 | 1975 | 12 780 kW | 13 040 Btu/kw-hr | 15 960 kW | 12 390 Btu/kw-hr | 9.8 |
| EAS-135/1 | 1977 | 13 970 kW | 12 920 Btu/kw-hr | 16 684 kW | 12 448 Btu/kw-hr | 9.5 |
| EAS-135/2 | 1979 | 14 190 kW | 12 780 Btu/kw-hr | 17 043 kW | 12 130 Btu/kw-hr | 9.5 |
| ESP-1 | 1980 | 11 200 kW | 10 800 Btu/kw-hr | 13 230 kW | 10 410 Btu/kw-hr | --- |
| ERB-124 | 1980 | 20 350 kW | 10 480 Btu/kw-hr | 22 600 kW | 10 270 Btu/kw-hr | 20.0 |
| EO-1C | 1972 | 22 940 kW | 11 458 Btu/kw-hr | 27 329 kW | 11 224 Btu/kw-hr | 11.0 |
| DEO-1C | 1972 | 45 880 kW | 11 458 Btu/kw-hr | 55 658 kW | 11 224 Btu/kw-hr | 11.0 |

# ELECTRIC POWER

| Air Flow (lb/sec) | Turbine Speed | Turbine Inlet Temp | Exhaust Temp | Dry Weight (Approx) | Length x Width x Height (Approx) | Comments and Footnotes |
|---|---|---|---|---|---|---|
| at peak load | output shaft speed | at peak rating | | | | |
| | | | | | | **Centrax Gas Turbine** |
| 11 lb/sec | 22 000 rpm | 800°C | 500°C | 11 200 lb | 13 x 5 x 6 ft | dual fuel packaged set |
| 11 lb/sec | 22 000 rpm | 880°C | 550°C | 13 440 lb | 15 x 5 x 9 ft | distillate packaged set |
| 11 lb/sec | 22 000 rpm | 880°C | 550°C | 11 200 lb | 13 x 4 x 6 ft | distillate packaged set |
| 12 lb/sec | 22 000 rpm | 880°C | 535°C | 11 200 lb | 13 x 4 x 6 ft | distillate packaged set |
| 34 lb/sec | 14 210 rpm | 1037°C | 550°C | 30 000 lb | 22 x 6 x 6 ft | Allison 501-KB |
| 42 lb/sec | 11 500 rpm | *830°C | 605°C | 35 000 lb | 22 x 6 x 6 ft | Allison 570-KA |
| | | *power turbine inlet temp | | | | |
| | | | | | | **Curtiss-Wright Power Systems** |
| 144 lb/sec | 3600/3000 | *2150°F | 888°F | 78 000 lb | 21 x 11 x 11 ft | LM2500 gas generator |
| 236 lb/sec | 3600/3000 | *1990°F | 995°F | 93 000 lb | 30 x 11 x 11 ft | Olympus C gas generator |
| 472 lb/sec | 3600/3000 | *1990°F | 995°F | 186 000 lb | 2 [30 x 11 x 11] | 2 Olympus C gas generators |
| | | *est | | | | |
| | | | | | | **Detroit Diesel Allison** |
| 3 lb/sec | 3 600 rpm | 1875°F | 543°F | 1 800 lb | 4 x 3 x 3.5 ft | 1800 rpm available |
| 33 lb/sec | 13 820 rpm | 1970°F | 950°F | 1 270 lb | 7.5 x 3 x 2.5 ft | |
| 42 lb/sec | 11 500 rpm | *1562°F | 1150°F | 1 350 lb | 6 x 2.5 x 2.5 ft | |
| 43 lb/sec | 11 500 rpm | *1535°F | 1070°F | 1 490 lb | 6.3 x 3 x 3 ft | under development |
| | | *power turbine inlet temp | | | | |
| | | | | | | **Fiat Aviazione** |
| 34 lb/sec | 3000/3600 | *1320°F | 940°F | 2 260 lb | 10 x 3.8 x 3.8 ft | all ratings gas |
| 125 lb/sec | 3 000 rpm | *1145°F | 775°F | 10 300 lb | 21 x 7 x 7 ft | fuel; at electric |
| 144 lb/sec | 3000/3600 | *1410°F | 920°F | 10 300 lb | 21 x 7 x 7 ft | gen terminals |
| | | *power turbine inlet temp | | | | |
| | | | | | | **Fiat TTG** |
| 134 lb/sec | 6 000 rpm | ------- | 831°F | 39 000 lb | 25 x 7 x 7 ft | gas fuel ratings |
| 267 lb/sec | 4 850 rpm | ------- | 786°F | 88 000 lb | 25 x 10 x 10 ft | |
| 355 lb/sec | 4 920 rpm | ------- | 1018°F | 130 000 lb | 30 x 10 x 10 ft | |
| 851 lb/sec | 3 000 rpm | ------- | 1031°F | 346 000 lb | 47 x 14 x 20 ft | |
| | | | | | | **Garrett Corp** |
| 7.9 lb/sec | 1500/1800 | 1765°F | 930°F | 1 750 lb | 4 x 3 x 3 ft | liquid fuel |
| 7.9 lb/sec | 1500/1800 | 1765°F | 930°F | 1 750 lb | 4 x 3 x 3 ft | gas fuel |
| 43.0 lb/sec | 1500/1800 | 1900°F | 880°F | 7 500 lb | 11 x 4 x 4 ft | |
| | | | | | | **GEC Gas Turbines** |
| 100 lb/sec | 5100/6300 | 1481°F | 860°F | 105 000 lb | 29 x 11 x 19 ft | all ratings at electric |
| 742 lb/sec | 3 000 rpm | ------- | ------- | 360 000 lb | 47 x 13 x 15 ft | gen terminals |
| 168 lb/sec | 6 500 rpm | 1652°F | 848°F | 45 000 lb | 24 x 11 x 10 ft | Avon 1533 gasifier |
| 168 lb/sec | 6 500 rpm | 1787°F | 913°F | 45 000 lb | 24 x 11 x 10 ft | Avon 1534 gasifier |
| 180 lb/sec | 6 500 rpm | 1823°F | 937°F | 45 000 lb | 24 x 11 x 10 ft | Avon 1535 gasifier |
| 180 lb/sec | 6 500 rpm | 1825°F | 926°F | 45 000 lb | 24 x 11 x 10 ft | Avon 1535 gasifier |
| ------- | 6 500 rpm | ------- | ------- | 48 000 lb | 16 x 11 x 10 ft | Spey gasifier |
| 203 lb/sec | 6 500 rpm | ------- | ------- | 52 000 lb | 22 x 11 x 10 ft | RB-211-24B gasifier |
| 242 lb/sec | 3 600 rpm | 1980°F | 995°F | 56 000 lb | 22 x 11 x 11 ft | Olympus C gasifier |
| 484 lb/sec | 3 600 rpm | 1980°F | 995°F | 392 000 lb | 90 x 13 x 13 ft | weight and size with gen |

(continued next page)

| series designation | 1st year unit was (will be) available | 6000 hrs a year or more | lower heating value (LHV) | good for up to 2000 hrs a year | lower heating value (LHV) | overall compressor |
|---|---|---|---|---|---|---|
| **Model** | **Year** | **ISO Base Rating (kW)** | **Heat Rate (Btu/kw-hr)** | **ISO Peak Rating (kW)** | **Heat Rate (Btu/kw-hr)** | **Pressure Ratio** |
| **GEC Gas Turbines (continued)** | | | | | | |
| ELM-125/2 | 1979 | 18 820 kW | 9 971 Btu/kw-hr | 20 650 kW | 9 910 Btu/kw-hr | 17.2 |
| ELM-125/6 | 1978 | 20 060 kW | 9 452 Btu/kw-hr | 21 890 kW | 9 424 Btu/kw-hr | 17.2 |
| ELM-150 | 1979 | 31 000 kW | 9 437 Btu/kw-hr | 36 090 kW | 9 280 Btu/kw-hr | 30.0 |
| DELM-150 | 1979 | 62 000 kW | 9 437 Btu/kw-hr | 72 180 kW | 9 280 Btu/kw-hr | 30.0 |
| EO-2 | 1969 | 29 400 kW | 13 260 Btu/kw-hr | 41 000 kW | 12 300 Btu/kw-hr | 10.9 |
| DEO-2 | 1976 | 58 800 kW | 13 260 Btu/kw-hr | 82 000 kW | 12 300 Btu/kw-hr | 10.4 |
| **General Electric GT Div** | | | | | | |
| G3142 | 1949 | 10 200 kW | 13 540 Btu/kw-hr | ------- | ------- | 7.2 |
| G3132R | 1951 | 9 750 kW | 10 520 Btu/kw-hr | ------- | ------- | 7.2 |
| G5261 | 1958 | 18 900 kW | 13 400 Btu/kw-hr | ------- | ------- | 8.1 |
| G5361 | 1958 | 25 420 kW | 12 230 Btu/kw-hr | ------- | ------- | 10.2 |
| PG5361 | 1961 | 24 800 kW | 12 450 Btu/kw-hr | 26 620 kW | 12 340 Btu/kw-hr | 10.2 |
| G7111 | 1977 | 76 900 kW | 10 590 Btu/kw-hr | ------- | ------- | 11.5 |
| PG7111 | 1976 | 75 000 kW | 10 790 Btu/kw-hr | 81 150 kW | 10 750 Btu/kw-hr | 11.5 |
| G9151 | 1976 | 109 300 kW | 10 700 Btu/kw-hr | ------- | ------- | 9.6 |
| PG9151 | 1976 | 106 700 kW | 10 850 Btu/kw-hr | 115 900 kW | 10 810 Btu/kw-hr | 9.6 |
| LM2500 | 1979 | 20 100 kW | 9 800 Btu/kw-hr | ------- | ------- | 18.0 |
| PG6521 | 1978 | 35 860 kW | 11 280 Btu/kw-hr | 38 950 kW | 11 180 Btu/kw-hr | 11.5 |
| G6521 | 1978 | 36 730 kW | 11 120 Btu/kw-hr | ------- | ------- | 11.5 |
| **Hispano-Suiza** | | | | | | |
| THM1103 | 1979 | 4 400 kW | 15 160 Btu/kw-hr | ------- | ------- | 6.7 |
| THM1103R | 1979 | 4 200 kW | *11 185 Btu/kw-hr | ------- | ------- | 6.7 |
| THM1203 | 1971 | 5 600 kW | 14 214 Btu kw-hr | ------- | ------- | 7.6 |
| THM1203R | 1971 | 5 400 kW | *10 660 Btu/kw-hr | ------- | ------- | 7.6 |
| THM1304 | 1978 | 8 100 kW | 12 640 Btu/kw-hr | ------- | ------- | 9.2 |
| THM1304R | 1978 | 7 830 kW | *10 660 Btu/kw-hr | ------- | ------- | 9.2 |
| THM1304-10 | 1983 | 10 000 kW | 11 764 Btu/kw-hr | ------- | ------- | 10.7 |
| THM1304-10R | 1983 | 9 700 kW | *10 034 Btu/kw-hr | ------- | ------- | 10.7 |
| | | | *heat rate with regenerator | | | |
| **Hitachi** | | | | | | |
| G3142 | 1966 | 10 200 kW | 13 540 Btu/kw-hr | ------- | ------- | 7.1 |
| G5341 | 1967 | 24 050 kW | 12 310 Btu/kw-hr | 25 950 kW | 12 180 Btu/kw-hr | 10.0 |
| G6501B | 1982 | 36 400 kW | 11 050 Btu/kw-hr | ------- | ------- | 11.5 |
| G7821 | 1972 | 60 000 kW | 10 960 Btu/kw-hr | 66 300 kW | 10 860 Btu/kw-hr | 9.7 |
| G7111 | 1982 | 78 400 kW | 10 510 Btu/kw-hr | ------- | ------- | 11.5 |
| G9141 | 1982 | 105 600 kW | 10 790 Btu/kw-hr | ------- | ------- | 9.6 |
| PG5341 | 1967 | 23 450 kW | 12 490 Btu/kw-hr | 25 300 kW | 12 370 Btu/kw-hr | 10.2 |
| PG6501B | 1982 | 35 600 kW | 11 180 Btu/kw-hr | 38 800 kW | 11 070 Btu/kw-hr | 11.5 |
| PG7821 | 1972 | 58 500 kW | 11 130 Btu/kw-hr | 64 600 kW | 11 030 Btu/kw-hr | 9.7 |
| PG7111 | 1982 | 76 500 kW | 10 690 Btu/kw-hr | 82 800 kW | 10 640 Btu/kw-hr | 11.5 |
| PG9111 | 1980 | 82 600 kW | 11 210 Btu/kw-hr | 91 600 kW | 11 090 Btu/kw-hr | 9.6 |
| PG9141 | 1982 | 102 700 kW | 10 900 Btu/kw-hr | 111 300 kW | 10 860 Btu/kw-hr | 9.6 |

# ELECTRIC POWER

| Air Flow (lb/sec) (at peak load) | Turbine Speed (output shaft speed) | Turbine Inlet Temp (at peak rating) | Exhaust Temp (at peak rating) | Dry Weight (Approx) | Length x Width x Height (Approx) | Comments and Footnotes |
|---|---|---|---|---|---|---|
| | | | | | | **GEC Gas Turbines (continued)** |
| 143 lb/sec | 6 500 rpm | ------- | 942°F | 47 000 lb | 18 x 11 x 10 ft | LM2500 gasifier |
| 144 lb/sec | 3000/3600 | ------- | 922°F | 47 000 lb | 18 x 11 x 10 ft | LM2500 gasifier |
| 272 lb/sec | 3000/3600 | ------- | 783°F | 63 000 lb | 29 x 11 x 11 ft | LM5000 gasifier |
| 544 lb/sec | 3000/3600 | ------- | 783°F | 530 000 lb | 85 x 13 x 13 ft | weight and size with gen. |
| 484 lb/sec | 3000/3600 | 1735°F | 896°F | 300 000 lb | 58 x 24 x 11 ft | 2 Olympus B: 1 power turbine |
| 968 lb/sec | 3000/3600 | 1735°F | 896°F | 615 700 lb | 98 x 23 x 12 ft | 4 Olympus B: 2 power turbines |
| | | | | | | **General Electric GT Div** |
| 113 lb/sec | 6 500 rpm | 1730°F | 979°F | 242 000 lb | 63 x 11 x 12 ft | all ratings on |
| 113 lb/sec | 6 500 rpm | 1750°F | 668°F | 242 000 lb | 63 x 11 x 12 ft | distillate fuel |
| 213 lb/sec | 5 100 rpm | 1720°F | 955°F | 318 000 lb | 68 x 10 x 12 ft | |
| 269 lb/sec | 5 105 rpm | 1755°F | 916°F | 398 000 lb | 77 x 11 x 12 ft | |
| 267 lb/sec | 5 105 rpm | 1825°F | 966°F | 570 000 lb | 115 x 19 x 35 ft | packaged plant |
| 590 lb/sec | 3 600 rpm | 2020°F | 999°F | 587 000 lb | 70 x 29 x 13 ft | |
| 587 lb/sec | 3 600 rpm | 2120°F | 1063°F | 1 070 000 lb | 132 x 71 x 31 ft | packaged plant |
| 876 lb/sec | 3 000 rpm | 1985°F | 968°F | 650 000 lb | 83 x 15 x 19 ft | 50-Hz machine |
| 876 lb/sec | 3 000 rpm | 2085°F | 1034°F | 1 425 000 lb | 140 x 48 x 38 ft | packaged plant |
| 144 lb/sec | 3 600 rpm | 2167°F | 940°F | 220 000 lb | 53 x 11 x 13 ft | |
| 301 lb/sec | 5 105 rpm | 2120°F | 1083°F | 700 000 lb | 123 x 24 x 34 ft | packaged plant |
| 304 lb/sec | 5 105 rpm | 2020°F | 1017°F | 528 000 lb | 89 x 11 x 12 ft | |
| | | | | | | **Hispano-Suiza** |
| 64 lb/sec | 7 800 rpm | 1607°F | 905°F | 15 420 kg | 8.2 x 2.4 x 3.2 m | size with baseplate & aux |
| 64 lb/sec | 7 800 rpm | 1607°F | ------- | 15 420 kg | 8.2 x 2.4 x 3.2 m | but without stack, filter, |
| 75 lb/sec | 7 800 rpm | 1661°F | 928°F | 16 520 kg | 8.2 x 2.4 x 3.2 m | silencer or regenerator |
| 75 lb/sec | 7 800 rpm | 1661°F | ------- | 16 520 kg | 8.2 x 2.4 x 3.2 m | |
| 93 lb/sec | 8 000 rpm | 1741°F | 914°F | 23 130 kg | 9.2 x 2.4 x 3.2 m | |
| 93 lb/sec | 8 000 rpm | 1741°F | ------- | 23 130 kg | 9.2 x 2.4 x 3.2 m | |
| 106 lb/sec | 8 000 rpm | 1832°F | 880°F | 23 200 kg | 9.2 x 2.4 x 3.2 m | |
| 106 lb/sec | 8 000 rpm | 1832°F | ------- | 23 200 kg | 9.2 x 2.4 x 3.2 m | |
| | | | | | | **Hitachi** |
| 116 lb/sec | 6 500 rpm | ------- | ------- | 242 000 lb | 64 x 11 x 13 ft | all ratings on |
| 261 lb/sec | 5 105 rpm | 1805°F | 954°F | 398 000 lb | 78 x 11 x 13 ft | natural gas fuel |
| 302 lb/sec | 5 105 rpm | 1985°F | 989°F | 400 000 lb | 73 x 11 x 12 ft | |
| 530 lb/sec | 3 600 rpm | 1950°F | 1016°F | 587 000 lb | 70 x 29 x 14 ft | |
| 609 lb/sec | 3 600 rpm | 2020°F | 1038°F | 587 000 lb | 70 x 29 x 14 ft | |
| 760 lb/sec | 3 000 rpm | 1955°F | 953°F | 650 000 lb | 83 x 15 x 19 ft | |
| 258 lb/sec | 5 105 rpm | 1805°F | 958°F | 570 000 lb | 116 x 19 x 36 ft | |
| 302 lb/sec | 5 105 rpm | 2085°F | 1055°F | 700 000 lb | 124 x 24 x 34 ft | |
| 524 lb/sec | 3 600 rpm | 1950°F | 1023°F | 1 095 000 lb | 117 x 62 x 32 ft | |
| 609 lb/sec | 3 600 rpm | 2120°F | 1045°F | 1 070 000 lb | 133 x 71 x 31 ft | |
| 753 lb/sec | 3 000 rpm | 1950°F | 1020°F | 1 425 000 lb | 140 x 48 x 34 ft | |
| 753 lb/sec | 3 000 rpm | 2055°F | 1020°F | 1 425 000 lb | 140 x 48 x 38 ft | |

| Model (series designation) | Year (1st year unit was (will be) available) | ISO Base Rating (kW) (6000 hrs a year or more) | Heat Rate (Btu/kw-hr) (lower heating value (LHV)) | ISO Peak Rating (kW) (good for up to 2000 hrs a year) | Heat Rate (Btu/kw-hr) (lower heating value (LHV)) | Pressure Ratio (overall compressor) |
|---|---|---|---|---|---|---|
| **Hitachi Zosen** | | | | | | |
| GT10 | 1980 | 3 066 kW | 12 844 Btu/kw-hr | 3 713 kW | 12 200 Btu/kw-hr | 9.3 |
| GT15 | 1980 | 4 512 kW | 12 104 Btu/kw-hr | 5 019 kW | 12 157 Btu/kw-hr | 11.2 |
| GT35 | 1979 | 12 300 kW | 11 960 Btu/kw-hr | 14 000 kW | 11 780 Btu/kw-hr | 12.0 |
| **Ishikawajima-Harima Heavy Industries** | | | | | | |
| IGT-90 | 1972 | 45 kW | 42 500 Btu/kw-hr | ------- | ------- | 2.8 |
| IM100-2G | 1972 | 920 kW | 17 000 Btu/kw-hr | ------- | ------- | 8.4 |
| IM100-4G | 1974 | 1 110 kW | 15 000 Btu/kw-hr | ------- | ------- | 8.4 |
| IM300/310 | 1967 | 1 870 kW | 14 200 Btu/kw-hr | ------- | ------- | 12.5 |
| IM400 | 1979 | 2 800 kW | 13 800 Btu/kw-hr | ------- | ------- | 9.3 |
| IM1500 | 1968 | 10 000 kW | 14 000 Btu/kw-hr | 11 000 kW | 13 760 Btu/kw-hr | 12.6 |
| IM2500 | 1976 | 20 000 kW | 9 580 Btu/kw-hr | 21 800 kW | 9 550 Btu/kw-hr | 18.1 |
| IM5000 | 1978 | 35 400 kW | 9 210 Btu/kw-hr | 38 100 kW | 9 080 Btu/kw-hr | 28.8 |
| Type 3 | 1977 | 5 040 kW | 13 300 Btu/kw-hr | ------- | ------- | 8.1 |
| Type 7 | 1976 | 10 050 kW | 14 200 Btu/kw-hr | ------- | ------- | 7.5 |
| **John Brown Engineering** | | | | | | |
| MS3002 | 1949 | 10 450 kW | 13 320 Btu/kw-hr | ------- | ------- | 7.2 |
| MS5001 | 1958 | 25 890 kW | 12 110 Btu/kw-hr | 27 750 kW | 12 020 Btu/kw-hr | 10.2 |
| MS6001 | 1978 | 37 500 kW | 11 000 Btu/kw-hr | 40 680 kW | 10 930 Btu/kw-hr | 11.5 |
| MS7001E | 1976 | 78 400 kW | 10 510 Btu/kw-hr | 84 800 kW | 10 480 Btu/kw-hr | 11.5 |
| MS9001B | 1976 | 87 400 kW | 10 820 Btu/kw-hr | 96 900 kW | 10 700 Btu/kw-hr | 9.6 |
| MS9001E | 1979 | 111 400 kW | 10 620 Btu/kw-hr | 121 420 kW | 10 610 Btu/kw-hr | 11.4 |
| LM2500-20 | 1978 | 12 850 kW | 10 041 Btu/kw-hr | ------- | ------- | 14.8 |
| LM2500-30 | 1978 | 20 100 kW | 9 800 Btu/kw-hr | ------- | ------- | 18.0 |
| **Kawasaki Heavy Industries** | | | | | | |
| S1A-01 | 1977 | ------- | ------- | 150 kW | 23 300 Btu/kw-hr | 8.5 |
| S1A-02 | 1978 | 180 kW | 21 800 Btu/kw-hr | 180 kW | 21 800 Btu/kw-hr | 9.0 |
| S1B-02 | 1981 | 180 kW | 21 800 Btu/kw-hr | 180 kW | 21 800 Btu/kw-hr | 9.0 |
| S1T-02 | 1978 | 350 kW | 22 600 Btu/kw-hr | 350 kW | 22 600 Btu/kw-hr | 9.0 |
| S2A-01 | 1979 | 600 kW | 17 980 Btu/kw-hr | 600 kW | 17 980 Btu/kw-hr | 9.0 |
| M1A-01 | 1978 | 1 000 kW | 17 600 Btu/kw-hr | 1 000 kW | 17 600 Btu/kw-hr | 8.0 |
| M1A-03 | 1982 | 1 200 kW | 17 900 Btu/kw-hr | 1 200 kW | 17 900 Btu/kw-hr | 9.1 |
| M1T-01 | 1979 | 2 000 kW | 17 800 Btu/kw-hr | 2 000 kW | 17 800 Btu/kw-hr | 8.0 |
| Super KTF25 | 1977 | 1 835 kW | 15 041 Btu/kw-hr | 2 163 kW | 14 757 Btu/kw-hr | 6.5 |
| Super KTF35 | 1977 | 2 550 kW | 12 941 Btu/kw-hr | 3 000 kW | 12 715 Btu/kw-hr | 7.6 |
| SK30 | 1977 | 23 500 kW | 11 200 Btu/kw-hr | 28 000 kW | 10 950 Btu/kw-hr | ---- |
| SK60 | 1977 | 47 000 kW | 11 200 Btu/kw-hr | 56 000 kW | 10 950 Btu/kw-hr | ---- |
| **Klockner-Humboldt-Deutz** | | | | | | |
| TF40 | 1976 | 2 950 kW | 12 880 Btu/kw-hr | 3 312 kW | 12 760 Btu/kw-hr | 8.4 |
| T216 | 1955 | 80 kW | 36 520 Btu/kw-hr | ------- | ------- | 2.8 |
| MI990 | 1978 | 3 960 kW | 12 080 Btu/kw-hr | 4 175 kW | ------- | 12.0 |

# ELECTRIC POWER

| Air Flow (lb/sec) (at peak load) | Turbine Speed (output shaft speed) | Turbine Inlet Temp (at peak rating) | Exhaust Temp (at peak rating) | Dry Weight (Approx) | Length x Width x Height (Approx) | Comments and Footnotes |
|---|---|---|---|---|---|---|
| | | | | | | **Hitachi Zosen** |
| 33 lb/sec | 13 820 rpm | 1970°F | 950°F | 1 270 lb | 7.5 x 3 x 2.5 ft | Allison 501-KB |
| 42 lb/sec | 11 500 rpm | *1562°F | 1150°F | 1 350 lb | 6 x 2.5 x 2.5 ft | Allison 570-KA |
| 185 lb/sec | 3000/3600 | 1560°F | 750°F | ------- | 46 x 11 x 10 ft | Stal-Laval GT35B |
| | | *power turbine inlet temp | | | | |
| | | | | | | **Ishikawajima-Harima Heavy Industries** |
| 2 lb/sec | 42 750 rpm | ------- | 630°C | 65 kg | .8 x .5 x .8 m | ratings at gen terminals |
| 13 lb/sec | 19 500 rpm | ------- | 495°C | 160 kg | 1.4 x .5 x .6 m | on all machines listed |
| 14 lb/sec | 19 500 rpm | ------- | 485°C | 200 kg | 1.7 x .6 x .7 m | |
| 24 lb/sec | 13 600 rpm | ------- | 450°C | 365 kg | 2.1 x .6 x 1.1 m | |
| 33 lb/sec | 13 820 rpm | 982°C | 555°C | 580 kg | 2.3 x .9 x .8 m | |
| 161 lb/sec | 3000/3600 | *930°C | 450°C | 16 500 kg | 8.0 x 2.8 x 2.8 m | |
| 149 lb/sec | 3000/3600 | *1170°C | 530°C | 4 700 kg | 6.6 x 2.1 x 2.1 m | |
| 289 lb/sec | 3000/3600 | *1160°C | 440°C | 40 000 kg | 8.7 x 3.5 x 3.5 m | twin sets available |
| 61 lb/sec | 10 000 rpm | 955°C | 493°C | 27 000 kg | 7.4 x 3.1 x 2.7 m | |
| 137 lb/sec | 6 400 rpm | 925°C | 480°C | 63 000 kg | 11.7 x 3.6 x 3.8 m | |
| | | *est | | | | |
| | | | | | | **John Brown Engineering** |
| 115 lb/sec | 6 500 rpm | 1730°F | 979°F | 250 000 lb | 63 x 11 x 12 ft | performance gross at |
| 269 lb/sec | 5 105 rpm | 1825°F | 960°F | 380 000 lb | 71 x 11 x 12 ft | gen terminals on |
| 304 lb/sec | 5 100 rpm | 2120°F | 1075°F | 420 000 lb | 76 x 11 x 12 ft | gas fuel |
| 604 lb/sec | 3 600 rpm | 2120°F | 1070°F | 710 000 lb | 100 x 13 x 14 ft | weights, dimensions |
| 760 lb/sec | 3 000 rpm | 1950°F | 1015°F | 1 000 000 lb | 110 x 15 x 23 ft | for turbo-alternator |
| 884 lb/sec | 3 000 rpm | 2085°F | 1030°F | 1 050 000 lb | 105 x 15 x 23 ft | unit only |
| 125 lb/sec | 3 000 rpm | *1145°F | 775°F | 220 000 lb | 53 x 11 x 13 ft | |
| 144 lb/sec | 3 600 rpm | *1410°F | 940°F | 220 000 lb | 53 x 11 x 13 ft | |
| | | *power turbine inlet temp | | | | |
| | | | | | | **Kawasaki Heavy Industries** |
| 3.9 lb/sec | 1500/1800 | ------- | 470°C | 390 kg | 1.0 x 1.0 x 0.8 m | |
| 3.9 lb/sec | 1500/1800 | ------- | 505°C | 390 kg | 1.0 x 1.0 x 0.8 m | |
| 3.9 lb/sec | 3000/3600 | ------- | 505°C | 240 kg | 1.0 x 0.8 x 0.9 m | |
| 7.7 lb/sec | 1500/1800 | ------- | 505°C | 650 kg | 1.2 x 1.1 x 1.1 m | |
| 10.6 lb/sec | 1500/1800 | ------- | 490°C | 1 300 kg | 1.6 x 1.1 x 1.2 m | |
| 16.9 lb/sec | 1500/1800 | ------- | 500°C | 2 450 kg | 2.1 x 1.4 x 1.4 m | |
| 19.8 lb/sec | 1500/1800 | ------- | 520°C | 2450 kg | 2.1 x 1.4 x 1.4 m | |
| 33.9 lb/sec | 1500/1800 | ------- | 500°C | 5 200 kg | 2.1 x 2.0 x 1.6 m | |
| 20.7 lb/sec | 1500/1800 | 965°C | 580°C | 610 kg | 1.3 x 0.9 x 1.1 m | |
| 26.4 lb/sec | 1500/1800 | 995°C | 610°C | 670 kg | 1.3 x 0.9 x 1.1 m | |
| ------- | 3000/3600 | ------- | 530°C | 475 000 lb | 79 x 53 x 53 ft | packaged plant |
| ------- | 3000/3600 | ------- | 530°C | 930 000 lb | 128 x 53 x 53 ft | packaged plant |
| | | | | | | **Klockner-Humboldt-Deutz** |
| 28 lb/sec | 14 630 rpm | 1940°F | 1130°F | 1 486 lb | 4 x 3 x 3.5 ft | |
| 2 lb/sec | 50 000 rpm | ------- | 1200°F | 200 lb | 2.7 x 2 x 2.7 ft | |
| 43 lb/sec | 7 20 rpm | 1900°F | 880°F | 7 000 lb | 11 x 4 x 4 ft | |

| Model (series designation) | Year (1st year unit was (will be) available) | ISO Base Rating (kW) (6000 hrs a year or more) | Heat Rate (Btu/kw-hr) (lower heating value (LHV)) | ISO Peak Rating (kW) (good for up to 2000 hrs a year) | Heat Rate (Btu/kw-hr) (lower heating value (LHV)) | Pressure Ratio (overall compressor) |
|---|---|---|---|---|---|---|
| **Kongsberg Gas Turbines and Power Systems** | | | | | | |
| KG2-3C | 1968 | 1 470 kW | 19 800 Btu/kw-hr | 1 670 kW | 19 200 Btu/kw-hr | 4.0 |
| KG2-3D | 1976 | ------- | ------- | 1 890 kW | 19 210 Btu/kw-hr | 4.0 |
| KG 5 | 1976 | 2 960 kW | 14 600 Btu/kw-hr | 3 260 kW | 14 500 Btu/kw-hr | 6.3 |
| KG 831 | 1976 | 530 kW | 15 570 Btu/kw-hr | 596 kW | 15 050 Btu/kw-hr | 11.0 |
| KG 831 | 1976 | 548 kW | 15 330 Btu/kw-hr | 596 kW | 14 960 Btu/kw-hr | 11.0 |
| **Kraftwerk Union** | | | | | | |
| V93.2 | 1971 | 74 460 kW | 11 790 Btu/kw-hr | 81 300 kW | 11 730 Btu/kw-hr | 10.0 |
| V94.2 | 1974 | 118 480 kW | 10 970 Btu/kw-hr | 128 770 kW | 10 920 Btu/kw-hr | 10.0 |
| **Kvaerner Brug** | | | | | | |
| G5341 | 1958 | 24 700 kW | 12 230 Btu/kw-hr | 26 500 kW | ------- | 10.2 |
| PG5341 | 1961 | 24 000 kW | 12 460 Btu/kw-hr | 25 920 kW | 12 350 Btu/kw-hr | 10.2 |
| G6001 | 1979 | 31 800 kW | 11 250 Btu/kw-hr | 33 300 kW | ------- | 11.5 |
| G6501B | 1979 | 35 650 kW | 11 170 Btu/kw-hr | 38 800 kW | ------- | 11.5 |
| G9001 | 1979 | 105 600 kW | 10 700 Btu/kw-hr | 114 000 kW | 10 860 Btu/kw-hr | 9.6 |
| PG9001 | 1979 | 102 700 kW | 10 900 Btu/kw-hr | 111 300 kW | 10 860 Btu/kw-hr | 9.6 |
| LM2500 | 1978 | 20 100 kW | 9 800 Btu/kw-hr | ------- | ------- | 18.0 |
| LM2500 | 1978 | 19 730 kW | 9 990 Btu/kw-hr | ------- | ------- | 18.0 |
| **Mitsubishi Heavy Industries** | | | | | | |
| MW-101L | 1968 | 8 710 kW | 14 960 Btu/kw-hr | 9 250 kW | 14 640 Btu/kw-hr | 8.0 |
| MW-191 | 1963 | 17 790 kW | 13 440 Btu/kw-hr | 19 100 kW | 13 220 Btu/kw-hr | 7.0 |
| MW-251 | 1964 | 36 840 kW | 11 800 Btu/kw-hr | 39 980 kW | 11 520 Btu/kw-hr | 12.0 |
| MW-501B | 1973 | 83 620 kW | 10 850 Btu/kw-hr | 90 600 kW | 10 720 Btu/kw-hr | 12.0 |
| MW-501D | 1980 | 100 000 kW | 10 470 Btu/kw-hr | 107 900 kW | 10 400 Btu/kw-hr | 14.0 |
| MW-701B | 1976 | 90 830 kW | 11 160 Btu/kw-hr | 97 780 kW | 11 070 Btu/kw-hr | 11.0 |
| MW-701D | 1981 | 121 800 kW | 10 670 Btu/kw-hr | 130 400 kW | 10 580 Btu/kw-hr | 14.0 |
| **Mitsui Engineering & Shipbuilding** | | | | | | |
| SB30C | 1973 | 4 870 kW | 14 073 Btu/kw-hr | 5 110 kW | 13 935 Btu/kw-hr | 6.9 |
| SB60C-M | 1981 | 13 070 kW | 11 123 Btu/kw-hr | 14 140 kW | 10 986 Btu/kw-hr | 13.2 |
| SB90C | 1969 | 16 170 kW | 13 364 Btu/kw-hr | 16 810 kW | 13 299 Btu/kw-hr | 6.9 |
| **Noel Penny Turbines** | | | | | | |
| NPT 403 | 1976 | 294 kW | 20 000 Btu/kw-hr | 318 kW | 19 050 Btu/kw-hr | 5.8 |
| NPT 603 | 1979 | 455 kW | 16 500 Btu/kw-hr | 492 kW | 16 300 Btu/kw-hr | 5.2 |
| **Nuovo Pignone** | | | | | | |
| G1602 | 1972 | 4 100 kW | 14 300 Btu/kw-hr | ------- | ------- | 7.9 |
| G1522 | 1972 | 3 575 kW | 15 240 Btu/kw-hr | ------- | ------- | 7.0 |
| G3142 | 1949 | 10 200 kW | 13 540 Btu/kw-hr | ------- | ------- | 7.1 |
| PG5341 | 1958 | 24 000 kW | 12 460 Btu/kw-hr | 25 920 kW | 12 350 Btu/kw-hr | 10.2 |
| PG6501 | 1978 | 34 850 kW | 11 310 Btu/kw-hr | 38 000 kW | 11 190 Btu/kw-hr | 11.5 |
| PG9141 | 1976 | 102 700 kW | 10 900 Btu/kw-hr | 111 300 kW | 10 860 Btu/kw-hr | 9.6 |
| LM2500 | 1979 | 20 100 kW | 9 800 Btu/kw-hr | ------- | ------- | 18.0 |

# ELECTRIC POWER

| Air Flow (lb/sec) at peak load | Turbine Speed output shaft speed | Turbine Inlet Temp at peak rating | Exhaust Temp at peak rating | Dry Weight (Approx) | Length x Width x Height (Approx) | Comments and Footnotes |
|---|---|---|---|---|---|---|
| | | | | | | **Kongsberg Gas Turbines and Power Systems** |
| 29 lb/sec | 1500/1800 | ------- | 603°C | 2 705 kg | 2.0 x 1.6 x 2.2 m | |
| 30 lb/sec | 1500/1800 | ------- | 638°C | 2 705 kg | 2.0 x 1.6 x 2.2 m | max standby rating |
| 46 lb/sec | 1500/1800 | ------- | 494°C | 3 760 kg | 3.7 x 1.9 x 2.2 m | |
| 7.7 lb/sec | 1500/1800 | ------- | 525°C | 770 kg | 1.6 x 1.0 x 0.9 m | liquid fuel rating |
| 7.7 lb/sec | 1500/1800 | ------- | 522°C | 770 kg | 1.6 x 1.0 x 0.9 m | gas fuel rating |
| | | | | | | **Kraftwerk Union** |
| 763 lb/sec | 3 000 rpm | 980°C | 526°C | 238 000 kg | 15 x 11 x 8 m | ratings at gen terminals, |
| 1072 lb/sec | 3 000 rpm | 1010°C | 532°C | 280 000 kg | 18 x 12 x 9 m | gas fuel on LHV 50067 kJ/kg |
| | | | | | | **Kvaerner Brug** |
| 266 lb/sec | 5 105 rpm | 1805°F | 955°F | 180 000 kg | 77 x 11 x 12 ft | distillate fuel |
| 266 lb/sec | 5 105 rpm | 1805°F | 955°F | 194 000 kg | 87 x 11 x 12 ft | ratings; higher |
| 300 lb/sec | 5 100 rpm | 1950°F | 967°F | ------- | ------- | on gas. |
| 300 lb/sec | 5 100 rpm | 2085°F | 1085°F | 240 000 kg | 87 x 11 x 12 ft | |
| 868 lb/sec | 3 000 rpm | 2055°F | 1020°F | 295 000 kg | 83 x 15 x 19 ft | 50-Hz machine |
| 868 lb/sec | 3 000 rpm | 2055°F | 1020°F | ------- | ------- | |
| 144 lb/sec | 3 600 rpm | ------- | ------- | 100 000 kg | 50 x 11 x 14 ft | packaged w/o duct |
| 144 lb/sec | 3 000 rpm | ------- | ------- | 100 000 kg | 50 x 11 x 14 ft | 50-Hz machine |
| | | | | | | **Mitsubishi Heavy Industries** |
| 145 lb/sec | 6 000 rpm | ------- | 806°F | 40 000 kg | 8.0 x 3.2 x 2.9 m | gas fuel ratings |
| 269 lb/sec | 4 912 rpm | ------- | 784°F | 85 000 kg | 10.9 x 3.7 x 3.5 m | |
| 347 lb/sec | 4 984 rpm | ------- | 1008°F | 100 000 kg | 12.4 x 3.7 x 3.7 m | |
| 782 lb/sec | 3 600 rpm | ------- | 952°F | 143 000 kg | 11.6 x 5.7 x 4.4 m | |
| 779 lb/sec | 3 600 rpm | ------- | 1029°F | 143 000 kg | 11.6 x 5.7 x 4.4 m | |
| 863 lb/sec | 3 000 rpm | ------- | 937°F | 170 000 kg | 12.5 x 5.2 x 5.2 m | |
| 971 lb/sec | 3 000 rpm | ------- | 1027°F | 170 000 kg | 12.5 x 5.2 x 5.2 m | |
| | | | | | | **Mitsui Engineering & Shipbuilding** |
| 62 lb/sec | 9 410 rpm | 951°C | 535°C | 28 300 kg | 7.8 x 3.0 x 4.3 m | weight dimensions |
| 134 lb/sec | 5 680 rpm | 1031°C | 480°C | 54 700 kg | 13.0 x 3.5 x 5.5 m | include base plate |
| 190 lb/sec | 5 475 rpm | 951°C | 530°C | 53 000 kg | 13.0 x 3.5 x 6.0 m | and accessories |
| | | | | | | **Noel Penny Turbines** |
| 5.0 lb/sec | 1500/3600 | 1712°F | 1080°F | 335 lb | 4.5 x 1.4 x 1.4 ft | |
| 5.4 lb/sec | 1500/3600 | 1880°F | 1265°F | 450 lb | 4.4 x 1.4 x 1.4 ft | |
| | | | | | | **Nuovo Pignone** |
| 53 lb/sec | 10 290 rpm | *1700°F | 990°F | 92 500 lb | 19 x 8 x 11 ft | ratings on distillate; |
| 45 lb/sec | 10 290 rpm | *1700°F | 1010°F | 92 500 lb | 19 x 8 x 10 ft | slightly higher on gas |
| 115 lb/sec | 6 500 rpm | *1730°F | 980°F | 242 000 lb | 63 x 11 x 12 ft | |
| 266 lb/sec | 5 100 rpm | *1805°F | 955°F | 398 000 lb | 77 x 11 x 12 ft | |
| 302 lb/sec | 5 100 rpm | *2085°F | 955°F | 500 000 lb | 67 x 11 x 12 ft | |
| 868 lb/sec | 3 000 rpm | *2055°F | 1020°F | 1 015 000 lb | 95 x 11 x 22 ft | |
| 135 lb/sec | 3 600 rpm | *2167°F | 940°F | 220 000 lb | 53 x 11 x 12 ft | |
| | | *est | | | | |

| Model (series designation) | Year (1st year unit was (will be) available) | ISO Base Rating (kW) (6000 hrs a year or more) | Heat Rate (Btu/kw-hr) (lower heating value (LHV)) | ISO Peak Rating (kW) (good for up to 2000 hrs a year) | Heat Rate (Btu/kw-hr) (lower heating value (LHV)) | Pressure Ratio (overall compressor) |
|---|---|---|---|---|---|---|
| **Pratt & Whitney Aircraft Canada** | | | | | | |
| ST6J70 | 1968 | 350 kW | 17 827 Btu/kw-hr | 430 kW | 16 974 Btu/kw-hr | 6.2 |
| ST6J77 | 1972 | 380 kW | 17 684 Btu/kw-hr | 620 kW | 16 090 Btu/kw-hr | 6.9 |
| CFT4C-3F | 1976 | 29 820 kW | 10 880 Btu/kw-hr | 32 180 kW | 10 760 Btu/kw-hr | 13.8 |
| P&WC | 1978 | 59 950 kW | 10 820 Btu/kw-hr | 64 700 kW | 10 700 Btu/kw-hr | 13.8 |
| **Prvni Brnenska Strojirna** | | | | | | |
| ST1.5 | 1972 | ------- | ------- | 1 500 kw | 22 600 Btu/kw-hr | 3.9 |
| ST9 | 1970 | 9 500 kW | 16 500 Btu/kw-hr | 9 500 kW | 16 500 Btu/kw-hr | 5.0 |
| ST15RD | 1973 | 15 000 kW | ------- | 15 000 kW | ------- | --- |
| ST25 | 1970 | 26 600 kW | 13 200 Btu/kw-hr | 26 600 kW | 13 200 Btu/kw-hr | 17.0 |
| ST30RD | 1973 | 30 000 kW | ------- | 30 000 kW | ------- | --- |
| **Rolls-Royce** | | | | | | |
| Spey | 1976 | 11 900 kW | 10 240 Btu/kw-hr | 13 800 kW | 10 050 Btu/kw-hr | 19.0 |
| Avon | 1964 | 16 300 kW | 11 790 Btu/kw-hr | 18 200 kW | 11 540 Btu/kw-hr | 10.0 |
| RB 211 | 1974 | 23 800 kW | 9 840 Btu/kw-hr | 25 700 kW | 9 720 Btu/kw-hr | 19.2 |
| Olympus | 1962 | 28 100 kW | 11 150 Btu/kw-hr | 28 900 kW | 11 120 Btu/kw-hr | 11.0 |
| SK15 | 1981 | 16 300 kW | 11 790 Btu/kw-hr | 18 200 kW | 11 540 Btu/kw-hr | 10.0 |
| SK25 | 1981 | 23 800 kW | 9 840 Btu/kw-hr | 25 700 kW | 9 720 Btu/kw-hr | 19.2 |
| SK30 | 1977 | 28 100 kW | 11 150 Btu/kw-hr | 28 900 kW | 11 120 Btu/kw-hr | 11.0 |
| SK50 | 1981 | 47 600 kW | 9 840 Btu/kw-hr | 51 400 kW | 9 720 Btu/kw-hr | 19.2 |
| SK60 | 1977 | 56 200 kW | 11 150 Btu/kw-hr | 57 800 kW | 11 120 Btu/kw-hr | 11.0 |
| **Ruston Gas Turbines** | | | | | | |
| TA1750 | 1969 | 1 269 kW | 20 410 Btu/kw-hr | ------- | ------- | 4.3 |
| TA2500 | 1978 | 1 745 kW | 17 010 Btu/kw-hr | ------- | ------- | 4.9 |
| TB5000 | 1977 | 3 460 kW | 14 150 Btu/kw-hr | ------- | ------- | 7.0 |
| Tornado | 1982 | 6 030 kW | 11 450 Btu/kw-hr | ------- | ------- | 12.1 |
| **Solar Turbines** | | | | | | |
| Saturn | 1960 | 800 kW | 16 250 Btu/kw-hr | ------- | ------- | 6.2 |
| Saturn | 1960 | ------- | ------- | *900 kW | 15 650 Btu/kw-hr | 6.2 |
| Centaur | 1968 | 2 800 kW | 13 930 Btu/kw-hr | *3 130 kW | 13 750 Btu/kw-hr | 9.0 |
| Mars | 1977 | 7 400 kW | 11 130 Btu/kw-hr | *9 130 kW | 10 470 Btu/kw-hr | 16.0 |
| | | | | *standby rating | | |
| **Stal-Laval Turbin** | | | | | | |
| GT35B | 1968 | 12 600 kW | 11 960 Btu/kw-hr | 14 500 kW | 11 650 Btu/kw-hr | 11.0 |
| PP3-Avon 1533 | 1966 | 14 300 kW | 12 220 Btu/kw-hr | 15 400 kW | 12 200 Btu/kw-hr | 10.0 |
| GT120C | 1959 | 70 000 kW | 11 960 Btu/kw-hr | 75 000 kW | 11 780 Btu/kw-hr | 15.0 |
| GT200 | 1977 | 75 000 kW | 10 300 Btu/kw-hr | 85 000 kW | 10 120 Btu/kw-hr | 16.0 |
| **Sulzer Brothers** | | | | | | |
| 3 | 1976 | 5 800 kW | 12 640 Btu/kw-hr | 6 100 kW | 12 540 Btu/kw-hr | 8.9 |
| R3 | 1976 | 5 570 kW | 10 340 Btu/kw-hr | 5 850 kW | 10 240 Btu/kw-hr | 9.0 |
| 7 | 1970 | 10 700 kW | 13 430 Btu/kw-hr | 11 200 kW | 13 280 Btu/kw-hr | 7.6 |
| R7 | 1970 | 10 300 kW | 10 630 Btu/kw-hr | 10 800 kW | 10 500 Btu/kw-hr | 7.7 |

# ELECTRIC POWER

| Air Flow (lb/sec) (at peak load) | Turbine Speed (output shaft speed) | Turbine Inlet Temp (at peak rating) | Exhaust Temp (at peak rating) | Dry Weight (Approx) | Length x Width x Height (Approx) | Comments and Footnotes |
|---|---|---|---|---|---|---|
| | | | | | | **Pratt & Whitney Aircraft Canada** |
| 6.0 lb/sec | 1 800 rpm | 1650°F | 973°F | 350 lb | 5 x 2 x 2 ft | |
| 6.9 lb/sec | 1 800 rpm | 1800°F | 1065°F | 369 lb | 5 x 2 x 2 ft | |
| 306 lb/sec | 3 600 rpm | 1970°F | 877°F | 580 000 lb | 60 x 20 x 40 ft | |
| 611 lb/sec | 3 600 rpm | 1970°F | 877°F | 860 000 lb | 102 x 20 x 40 ft | height w. tallest stack |
| | | | | | | **Prvni Brneska Strojirna** |
| 33 lb/sec | 14 250 rpm | 1472°F | 968°F | 17 650 lb | 8 x 7 x 7 ft | with 1500rpm gearbox & base |
| 176 lb/sec | 3 000 rpm | 1360°F | 805°F | 275 000 lb | 35 x 21 x 21 ft | |
| ------- | 3 000 rpm | ------- | ------- | 73 500 lb | 36 x 10 x 11 ft | gasifier & power turbine |
| 304 lb/sec | 3 000 rpm | 1380°F | 735°F | 475 000 lb | 38 x 30 x 28 ft | |
| ------- | 3 000 rpm | ------- | ------- | 308 650 lb | 89 x 12 x 12 ft | elec gen & 2x gasifers |
| | | | | | | **Rolls Royce** |
| 128 lb/sec | ------- | ------- | ------- | 3 250 lb | 9 x 4 x 5 ft | gas generator only |
| 181 lb/sec | ------- | ------- | ------- | 3 560 lb | 11.5 x 3 x 3 ft | gas generator only |
| 207 lb/sec | ------- | ------- | ------- | 5 740 lb | 10 x 5 x 5 ft | gas generator only |
| 240 lb/sec | ------- | ------- | ------- | 4 800 lb | 12 x 3.6 x 3.6 ft | gas generator only |
| 181 lb/sec | 3000/3600 | *710°C | 500°C | 376 000 lb | 60 x 14 x 24 ft | Avon packaged set |
| 207 lb/sec | 3000/3600 | ------- | ------- | 380 000 lb | 60 x 14 x 24 ft | RB 211 packaged set |
| 240 lb/sec | 3000/3600 | *805°C | 555°C | 515 000 lb | 76 x 14 x 49 ft | Olympus packaged set |
| 414 lb/sec | 3000/3600 | ------- | ------- | 620 000 lb | 93 x 14 x 24 ft | 2x RB 211 package |
| 480 lb/sec | 3000/3600 | *805°C | 555°C | 840 000 lb | 118 x 14 x 49 ft | 2x Olympus package |
| | | *power turbine inlet temp | | | | |
| | | | | | | **Ruston Gas Turbines** |
| 25 lb/sec | 6 000 rpm | 1516°F | 979°F | 22 500 lb | 16 x 8 x 8.5 ft | |
| 28 lb/sec | 7 950 rpm | 1564°F | 944°F | 20 000 lb | 19 x 8 x 8.5 ft | |
| 46 lb/sec | 7 950 rpm | 1633°F | 944°F | 30 000 lb | 19 x 8 x 8 ft | |
| 60 lb/sec | 11 085 rpm | 1832°F | 837°F | 55 000 lb | 24 x 8 x 9.5 ft | weight w. gearbox |
| | | | | | | **Solar Turbines** |
| 12.8 lb/sec | 22 300 rpm | 1360°F | 830°F | 20 250 lb | 21 x 6 x 7 ft | turbine inlet and |
| 12.8 lb/sec | 22 300 rpm | 1405°F | 870°F | 11 045 lb | 14 x 4 x 5.5 ft | exhaust temps at |
| 38 lb/sec | 14 950 rpm | 1612°F | 813°F | 40 000 lb | 24 x 7 x 8.5 ft | continuous rating |
| 80 lb/sec | 8 000 rpm | 1825°F | 780°F | 140 000 lb | 49 x 8 x 9 ft | |
| | | | | | | **Stal-Laval Turbin** |
| 186 lb/sec | 3000/3600 | 1560°F | 740°F | 85 000 lb | 46 x 11 x 10 ft | |
| 180 lb/sec | 3000/3600 | 1650°F | 860°F | ------- | 43 x 11 x 10 ft | Avon 1533 gas generator |
| 790 lb/sec | 3 000 rpm | 1530°F | 635°F | ------- | 75 x 42 x 42 ft | burns residual fuel |
| 770 lb/sec | 3 000 rpm | ------- | ------- | 250 000 lb | 85 x 30 x 28 ft | packaged plant design |
| | | | | | | **Sulzer Brothers** |
| 158 lb/sec | 10 600 rpm | 970°C | 505°C | 60 000 lb | 24 x 10 x 9 ft | |
| 166 lb/sec | 10 600 rpm | 970°C | 391°C | 60 000 lb | 24 x 10 x 9 ft | |
| 140 lb/sec | 6 400 rpm | 925°C | 493°C | 138 000 lb | 38 x 12 x 13 ft | |
| 140 lb/sec | 6 400 rpm | 925°C | 342°C | 138 000 lb | 38 x 12 x 13 ft | |

**(continued next page)**

| Model (series designation) | Year (1st year unit was (will be) available) | ISO Base Rating (kW) (6000 hrs a year or more) | Heat Rate (Btu/kw-hr) (lower heating value (LHV)) | ISO Peak Rating (kW) (good for up to 2000 hrs a year) | Heat Rate (Btu/kw-hr) (lower heating value (LHV)) | Pressure Ratio (overall compressor) |
|---|---|---|---|---|---|---|
| **Sulzer Brothers (continued)** | | | | | | |
| 10 | 1981 | 20 700 kW | 10 310 Btu/kw-hr | ------- | ------- | 13.5 |
| PRIMO 12 | 1978 | 12 200 kW | 12 060 Btu/kw-hr | 15 800 kW | 11 600 Btu/kw-hr | 8.8 |
| PRIMO 14 | 1978 | 15 500 kW | 11 680 Btu/kw-hr | 17 400 kW | 11 450 Btu/kw-hr | 8.8 |
| PRIMO 15 | 1979 | 16 200 kW | 11 680 Btu/kw-hr | 18 000 kW | 11 450 Btu/kw-hr | 9.0 |
| **Thomassen Holland** | | | | | | |
| TF10 | 1983 | 6 700 kW | 8 120 Btu/kw-hr | ------- | ------- | 8.5 |
| G3142J | 1949 | 10 450 kW | 13 320 Btu/kw-hr | ------- | ------- | 7.2 |
| G3142R | 1951 | 10 000 kW | 10 370 Btu/kw-hr | ------- | ------- | 7.2 |
| G5341P | 1958 | 25 350 kW | 12 020 Btu/kw-hr | ------- | ------- | 10.2 |
| PG5341P | 1961 | 24 500 kW | 12 310 Btu/kw-hr | 26 450 kW | 12 220 Btu/kw-hr | 10.2 |
| G6441A | 1979 | 32 450 kW | 11 150 Btu/kw-hr | ------- | ------- | 11.1 |
| PG6441A | 1979 | 31 750 kW | 11 280 Btu/kw-hr | 34 750 kW | 11 210 Btu/kw-hr | 11.1 |
| PG6441B | 1981 | 34 140 kW | 11 280 Btu/kw-hr | 36 900 kW | 11 250 Btu/kw-hr | 11.5 |
| G7101E | 1979 | 76 400 kW | 10 510 Btu/kw-hr | ------- | ------- | 11.5 |
| PG7101E | 1979 | 74 400 kW | 10 690 Btu/kw-hr | 80 500 kW | 10 640 Btu/kw-hr | 11.5 |
| G9141E | 1979 | 107 500 kW | 10 620 Btu/kw-hr | ------- | ------- | 9.6 |
| PG9141E | 1979 | 104 500 kW | 10 820 Btu/kw-hr | 113 300 kW | 10 790 Btu/kw-hr | 9.6 |
| **Toshiba** | | | | | | |
| Type 9 | 1970 | 34 500 kW | 12 010 Btu/kw-hr | 37 500 kW | 11 890 Btu/kw-hr | 8.8 |
| Type 11 | 1971 | 72 500 kW | 10 760 Btu/kw-hr | 78 900 kW | 10 660 Btu/kw-hr | 11.0 |
| Type 13 | 1970 | 88 900 kW | 10 730 Btu/kw-hr | 97 300 kW | 10 580 Btu/kw-hr | 11.8 |
| **Turbine Industrie** | | | | | | |
| Doller 1202 | 1978 | 900 kW | 19 300 Btu/kw-hr | 900 kW | 19 300 Btu/kw-hr | 6.0 |
| **Turbomeca** | | | | | | |
| Astazou IV | 1972 | 320 kW | 18 170 Btu/kw-hr | 340 kW | 17 670 Btu/kw-hr | 5.3 |
| Astazou XIV | 1970 | 430 kW | 16 200 Btu/kw-hr | 500 kW | 15 100 Btu/kwhr | 7.5 |
| Bastan VI | 1968 | 615 kW | 17 550 Btu/kw-hr | 685 kW | 16 700 Btu/kw-hr | 5.7 |
| Bastan VII | 1972 | 870 kW | 14 700 Btu/kw-hr | 1 050 kW | 13 800 Btu/kw-hr | 7.1 |
| Bi Bastan VI | 1978 | 1 160 kW | 17 930 Btu/kw-hr | 1 295 kW | 17 080 Btu/kw-hr | 5.7 |
| Bi Bastan VII | 1978 | 1 645 kW | 15 500 Btu/kw-hr | 1 720 kW | 15 260 Btu/kw-hr | 7.1 |
| **UTC Power Systems** | | | | | | |
| FT4C-3F | 1974 | 30 560 kW | 10 920 Btu/kw-hr | 33 000 kW | 10 834 Btu/kw-hr | 13.8 |
| TP4-2(C3F) | 1974 | 61 440 kW | 10 865 Btu/kw-hr | 66 320 kW | 10 780 Btu/kw-hr | 13.8 |
| **USSR Energomachexport** | | | | | | |
| GTG-6 | 1960 | 6 300 kW | 14 720 Btu/kw-hr | 7 200 kW | ------- | 6.0 |
| GTG-10 | 196? | 10 000 kW | 11 960 Btu/kw-hr | 13 000 kW | ------- | 6.0 |
| GTG-16 | 1962 | 16 000 kW | 11 960 Btu/kw-hr | 20 000 kW | ------- | 11.5 |
| GTG-25 | 1971 | 25 000 kW | 11 590 Btu/kw-hr | 30 000 kW | ------- | 12.5 |
| GT-35 | 196? | 31 000 kW | 14 352 Btu/kw-hr | 40 000 kW | ------- | 6.5 |
| GT-100-3 | 1970 | 95 000 kW | 11 830 Btu/kw-hr | 115 000 kW | 10 600 Btu/kw-hr | 26.5 |

# ELECTRIC POWER

| Air Flow (lb/sec) (at peak load) | Turbine Speed (output shaft speed) | Turbine Inlet Temp (at peak rating) | Exhaust Temp (at peak rating) | Dry Weight (Approx) | Length x Width x Height (Approx) | Comments and Footnotes |
|---|---|---|---|---|---|---|
| **Sulzer Brothers (continued)** | | | | | | |
| 158 lb/sec | 7 700 rpm | ------- | 507°C | 61 700 lb | 26 x 10 x 11 ft | |
| 165 lb/sec | 6 500 rpm | 810°C | 403°C | 62 000 lb | 40 x 14 x 13 ft | Avon 1533 gas generator |
| 171 lb/sec | 6 500 rpm | 933°C | 464°C | 62 000 lb | 40 x 14 x 13 ft | Avon 1534 gas generator |
| 172 lb/sec | 6 500 rpm | 955°C | 478°C | 62 000 lb | 40 x 14 x 13 ft | Avon 1535 gas generator |
| **Thomassen Holland** | | | | | | |
| 55 lb/sec | 9 200 rpm | 1965°F | 490°F | 96 000 lb | 32 x 10 x 12 ft | |
| 115 lb/sec | 6 500 rpm | 1730°F | 979°F | 242 000 lb | 63 x 11 x 12 ft | peak ratings on request |
| 115 lb/sec | 6 500 rpm | 1750°F | 668°F | 242 000 lb | 63 x 11 x 12 ft | |
| 269 lb/sec | 5 100 rpm | 1730°F | 901°F | 398 000 lb | 77 x 11 x 12 ft | |
| 266 lb/sec | 5 100 rpm | 1805°F | 955°F | 570 000 lb | 115 x 19 x 35 ft | packaged plant |
| 304 lb/sec | 5 100 rpm | 1850°F | 901°F | 202 000 lb | ------- | |
| 301 lb/sec | 5 100 rpm | 1950°F | 967°F | 700 000 lb | 123 x 24 x 34 ft | packaged plant |
| 301 lb/sec | 5 100 rpm | 2085°F | 1041°F | 700 000 lb | 123 x 24 x 34 ft | packaged plant |
| 604 lb/sec | 3 600 rpm | 1985°F | 977°F | 587 000 lb | 70 x 29 x 13 ft | |
| 599 lb/sec | 3 600 rpm | 2085°F | 1045°F | 1 070 000 lb | 132 x 71 x 31 ft | packaged plant |
| 876 lb/sec | 3 000 rpm | 1955°F | 953°F | 650 000 lb | 83 x 15 x 19 ft | |
| 868 lb/sec | 3 000 rpm | 2055°F | 1020°F | 1 425 000 lb | 148 x 48 x 38 ft | packaged plant |
| **Toshiba** | | | | | | |
| 355 lb/sec | 4473/4485 | ------- | 545°C | 79 000 kg | 10 x 3.2 x 7.6 m | ratings at output |
| 639 lb/sec | 3600 rpm | ------- | 550°C | 165 000 kg | 11.2 x 7.2 x 5.1 m | shaft of GT, |
| 825 lb/sec | 3000 rpm | ------- | 526°C | 270 000 kg | 14.5 x 4.6 x 9.8 m | distillate fuel |
| **Turbine Industrie** | | | | | | |
| 15.5 lb/sec | 1 500 rpm | 870°C | 525°C | 1 500 kg | 2.6 x 1.2 x 2 m | output speeds to 17,000 rpm |
| **Turbomeca** | | | | | | |
| 5.5 lb/sec | 1500/1800 | 845°C | 485°C | 310 kg | 4 x 2 x 2 ft | |
| 7.3 lb/sec | 1 500 rpm | 870°C | 450°C | ------- | 6 x 2 x 3 ft | |
| 10.1 lb/sec | 1500/1800 | 870°C | 535°C | 270 kg | 6 x 3 x 3 ft | |
| 13.3 lb/sec | 1500/1800 | 880°C | 485°C | 290 kg | 6 x 3 x 3 ft | |
| 20.2 lb/sec | 1500/1800 | 870°C | 535°C | 1 400 kg | 7 x 4 x 3 ft | |
| 30 lb/sec | 1500/1800 | 800°C | 485°C | 1 450 kg | 8 x 4 x 3 ft | |
| **UTC Power Systems** | | | | | | |
| 306 lb/sec | 3 600 rpm | 1984°F | 877°F | 30 600 lb | 27 x 10 x 17 ft | |
| 611 lb/sec | 3 600 rpm | 1984°F | 877°F | 360 tons | 102 x 20 x 40 ft | height w. tallest stack |
| **USSR Energomachexport** | | | | | | |
| 100 lb/sec | 3000/3600 | 760°C | 430°C | 47 000 kg | 12 x 3 x 4 m | |
| 190 lb/sec | 3000/3600 | ------- | 315°C | 56 000 kg | 9 x 3 x 3 m | |
| 188 lb/sec | 3000/3600 | 900°C | 430°C | 72 000 kg | 13 x 3 x 4 m | |
| 380 lb/sec | 3000/3600 | ------- | 385°C | 63 000 kg | 13 x 3 x 4 m | |
| 468 lb/sec | 3 000 rpm | 770°C | 435°C | 240 000 kg | ------- | |
| 980 lb/sec | 3 000 rpm | 750°C | 395°C | 600 000 kg | 24 x 6 x 6.6 m | designed with intercooling |

| Model<br>(series designation) | Year<br>(1st year unit was (will be) available) | ISO Base Rating (kW)<br>(6000 hrs a year or more) | Heat Rate (Btu/kw-hr)<br>(lower heating value (LHV)) | ISO Peak Rating (kW)<br>(good for up to 2000 hrs a year) | Heat Rate (Btu/kw-hr)<br>(lower heating value (LHV)) | Pressure Ratio<br>(overall compressor) |
|---|---|---|---|---|---|---|
| **Westinghouse Canada** | | | | | | |
| W101G | 1961 | 8 300 kW | 15 010 Btu/kw-hr | ------- | ------- | 6.6 |
| W191G | 1960 | 18 040 kW | 13 220 Btu/kw-hr | 19 350 kW | 13 010 Btu/kw-hr | 7.9 |
| W101PG | 1966 | 8 080 kW | 15 340 Btu/kw-hr | ------- | ------- | 6.6 |
| W191PG | 1966 | 17 700 kW | 13 390 Btu/kw-hr | 19 000 kW | 13 210 Btu/kw-hr | 7.9 |
| W191PG Super | 1980 | 19 250 kW | 12 930 Btu/kw-hr | ------- | ------- | 8.6 |
| W301G | 1961 | 32 300 kW | 13 060 Btu/kw-hr | 34 400 kW | 12 870 Btu/kw-hr | 6.8 |
| W301G Super | 1978 | 35 125 kW | 12 600 Btu/kw-hr | ------- | ------- | 7.5 |
| **Westinghouse Electric Combustion Turbine Systems** | | | | | | |
| W251B 60Hz | 1974 | 38 850 kW | 11 980 Btu/kw-hr | 41 960 kW | 11 845 Btu/kw-hr | 12.0 |
| W251B 50Hz | 1974 | 38 850 kW | 11 980 Btu/kw-hr | 41 960 kW | 11 845 Btu/kw-hr | 12.0 |
| W501D 60Hz | 1975 | 96 440 kW | 10 660 Btu/kw-hr | 104 220 kW | 10 580 Btu/kw-hr | 14.2 |
| W501D 50Hz | 1981 | 95 330 kW | 10 790 Btu/kw-hr | 102 970 kW | 10 710 Btu/kw-hr | 14.2 |
| **Norwalk-Turbo** | | | | | | |
| TG-7 | 1975 | 597 kW | 15 400 Btu/kw-hr | 660 kW | 14 070 Btu/kw-hr | 7.0 |
| TG-9 | 1975 | 665 kW | 16 030 Btu/kw-hr | 740 kW | 13 720 Btu/kw-hr | 9.0 |
| TG-14 | 1975 | 1 194 kW | 15 400 Btu/kw-hr | 1 320 kW | 14 070 Btu/kw-hr | 7.0 |
| TG-18 | 1975 | 1 330 kW | 16 030 Btu/kw-hr | 1 480 kW | 13 720 Btu/kw-hr | 9.0 |
| TG-25 | 1975 | 1 736 kW | 14 773 Btu/kw-hr | 2 083 kW | 14 376 Btu/kw-hr | -- |
| TG-40 | 1975 | 2 778 kW | 12 713 Btu/kw-hr | 3 055 kW | 12 563 Btu/kw-hr | 8.4 |
| TG-60 | 1981 | 3 611 kW | 11 171 Btu/kw-hr | 4 166 kW | 11 067 Btu/kw-hr | 15.0 |

# ELECTRIC POWER

| Air Flow (lb/sec) at peak load | Turbine Speed output shaft speed | Turbine Inlet Temp at peak rating | Exhaust Temp at peak rating | Dry Weight (Approx) | Length x Width x Height (Approx) | Comments and Footnotes |
|---|---|---|---|---|---|---|
| | | | | | | **Westinghouse Canada** |
| 134 lb/sec | 6 000 rpm | 1450°F | 820°F | 80 000 lb | 27 x 9 x 11 ft | 17% less fuel regen |
| 273 lb/sec | 4 912 rpm | 1450°F | 773°F | 170 550 lb | 36 x 10 x 12 ft | 12% less fuel regen |
| 134 lb/sec | 6 000 rpm | 1450°F | 820°F | ------- | ------- | packaged powerplant |
| 270 lb/sec | 4 912 rpm | 1450°F | 773°F | ------- | ------- | packaged powerplant |
| 297 lb/sec | 4 912 rpm | 1450°F | 758°F | ------- | ------- | supercharged |
| 438 lb/sec | 3 600 rpm | 1450°F | 800°F | 233 700 lb | 44 x 14 x 15 ft | 60 Hz; geared for 50 Hz |
| 481 lb/sec | 3 600 rpm | 1450°F | 785°F | 233 700 lb | 44 x 14 x 15 ft | supercharged |
| | | | | | | **Westinghouse Electric Combustion Turbine Systems** |
| 363 lb/sec | 5 015 rpm | *2066°F | 1101°F | 826 000 lb | 116 x 33 x 26 ft | distillate fuel ratings; |
| 363 lb/sec | 5 011 rpm | *2066°F | 1101°F | 826 000 lb | 116 x 33 x 26 ft | slightly higher on gas; |
| 778 lb/sec | 3 600 rpm | *2066°F | 1026°F | 1 600 000 lb | 132 x 60 x 48 ft | packaged plants |
| 778 lb/sec | 3 600 rpm | *2066°F<br>*est | 1026°F | 1 900 000 lb | 153 x 60 x 49 ft | |
| | | | | | | **Norwalk-Turbo** |
| 8.2 lb/sec | 33 000 rpm | ------- | ------- | 8 800 lb | 15 x 5 x 6 ft | ST6L-795 driver |
| 8.5 lb/sec | 28 500 rpm | ------- | ------- | 8 800 lb | 15 x 5 x 6 ft | ST66-813 driver |
| 16.4 lb/sec | 33 000 rpm | ------- | ------- | 18 000 lb | 15 x 10 x 6 ft | ST6L-795 driver (tandem) |
| 17.0 lb/sec | 28 500 rpm | ------- | ------- | 18 000 lb | 15 x 10 x 6 ft | ST6L-813 driver (tandem) |
| 23.0 lb/sec | 14 500 rpm | ------- | 1150°F | 15 000 lb | 19 x 6 x 7 ft | TF-25 driver |
| 28.0 lb/sec | 15 400 rpm | ------- | 1130°F | 15 000 lb | 22 x 6 x 7 ft | TF-40 driver |
| 36.0 lb/sec | 7 000 rpm | ------- | 1053°F | 21 000 lb | 23 x 6 x 7 ft | LM-500 driver |

## COMBINED CYCLE PLANTS

# GAS TURBINE PERFORMANCE SPECIFICATIONS

| Power Plant (series designation) | Year (1st year unit was (will be) available) | ISO Base Load Output (#2 Oil) (6000 hrs a year or more) Net Plant | Heat Rate | ISO Peak Load Output (#2 Oil) (good for up to 2000 hrs a year) Net Plant | Heat Rate |
|---|---|---|---|---|---|
| **ACEC** | | | | | |
| COCY 50 | 1979 | 51 100 kW | 8 520 Btu/kw-hr | 54 400 kW | 8 470 Btu/kw-hr |
| COCY 100 | 1977 | 94 500 kW | 8 640 Btu/kw-hr | 99 200 kW | 8 410 Btu/kw-hr |
| COCY 130 | 1977 | 128 300 kW | 7 990 Btu/kw-hr | 134 700 kW | 7 770 Btu/kw-hr |
| COCY 140 | 1979 | 143 000 kW | 7 440 Btu/kw-hr | 153 000 kW | 7 430 Btu/kw-hr |
| **AEG-Kanis** | | | | | |
| various highly standardized combined cycle power plants for MS 3002, 5001, 6001, 7001 and 9001 for generation of electricity and/or district heating. | | | | | |
| **Alsthom-Atlantique** | | | | | |
| VEGA 105P | 1976 | 33 100 kW | 8 850 Btu/kw-hr | 36 000 kW | 8 570 Btu/kw-hr |
| VEGA 205P | 1976 | 69 000 kW | 8 575 Btu/kw-hr | ------- | ------- |
| VEGA 206B | 1978 | 102 300 kW | 8 030 Btu/kw-hr | ------- | ------- |
| VEGA 209B | 1976 | 237 000 kW | 7 900 Btu/kw-hr | ------- | ------- |
| VEGA 209E | 1978 | 290 000 kW | 7 791 Btu/kw-hr | ------- | ------- |
| VEGA109E | 1978 | 144 200 kW | 7 845 Btu/kw-hr | ------- | ------- |
| **BBC Brown, Boveri & Co Ltd<br>Brown, Boveri & Cie AG<br>Brown Boveri Turbomachinery Inc** | | | | | |
| KA9-1 | 1977 | 51 300 kW | 8 061 Btu/kw-hr | 55 900 kW | 7 980 Btu/kw-hr |
| KA9-2 | 1977 | 104 400 kW | 7 921 Btu/kw-hr | 113 700 kW | 7 842 Btu/kw-hr |
| KA9-3 | 1979 | 157 500 kW | 7 874 Btu/kw-hr | 171 600 kW | 7 795 Btu/kw-hr |
| KA9-4 | 1979 | 210 800 kW | 7 846 Btu/kw-hr | 229 700 kW | 7 768 Btu/kw-hr |
| KA11-1 | 1980 | 105 200 kW | 7 399 Btu/kw-hr | 114 600 kW | 7 325 Btu/kw-hr |
| KA11-2 | 1980 | 212 300 kW | 7 329 Btu/kw-hr | 231 400 kW | 7 256 Btu/kw-hr |
| KA11-3 | 1980 | 318 500 kW | 7 329 Btu/kw-hr | 347 100 kW | 7 256 Btu/kw-hr |
| KA11-4 | 1980 | 427 900 kW | 7 335 Btu/kw-hr | 462 400 kW | 7 262 Btu/kw-hr |
| KA13-1 | 1981 | 128 000 kW | 7 413 Btu/kw-hr | 139 500 kW | 7 339 Btu/kw-hr |
| KA13-2 | 1981 | 258 200 kW | 7 349 Btu/kw-hr | 281 400 kW | 7 276 Btu/kw-hr |
| KA13-3 | 1981 | 385 300 kW | 7 384 Btu/kw-hr | 419 900 kW | 7 310 Btu/kw-hr |
| KA13-4 | 1981 | 512 500 kW | 7 401 Btu/kw-hr | 558 600 kW | 7 327 Btu/kw-hr |
| **Curtiss-Wright Power Systems** | | | | | |
| TEC 150 | 1978 | 150 000 kW | 9 100 Btu/kw-hr | 165 000 kW | 9 000 Btu/kw-hr |
| TEC 165 | 1983 | 160 000 kW | 8 525 Btu/kw-hr | ------- | ------- |
| TEC 200 | 1978 | 210 000 kW | 9 300 Btu/kw-hr | 230 000 kW | 9 200 Btu/kw-hr |
| TEC 330 | 1984 | 320 000 kW | 8 525 Btu/kw-hr | ------- | ------- |
| TEC 500 | 1985 | 480 000 kW | 8 525 Btu/kw-hr | ------- | ------- |
| TEC 750 | 1988 | 750 000 kW | 7 210 Btu/kw-hr | ------- | ------- |
| **Fiat TTG** | | | | | |
| CC 50 | 1978 | 53 700 kW | 8 115 Btu/kw-hr | ------- | ------- |
| CC 100 | 1978 | 107 400 kW | 8 115 Btu/kw-hr | ------- | ------- |
| CC 150 | 1978 | 126 000 kW | 8 035 Btu/kw-hr | ------- | ------- |
| CC 250 | 1978 | 252 000 kW | 8 035 Btu/kw-hr | ------- | ------- |

output shaft speed(s) — fired steam boiler

# COMBINED CYCLE

| Electrical Frequency | Boiler Fuel | Gas Turbine Output | Steam Turbine Output | Number Of Gas Turbines | Comments and Footnotes |
|---|---|---|---|---|---|
| **ACEC** | | | | | |
| 50 Hz | non-fired | 36 500 kW | 14 600 kW | one W251 | ratings at generator |
| 50 Hz | non-fired | 67 000 kW | 27 500 kW | two W251 | terminals of plant |
| 50 Hz | non-fired | 91 300 kW | 37 000 kW | one W1101 | (GT & ST), without |
| 50 Hz | non-fired | 91 700 kW | 51 300 kW | one W1101 | inlet and exhaust losses |
| **AEG-Kanis** | | | | | |
| **Alsthom-Atlantique** | | | | | |
| 50/60 Hz | non-fired | 22 400 kW | 10 700 kW | one PG5341 | fired boiler cycle and peak |
| 50/60 Hz | non-fired | 47 350 kW | 22 100 kW | two PG5341 | rating upon request; all |
| 50/60 Hz | non-fired | 72 800 kW | 30 200 kW | two PG6441 | ratings on natural gas; |
| 50 Hz | non-fired | 168 900 kW | 70 000 kW | two PG9111 | ratings available upon |
| 50 Hz | non-fired | 209 400 kW | 87 500 kW | two PG9141` | request for light, heavy |
| 50 Hz | non-fired | 104 700 kW | 40 000 kW | one PG9141 | distillate and residual fuel. |
| **BBC Brown, Boveri & Co Ltd<br>Brown, Boveri & Cie AG<br>Brown Boveri Turbomachinery Inc** | | | | | |
| 50/60 Hz | non-fired | 31 870 kW | 19 430 kW | one Type 9 | all ratings at gen terminals on |
| 50/60 Hz | non-fired | 63 740 kW | 40 660 kW | two Type 9 | distillate fuel, dual pressure HRSG, |
| 50/60 Hz | non-fired | 95 610 kW | 61 890 kW | three Type 9 | one single pressure steam turbine |
| 50/60 Hz | non-fired | 127 480 kW | 83 320 kW | four Type 9 | at 0.04 bar condenser pressure. |
| 60 Hz | non-fired | 68 400 kW | 36 800 kW | one Type 11 | all plants available in following |
| 60 Hz | non-fired | 136 800 kW | 75 500 kW | two Type 11 | versions:water or air cooled; |
| 60 Hz | non-fired | 205 200 kW | 113 300 kW | three Type 11 | dual fuel gas/oil; heavy or crude |
| 60 Hz | non-fired | 273 600 kW | 150 700 kW | four Type 11 | oil fired; w. fired HRSG. |
| 50 Hz | non-fired | 84 700 kW | 43 300 kW | one Type 13 | for cogeneration of heat |
| 50 Hz | non-fired | 169 400 kW | 88 800 kW | two Type 13 | and power for industry, |
| 50 Hz | non-fired | 254 100 kW | 131 209 kW | three Type 13 | seawater desalination or |
| 50 Hz | non-fired | 338 800 kW | 173 700 kW | four Type 13 | district heating. |
| **Curtiss-Wright Power Systems** | | | | | |
| 50/60 Hz | non-fired | 89 000 kW | 61 000 kW | four | |
| 50/60 Hz | non-fired | 97 500 kW | 62 500 kW | two | |
| 50/60 Hz | non-fired | 89 000 kW | 121 000 kW | four | |
| 50/60 Hz | non-fired | 195 000 kW | 125 000 kW | four | |
| 50/60 Hz | non-fired | 290 000 kW | 190 000 kW | six | |
| 50/60 Hz | non-fired | 530 000 kW | 220 000 kW | four | |
| **Fiat TTG** | | | | | |
| 50/60 Hz | non-fired | 36 300 kW | 17 400 kW | one TG.20 | ratings on natural gas fuel |
| 50/60 Hz | non-fired | 72 600 kW | 34 800 kW | two TG.20 | |
| 50 Hz | non-fired | 88 000 kW | 38 000 kW | one TG.50 | |
| 50 Hz | non-fired | 176 000 kW | 76 000 kW | two TG.50 | |

| Power Plant (series designation) | Year (1st year unit was (will be) available) | ISO Base Load Output (#2 Oil) (6000 hrs a year or more) Net Plant | Heat Rate | ISO Peak Load Output (#2 Oil) (good for up to 2000 hrs a year) Net Plant | Heat Rate |
|---|---|---|---|---|---|
| **GEC Gas Turbines** | | | | | |
| EM610 | 1975 | 118 200 kW | 7 790 Btu/kw-hr | ------- | ------- |
| **General Electric GT Div** | | | | | |
| STAG 109E | 1979 | 140 400 kW | 7 940 Btu/kw-hr | ------- | ------- |
| STAG 209E | 1979 | 282 300 kW | 7 890 Btu/kw-hr | ------- | ------- |
| STAG 309E | 1979 | 423 800 kW | 7 890 Btu/kw-hr | ------- | ------- |
| STAG 409E | 1979 | 564 700 kW | 7 890 Btu/kw-hr | ------- | ------- |
| STAG 205P | 1979 | 67 400 kW | 8 700 Btu/kw-hr | ------- | ------- |
| STAG 405P | 1977 | 138 400 kW | 8 700 Btu/kw-hr | ------- | ------- |
| STAG 206B | 1979 | 96 200 kW | 8 070 Btu/kw-hr | ------- | ------- |
| STAG 406B | 1979 | 192 400 kW | 8 070 Btu/kw-hr | ------- | ------- |
| STAG 107E | 1977 | 99 300 kW | 7 700 Btu/kw-hr | ------- | ------- |
| STAG 407E | 1979 | 396 700 kW | 7 650 Btu/kw-hr | ------- | ------- |
| STAG 607E | 1979 | 595 600 kW | 7 630 Btu/kw-hr | ------- | ------- |
| **Hitachi** | | | | | |
| 205 | 1982 | 68 300 kW | 9 100 Btu/kw-hr | 74 400 kW | 8 930 Btu/kw-hr |
| 107B | 1982 | 86 000 kW | 8 430 Btu/kw-hr | 97 000 kW | 8 230 Btu/kw-hr |
| 207B | 1982 | 174 000 kW | 8 350 Btu/kw-hr | 195 000 kW | 8 150 Btu/kw-hr |
| 307B | 1982 | 262 000 kW | 8 350 Btu/kw-hr | 293 000 kW | 8 150 Btu/kw-hr |
| 407B | 1982 | 351 000 kW | 8 290 Btu/kw-hr | 393 000 kW | 8 110 Btu/kw-hr |
| 107E | 1982 | 106 000 kW | 7 700 Btu/kw-hr | 115 000 kW | 7 940 Btu/kw-hr |
| 207E | 1982 | 214 000 kW | 7 650 Btu/kw-hr | 233 000 kW | 7 850 Btu/kw-hr |
| 307E | 1982 | 323 000 kW | 7 600 Btu/kw-hr | 352 000 kW | 7 800 Btu/kw-hr |
| 407E | 1982 | 430 000 kW | 7 600 Btu/kw-hr | 469 000 kW | 7 800 Btu/kw-hr |
| 109B | 1980 | 123 000 kW | 8 410 Btu/kw-hr | 138 000 kW | 8 210 Btu/kw-hr |
| 209B | 1982 | 248 000 kW | 8 350 Btu/kw-hr | 278 000 kW | 8 130 Btu/kw-hr |
| 309B | 1982 | 373 000 kW | 8 330 Btu/kw-hr | 418 000 kW | 8 110 Btu/kw-hr |
| 109E | 1982 | 155 000 kW | 7 700 Btu/kw-hr | ------- | ------- |
| 209E | 1982 | 311 100 kW | 7 650 Btu/kw-hr | ------- | ------- |
| 409E | 1982 | 626 600 kW | 7 600 Btu/kw-hr | ------- | ------- |
| **John Brown Engineering** | | | | | |
| 103J | 1979 | 14 500 kW | 9 180 Btu/kw-hr | ------- | ------- |
| 105P | 1978 | 35 500 kW | 8 760 Btu/kw-hr | ------- | ------- |
| 106B | 1980 | 51 000 kW | 8 000 Btu/kw-hr | ------- | ------- |
| 205P | 1980 | 71 700 kW | 8 670 Btu/kw-hr | ------- | ------- |
| 306B | 1980 | 154 000 kW | 7 950 Btu/kw-hr | ------- | ------- |
| 107E | 1980 | 110 000 kW | 7 410 Btu/kw-hr | ------- | ------- |
| 109B | 1979 | 123 000 kW | 7 600 Btu/kw-hr | ------- | ------- |
| 109E | 1980 | 154 000 kW | 7 500 Btu/kw-hr | ------- | ------- |
| 307E | 1980 | 334 000 kW | 7 330 Btu/kw-hr | ------- | ------- |
| 209B | 1980 | 247 000 kW | 7 570 Btu/kw-hr | ------- | ------- |
| 209E | 1980 | 309 000 kW | 7 450 Btu/kw-hr | ------- | ------- |

# COMBINED CYCLE

output shaft speed(s) — fired steam boiler

| Electrical Frequency | Boiler Fuel | Gas Turbine Output | Steam Turbine Output | Number Of Gas Turbines | Comments and Footnotes |
|---|---|---|---|---|---|
| | | | | | **GEC Gas Turbines** |
| 50 Hz | non-fired | ------- | ------- | one EM610 | one, two and three gas turbine based modules available |
| | | | | | **General Electric GT Div** |
| 50 Hz | non-fired | ------- | ------- | one MS9001E | net heat rate on LHV of distillate on all plants listed |
| 50 Hz | non-fired | ------- | ------- | two MS9001E | |
| 50 Hz | non-fired | ------- | ------- | three MS9001E | |
| 50 Hz | non-fired | ------- | ------- | four MS9001E | |
| 50/60 Hz | non-fired | ------- | ------- | two MS5001P | |
| 50/60 Hz | non-fired | ------- | ------- | four MS5001P | |
| 50/60 Hz | non-fired | ------- | ------- | two MS6001B | |
| 50/60 Hz | non-fired | ------- | ------- | four MS6001B | |
| 60 Hz | non-fired | ------- | ------- | one MS7001E | |
| 60 Hz | non-fired | ------- | ------- | four MS7001E | |
| 60 Hz | non-fired | ------- | ------- | six MS7001E | |
| | | | | | **Hitachi** |
| 50/60 Hz | non-fired | 46 000 kW | 22 300 kW | two PG5341 | |
| 60 Hz | non-fired | 59 000 kW | 27 000 kW | one PG7821 | |
| 60 Hz | non-fired | 117 000 kW | 57 000 kW | two PG7821 | |
| 60 Hz | non-fired | 176 000 kW | 86 000 kW | three PG7821 | |
| 60 Hz | non-fired | 235 000 kW | 110 000 kW | four PG7821 | |
| 60 Hz | non-fired | 73 000 kW | 33 000 kW | one PG7981 | |
| 60 Hz | non-fired | 147 000 kW | 67 000 kW | two PG7981 | |
| 60 Hz | non-fired | 220 000 kW | 103 000 kW | three PG7981 | |
| 60 Hz | non-fired | 293 000 kW | 137 000 kW | four PG7981 | |
| 50 Hz | non-fired | 83 000 kW | 40 000 kW | one PG9111 | |
| 50 Hz | non-fired | 165 000 kW | 83 000 kW | two PG9111 | |
| 50 Hz | non-fired | 248 000 kW | 125 000 kW | three PG9111 | |
| 50 Hz | non-fired | 102 600 kW | 52 400 kW | one PG9141 | |
| 50 Hz | non-fired | 205 200 kW | 105 900 kW | two PG9141 | |
| 50 Hz | non-fired | 410 400 kW | 216 200 kW | four PG9141 | |
| | | | | | **John Brown Engineering** |
| 50/60 Hz | non-fired | ------- | ------- | one MS3002J | single pressure level heat recovery steam generator, gas fuel |
| 50/60 Hz | non-fired | ------- | ------- | one MS5001P | |
| 50/60 Hz | non-fired | ------- | ------- | one MS6001B | |
| 50/60 Hz | non-fired | ------- | ------- | two MS5001P | |
| 50/60 Hz | non-fired | ------- | ------- | three MS6001B | |
| 60 Hz | non-fired | ------- | ------- | one MS7001E | dual pressure level heat recovery boiler |
| 50 Hz | non-fired | ------- | ------- | one MS9001B | |
| 50 Hz | non-fired | ------- | ------- | one MS9001E | |
| 60 Hz | non-fired | ------- | ------- | three MS7001E | |
| 50 Hz | non-fired | ------- | ------- | two MS9001B | |
| 50 Hz | non-fired | ------- | ------- | two MS9001E | |

| Power Plant (series designation) | Year (1st year unit was (will be) available) | ISO Base Load Output (#2 Oil) (6000 hrs a year or more) Net Plant | Heat Rate | ISO Peak Load Output (#2 Oil) (good for up to 2000 hrs a year) Net Plant | Heat Rate |
|---|---|---|---|---|---|
| **Kraftwerk Union** | | | | | |
| GUD 17.14 | 1977 | 170 000 kW | 7 588 Btu/kw-hr | ------- | ------- |
| GUD 35.24 | 1977 | 340 000 kW | 7 587 Btu/kw-hr | ------- | ------- |
| GUD 70.44 | 1977 | 680 000 kW | 7 586 Btu/kw-hr | ------- | ------- |
| **Mitsubishi Heavy Industries** | | | | | |
| MPCP1(501D) | 1981 | 144 000 kW | 7 260 Btu/kw-hr | 157 900 kW | 7 100 Btu/kw-hr |
| MPCP2(501D) | 1981 | 289 900 kW | 7 215 Btu/kw-hr | 317 900 kW | 7 050 Btu/kw-hr |
| MPCP3(501D) | 1981 | 435 500 kW | 7 200 Btu/kw-hr | 477 600 kW | 7 040 Btu/kw-hr |
| MPCP4(501D) | 1981 | 581 000 kW | 7 200 Btu/kw-hr | 637 200 kW | 7 040 Btu/kw-hr |
| MPCP1(701D) | 1981 | 175 800 kW | 7 390 Btu/kw-hr | 191 800 kW | 7 190 Btu/kw-hr |
| MPCP2(701D) | 1981 | 353 400 kW | 7 360 Btu/kw-hr | 385 500 kW | 7 160 Btu/kw-hr |
| MPCP3(701D) | 1981 | 530 900 kW | 7 350 Btu/kw-hr | 579 200 kW | 7 150 Btu/kw-hr |
| MPCP4(701D) | 1981 | 708 200 kW | 7 340 Btu/kw-hr | 772 600 kW | 7 140 Btu/kw-hr |
| **Pratt & Whitney Aircraft Canada** | | | | | |
| TSP-2 | 1980 | 81 150 kW | 8 385 Btu/kw-hr | ------- | ------- |
| TSP-4 | 1980 | 162 950 kW | 8 350 Btu/kw-hr | ------- | ------- |
| TSP-8 | 1980 | 326 580 kW | 8 335 Btu/kw-hr | ------- | ------- |
| **Prvni Brnenska Strojirna** | | | | | |
| STGT 9 | 1974 | 11 700 kW | 8 100 Btu/kw-hr | 11 700 kW | 8 100 Btu/kw-hr |
| **Rolls-Royce** | | | | | |
| SK15-CS | 1981 | 24 670 kW | 7 960 Btu/kw-hr | ------- | ------- |
| SK25-CS | 1981 | 32 750 kW | 7 100 Btu/kw-hr | ------- | ------- |
| SK30-CS | 1977 | 44 220 kW | 7 100 Btu/kw-hr | ------- | ------- |
| SK50-CS | 1981 | 65 500 kW | 7 100 Btu/kw-hr | ------- | ------- |
| SK60-CS | 1977 | 88 440 kW | 7 100 Btu/kw-hr | ------- | ------- |
| **Solar Turbines** | | | | | |
| Centaur CC | 1981 | 4 013 kW | 9 172 Btu/kw-hr | ------- | ------- |
| Mars CC | 1981 | 10 920 kW | 7 767 Btu/kw-hr | ------- | ------- |
| **Stal-Laval Turbin** | | | | | |
| Gasteam 125 | 1977 | 113 000 kW | 7 500 Btu/kw-hr | 122 000 kW | 7 420 Btu/kw-hr |
| Gasteam 250 | 1977 | 228 000 kW | 7 420 Btu/kw-hr | 246 000 kW | 7 340 Btu/kw-hr |
| **Sulzer Brothers** | | | | | |
| Turbotur 205 | 1974 | 8 100 kW | 8 220 Btu/kw-hr | ------- | ------- |
| Turbotur 305 | 1974 | 16 300 kW | 8 210 Btu/kw-hr | ------- | ------- |
| Turbotur 210 | 1974 | 17 000 kW | 8 340 Btu/kw-hr | ------- | ------- |
| Turbotur 310 | 1974 | 34 000 kW | 8 330 Btu/kw-hr | ------- | ------- |
| Turbotur 214 | 1978 | 20 400 kW | 8 030 Btu/kw-hr | ------- | ------- |
| Turbotur 314 | 1978 | 41 000 kW | 8 010 Btu/kw-hr | ------- | ------- |
| Turbotur 225 | 1978 | 39 300 kW | 7 320 Btu/kw-hr | ------- | ------- |
| Turbotur 325 | 1978 | 78 800 kW | 7 300 Btu/kw-hr | ------- | ------- |

# COMBINED CYCLE

| Electrical Frequency (output shaft speed(s)) | Boiler Fuel (fired steam boiler) | Gas Turbine Output | Steam Turbine Output | Number Of Gas Turbines | Comments and Footnotes |
|---|---|---|---|---|---|
| | | | | | **Kraftwerk Union** |
| 50 Hz | non-fired | *116 470 kW | 60 310 kW | one | net plant output on natural |
| 50 Hz | non-fired | *232 940 kW | 120 620 kW | two | gas LHV 50056 kJ/kg, w. |
| 50 Hz | non-fired | *465 880 kW | 241 240 kW | four | duct losses, condenser pressure |
| | | *ISO w/o duct losses | | | 0.043 bar. |
| | | | | | **Mitsubishi Heavy Industries** |
| 60 Hz | non-fired | 98 700 kW | 45 300 kW | one MW-501D | all ratings on |
| 60 Hz | non-fired | 197 400 kW | 92 500 kW | two MW-501D | natural gas fuel |
| 60 Hz | non-fired | 296 100 kW | 139 400 kW | three MW-501D | at electric generator |
| 60 Hz | non-fired | 394 800 kW | 186 200 kW | four MW-501D | terminals. |
| 50 Hz | non-fired | 119 600 kW | 56 200 kW | one MW-701D | all heat rates LHV |
| 50 Hz | non-fired | 239 200 kW | 114 200 kW | two MW-701D | of natural gas fuel |
| 50 Hz | non-fired | 358 800 kW | 172 100 kW | three MW-701D | |
| 50 Hz | non-fired | 478 400 kW | 229 800 kW | four MW-701D | |
| | | | | | **Pratt & Whitney Aircraft Canada** |
| 60 Hz | non-fired | 57 250 kW | 23 900 kW | two CFT4C-3F | 50 Hz units available |
| 60 Hz | non-fired | 114 500 kW | 48 450 kW | four CFT4C-3F | |
| 60 Hz | non-fired | 229 000 kW | 97 580 kW | eight CFT4C-3F | |
| | | | | | **Prvni Brnenska Strojirna** |
| 50 Hz | gas-fired | 7 000 kW | 4 700 kW | one ST9 | two boilers 2x50 tons/hr for process steam |
| | | | | | **Rolls-Royce** |
| 50/60 Hz | non-fired | 16 000 kW | 8 760 kW | one Avon | |
| 50/60 Hz | non-fired | 22 750 kW | 10 000 kW | one RB 211 | |
| 50/60 Hz | non-fired | 27 500 kW | 16 720 kW | one Olympus | higher plant outputs |
| 50/60 Hz | non-fired | 45 500 kW | 20 000 kW | two RB 211 | available with multiple units |
| 50/60 Hz | non-fired | 55 000 kW | 33 440 kW | two Olympus | on all plants listed |
| | | | | | **Solar Turbines** |
| ---- | non-fired | 2 945 kW | 1 017 kW | one Centaur | all ratings on gas fuel; shaft kW, |
| ---- | non-fired | 8 128 kW | 2 731 kW | one Mars | avail. for mech drive and electric power |
| | | | | | **Stal-Laval Turbin** |
| 50 Hz | non-fired | 80 000 kW | 34 000 kW | one GT200 | gas fuel rating |
| 50 Hz | non-fired | 160 000 kW | 70 000 kW | two GT200 | gas fuel rating |
| | | | | | **Sulzer Brothers** |
| 50/60 Hz | non-fired | 5 300 kW | 2 800 kW | one Type 3 | units avail w. |
| 50/60 Hz | non-fired | 10 600 kW | 5 700 kW | two Type 3 | supplementary firing, |
| 50/60 Hz | non-fired | 10 580 kW | 6 420 kW | one Type 7 | other gas turbines |
| 50/60 Hz | non-fired | 21 160 kW | 12 840 kW | two Type 7 | |
| 50/60 Hz | non-fired | 13 700 kW | 6 700 kW | one PRIMO 14 | |
| 50/60 Hz | non-fired | 27 400 kW | 13 600 kW | two PRIMO 14 | |
| 50/60 Hz | non-fired | 25 400 kW | 13 900 kW | one Olympus | |
| 50/60 Hz | non-fired | 50 800 kW | 28 000 kW | two Olympus | |

| Power Plant (series designation) | Year (1st year unit was (will be) available) | ISO Base Load Output (#2 Oil) (6000 hrs a year or more) Net Plant | Heat Rate | ISO Peak Load Output (#2 Oil) (good for up to 2000 hrs a year) Net Plant | Heat Rate |
|---|---|---|---|---|---|
| **Thomassen Holland** | | | | | |
| STEG 105P | 1972 | 33 400 kW | 8 740 Btu/kw-hr | ------- | ------- |
| STEG 205P | 1972 | 67 200 kW | 8 720 Btu/kw-hr | ------- | ------- |
| STEG 106B | 1980 | 47 800 kW | 8 120 Btu/kw-hr | ------- | ------- |
| STEG 206B | 1980 | 95 900 kW | 8 100 Btu/kw-hr | ------- | ------- |
| STEG 107E | 1979 | 99 300 kW | 7 700 Btu/kw-hr | ------- | ------- |
| STEG 207E | 1979 | 198 800 kW | 7 660 Btu/kw-hr | ------- | ------- |
| STEG 109E | 1979 | 140 400 kW | 7 940 Btu/kw-hr | ------- | ------- |
| STEG 209E | 1979 | 282 300 kW | 7 890 Btu/kw-hr | ------- | ------- |
| **Turbotecnica** | | | | | |
| CC-205P | 1981 | 67 500 kW | 8 800 Btu/kw-hr | ------- | ------- |
| CC-206B | 1981 | 99 000 kW | 7 960 Btu/kw-hr | ------- | ------- |
| CC-209E | 1981 | 293 000 kW | 7 650 Btu/kw-hr | ------- | ------- |
| **UTC Power Systems** | | | | | |
| TSP-2 | 1980 | 81 150 kW | 8 385 Btu/kw-hr | ------- | ------- |
| TSP-4 | 1980 | 162 950 kW | 8 350 Btu/kw-hr | ------- | ------- |
| TSP-8 | 1977 | 326 580 kW | 8 335 Btu/kw-hr | ------- | ------- |
| **Westinghouse Canada** | | | | | |
| Mini Pace | 198? | 53 445 kW | 9 800 Btu/kw-hr | ------- | ------- |
| **Westinghouse Electric Combustion Turbine Systems** | | | | | |
| PACE 2511 | 1981 | 55 200 kW | 8 300 Btu/kw-hr | ------- | ------- |
| PACE 2512 | 1981 | 110 400 kW | 8 300 Btu/kw-hr | ------- | ------- |
| PACE 2513 | 1981 | 165 600 kW | 8 300 Btu/kw-hr | ------- | ------- |
| PACE 2514 | 1981 | 220 800 kW | 8 300 Btu/kw-hr | ------- | ------- |
| PACE 5011 | 1981 | 134 700 kW | 7 630 Btu/kw-hr | ------- | ------- |
| PACE 5012 | 1981 | 269 400 kW | 7 630 Btu/kw-hr | ------- | ------- |
| PACE 5013 | 1981 | 404 100 kW | 7 630 Btu/kw-hr | ------- | ------- |
| PACE 5014 | 1981 | 538 800 kW | 7 630 Btu/kw-hr | ------- | ------- |
| PACE 5015 | 1981 | 673 500 kW | 7 630 Btu/kw-hr | ------- | ------- |

# COMBINED CYCLE

| Electrical Frequency (output shaft speed(s)) | Boiler Fuel (fired steam boiler) | Gas Turbine Output | Steam Turbine Output | Number Of Gas Turbines | Comments and Footnotes |
|---|---|---|---|---|---|
| **Thomassen Holland** | | | | | |
| 50/60 Hz | non-fired | ------- | ------- | one PG5341 | various other combined cycle plants available |
| 50/60 Hz | non-fired | ------- | ------- | two PG5341 | |
| 50 Hz | non-fired | ------- | ------- | one PG6441 | |
| 50 Hz | non-fired | ------- | ------- | two PG6441 | |
| 60 Hz | non-fired | ------- | ------- | one PG7891 | |
| 60 Hz | non-fired | ------- | ------- | two PG7891 | |
| 50 Hz | non-fired | ------- | ------- | one PG9141 | |
| 50 Hz | non-fired | ------- | ------- | two PG9141 | |
| **Turbotecnica** | | | | | |
| 50 Hz | non-fired | 47 000 kW | 20 500 kW | two G5341 | |
| 50 Hz | non-fired | 69 000 kW | 30 000 kW | two G6501 | |
| 50 Hz | non-fired | 204 000 kW | 89 000 kW | two G9141 | |
| **UTC Power Systems** | | | | | |
| 60 Hz | non-fired | 57 250 kW | 23 900 kW | two FT4C | 50 Hz units available |
| 60 Hz | non-fired | 114 500 kW | 48 450 kW | four FT4C | |
| 60 Hz | non-fired | 229 000 kW | 97 580 kW | eight FT4C | |
| **Westinghouse Canada** | | | | | |
| 50/60 Hz | bunker | 35 244 kW | 18 200 kW | two W191 | 105°F and sea level; heat rate - LHV; both gas turbines and boiler fired on residual oil. |
| **Westinghouse Electric Combustion Turbine Systems** | | | | | |
| 50/60 Hz | non-fired | 36 800 kW | 18 400 kW | one W251B | ISO ratings; net heat rate on LHV of distillate; nominal performance |
| 50/60 Hz | non-fired | 73 500 kW | 36 900 kW | two W251B | |
| 50/60 Hz | non-fired | 110 300 kW | 55 300 kW | three W251B | |
| 50/60 Hz | non-fired | 147 100 kW | 73 700 kW | four W251B | |
| 50/60 Hz | non-fired | 93 800 kW | 40 900 kW | one W501D | supplementary firing, optimized configurations available on all models listed |
| 50/60 Hz | non-fired | 187 500 kW | 81 900 kW | two W501D | |
| 50/60 Hz | non-fired | 281 300 kW | 122 800 kW | three W501D | |
| 50/60 Hz | non-fired | 375 000 kW | 163 800 kW | four W501D | |
| 50/60 Hz | non-fired | 468 800 kW | 204 800 kW | five W501D | |

# INDEX

## ABOUT THE AUTHOR

Charles Butler has provided engineering, design, and financial analysis for cogeneration projects since 1975. He holds seven professional engineering registrations in addition to B.S. and M.S. degrees in mechanical engineering. Mr. Butler is currently a member of the American Society of Mechanical Engineers and chairman of the Northern California Regional Chapter of the International Cogeneration Society. He frequently lectures on cogeneration for the Electric Power Institute, the California Energy Commission, and the Pacific Coast Electrical Association.

## Lower Electric Rates May Slow Shift to Cogenera

# Utilities Act to Keep Bi

By GREG JOHNSON, *Times Staff Writer*

San Diego Gas & Electric executives snapped to attention earlier this month when the Navy, upset that its 1986 electric bill in San Diego will hit $80 million, threatened to abandon SDG&E and build its own electric generation plant.

Hoping to keep its single largest electric customer from weighing anchor, SDG&E countered with a rate package that would match or beat the Navy's predicted annual cost savings of $15 million.

SDG&E's plight is a graphic example of how, when the economics are right, heavy electric users will turn off high-priced utility power and generate their own electricity.

Navy officials said they would study the lower rates, but insisted that they would "move forward on . . . a more cost effective means for providing steam and electricity" to the Navy's six shore establishments that ring San Diego Bay.

Aware that negotiations with the Navy might prove fruitless, SDG&E also began a push for "mitigating legislation" to prohibit major federal government installations from abandoning the nation's utilities.

The Navy initially turned to cogeneration more than 10 years ago to supply inexpensive steam for its San Diego shore operations. It now receives steam generated by three cogeneration plants owned and operated by a third party that turns a profit by selling electricity to SDG&E.

permission to negotiate long-term, lower-cost electric contracts with larger electric customers that otherwise would install cogeneration plants and abandon PG&E.

However, letting utilities and customers set prices "raises anti-trust questions or the possibility of discriminatory rates," Hamrin warned. "If I were ABC Widget Co. and found out that Ace Co. is paying lower electric rates to the utility than I am, I'd be bent out of shape [because] one of the utility's charges is to provide non-discriminatory rates."

Utilities defend the proposed special rates, arguing that they "reflect the true costs of service to our larger customers," Martin said. "If we can make the sale at cost plus something a bit less than our retail charge, we can make it uneconomic for [customers] to install a cogeneration plant."

Utilities won the ability to negotiate contracts with their natural gas customers several years ago when large industrial companies began switching from utility-supplied natural gas to free-market natural gas, fuel oil, butane and propane, Martin said.

### Negotiated Deal

PG&E successfully negotiated a gas contract with Chevron when the utility's largest natural gas customer threatened to take its business elsewhere.

"Several years ago we realized

tion

# Customers

"We need to determine whether or not the rate-making decision process fits the era we're in."

California's existing rate-making scheme, which was created in the midst of an energy crisis, pushes utilities to reduce electric and gas sales in the name of conservation, said Mike Florio, an attorney with San Francisco-based Toward Utility Rate Normalization.

"The economics have changed remarkably over the last five years," Florio said. "We've gone from an era of [energy] shortage to an era of surplus."

Interestingly, Applied Energy, a former SDG&E subsidiary, installed and operated the Navy's first cogeneration plant more than 10 years ago during an energy shortage. The utility championed the technology because cogeneration plants provided steam for the Navy's industrial processes and generated relatively cheap electricity that SDG&E purchased.

Cogeneration plants now generate 140 megawatts of SDG&E's generation capacity and will supply nearly half of the 1,000 megawatts of additional generating capacity that SDG&E will need to meet customer demand in 1995.

Most cogeneration plants being built in the state generate 20 megawatts or less of electricity, according to Claudia Barker, a public information officer with the California Energy Commission. The commission exercises control over plants that produce 50 mega-

posals, including a 395 megawatt facility planned by Atlantic Richfield Co. in Los Angeles.

Although utilities have relied upon cogeneration to meet electric demand, they "now see their historical monopoly [on power generation] being destroyed," Florio said. "Cogeneration has seen a lot more development than anyone every expected because between tax credits and other incentives, it really took off."

It is unlikely that Congress would prohibit industrial and commercial customers from developing cheaper sources of power.

"I don't think the compact that the citizens of this country have signed with their utilities is an all-events situation," observed Robert E. Burt, a Sacramento-based energy consultant for the California Assn. of Manufacturers. "No business that has built a given capacity to serve customers is automatically entitled to have a guaranteed market."